Advances in Physicochemical Properties of Biopolymers

(Part 1)

Edited by:

Martin Masuelli & Denis Renard

Instituto de Física Aplicada-CONICET, Universidad Nacional de San Luis, Chacabuco 917, CP 5700, San Luis, Argentina

Advances in Physicochemical Properties of Biopolymers *(Part 1)*

Editors: Martin Masuelli & Denis Renard

ISBN (eBook): 978-1-68108-453-4

ISBN (Print): 978-1-68108-454-1

First published in 2017.

General:

1. Any dispute or claim arising out of or in connection with this License Agreement or the Work (including non-contractual disputes or claims) will be governed by and construed in accordance with the laws of the U.A.E. as applied in the Emirate of Dubai. Each party agrees that the courts of the Emirate of Dubai shall have exclusive jurisdiction to settle any dispute or claim arising out of or in connection with this License Agreement or the Work (including non-contractual disputes or claims).
2. Your rights under this License Agreement will automatically terminate without notice and without the need for a court order if at any point you breach any terms of this License Agreement. In no event will any delay or failure by Bentham Science Publishers in enforcing your compliance with this License Agreement constitute a waiver of any of its rights.
3. You acknowledge that you have read this License Agreement, and agree to be bound by its terms and conditions. To the extent that any other terms and conditions presented on any website of Bentham Science Publishers conflict with, or are inconsistent with, the terms and conditions set out in this License Agreement, you acknowledge that the terms and conditions set out in this License Agreement shall prevail.

Bentham Science Publishers Ltd.
Executive Suite Y - 2
PO Box 7917, Saif Zone
Sharjah, U.A.E.
Email: subscriptions@benthamscience.org

BENTHAM
SCIENCE

CONTENTS

PREFACE

The objective of this ebook is to provide to the readers the most recent state-of-the-art on physicochemical properties of biopolymers and their related end-uses applications. Biopolymers are usually described as polymers produced in a natural way by living species. Their molecular backbones are composed of repeating units of saccharide, nucleic acids, or amino acids and sometimes various additional side chains contributing also to their functionalities.

If the largest part of biopolymers is extracted from biomass, such as polysaccharides from cellulose and proteins from collagen, milk or wheat, biopolymers can also be produced from monomers using conventional chemical processes as polylactic acid, or directly in microorganisms or genetically modified organisms, as polyhydroxyalkanoates. The genetic manipulation of microorganisms brings a tremendous potentiality for the biotechnological production of biopolymers with tailored properties quite suitable for high-value medical applications such as tissue engineering and drug delivery.

Biopolymers from renewable sources, on the contrary display structural complexity and natural variability that need to be deeply studied and characterized before probing into the structure-function relationships for further applications. Research on natural polymers has focused on developing more environmentally friendly applications to reduce pollution caused by non-biodegradable material. Historically, biopolymers were mainly used by mankind as food, or for making clothing and furniture. Since the industrial time, fossil fuels such as oil are the greatest source in the development and manufacture of almost every commercial product, such as the plastic, which is currently used at a very large scale. But these fuels are not unlimited resources, and environmental concerns over all aspects of using fossil fuels for production and energy must be taken into account. We must act in a sustainable manner, which means that the resources must be consumed at a rate such that they can be restored by natural cycles of our planet [1].

Therefore, in recent years, the renewable nature of biopolymers leads them to a renaissance and a new considerable interest by industry due to the unique properties, including biodegradability, biocompatibility and nontoxicity, of biopolymers. To fulfil all these different functions, biopolymers must exhibit rather diverse properties. They must very specifically interact with a large variety of different substances, components and materials, and often they must have extraordinarily high affinities to them. Finally, they must have a high strength. Some of these properties are utilized directly or indirectly for various applications. This and the possibility to produce them from renewable resources, as living matter mostly does, make biopolymers interesting candidates to industry [2]. As a consequence of their properties, these biopolymers derived from natural products have found a place of choice in areas as diverse as effluent treatment, papermaking, chemical, food, cosmetic, pharmaceutical, petroleum and textile industries, as well as in analytical chemistry (biosensors) and molecular biology. However, biopolymers have to compete with polymers derived from fossil fuel not only because of their functional properties but also in terms of cost. In this respect, biopolymers are competitive when the price of oil is high and the price of feedstocks, such as starch from corn, is low.1 The continuing development of new and existing biopolymers will enable these materials to help supplement the increasing global demand for biopolymers-based products and to develop new markets with their niche applications.

The most common biopolymers used for industrial applications and thoroughly considered in this ebook are polysaccharides from plant, algal, microbial and animal origins such as starch, cellulose, lignin, arabinoxylans, sulfated polysaccharides from seaweeds, galactomannans and xyloglucans from brazilian seeds, chitin and its derivative chitosan. Natural gums such as mesquite, tara and arabic gums are also widely used in food and non-food industry and are discussed in this ebook. Animal and plant proteins such as collagen, gelatin, albumin, dairy proteins and wheat, corn and soy proteins are also considered as sources of proteins for biomedical, microencapsulation and plastic foams applications. Nucleic acids such as DNA and RNA and their related applications in genetic engineering for instance are not considered in this ebook.

This ebook presents a comprehensive review and compile information on biopolymers in 27 chapters covering from isolation and production, properties and applications, modification, and relevant analytical methods to reveal the structure and properties of some biopolymers.

Authors write this ebook from Argentina, France, Mexico, Spain, Iran, Brazil, Egypt, Turkey, Venezuela, India, Russia, Portugal, New Zealand and Malaysia. This ebook has tried to arrange the ebook chapters in a subject order to make it easier for the readers to find what they need. However, the reader can still find information on the same subject in more than one Section.

Section A (Part 1), which includes one chapter, is mainly an introduction to biopolymers. It includes concepts and molecular weight determination.

Section B (Part 1), which includes twelve chapters, refers to some physical chemistry determinations of biopolymers.

Section C (Part 1), which consists of two chapters, deals with studies on hydrodynamic properties of biopolymers.

Section D (Part 1), which consists of one chapter, refers to theoretical models for biopolymers.

Section A (Part 2), which includes four chapters, refers to special cases of polysaccharides separation and purification.

Section B (Part 2), which includes seven chapters, deals with applications of biopolymers/hydrogels in drug delivery systems, biomaterials, biothermoplastics, bio(nano)composites, bionanostructures, biocapsules, bioadsorbents, bioelectrospinning and biopackaging. This section deserves a special attention because it forms a fascinating interdisciplinary area that brings together biology, chemistry, materials science and (nano)-technology.

This ebook is expected to be of help to many graduate and post-graduate students, professors, scientists, pharmacists, engineers and other experts in a variety of disciplines, both academic and industrial, dedicated to the determination of polymers and biopolymers properties. This ebook may not only support research and development, but also be suitable for teaching. The audience will benefit with an excellent review offering advanced knowledge about technical determinations and physicochemical properties of macromolecules, a thorough knowledge of hydrodynamics and different methods of characterization. Readers will find in this ebook a triple deal, including educational, scientific and industrial applications.

The first main objective of this e-book is therefore to highlight the progress in different techniques of molecular weight determinations and physicochemical properties of biopolymers. The last two decades have seen a number of significant advances in the methodology for evaluating the molecular weight distributions of polydisperse macromolecular systems in solution at the molecular level. These advances have centered on the coupling of chromatographic or membrane based fractionation procedures with multiple detectors on line such as multi-angle laser light scattering, refractive index, UV-Vis absorbance and intrinsic viscosity detection systems. Recent advances in SEC-MALLS (size exclusion chromatography coupled to multi-angle laser light scattering) and FFF-MALLS (field flow fractionation coupled on line to MALLS) applied to complex polymers from renewable resources are therefore presented in this e-book. Beyond molecular charcaterization using HPSEC-A4F-MALLS technique, tremendous efforts were made these last years to elucidate the structural variability and complexity of polysaccharides using matrix-assisted laser-desorption ionization (MALDI) and electrospray ionization (ESI) mass spectrometry coupled or not to nuclear magnetic resonance (NMR) spectroscopy. One chapter of this ebook in section B considers the sequence, interresidue linkage position and substitution pattern of sulfated polysaccharides after enzymatic hydrolyses.

The most widely used method for the dynamic characterization of macromolecules in solution is the capillary viscometry, as it is a simple and economic method. Although in literature there is much information on hydrodynamic measurements from intrinsic viscosity determinations, very few of them evaluate the conformation of different biopolymers. The importance of this type of study lies in the analysis of the polysaccharides or proteins behaviour in industrial processes and product quality control after extraction and purification. These physicochemical studies help to elucidate the chemical structure, macromolecular conformation and the ability biopolymers have to form gels, films, agglomerates, *etc.* A particular attention is paid in this ebook on the intrinsic viscosity determination of proteins and strong synthetic polyelectrolytes for which theoretical models always need to be implemented in order to get reliable dynamic structural informations.

The ebook also focuses on the structural analyses at the mesoscopic scale using mechanical analyses, microscopy, small angle scattering and free volume measurements and different applications related to biopolymers such as biomaterials, microcapsules, biothermoplastics, nanostructured biocomposites, super-absorbents, bioelectrospinning, biopolymers-based dermal and transdermal drug delivery systems, and biopackaging. All these applications using biopolymers aim to provide a means to reduce dependence on fossil fuels, and decrease the environmental impact of non-biodegradable materials. The main challenge to overcome with biopolymers-based materials is the control of biopolymer-biopolymer interactions, a challenge always present and discussed throughout the ebook by authors.

To conclude, the content of this ebook will bring its readers a basic understanding of the physical chemistry of biopolymers, but also the latest findings about new macromolecules recently discovered and published. Theoretical aspects of computational structural description of biopolymers are also thoroughly described. Therefore, this ebook will appeal to different readers as a great source of knowledge about the science of biopolymers.

Biopolymers Audience

- Separation, purification, characterization of biopolymers
- Hydrodynamic, molecular weight, size, shape, conformation
- Macromolecular assembly
- Molecular design and bio-nanotechnology
- Biopolymer processing and degradation
- Experimental and theoretical studies of biopolymer structures
- Three-dimensional structures of biopolymers determined by X-ray, neutrons, NMR

- Interactions and thermodynamics
- Food biocolloids
- Structure and function
- Preparation and characterization of novel biomaterials
- Capsules and microcapsules
- Biocatalysis
- Biopolymers for bioremediation
- Thin films, membranes & packaging

REFERENCES

[1] Thomas, Sabu; Durand, Dominique; Jyotishkumar, P.; Chassenieux, Christophe, editors. Handbook of Biopolymer-Based Materials: From Blends and Composites to Gels and Complex Networks. 1st. Wiley-VCH Verlag Gmbh and Co. KGaA; 2013.

[2] Steinbüchel, Alexander, editor. Biopolymers. 10. Wiley-VCH, Weinheim; 2003.

Denis Renard
Unité de Recherches Biopolymères,
Interactions, Assemblages,
INRA, Rue de la Géraudière,
France

ACKNOWLEDGEMENTS

We first would like to thanks all the authors contributing to the realization of this ebook.

We would like to thank the Agronomic Research National Institute, France and the Área de Química Física - Departamento de Química - Facultad de Química Bioquímica y Farmacia - Universidad Nacional de San Luis – Instituto de Física Aplicada – CONICET, Argentina for giving us the time and opportunity for the writing and edition work of this ebook.

We are grateful to the editing team at Bentham eBooks for their help, especially Humaira Hashmi and Salma Sarfaraz.

Finally, we wish to express our gratitude to our spouses, Maria Gabriela and Marie-Christine, for their patience and support during the long hours of writing, reviewing, commenting, editing of this ebook.

List of Contributors

Alberto Tecante	Departamento de Alimentos y Biotecnología, Facultad de Química, Universidad Nacional Autónoma de México, Cd. Universitaria, D.F., 04510, México
Agnès Rolland-Sabaté	UR1268 Biopolymères Interactions Assemblages, INRA, F-44316 Nantes, France
Alma R. Toledo-Guillén	Biopolymers Group, Research Center for Food and Development, (C.I.A.D., A.C.), Hermosillo, Sonora 83304, Mexico
Alexander S. Gubarev	St. Petersburg State University, Saint Petersburg, Russian Federation
Asmita Sen Gupta	Physics Department, Visva-Bharati Central University, Santiniketan-731235, W.B, India
C. Sanchez	UMR1208 Ingénierie des Agropolymères et Technologies Emergentes, INRA-Montpellier SupAgro-CIRAD-Université Montpellier, 2 Place Pierre Viala, F-34060 Montpellier, France
Ciria G. Figueroa-Soto	Biochemistry of proteins, CCA, Research Center for Food and Development, Hermosillo, Sonora, México
Denis Renard	INRA, UR1268 Biopolymères Interactions Assemblages, F-44300 Nantes Cedex, France
Elisa M. Valenzuela-Soto	Biochemistry of proteins, CCA, Research Center for Food and Development, Hermosillo, Sonora, México
Elizabeth Carvajal-Millan	Biopolymers, CTAOA, Research Center for Food and Development, Hermosillo, Sonora, México
F. Guillermo Díaz Baños	Departamento de Química Física. Universidad de Murcia. Campus de Espinardo, 30100. Murcia, Spain
Frédéric Bonfils	CIRAD, UMR Ingénierie des Agropolymères et Technologies Emergentes (IATE), F-34398 Montpellier, France
Georges M. Pavlov	Institute of Macromolecular Compounds, Russian Academy of Science, Russian Federation St. Petersburg State University, Saint Petersburg, Russian Federation
Guillermo J. Copello	Universidad de Buenos Aires (UBA), Facultad de Farmacia y Bioquímica, Cátedra de Química Analítica Instrumental, IQUIMEFA (UBA-CONICET), Junín 956, C1113AAD Buenos Aires, Argentina
J. Ginés Hernández Cifre	Departamento de Química Física. Universidad de Murcia. Campus de Espinardo, 30100. Murcia, Spain
Jaime Lizardi-Mendoza	Biopolymers Group, Research Center for Food and Development, (C.I.A.D., A.C.), Hermosillo, Sonora 83304, Mexico
Joaquín A. González	Universidad de Buenos Aires (UBA), Facultad de Farmacia y Bioquímica, Cátedra de Química Analítica Instrumental, IQUIMEFA (UBA-CONICET), Junín 956, C1113AAD Buenos Aires, Argentina
Karim Chelbi	CIRAD, UMR Ingénierie des Agropolymères et Technologies Emergentes (IATE), F-34398 Montpellier, France

Kelvia Álvarez	Department of Analytical Chemistry, Faculty of Pharmacy, Central University of Venezuela (UCV), PO Box 40109, Caracas 1040-A, Venezuela Institute of Food Science and Technology, Faculty of Sciences, Central University of Venezuela (UCV), PO Box 47097, Caracas 1041-A, Venezuela
Laura Patricia Martínez-Padilla	Laboratorio de Propiedades Reológicas y Funcionales en Alimentos, Facultad de Estudios Superiores Cuautitlán, Universidad Nacional Autónoma de México, México
Lucía Famá	LPMC, IFIBA-CONICET, Dep. Physics, Faculty of Exact and Natural Sciences, University of Buenos Aires, University City (1428), Pab. 1, Buenos Aires, Argentina
Madhav Yadav	Sustainable Biofuels and Coproducts Research Unit. Eastern Regional Research Center, Agricultural Research Service, US Department of Agriculture, Wyndmoor, PA, USA
Martin Alberto Masuelli	Instituto de Física Aplicada-CONICET, Universidad Nacional de San Luis, Chacabuco 917, CP 5700, San Luis, Argentina
Margarite Rinaudo	Biomaterials Applications, 6, rue Lesdiguiéres Grenoble, 38000, France
M. C. Ortiz-Tafoya	Departamento de Alimentos y Biotecnología, Facultad de Química, Universidad Nacional Autónoma de México, Cd. Universitaria, D.F., 04510, México
María C. Matulewicz	Universidad de Buenos Aires, Facultad de Ciencias Exactas y Naturales, Departamento de Química Orgánica, Consejo Nacional de lnvestigaciones Científicas y Técnicas- Centro de Investigación de Hidratos de Carbono (CIHIDECAR, CONICET-UBA), Ciudad Universitaria-Pabellón 2, C1428EHA Buenos Aires, Argentina
Marina Ciancia	Universidad de Buenos Aires, Facultad de Ciencias Exactas y Naturales, Departamento de Química Orgánica, Consejo Nacional de lnvestigaciones Científicas y Técnicas- Centro de Investigación de Hidratos de Carbono (CIHIDECAR, CONICET-UBA), Ciudad Universitaria-Pabellón 2, C1428EHA Buenos Aires, Argentina Universidad de Buenos Aires, Facultad de Agronomía, Departamento de Biología Aplicada y Alimentos, Cátedra de Química de Biomoléculas, Av. San Martín 4453, C1417DSE Buenos Aires, Argentina
María Emilia Villanueva	Universidad de Buenos Aires (UBA), Facultad de Farmacia y Bioquímica, Cátedra de Química Analítica Instrumental, IQUIMEFA (UBA-CONICET), Junín 956, C1113AAD Buenos Aires, Argentina
Mayra A. Méndez-Encinas	Biopolymers, Research Center for Food and Development, Hermosillo, Sonora, México
Michel Martin	Ecole Supérieuure de Physique et de Chimie Industrielles, Laboratoire de Physique et Mécanique des Milieux Hétérogènes (PMMH, UMR 7636 CNRS, ESPCI-ParisTech, Université Pierre et Marie Curie, Université Paris-Diderot), 10 rue Vauquelin, 75231 Paris Cedex 05, France
Mohammad R. Kasaai	Department of Food Science and Technology, Sari Agricultural Sciences and Natural Resources University, Khazar Abad road, Km. 9, P.O. Box, 578, Sari, Mazandaran, Iran

Orlando Tortoledo-Ortiz Instrumental Analysis, CN, Research Center for Food and Development, Hermosillo, Sonora, México

Serge Battu EA3842 Homéostasie et Pathologies Cellulaires, Laboratoire de Chimie Analytique, Faculté de Pharmacie, Univeristé de Limoges, France

Tomy J. Gutiérrez Department of Analytical Chemistry, Faculty of Pharmacy, Central University of Venezuela (UCV), PO Box 40109, Caracas 1040-A, Venezuela
Institute of Food Science and Technology, Faculty of Sciences, Central University of Venezuela (UCV), PO Box 47097, Caracas 1041-A, Venezuela
Composite Materials Group (CoMP), Research Institute of Materials Science and Technology (INTEMA), Faculty of Engineering, University of Mar del Plata and National (UNdMP) and National Council of Scientific and Technical Research (CONICET), Colón 10850, Mar del Plata 7600, Buenos Aires, Argentina

Yolanda L. López-Franco Biopolymers Group, Research Center for Food and Development, (C.I.A.D., A.C.), Hermosillo, Sonora 83304, Mexico

CHAPTER 1

Molecular Weight and Molecular Weight Distribution for Biopolymers

Mohammad R. Kasaai*

Department of Food Science and Technology, Sari Agricultural Sciences and Natural Resources University, Khazar Abad road, Km. 9, Sari, Mazandaran, Iran

Abstract: In this chapter, molecular weight (M), and molecular weight distribution (MWD), of polymers with emphasis on M and MWD of biopolymers, *e.g.*, carbohydrate polymers, proteins, deoxyribonucleic acid, DNA, and ribonucleic acid, RNA, are reviewed. The M and MWD of biopolymers are compared with those of synthetic polymers. The following conclusions are drawn. (1) Unlike simple low molecular substances, most polymers do not have unique molecular weights. Practically, no polymer exists whose molecules are all strictly of the same size or have the same degree of polymerization. Thus, polymers are more or less heterogeneous with respect to their molecular weights. (2) The concept of average molecular weight is used for polymers. (3) Different numerical values for molecular weights of polymers have already been defined as average molecular weights (M_n, M_w, M_z, and M_v), depending on the methods by which they are measured. (4) The average values vary in the following order: $M_n < M_v < M_w < M_z < M_{z+1}$. The disparity between average molecular weights provides a measure of the degree of heterogeneity, *i.e.* dispersity, in the molecular weight distribution. (5) The constitution of a polymer as well as the MWD may be described either by a set of different average molecular weights, the ratios of two different types of average molecular weights, or by the distribution functions *via* graphical presentation and (6) Polysaccharides in a similar way to synthetic polymers are polydisperse polymers, whereas proteins, DNA, and RNA, are mostly monodisperse macromolecules.

Keywords: Biopolymers, Dispersity, DNA, Heterogeneity, Molecular weight, Molecular weight distribution, Polysaccharides, Proteins, RNA.

* **Corresponding author Mohammad R. Kasaai**: *Department of Food Science and Technology, Sari Agricultural Sciences and Natural Resources University, Khazar Abad road, Km. 9, Sari, Mazandaran, Iran; E-mail:*reza_kasaai@hotmail.com

Martin Masuelli & Denis Renard (Eds.)

INTRODUCTION

A macromolecule, known as a giant molecule or a polymer, is a chemical species, composed of a long chain with a regularly repeating unit, a high molecular weight and a high molecular size [1, 2]. The unit for molecular weight is usually the Dalton (Da); one Dalton is equal to one atomic mass unit. Symbols and parameters appearing in this chapter are given in Table **1**. Macromolecules are divided into natural and man- made polymers. The latter are known as synthetic polymers [3].

Table 1. Constants and symbols used in this chapter.

Symbol	Definition
M	Molecular weight
M_w	Weight-average molecular weight
M_n	Number-average molecular weight
M_z	Z-average (centrifuge average) molecular weight
M_{z+1}	(z+1)- average molecular weight
M_v	Viscosity-average molecular weight
[η]	Intrinsic viscosity in a solvent
K	Intercept for MHS equation
a	Exponent for MHS equation,
MHS	Mark-Houwink –Sakurada
q_{MHS}	Dispersity correction factor
$f_w(X)$	Cumulative weight fraction
X	Parameter characterizing the chain length
W(M)	Weight distribution function
N(M)	Number distribution function
a_1	Positive adjustable parameter
b_1	Positive adjustable parameter
μ	Mean value
σ_n	Standard deviation
K_{av}	Distribution coefficient
V_R	Solute elution volumen
V_0	Void volumen
V_C	Total bed volumen
b	Constant, it is an empirical polynomial function of MHS exponent *a*

(Table 1) contd.....

Symbol	Definition
c	Constant, it is an empirical polynomial function of MHS exponent *a*
N_{Av}	Avogadro's constant
S	Sedimentation constant
D	Diffusion Constant

This chapter focuses on natural polymers, also known as biopolymers.

Natural polymers, are produced by biosynthesis in nature, whereas synthetic polymers are made and their synthesis controlled by human beings. Biopolymers may be classified into proteins, nucleic acids, polysaccharides, and others. In this chapter, three main groups; polysaccharides, proteins and nucleic acids, which play important roles in all biological phenomena and processes are discussed [4 - 6]. A wide variety of natural polymers relevant to the field of biomaterials, is derived from plants and animals [5, 7]. Generally, biopolymers consist of carbohydrate polymers, proteins, deoxyribonucleic acids (DNA), and ribonucleic acids (RNA). They are fundamental to the biological substance of life [5]. Abbreviations and expressions are listed in Table **2**.

Table 2. Abbreviations used in this chapter.

Abbreviation	Expression, Term
CLND	Chemi-Luminescent Nnitrogen Detector
Da	Dalton
DI	Dispersity Index
DNA	Deoxyribonucleic Acid
DRI	Differential Refractive Index
GPC	Gel Permeation Chromatography
IUPAC	International Union of Pure and Applied Chemistry
RNA	Ribonucleic Acid
SEC	Size Exclusion Chromatography
MS	Mass Spectrometry
MWD	Molecular Weight Distribution

GENERAL CONSIDERATIONS

Carbohydrate macromolecules are known as polysaccharides. Monosaccharide units join together *via* glucosidic linkages and form a polysaccharide [7]. Proteins are linear polymers formed by linking the α-carboxyl group of one amino acid to the α-amino group of another amino acid with a peptide bond. A polypeptide chain consists of a regularly repeating part of amino acids joined by peptide bonds. Most of natural polypeptide chains contain between 50 and 2000 amino acid residues and are commonly referred to as proteins. The M_w of most proteins lies between 5.5 kDa and 220 kDa [7]. Polysaccharides can be defined as linear or branched macromolecules formed by many monosaccharide units linked by glycosidic bonds. These biopolymers, sometimes also called glycans, can be classified as homo-polysaccharides, *i.e.* homopolymers, which consist of monosaccharide units, and hetero-polysaccharides, *i.e.* copolymers which consist of two or more different monosaccharide units [8]. The glycosidic bonds can be α or β (1→4, 1→6, 1→3, for instance). Depending on their functions, they can be also classified as structural, such as cellulose and chitin, and storage, like starch, inulin and glycogen [9].

Proteins, DNA, and RNA are linear polymers. DNA and RNA are nucleotide polymers and called nucleic acids [10]. However, proteins are more complex than DNA and RNA. Proteins are formed from a selection of 20 building blocks, called amino acids, whereas DNA and RNA are formed from four monomer units; nucleotide units [7]. Proteins, DNA and RNA with different types of monomers are also classified as copolymers [5]. The structure of a protein depends on the sequence in which individual amino acids are connected. All proteins are polypeptides. A protein has a polyamide backbone with different side chains attached to the backbone. A nucleic acid has an alternating sugar-phosphate backbone with a different amine base attached to it [11]. The structure of a nucleic acid depends on the sequence of individual nucleotides [11].

DNA and RNA consist of a large number of linked nucleotides, each composed of a sugar, a phosphate, and a base. Sugars linked by phosphates form a common backbone, whereas the bases vary among four different kinds [7]. RNA like DNA are long, linear (long un-branched) polymers consisting of nucleotides joined by 3 to 5 phosphodiesters. In both DNA and RNA, the heterocyclic amino base is bonded to C1' of the sugar, whereas the phosphoric acid is bonded by a phosphate ester linkage to the C5' sugar position [11].

The structure of RNA differs from that of DNA in two respects. The sugar units in RNA are riboses, and the sugar in DNA is 2'- deoxyriboses [7, 11]. The other difference is that one of the four major bases in RNA is uracil (U) instead of thymine (T). Thus, four monomer units in DNA are adenine, cytosine, guanine, and thymine, whereas in RNA adenine, cytosine, guanine, and uracil [7]. The sugars in nucleic acids are linked to one another by phosphor-diester bridges (with negative charges). The chain of sugars linked by phosphodiester bridges is referred to the backbone of the nucleic acids. Whereas the backbone is constant in

DNA and RNA, the bases vary from one monomer to the next.

Though chemically DNA and RNA are similar, DNA and RNA differ in size. Molecules of DNA are enormous. They have molecular weights of up to 150 billion and length of up to 12 cm when stretched out, and they are found mostly in the nucleus of cells [11]. In contrast, molecules of RNA are much smaller as low as 35 kDa in molecular weight and are found mostly outside the cell nucleus [11].

CONCEPT OF AVERAGE MOLECULAR WEIGHT AND THEIR DETERMINATION

In this chapter, molecular weight as a dimensionless term is used instead of molar mass which is defined as the mass of 1 mole of a substance, $g.mol^{-1}$ in SI units. The M is the most characteristic feature of polymeric compounds. Unlike simple low molecular substances, polymers do not have unique molecular weights. Practically there are no polymers whose molecules are all strictly of the same size or have the same degree of polymerization [12], due to the random nature of polymerization reactions [13]. Most macromolecules consist of a mixture of chains of different number of units. Hence, one uses the term average molecular weight when describing the M of these macromolecules [14], and that is why in polymer science and technology, the concept of average molecular weight is used. A series of polymeric compounds of the same chemical structure differing only in molecular weight is known as a polymer-homologous series [12].

Different methods for calculation of average molecular weights are used, resulting in different numerical values, depending on the methods by which they are measured. Thus, different numerical values for molecular weights of polymers have already been defined as average molecular weights (M_n, M_w, M_z, M_v). Generally, the average values vary in the following order ($M_n < M_v < M_w < M_z < M_{z+1}$). The number-average molecular weight (M_n) of a polydisperse polymer does not coincide with its weight-average (M_w) value, and the value of M_w is always greater than M_n. The mean average molecular weight, M_m, is calculated using the following equation [15]:

$$\overline{M_m} = \overline{M_n}\,\overline{M_w} \tag{1}$$

Various procedures and methods have been used for the determination of different average molecular weights, in which various kinds of averages can be determined experimentally [13]. There is no difference in the behavior, study, or testing of natural and synthetic polymers. Techniques suitable for application to synthetic polymers are equally applicable to the study and behavior of natural polymers [14]. Methods and procedures for determining average-molecular weights, the distribution and DI for biopolymers are mostly similar to synthetic polymers.

The calculation procedure of different average-molecular weights in size exclusion chromatography (SEC), may be performed as follows. The height of chromatogram of each fraction ($h_{i,}$), is proportional to its weight fraction, w_i, which in turn is proportional to the product of number of molecules, n_i, and their molecular weight, M_i according to:

$$h_i = k.w_i = k.n_1.M_1 \tag{2}$$

Where k is a constant and represents the relationship between the signal height and its weight fraction [16 - 18]. Practically, each chromatogram is divided into several equal segments and the corresponding heights are determined. Average-molecular weights ($\bar{M}_n$, $\bar{M}_v$, $\bar{M}_w$, and $\bar{M}_z$) are expressed as summation of individual segments and calculated from the following equations:

$$\overline{Mn} = \frac{\sum_{i=1} wi}{\sum_{i=1} ni} = \frac{\sum_{i=1} ni\,Mi}{\sum_{i=1} ni} = \frac{\sum_{i=1} hi}{\sum_{i=1} \left(hi / Mi \right)} = \tag{4}$$

$$\overline{Mw} = \frac{\sum_{i=1} wi\,Mi}{\sum_{i=1} wi} = \frac{\sum_{i=1} ni\,Mi^2}{\sum_{i=1} ni\,Mi} = \frac{\sum_{i=1} hi\,Mi}{\sum_{i=1} hi} \tag{4}$$

$$\overline{Mz} = \frac{\sum_{i=1} wi\,Mi^2}{\sum_{i=1} wi\,Mi} = \frac{\sum_{i=1} ni\,Mi^3}{\sum_{i=1} ni\,Mi^2} = \frac{\sum_{i=1} hi\,Mi^2}{\sum_{i=1} hi\,Mi} \tag{5}$$

$$\overline{M_v} = \left(\frac{\sum_{i=1} w_i M_i^a}{\sum_{i=1} w_i} \right)^{1/a} = \left(\frac{\sum_{i=1} n_i M_i^{1+a}}{\sum_{i=1} n_i M_i} \right)^{1/a} = \left(\frac{\sum_{i=1} h_i M_i^a}{\sum_{i=1} h_i} \right)^{1/a} \tag{6}$$

where h_i and M_i are the peak height and molecular weight of compound i. The value of h_i is read directly from the chromatogram. One can also calculate the average-molecular weight, using the area for compound i,

$$\overline{Mn} = \frac{\sum_{i=1} wi}{\sum_{i=1} ni} = \frac{\sum_{i=1} ni\,Mi}{\sum_{i=1} ni} = \frac{\sum_{i=1} Ai}{\sum_{i=1} \left(Ai / Mi \right)} \tag{7}$$

$$\overline{Mw} = \frac{\sum_{i=1} wi\,Mi}{\sum_{i=1} wi} = \frac{\sum_{i=1} ni\,Mi^2}{\sum_{i=1} ni\,Mi} = \frac{\sum_{i=1} Ai\,Mi}{\sum_{i=1} Ai} \tag{8}$$

$$\overline{Mz} = \frac{\sum_{i=1} wi\,Mi^2}{\sum_{i=1} wi\,Mi} = \frac{\sum_{i=1} ni\,Mi^3}{\sum_{i=1} ni\,Mi^2} = \frac{\sum_{i=1} AiM_i^2}{\sum_{i=1} AiM_i} \tag{9}$$

$$\overline{M_v} = \left(\frac{\sum_{i=1} w_i M_i^a}{\sum_{i=1} w_i} \right)^{1/a} = \left(\frac{\sum_{i=1} n_i M_i^{1+a}}{\sum_{i=1} n_i M_i} \right)^{1/a} = \left(\frac{\sum_{i=1} A_i M_i^a}{\sum_{i=1} A_i} \right)^{1/a} \tag{10}$$

where A_i is the peak area. The two methods seem to give the same values for average molecular weights. However, the area method is much more accurate and reliable due to the two following reasons: (a) if the chromatogram is symmetrical, one can treat it as a Gaussian distribution, *i.e.* the area method is just as good as the height method, actually, the two methods are identical in this case; (b) if the chromatogram is not strictly Gaussian [15]. The maximum height does not necessarily represent the median and the area method would minimize the error.

$$M_n = \frac{\int_0^\infty MN(M)dM}{\int_0^\infty N(M)dM} \tag{11}$$

$$M_W = \frac{\int_0^{\infty} M.W(M)dM}{\int_0^{\infty} W(M)dM} = \frac{\int_0^{\infty} M^2 N(M)dM}{\int_0^{\infty} M.N(M)dM} \quad \textbf{(12)}$$

$$M_z = \frac{\int_0^{\infty} M^2 W(M)dM}{\int_0^{\infty} M.W(M)dM} = \frac{\int_0^{\infty} M^3 N(M)dM}{\int_0^{\infty} M^2.N(M)dM} \quad \textbf{(13)}$$

$$M_v = \left(\frac{\int_0^{\infty}(M^a W(M)dM)}{\int_0^{\infty}(W(M)dM)}\right)^{1/a} = \left(\frac{\int_0^{\infty} M^{a+1} N(M)dM}{\int_0^{\infty} M.N(M)dM}\right)^{1/a} \quad \textbf{(14)}$$

Alternatively, the average molecular weights can be expressed in integral forms as follows [19]:

The M_w is particularly sensitive to the presence of larger species, whereas the M_n is sensitive to the presence of lower molecular weights.The value of viscosity-average molecular weight (M_v) is different from other average molecular weights [2]. It is not a unique value, and varies from one solvent to another, since it is a function of *a*, *K*, and [η] [20, 21]. The value of M_v can be obtained experimentally. It can be calculated once the values of the constants are determined in a polydisperse polymer samples. The viscometry method is not an absolute technique, since the M_v value must be calibrated using an absolute molecular weight determination technique [14]. One can employ an absolute method as a calibrated technique to calibrate the molecular weight of a sample obtained from a relative method [14]. Light scattering, ultra-centrifugation, *i.e.* sedimentation, and collective methods, *e.g.* osmometry, end-group determinations are absolute methods for molecular weight determinations [4, 8, 14, 22 - 24], because there is a direct mathematical connection between molecular weight and the particular property used to determine it [14].

Electrophoresis is generally used to determine molecular weight of biopolymers particularly for proteins, DNA and RNA. Electrophoresis usually separates the components by molecular size as well as by electro-phoretic mobility [4]. Electrophoresis is a relative method for molecular weight determination. Thus, a series of standard samples with definite molecular weights is required to determine molecular weight and MWD of unknown samples [14, 24].

CONCEPT OF DISPERSITY AND MOLECULAR WEIGHT DISTRIBUTION

The constitution of a polymer system is described either by a set of different average molecular weights or by the distribution [1]. Nearly all synthetic macromolecules are polydisperse polymers, due to the random nature of the polymerization reactions by which they are formed [4]. In nature, some macromolecules occur naturally as polydisperse samples. Therefore, large molecules of a polymer sample may also contain molecules relatively smaller and larger than the intermediate size. Hence, all polymers are more or less heterogeneous with respect to molecular weight.

$$M_n < M_v < M_w < M_z < M_{z+1} \qquad \textbf{(15)}$$

The degree of dispersity *i.e.* dispersity index (DI) or dispersity is a new term for the previous terms "polydispersity" and for the original term "poly-molecularity" [8, 25]. The new term was given by the International Union of Pure and Applied Chemistry (IUPAC). In this chapter and in the following, in order to avoid confusion, the term "dispersity" replaces polydispersity. The DI is generally expressed as the ratios of two different types of average molecular weights. Various average molecular weights will have the same value if the polymer is perfectly monodisperse, *i.e.*, if all the molecules are of the same molecular weight. Otherwise, the averages vary in the following order as given in the following equation as well as it is illustrated in Fig. (**1**) [13, 23, 24].

The disparity between average molecular weights provides a measure of the degree of heterogeneity in the molecular weight distribution [2]. The z – and (z+1) averages are important for very broad polydisperse polymers [26].

The molecular weight distribution (MWD), is generally expressed as the ratios of two different types of average molecular weights. Thus, it is necessary to determine different average molecular weights. It is useful to calculate two dispersity indices (DI_1, DI_2) as follows:

$$DI_1 = I_1 = M_w / M_n \qquad \textbf{(16)}$$

$$I_2 = I_2 = M_z / M_w \qquad \textbf{(17)}$$

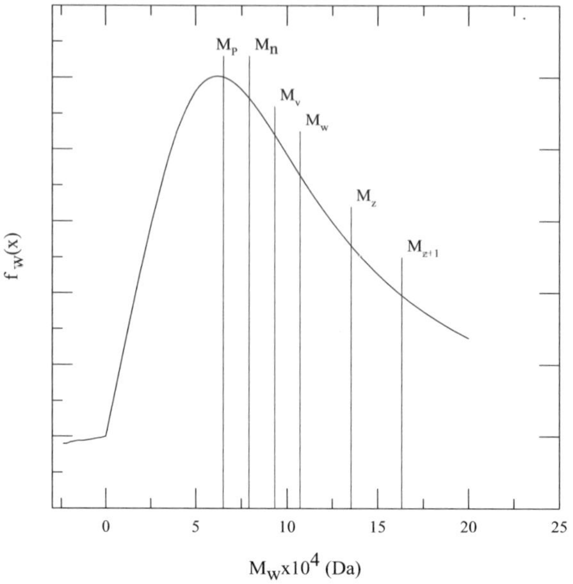

Fig. (1). A typical molecular weight distribution profile with a semi-qualitative comparison of different average molecular weights.

The two indices can be used to evaluate the level of heterogeneity of different samples of a polymer [27]. The DI_2 is sensitive to the presence of the high molecular species. The DI is usually obtained from the ratio of M_w/M_n [2, 8, 13]. For mono-dispersed polymers the indices are close to unity, and the higher the indices, the greater the difference between the average weights and the larger the distribution [27]. In the case of monodisperse polymers, the average molecular weights are determined by different methods, likely to coincide and reach an identical value [12]. For instance, in the mono-disperse system, $M_w = M_n$. The molecular weights of samples with a wide MWD may differ by a factor of more than two.

The constitution of a polymer may be described either by a set of different average molecular weights or by the distribution function itself [1]. The MWD can be expressed in term of the number-distribution function [N(M)] or the weight-distribution function [W(M)]. The value of N(M) is the relative number of chains of molecular weight, and W(M) is the relative weight of chains of molecular weight. One can choose the normalization constant: (1) the total number of chains $\int_0^\infty N(M)dM = 1$; (2) $\int_0^\infty M\, N(M).dM = 1$; or (3)

$\int_0^\infty M^2 N(M)\, dM = 1$. Normalization has no physical significance; thus it is essential to ensure that any inference drawn from an experimental MWD does no depend on normalization [19]. Dispersity may be illustrated graphically [26]. The graphical presentation is used for qualitative and quantitative evaluation of dispersity.

Chemically heterogeneous macromolecules are polymers that contain units of different composition in the same chain. For example, some of the units may be completely esterified, while others may contain free hydroxyl groups. The chemical composition of such polymers is conventionally characterized by the average percentage content of their functional groups, *e.g.* acetyl [12]. Copolymers are more complex than most of homo-polymers. The chemical composition of a copolymer may not be uniform. Non-uniformity results in compositional heterogeneity [22], which is required to characterize the copolymers in terms of chemical compositional distribution, usually fractionation followed by chemical characterization of the fractions [22]. Chemical composition and size, *i.e.* molecular weight of a copolymer chain may vary [22]. Two polymers may have exactly the same or similar average molecular weights but very different MWDs [22].

Dispersity Correction Factor

The value of [η] varies with M_v, for a homologous series according to the Mark-Houwink-Sakurada (MHS) equation [2, 28]:

$$[\eta] = K.\ M_v{}^{a} \qquad \textbf{(18)}$$

Equation (1) can be rearranged and result in a modified MHS equation as follows:

$$[\eta] = K\, M_v{}^{a} = K\, (M_v / M_w)^{a} . M_w{}^{a} = K\, q_{MHS}\, M_w{}^{a} \qquad \textbf{(19)}$$

The value of q_{MHS} is a statistical function of MWD. It is a measure of the width of the MWD as well as the probability of molecular weight distribution curve. The value of q_{MHS} varies from one sample to another. It is a function of *a* and the average-molecular weights (M_v, M_w). It can be calculated using M_w, the ratio of M_v/M_w and the exponent *a*. The value of q_{MHS} is equal to $(M_v/M_w)^a$. Alternatively, the value of q_{MHS} can be calculated using a numerical method and average-molecular weights other than M_v, *e.g.* (M_n, M_w, M_z) according to [21, 29]:

$$q_{MHS} = (M_w / M_n)^{b}\, (M_z / M_w)^{c} \qquad \textbf{(20)}$$

On the whole, the correction factor, q_{MHS}, is a function of exponent *a*, and (M_n, M_v, M_w, M_z). The precision of the q_{MHS} value depends on the precision of both *a* and average-molecular weights.

Description of Some Models of Distribution

The MWD may be mono- bi-, tri-, or polymodal. In a mono-modal curve, the distribution profile exists as a maximum peak. A bimodal curve is often characteristic of a polymerization occurring under two different environments [14]. In a bimodal, the differential weight distribution function, W(r), has two maxima, often observed in polymer fractions obtained by fractionation [4]. The distribution function may be discrete, *i.e.* take on only certain specified values of the random variable(s), or continuous, *i.e.* take on any intermediate values of the random variables(s), in a given range. Most distributions in polymer science are intrinsically discrete, but it is often convenient to regard them as continuous or to use distribution functions that are inherently continuous [8]. Distribution functions may be integral forms or cumulative forms, *i.e.*, give the proportion of the population for which a random variable is less than or equal to a given value. Alternatively they may be a differential function or probability density functions, *i.e.*, give the proportion of the population for which the random variable is within an interval of its range [8]. The distribution function is a mathematical expression describing the distribution of molecular sizes. Discontinuous functions, frequency functions, give the distribution of statistical weights of M_i, whereas cumulative distribution functions give the summation over all statistical weights up to M_i [4]. The statistical functions may be the number (n_i), weight (w_i) or z (z_i)-fractions number, weight and z- average, respectively [4].

Number-Distribution Function

The function representing the relation between the mole fraction and molecular weight is called the number distribution function [N(M)]. It is obtained by plotting the mole fraction *versus* molecular weight [8, 23].

Weight-Distribution Function

The function representing the relation between the weight fraction and the molecular weight is called the differential weight distribution function [W(M)]. The differential molecular weight distribution curve is obtained by plotting the weight fraction *versus* molecular weight [8, 23].

Gaussian Distribution

Gaussian distribution is a statistical distribution of a chain end-to-end-distribution, which is symmetrical about the mean value as shown in Fig. (**2**) [4, 30]. It takes a bell-shaped profile, which is symmetrical about the mean value, where the peak of the bell is the mean μ, and the width is determined by the standard deviation σ. The following equation is used to calculate a Gaussian distribution [4, 30, 31]:

$$f_w(x) = \frac{1}{\sqrt{2\pi\sigma^2}} e^{\frac{(X-\mu)^2}{2\sigma^2}} \quad \textbf{(21)}$$

where the standard normal distribution occurs when $\mu = 0$, and $\sigma^2 = 1$. The quantity σ in the equation 21 acts as a curve-fitting parameter. It describes the width of the distribution, and thus, the deviation from the mean value. The standard deviation of the molar distribution of the degree of polymerization can be calculated from the number- and weight averages. The standard deviation for a Gaussian distribution is an absolute measure of the width of such distribution [30].

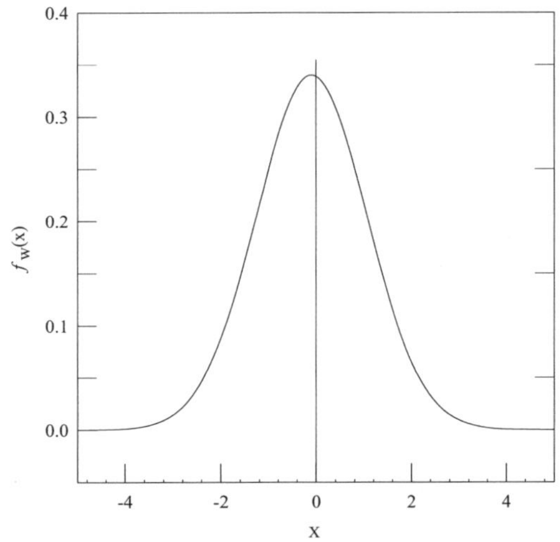

Fig. (2). A typical profile for a Gaussian distribution.

Normal distribution is an alternative name for Gaussian distribution. In mathematics, the Gaussian distribution is called normal distribution. In polymer science, the Schultz-Flory distribution is often called normal distribution [4, 30]. Gaussian distribution is used occasionally to describe the MWD of a narrow distribution polymer as an approximation of the Poisson distribution [4].

Logarithmic Normal Distribution

Logarithmic normal distribution is similar to Gaussian distribution, but with ln r replacing r, where r is the measure of the individual molecular sizes, *e.g.* degree of polymerization [4]. The differential logarithmic normal distribution has the same mathematical form as the Gaussian distribution with the small difference that the logarithm of the property occurs in place of the property itself, log X instead of X. In other words, the log-normal distribution is given by a Gaussian distribution with respect to log M. The distribution takes one of the following forms [4, 8, 30]:

$$f_w(x)dx = \frac{1}{a_1 x.\sqrt{\pi}}.\exp\left(-\frac{1}{a_1^2}\left[\ln\left(\frac{x}{b_1}\right)\right]^2\right)dx \qquad \textbf{(22)}$$

$$f_w(x) = \frac{1}{\sqrt{2\pi X.\sigma n^2}} e^{\frac{(\log M - \log \mu)^2}{2\sigma n^2}} \qquad \textbf{(23)}$$

where x is a parameter characterizing the chain length, such as relative molecular weight or degree of polymerization, a_1 and b_1 are positive adjustable parameters [8]. The values of log μ and σ_n in equation 23 are the mean and the standard deviation of the molar distribution, respectively. The curve is symmetric about logμ [31].

If the molar logarithmic normal distribution is plotted in terms of the weight fraction, the shape of the curve remains essentially unchanged. However, the maximum value for the distribution is not identical with the number- or weight-average degree of polymerization. The plot of log (integral molecular weight fraction) *versus* log M, or log (cumulative weight fraction) *versus* log M, results in a straight line when log scales for both X and Y are used [30]. The logarithmic normal distribution is more useful than the Gaussian distribution in describing broader distributions, especially those from which the lower molecular weight "tail" has been removed [4].

Poisson Distribution

A Poisson distribution occurs when a constant number of polymer chains begin to grow simultaneously, and when the addition of monomer units is random and occurs independently of other monomer units to the macromolecule that is growing. The Poisson distribution generally occurs in living polymers [30]. The discrete distribution with the differential weight-distribution function takes the following form [8]:

$$f_w(x) = \frac{x.e^{-a_1} a_1^{x-1}}{(a_1+1)(x-1)!} \quad \textbf{(24)}$$

where x is a parameter characterizing the chain length, such as relative molecular weight or degree of polymerization and a_1 is a positive adjustable parameter [8].

Most Probable Distribution

The most probable distribution describes the relative ratios of polymers with different lengths after a polymerization process, based on their relative probabilities of occurrence. The following equation is used to calculate the distribution:

$$f_w(x) = a_1^2.x(1-x)^{x-1} \quad \textbf{(25)}$$

where x is a variable characterizing the chain length, *e.g.* number average molecular weight, degree of polymerization, and a_l is an empirically-determined constant [4, 8]. The form of this distribution implies that shorter polymers are favored over longer ones. This type of distribution is occurs in poly-condensation and most free radical polymerization possesses. It is in contrast to Poisson distribution [30]. The most probable distribution often referred to is Flory distribution or Flory-Schulz distribution, where $M_w/M_n = 2$ [30]. Alternatively, the most probable distribution is described as follows:

$$M_n : M_v : M_w = 1/\ [(1+a)\Gamma(1+a)]^{1/a}/2 \quad \textbf{(26)}$$

where $\Gamma(1+a)]$ is the gamma function of $1 + a$. As the value of a varies from 0.50 to 1.00, M_v/M_n for this particular molecular weight distribution increases from 1.67 to 2.00 [2]. When $a = 1$, $M_v = M_w$. The results just cited for the most probable distribution can be extended to the broader conclusion that the M_v will always be considerably closer to M_w than to M_n for any distribution likely to be encountered in a high polymer [2].

Determination of Molecular Weight Distribution

There are several ways to measure MWD: (1) gel permeation chromatography (GPC) or size exclusion chromatography (SEC) has been used for the determination of molecular weight and molecular-weight distribution of polymers [12, 23, 32]; (2) fractionation of a polymer with a broad MWD into narrower MWD fractions and determination of molecular weight of the narrow fractions [33]. Fractionation is helpful in evaluating the true range of dispersity of polymers with a narrow MWD [34]; and (3) the molecular weight distribution curve of a polymer can be also obtained directly from the data on sedimentation of a disperse polymer sample using an ultracentrifugation procedure [12].

Size Exclusion Chromatography

SEC has been used for many decades to estimate M and MWD, and DI of biopolymers and synthetic polymers through the use of calibration curves between molecular weight and the distribution coefficient, K_{av}.

$$K_{av} = \frac{V_R - V_0}{V_C - V_0} \qquad \textbf{(27)}$$

where V_R, V_0, and V_C represent solute elution volume, void volume, and the total bed volume of fluid and SEC media combined, respectively [35]. For the determination of the retention volume, it is much better to use the area method, the center of mass, than that of the height method, the peak maximum [36]. The values of A_i or h_i and V_{Ri} are read directly from the chromatogram. The values of V_{Ri} can be converted into M_i.

The chromatographic method separates the molecules according to their sizes. The larger is the molecule the greater is the exclusion from various sized pores in the stationary phase material. Accordingly, the higher is the molecular weight, the lower is the elution volume. Fig. (**3**) shows three SEC chromatograms, A, B, and C, representing the highest, intermediate and the smallest macromolecules, respectively. Monodisperse polymer standards are required to translate elution volumes to molecular weights. From the SEC distribution curve, the various molecular weight averages may be also calculated [13, 16, 37].

The separation is based on the hydrodynamic volume of a polymer molecule. The hydrodynamic volume is proportional to the product, [η].M. The molecular weights of polymers do not correlate linearly with retention volumes, because retention volume is a function of effective hydrodynamic volumen, hydrodynamic volume [η]. M, correlates with retention volume, *i.e.* elution volume, Ve [22, 30]. In order to construct a universal calibration curve of the hydrodynamic volume of a polymer as a function of the elution volume; *Ve*: the intrinsic viscosity and SEC chromatograms of well-characterized mono-disperse polymer samples are measured [32]. Coupling of a SEC with an automatic capillary viscometer results in more accurate data for the dispersity indices than SEC alone [16, 38]. The coupling method also enables one to determine the resolution factor for a given SEC separation, column, system employed [38].

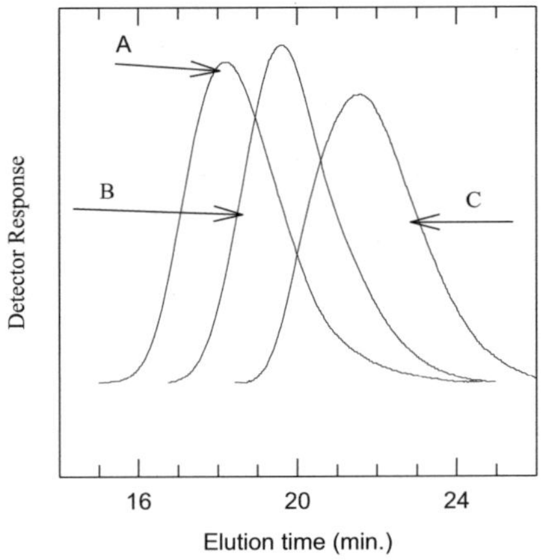

Fig. (3). Three SEC chromatograms, A, B, and C, representing the highest, intermediate and the smallest macromolecules, respectively.

According to the universal calibration theory, at a given elution volume two polymers 1 and 2 have the same hydrodynamic volume, [η].M [15]:

$$[\eta]=KM^{a} \quad \textbf{(28)}$$

$$[\eta]_1.M_1=[\eta]_2.M_2 \quad \textbf{(29)}$$

$$\log M_2=\frac{1}{1+a_2}.\log\frac{K_1}{K_2}+\frac{1+a_1}{1+a_2}.\log M_1 \quad \textbf{(30)}$$

Generally K_1 and a_1 are known in the literature, and K_2 and a_2 are either known in the literature or can be obtained by intrinsic-viscosity, osmotic-pressure or light-scattering techniques [15].

A direct measurement of the MWD can be carried out using SEC. SEC data often reported in terms of weight fractions, wt_i, corresponding to molecular weight, M_i. These data must be converted to a discrete log MWD [39]. For linear polymers, the viscosity MWD and the discrete log MWD function calculated from SEC data should be the same. Agreement is observed between the viscosity MWD and the SEC MWD [39]. In other words, MWD obtained from the viscosity procedure yields reliable results as well as SEC procedure [39].

Analyzing a complex mixture of unknown macromolecules is often challenging in GPC/SEC. GPC/SEC is used to separate, identify and characterize macromolecules with respect to their MWD. The precision and accuracy of the results depend on the selection of the proper separation columns. GPC/SEC separates macromolecules based on their hydrodynamic volume and, therefore, allows macromolecule chains with different lengths to be separated into small fractions. However, MWD determined by SEC can be influenced by a number of factors. Peak-broadening effects and incomplete resolution can give misleading information [22].

Qualitative determination of MWD is a difficult task. SEC instrument with suitable standard samples would result in a qualitative evaluation of MWD for polymer samples. However, quantitative evaluation of MWD is more difficult than that of a qualitative one. Combining SEC with light scattering or another absolute method for molecular weight method is an alternative solution for the quantitative evaluation of MWD [14, 24].

Generally, silica- and polymer-based materials are used as packing materials in SEC columns for determination of molecular weights and MWD of polymers. If silica-based materials were used to separate biopolymers such as pullulan, there should be interaction between silanol groups of silica-based packing materials and biopolymers. The interaction could perturb the validity of the calibration curve for the SEC process, and thus results in deviation for molecular weight data [21]. Generally, silica-based packing materials are chemically modified to remove the effect of the silanol groups, which tend to have a negative effect on biopolymer separations [40]. The polymer-based packaging materials have advantages over silica-based, due to lack or negligible negative effect on biopolymer separations.

Differential refractive index (DRI) is the most common means of mass detection for the MWD analysis of polymers by SEC [41]. A disadvantage of DRI, is that this detector only provides concentration information and no information about composition and heterogeneity. A chemiluminescent nitrogen detector (CLND) with SEC was developed to estimate average MWD of peptides and food grade protein hydrolysates, as well as protein hydrolysate-based food [42]. The

DRI/CLND SEC analysis can detect differences between two lots of a polymer which have similar MWD, but dissimilar chemical composition distributions [42]. Mass spectrometry (MS), provides structural information, differentiating molecules with small differences in molecular weight [43].

Pullulan is used as a standard in SEC, to determine the average molecular weights (M_n, M_w and M_z) and MWD for linear biopolymers by constructing a universal calibration curve. Up to date, there is no commercially available β-glucans or other similar polysaccharide standards that are comparable with pullulan with respects to the narrow dispersity. Generally, in order to have a good resolution as a function of molecular size, an appropriate solvent should be selected for the investigated polymers, polysaccharides or water-soluble polymers [21].

Fractionated dextrans have been also used as standard materials for M_w and MWD determinations of biopolymers and water-soluble polymers as well as for construction of universal calibration curve for evaluation of SEC results [44, 45].

Fractionation

Generally, the composition of a polymer substance is not homogeneous. MWD is a general feature for all synthetic polymers and polydisperse biopolymers. It is a consequence of the particular nature of the polymerization process by which synthetic polymers are made [46]. Biopolymers are usually formed in nature *via* biological/ biochemical processes. In nature, some of macromolecules occur commonly as polydisperse materials. Biopolymers are also susceptible to degradation under environmental conditions like temperature, humidity, oxygen, light, and others. Biopolymers may react with other biomaterials through biochemical and physical reactions. Thus, the size of biopolymer species formed in nature yields also polydisperse materials. However, some biopolymers may occur as relatively monodisperse samples.

Fractionation of a polymeric substance means separation of that substance into its different molecular species, using a suitable experimental technique, in order to obtain homogeneous fractions [46]. The disparity between different average molecular weights may be made small by careful fractionation [2]. Fractionation techniques separate polymers based on molecular weight or chemical composition. Practically, partial separations by molecular weight and chemical composition are often achieved simultaneously [22]. It is possible to separate and characterize a complex sample containing homo- and copolymer species based on chemical heterogeneity and molecular weight [22]. By fractionating a polymer and determining the molecular weights of each fraction, the MWD curve can be obtained. Differentiation of the integral curve gives the differential distribution curve. The basic characteristics of a differential curve are the position of its peak and its width. The broader the distribution curve the wider the molecular weight distribution [12].

Fractionation is an experimental procedure to separate a polymer sample containing different species based on their sizes or compositions [47]. The fractionation methods include fractional precipitation, fractional distribution between two phases, fractional dissolution, and fractional extraction [47]. In the fractional extraction, fractional solution or fractional elution, the polymer sample is successively extracted with a solvent, whose power gradually increases. The residues are removed at each stage and a series of fractions of increasing molecular weight are obtained from the solutions [4]. A fractional extraction method was more efficient than the conventional precipitation fraction in obtaining fractions with a narrow MWD [47]. In fractional precipitation of biopolymers, separation of different polymers from mixtures in aqueous solutions, *e.g.* for proteins, may be achieved by variation of pH, iso-electric precipitation, by variation of ionic strength, salting-in and salting-out, or by the use of organic solvents, often ethanol, at low temperatures to prevent denaturation [4]. The procedure of SEC (GPC) is also a method of a polymer fractionation. A series of fractions may be collected from the effluent, with gradually increasing their molecular weights. The MWD may be calculated from the chromatograms [4].

Sedimentation and Diffusion

Sedimentation

The technique of sedimentation equilibrium in the analytical ultracentrifuge can provide absolute sizes and size distribution information in terms of molecular weight averages and MWDs [42]. The MWD can be determined from a collection of fractions by employing a size or a composition characterization method [42].The sedimentation equilibrium method gives quantitative results [14, 24].

The size of particles is determined by the rate of their sedimentation. Dissolved particles with density ρ_2 travel through a solvent of density ρ_1 under the influence of a centrifugal field. Particles sediment in the direction of the centrifugal field, in a direction perpendicular to the axis of rotation. When $\rho_2 > \rho_1$ and the centrifugal field is strong enough as in an ultracentrifuge, the procedure may be performed when the polymer solution is much diluted, *i.e.* a very small amount of the polymer sample is used to prepare the solution [12, 30, 48]. Sedimentation works against diffusion caused by Brownian motion. With a sufficiently weak centrifugal field, relative to particle and density differences, a stage will be reached where the rate of sedimentation equals the rate of diffusion, and a state of sedimentation equilibrium occurs [30].

The sedimentation coefficient (S), like the diffusion coefficient (D), depends on concentration. Both of them must be extrapolated to infinite dilution for determination of molecular weight of a polymer, and molecular weight is obtained from the following equation:

$$M_w = \frac{R.T.S_0}{D_0(1-\bar{v}_2.d_0)} \quad \textbf{(31)}$$

where D_0 and S_0 are diffusion and sedimentation constants at infinite dilution, respectively [12]. In principle, it is possible to evaluate various averages of the sedimentation coefficient from the distribution of the concentration gradient, dC/dr, in the cell as follows [48]:

$$S_n = M_n \int_0^\infty W(M).M^{-1}.S(M)dM \quad \textbf{(32)}$$

$$S_w = \int_0^\infty W(M).S(M)dM \quad \textbf{(33)}$$

$$S_z = (1/M_W)\int_0^\infty W(M).M.S(M)dM \quad \textbf{(34)}$$

where $W(M) = (C/M)/\int_0^\infty C(M)dM = C(M)/C_0$ is the weight distribution of the polydisperse polymer simple in the cell. The S_n, S_w, and S_z are the number average, weight average and z-average of the sedimentation coefficient [47]. From these data M_n, M_w, M_z can be determined. *i.e.*, MWD can be obtained.

In many cases, migration of the maximum of the quantity dC/dr is determined [48]. In the ultracentrifugation method, the refractive index difference between polymer solution and solvent (dn), and the refractive index gradient (dn/dr) are determined [48]. With certain assumptions dn/dr can be related to the concentration gradient

$$dn/dr = R.dC/dr \quad \textbf{(35)}$$

where R is the specific refractive index increment. The dependence of S on the molecular weight can be given by a power law as follows [12, 48]:

$$S = K_S . M^{as} \quad \textbf{(36)}$$

where K_s and a_s are empirical constants for each polymer-solvent system at given values of temperature and pressure [12, 48]. The rate of sedimentation at definite time intervals can be measured by photo-metrically. The variation of concentration gradient during centrifugation is called the distribution curve or the sedimentation distribution curve [12]. Sedimentation in an ultracentrifuge is an absolute method to measure the molecular weight of a polymer, as no assumptions are made on macromolecular conformations [12].

Diffusion

The quantitative relationship between D, defined by Ficks' first law as the ratio of flow per unit area of substance to concentration gradient, - J/(dC/dx) and the size of a diffusion particles is determined theoretically by the Stokes- Einstein equation [12, 49]:

$$D = \frac{RT}{6\pi.\eta.r.N_A} \quad \textbf{(37)}$$

where η, r, N_A are the viscosity of the medium, the radius of diffusing particle and Avogadro's number, respectively. The molecular weight of a spherical molecule is given by:

$$M_w = \frac{R^3T^3d}{162.\pi^2.\eta^3.r..N_A^2.D^3} \quad \textbf{(38)}$$

If the molecule is not spherical, a correction is made [12]. The dependence of the diffusion coefficient on the molecular weight can be expressed by the following equation [12, 48]:

$$D = K_D . M^{aD} \quad \textbf{(39)}$$

where K_D and a_D are empirical constants for each polymer-solvent system [12, 48]. A similar consideration is taken on the sedimentation coefficient for a diffusion procedure. From measurements in a diffusion cell, one obtains the various averages of the diffusion coefficient from the distribution of the concentration gradient (dC/dr) in the cell:

$$D_n = M_n \int_0^\infty W(M).M^{-1}.D(M)dM \tag{40}$$

$$D_w = \int_0^\infty W(M).D(M)dM \tag{41}$$

$$D_z = (1/M_W) \int_0^\infty W(M).M.D(M)dM \tag{42}$$

where W(M)= $C(M)/C_0$ [48].

Molecular Weight Distribution and Dispersity of Polymers

Some naturally occurring polymers such as certain proteins and nucleic acids consist of molecules with a specific molecular weight and are monodisperse [14, 24]. Proteins are almost the only source of truly monodisperse polymers [50]. Nature makes all these molecules exactly alike [50]. Branching may occur, which broadens the MWD [50]. Other natural polymers, polysaccharides similar to most of synthetic polymers consist of molecules with different molecular weights and are polydisperse samples [14, 24].

CONCLUSIONS

Determination of both different average -molecular weights and molecular weight distribution using various experimental procedures for polymers, synthetic and natural, with particular attention on biopolymers is reviewed. Nearly all synthetic polymers and some biopolymers are polydisperse and can be described in terms of one or more MWD functions. Among a different family of biopolymers, polysaccharides are polydisperse, whereas proteins, DNA and RNA are monodisperse. Chemical composition, size or molecular weight of a homopolymer or a copolymer chain may vary. Two polymers may have exactly the same or similar average molecular weights but very different MWDs.

SEC is a reliable procedure for determination of the relative MWD. A series of standard polymer samples with definite molecular weights are required to determine molecular weight and MWD of unknown samples. The SEC method also makes possible a direct and simple determination of the resolution factor of the separation system employed.The combination of SEC with light scattering or another absolute method for molecular weight determination is an alternative solution for the quantitative evaluation of MWD and calculation of the DI, for polymers with both narrow and wide distributions.

CONFLICT OF INTEREST

The author (editor) declares no conflict of interest, financial or otherwise.

ACKNOWLEDGEMENTS

Declared none.

REFERENCES

[1] L. Mandelkern, *An Introduction to Macromolecule.* 2nd ed Springer-Verlag: New York, 1983. [http://dx.doi.org/10.1007/978-1-4612-5494-2]

[2] P.J. Flory, *Principles of Polymer Chemistry.* Cornell University Press: Ithaca, New York, 1953.

[3] B.D. Rather, A.S. Hoffman, and F.J. Schoen, *Biomedical Engineering Desk Reference.* Elsevier: Amsterdam, 2009.

[4] M. Alger, *Polymer Science Dictionary.* Chapman & Hall: London, 1999.

[5] F. den Hollander, *Random Polymers.* Springer-Verlag: Berlin, 2009. [http://dx.doi.org/10.1007/978-3-642-00333-2]

[6] A.Y. Grosberg, and A.R. Khokhlov, *Statistical Physics of Macromolecules.* AIP Press: New York, 1994.

[7] J.M. Grosberg, J.L. Tymoczko, and L. Stryer, *Biochemistry.* 5th ed W.H. Freeman and Company: New York, 2002.

[8] IUPAC, *Compendium of Polymer Terminology and Nomenclature.* RSC Publishing: Cambridge, UK, 2008.

[9] E.A. López-López, M.A. Hernández-Gallegos, M. Cornejo-Mazón, and H. Hernández-Sánchez, "Polysaccharide-based nanoparticles", In: *Food Nanoscience and Nanotechnology,* H. Hernandez-Sanchez, G.F. Gutierrez-Lopez, Eds., Springer: New York, 2015, pp. 59-68. [http://dx.doi.org/10.1007/978-3-319-13596-0_4]

[10] D.U. Stilverthorn, B.R. Johnson, W.C. Ober, C.W. Garrison, and A.C. Stiverthorn, *Human Physiology: An Integrated Approach.* 5th ed W.H. Freeman and Company: San Francisco, 2009.

[11] J. McMurry, *Organic Chemistry.* 5th ed Pearson: Benjamin Cummings: Boston, 2000, pp. 1150-1192.

[12] A. Tager, *Physical Chemistry of polymers.* 2nd ed Mir Publisher: Moscow, 1978.

[13] B.M. Mandal, *Fundamentals of Polymerization.* World Scientific: New Jersey, 2013, pp. 1-35. [http://dx.doi.org/10.1142/9789814322850_0001]

[14] C.E. Charraher, *Polymer Chemistry.* 7th ed. CRC, Taylor & Francis: Boca Raton, 2007.

[15] S.F. Sun, and E. Wong, "Modified method of universal calibration for the determination of molecular weight and molecular weight distribution", *J. Chromatogr. A,* vol. 208, no. 2, pp. 253-259, 1981. [http://dx.doi.org/10.1016/S0021-9673(00)81936-6]

[16] M.R. Kasaai, J. Arul, and G. Charlet, "Intrinsic viscosity-molecular weight relationship for chitosan", *J. Polym. Sci., B, Polym. Phys,* vol. 38, no. 19, pp. 2591-2598, 2000. [http://dx.doi.org/10.1002/1099-0488(20001001)38:19<2591::AID-POLB110>3.0.CO;2-6]

[17] J.F. Johson, "Chromatography", In: *Encyclopedia of Polymer Science and. Engineering.* vol. 3. 2nd ed. John Wiley & Sons: NewYork, 1985, pp. 511-518.

[18] A.C. Wu, W.A. Bough, E.C. Conrad, and K.E. Alden Jr, "Determination of molecular-weight distribution of chitosan by high-performance liquid chromatography", *J. Chromatogr. A,* vol. 128, no. 1, pp. 87-99, 1976.
[http://dx.doi.org/10.1016/S0021-9673(00)84034-0] [PMID: 993304]

[19] J.V. Castro, R.M. Ward, R.G. Gilbert, and M.A. Fitzgerald, "Measurement of the molecular weight distribution of debranched starch", *Biomacromolecules,* vol. 6, no. 4, pp. 2260-2270, 2005.
[http://dx.doi.org/10.1021/bm050041t] [PMID: 16004470]

[20] M.R. Kasaai, "Calculation of Mark-Houwink-Sakurada (MHS) Viscometric Constants for Chitosan in any Solvent-Temperature System using Experimental Reported Viscometric Constants Data", *Carbohydr. Polym,* vol. 68, no. 3, pp. 477-488, 2007.
[http://dx.doi.org/10.1016/j.carbpol.2006.11.006]

[21] M.R. Kasaai, "Intrinsic Viscosity–Molecular Weight Relationship and Hydrodynamic Volume for Pullulan", *J. Appl. Polym. Sci,* vol. 100, no. 6, pp. 4325-4332, 2006.
[http://dx.doi.org/10.1002/app.22324]

[22] W.J. Freeman, "Characterization of polymers", In: *Encyclopedia of polymer Science and Engineering,* H.F. Mark, Ed., vol. 3. John Wiley & Sons: New York, 1985, pp. 290-327.

[23] J.F. Rabek, *Experimental Methods in Polymer Chemistry.* John Wiley & Sons: New York, 1980.

[24] C.E. Charraher, *Introduction to polymer chemistry, CRC.* Taylor & Francis: Boca Raton, 2007.

[25] IUPAC, *Compendium of Chemical Technology.* (the Gold Book), RSC Publishing: Cambridge (UK), 1997.

[26] D. Berek, "Molecular characterization of synthetic polymers by means of liquid chromatography", In: *Physical chemistry of macromolecules,* C.H. Chan, C.H. Chia, S. Thomas, Eds., CRC Press, Taylor and Francis: Boca Raton, 2014, pp. 223-331.
[http://dx.doi.org/10.1201/b16706-13]

[27] A. Denuzière, N. Yagoubi, A. Baillet, and D. Ferrier, "An improved statistical parameter allowing elaborate comparison of polymer molecular weight distribution by gel filtration chromatography: Application to chitosan", *S.T.P. Pharma Sci,* vol. 5, no. 6, pp. 481-485, 1995.

[28] C. Tanford, *Physical Chemistry of Macromolecules.* John Wiley Press: New York, 1961, pp. 390-412.

[29] R.E. Bareiss, "Polymolecularity correction factor", In: *Polymer Handbook,* J. Brandrup, E.H. Immergut, E.A. Grulke, Eds., Wiley–Interscience: New York, 1999, pp. VII/215-VII/245.

[30] H.G. Elias, *Macromolecules. 1: Structure and Properties.* 2nd ed Plenum Press: New York, 1984, pp. 281-371.
[http://dx.doi.org/10.1007/978-1-4899-2809-2]

[31] S.E. Cabaniss, Q. Zhou, and P.A. Maurice, "P.A., Chin, Y-P., Aiken, "A log normal distribution model for the molecular weight of aquatic Fulvic acids", *Environ. Sci. Technol,* vol. 34, no. 6, pp. 1103-1109, 2000.
[http://dx.doi.org/10.1021/es990555y]

[32] M. Gaborieau, and P. Castignolles, "Size-exclusion chromatography (SEC) of branched polymers and polysaccharides", *Anal. Bioanal. Chem,* vol. 399, no. 4, pp. 1413-1423, 2011.
[http://dx.doi.org/10.1007/s00216-010-4221-7] [PMID: 20967430]

[33] R. Koningsveld, "Preparative and analytical aspects of polymer fractionation", In: *Adv. Polym.Sci.* vol. 7. Springer: Berlin, 1970, pp. 1-69.
[http://dx.doi.org/10.1007/BFb0051108]

[34] F.W. BillMeyer, and L.R. Sierbert, "Application of the summative-fractionation method to the determination of Mw/Mn for narrow-distribution polymers", In: *Adv. Chem. Ser.* vol. 125. American Chemical Society: Washington, 1973, pp. 9-16.

[35] C.J. Fee, and J.M. Van Alstine, "Prediction of the viscosity radius and the size exclusion chromatography behavior of PEGylated proteins", *Bioconjug. Chem,* vol. 15, no. 6, pp. 1304-1313, 2004.
[http://dx.doi.org/10.1021/bc049843w] [PMID: 15546197]

[36] W.W. Yau, J.J. Kirkland, and D.D. Bly, *Modern Size Exclusion Liquid Chromatography.* Wiley Inter-Science: New York, 1979.

[37] J.Z. Knaul, M.R. Kasaai, V.T. Bui, and K.A. Creber, "Characterization of deacetylated chitosan and chitosan molecular weight review", *Can. J. Chem,* vol. 76, no. 11, pp. 1699-1706, 1998.

[38] J. Janca, and S. Pokorny, "Coupling of a gel permeation chromatograph and automatic capillary viscometer: II. Method for the determination of the molecular weight distribution and of the resolution factor", *J Chromatogr,* vol. 134, vol. 2, pp. 273-278, 1977.

[39] P.M. Wood-Adams, and J.M. Dealy, "Using rheological data to determine the branching level in metallocene polyethylenes", *Macromolecules,* vol. 33, no. 20, pp. 7481-7488, 2000.
[http://dx.doi.org/10.1021/ma991534r]

[40] R.S. Cooley, and P.A. Schweitzer, "High performance liquid chromatography", In: *Handbook of Separation Techniques for Chemical Engineering, 3rd ed,* P.A. Schweitzer, Ed., McGraw-Hill, Inc: New York, 1997, pp. 1-669-1-679.

[41] F.J. Kolpaka, J.E. Brady, and E.M. Fujinari, "The determination of compositional and molecular weight distributions of cationic polymers using chemiluminescent nitrogen detection (CLND) in aqueous size exclusion chromatography", In: *Developments in Food Science: Instrumental Methods in Food and Beverage Analysis,* D.L. Wetzel, G. Charalambous, Eds., vol. 39. Elsevier: Amsterdam, 1998, pp. 467-473.
[http://dx.doi.org/10.1016/S0167-4501(98)80020-4]

[42] F.M. Groycoolea, and I.S. Chronakis, "Specific methods for the analysis of identity and purity of functional food polysaccharides", In: *Developments in Food Science. Instrumental Methods in Food and Beverage Analysis,* D.L. Wetzel, G. Charalambous, Eds., vol. 39. Elsevier: Amsterdam, 1998, pp. 99-140.
[http://dx.doi.org/10.1016/S0167-4501(98)80008-3]

[43] D.J. Burinsky, and F. Wang, "Mass spectral characterization", In: *Handbook of Iisolation and Characterization of Impurities in Pharmaceuticals,* S. Ahuja, K.M. Alsante, Eds., Academic Press: Amsterdam, 2003, pp. 249-299.

[44] Q. Wang, and S.W. Cui, "Understanding the physical properties of food carbohydrates", In: *Food carbohydrates,* S.W. Cui, Ed., CRC, Taylor & Francis Group: Boca Raton, 2005, pp. 161-217.

[45] M.R. Kasaai, "Dilute solution properties and degree of chain branching for dextran", *Carbohydr. Polym,* vol. 88, no. 1, pp. 373-381, 2012.
[http://dx.doi.org/10.1016/j.carbpol.2011.12.012]

[46] J.M. Barrales-Rienda, A. Bello, P. Bello, and G.M. Guzman, "Fractionation of polymers", In: *Polymer Handbook,* J. Brandrup, E. H. Immergut, E. A. Grulke, Eds., 4th ed Wiley–Interscience: New York, 1999, pp. VII/327-VII/496.

[47] F. Francuskiewicz, *Polymer Fractionation.* Springer: Berlin, 1994.
[http://dx.doi.org/10.1007/978-3-642-78704-1]

[48] M.D. Lechner, E. Nordmeier, and D.G. Steinmeier, "Sedimentation coefficients, diffusion coefficients, partial specific volumes, frictional ratios, and second virial coefficients of polymers in solution", In: *Polymer Handbook,* J. Brandrup, E. H. Immergut, E. A. Grulke, Eds., 4th ed Wiley–Interscience: New York, 1999, pp. VII/85-VII/213.

[49] G.D. Phillies, *Phenomenology of Polymer Solution Dynamics.* Cambridge University Press: New York, 2011.
[http://dx.doi.org/10.1017/CBO9780511843181]

[50] L.H. Sperling, *Introduction to Physical Polymer Science.* John Wiley & Sons: New York, 1992.

CHAPTER 2

Intrinsic Viscosity Bovine Serum Albumin in Aqueous Solutions: Temperature Influence on Mark-Houwink Parameters

Martin Alberto Masuelli[*,1] and **Jesica Gassmann**[1,2]

[1] *Instituto de Física Aplicada-CONICET, Universidad Nacional de San Luis, Chacabuco 917, CP 5700, San Luis, Argentina*

[2] *Policlínico Regional San Luis, San Luis, Argentina*

Abstract: Bovine serum albumin (BSA) in aqueous solution is scarcely studied, and the Mark-Houwink parameters from the intrinsic viscosity measurements have not been reported at different temperatures. This work discusses these with a simple calculation of the Mark-Houwink parameters of BSA in aqueous solution when the concentration ranged from 0.2 to 1.0% wt., and the temperature ranged from 20 to 45°C. The relationship between the concentration and intrinsic viscosity was determined according to different methods. It is well known that when the temperature increases, the intrinsic viscosity decreases. This is reflected in the stiffer chain curve with $d(\ln[\eta])/d(1/T)$ of -398.97 for A zone from 20-30ºC (gel zone), -2759.1 for B zone from 35-40ºC (active zone) and 5604.5 for C zone from 41-45ºC (denatured protein zone), the point of intersection between the zones A and B is 34.6ºC. The linear relation between the logarithmic of viscosity and reverse temperature is ΔE_{avf} with a value of 680 Cal/mol. Furthermore, this work proposes the determination of M-H parameters of a protein-water system and their thermodynamic implications in conformational changes.

Keywords: BSA, Intrinsic viscosity, Mark-Houwink parameters.

INTRODUCTION

Bovine serum albumin (BSA) is a serum albumin protein derived from cows; the most abundant plasma protein is a globular protein. BSA is a major contributor to oncotic pressure (also known as colloid osmotic pressure) of plasma, acting as a carrier for various substances. Albumin is a soluble, monomeric protein which

[*] **Corresponding author Martin A. Masuelli**: Instituto de Física Aplicada-CONICET, Universidad Nacional de San Luis, Chacabuco 917 (ZC: 5700), San Luis, Argentina; E-mail: masuelli@unsl.edu.ar

Martin Masuelli & Denis Renard (Eds.)

comprises about one-half of the blood serum proteins. Albumin functions primarily as a carrier protein for steroids, fatty acids, and thyroid hormones and plays an important role in stabilising the extracellular fluid volume.

Albumin is a globular non-glycosylated serum protein with a molecular weight of 66,500 g/mol. Albumin is synthesised in the liver as proalbumin, which has an N-terminal peptide that is removed before the nascent protein is released from the rough endoplasmic reticulum. The product, proalbumin, is in turn cleaved in the Golgi vesicles to produce the secreted albumin. Albumin (when ionised in water at pH 7.4, as found in the body) is negatively charged. The glomerular basement membrane is also negatively charged in the body; some studies suggest that this prevents the filtration of albumin in the urine. According to this theory, the charge plays a major role in the selective exclusion of albumin from the glomerular filtrate. A defect in this property results in nephrotic syndrome, leading to albumin loss in the urine. Nephrotic syndrome patients are sometimes given albumin to replace the lost albumin. Because smaller animals (for example rats) function at a lower blood pressure, they need less oncotic pressure to balance this, and thus need less albumin to maintain proper fluid distribution. The general structure of albumin is characterised by several long α (alpha) helices, which this allows it to maintain a relatively static shape, which is essential for regulating blood pressure. Serum albumin contains eleven distinct binding domains for hydrophobic compounds. One heme and six long-chain fatty acids can bind to serum albumin at the same time [1].

Table 1. Physical properties of BSA [8], from data sheet of SIGMA ALDRICH, Germany.

Number of amino acid residues	**Molecular weight (g/mol)**	**Stokes Radius (R_s)**	**Estimated α-helix, %**	**pH of 1% Solution**
583	66463 Da	3.48 nm	54	5.2-7
Isoelectric point in water at 25 °C	Dimensions	Intrinsic viscosity (cm^3/g)	Estimated β-form, %	Diffusion constant, $D_{20,W} \times 10^7$ (cm^2/s)
4.7	140 × 40 × 40 Å^3 (prolate ellipsoid where a = b < c)	4.13	18	5.9

In solution, BSA presents a versatile conformation modified by changes in pH, and ionic strength, which serves to characterise the structure, conditions and properties of BSA (see Table **1**). Conformational changes induced by pH are reversible [2]. Although there have been speculations on the possible function of each transition, its physiological meaning still remains unclear. Foster [2] classified conformers as: "E" for expanded, "F" for fast migration, "N" for normal dominant form at neutral pH, "B" for basic form and "A" for aged at alkaline pH. Molecular weight, intrinsic viscosity, and Mark-Houwink parameters of BSA are shown in Table **2**.

Table 2. Intrinsic viscosity at different temperature.

Solvent/Solution	pH	T (°C)	M_w (g/mol)	[η] (cm^3/g)	*a*	Reference
Water, 0.1M Sodium citrate	7.4	37		4.20	-	Hess & Cobure 1950 [9]
Water, 0.2 M NaCl	7	15 20 25 30	64540	4.12 4.17 4.18 4.06	-	Koenings & Perrings 1952 [10]
Water, 0.010 M KCl	-	25	65000	3.4-3.8	-	Tandford & Buzzel 1954 [11]
Water I=0.1	-	25	65000	3.8	-	Yang & Foster 1956 [12]
Water I=0.2, Na_3PO_4. Urea (8M), sodium borate (0.05 M). Sodium p-chloromercuribenzoate (0.005M)	7 8.3 9.9	30	66706	4.20 4.10 20.8-30.5	-	Charlwood 1955 [13]
Water, 0.15 NaCl	5	25	-	3.70	-	Tanford *et al.* 1955 [14]
Water I=0.1	2.03 3.06 3.73 2.83	60	-	5.1 8.6 10.7 4.8	-	Foster & Yang 1955 [15]
Water 0.5M KCl	4 5.13	25	67000	4.57 4.13	-	Loeb & Scheraga 1956 [16]
Water 0.15M NaCl	3 4.3 4.8 4.9 7.3 8.5 9.3 10.5	25	67000	6.50 3.74 3.76 3.71 3.68 4.17 4.00 4.00	-	Tandford & Bussel 1956 [17]
Formic acid, Formic acid-0.1M NaCl.	-	-	70000	31.8 21.6	-	Martin 1961 [18]
Water	3.6	25	66000	5.44	-	Bloomfield 1966 [19]

(Table 2) contd.....

Solvent/Solution	pH	T (°C)	M_W (g/mol)	[η] (cm^3/g)	*a*	Reference
Water, 6M Guanidyne hydrochloride, mercaptoethanol 0.1M	5.4	25	69000	52.20 52.20	0.6700 0.6600 0.67	Tanford 1966-67 [20, 21] 1966′
Water, Tris 0.05M NaCl 0.2M	8.15	25	66700	3.70	-	Squire *et al.* 1968 [22]
Water-Propanol I=0.03 10% 15% 20% 25%	-	25	-	3.8 4.1 4.4 8.5	-	Sun 1972 [23]
Na_3PO_4 0.06 M I=0.1	7.4	37	-	3.92-4.21	-	MacMillan 1974 [24]
Water	7	25	66267-66700	4.10	-	Peters 1985 [25]
Na_3PO_4 0.06M KCl 0.2M	7	25	68000	3.13-4	-	Khan 1986 [26]
Na_3PO_4 0.07M NaCl 0.2M	6.8	25	67000	5.2		Chikazumi & Ohta 1991 [27]
Water, Nature Denatured 0.01M KCl	7	25	66267	3.60 13.00	-	Richards 1993 [28]
0.06M Na_3PO_4	7	25		3.20-3.90		Saeed *et al.* 1993 [29]
Water	5.2	5 10 15 20 25 30 35 40 45	66000	6.6 6.3 6.1 5.9 5.8 5.7 5.65 5.63 5.6	-	Monkos 1996 [30]
Water	-	-	65000	3.7-3.9	-	Harding 1997 [31]
Water	7	20	66500	4.10	-	Ferrer *et al.* 2001 [32]
Water	6.7-7.3	23	65000	3.6	-	Lee & Tripathi 2005 [33]
Water	7.4	25	66500	4.86	0.5800	Curvale 2008-2009 [34, 35]
Na_3PO_4 0.01M	7.4	25	66000	4.00	-	Sharma *et al.* 2011 [36]

(Table 2) contd.....

Solvent/Solution	pH	T (°C)	M_W (g/mol)	[η] (cm^3/g)	*a*	Reference
Water	6.4	35 37 40 41	66500	4.85 4.59 4.42 4.10	-	Masuelli 2013 [37]
Water, I=0.02	3 4 5 6 7	-	67000	7.91-9.06 5.33-6.10 4.19-4.81 3.21-3.72 2.86-3.27	-	Sarangapani *et al.* 2013 [38]

The N-F transition implies the opening of the molecule by unfolding domain III [3, 4]. The F form is characterised by increased viscosity, lower solubility and loss of α-Helix content [5]. At pH <4, another BSA expansion results in the loss of the helicoidal structure connecting domain I with domains II and III. This expanded form is known as E, and shows a new increase in intrinsic viscosity and a 40 to 90 Å increment in the axis of the axial hydrodynamic radium [6]. At pH 9, albumin changes conformation to the B basic form. If the BSA solution is maintained at a constant pH 9, low ionic strength and 3°C for 3 or 4 days, another isomer known as form A appears [7].

Basic studies on proteins have focused on protein concentration, pH, ionic strength, polymeric additives, the dielectric properties of solvent and solvent mixtures, and effect of temperature. In the case of proteins and polysaccharides, solubility studies are closely related to studies of gelation that try to determine the temperature and concentration of gelation (T_{gel} and C_g) [39].

The conformational stability of a protein is determined by intramolecular factors and solvent interactions (hydration). Conformational changes (changes in functional activity) can be induced by changes in temperature, pressure, and solvent medium. The technological performance of proteins depends critically on conformation and hydration (water-holding capacity). Like other polymers, proteins can be characterised by their chain conformations. However, methods of polymer statistics cannot be applied to most proteins because they adopt specific (native) conformations under different physiological conditions. Since proteins are polyelectrolytes in water solutions, they are largely governed by electrostatic (ionic) interactions. In determining charge/charge interactions, pK_a, and pK_b values of individual amino acids play an important role [40].

Cohn [41, 42] studied the ionic strength and isoelectric point, which are related to the solubility of globular proteins, concluding that the solubility of proteins in the neighbourhood of their isoelectric points depends to a large extent upon the degree of their dissociation, with more soluble proteins being the more dissociated.

Monkos [30, 43 - 45] has made a detailed study on the viscosity and intrinsic viscosity of different globular proteins in terms of concentration and temperature, in order to identify a mathematical relationship that describes this phenomenon.

Curvale *et al.* [34, 35] determined the intrinsic viscosity of BSA in aqueous solutions and found an abnormal behaviour, according to measurements with the Huggins method; $[\eta]$ is an expression of the interaction between biopolymer and solvent which reflects the solvent's ability to swell the macromolecule. Thus, it can be seen that BSA has a very low $[\eta]$ value at pH 7.4, which explains the assigned globular shape.

Fan *et al.* [46] studied BSA solutions (2.5%, pH 7.2) which can undergo structural changes under heat treatment. They found that α-helices are transformed to random coils at 67°C, resulting in an increased rotation angle. With subsequent heating, the transformed random coils may once again transform to non-native β-sheets and restore the optical rotation angle. These two states are reversible. However, when the heating temperature increases to 69°C, the denatured BSA starts to transform into a rigid network and achieves an irreversible state.

Wetzel *et al.* [47] studied the structural alterations of albumin, their dependence on concentration and the role of free -SH groups in thermal denaturation, as well as the reversibility of thermally-induced structural changes. The helix content changes with heat treatment, giving rise to beta structures which are amplified when cooled and are correlated with the aggregation of albumin. With rising temperature and concentration, the proportion of beta structures and aggregates increases. At denaturation degrees of up to 20%, complete renaturation is possible in every case. The structure content is concentration-dependent, even at room temperature. It may be due to intermolecular interactions that induce additional alpha-helix structures, which are less stable than those stabilised by intramolecular interactions. Unfolding of the pocket containing the free -SH group of cysteine-34 enables disulphide bridges to be formed, leading to stable aggregates and irreversible structural alterations. At temperatures below 65-70ºC, oligomers are formed mainly *via* intermolecular beta structures.

Takeda *et al.* [48] studied the thermal denaturation of bovine serum albumin (BSA) at pH of 2.8 and 7.0 in the range of 2-65°C. They found, by the curve-fitting method of circular dichroism spectra, that the relative proportions of α-helix, β-structures, and disordered structures in the protein's conformation were determined as a function of temperature. With the increase of temperature at pH 7.0, the proportion of α-helix decreased above 30°C and those of β-structures and disordered structures increased in the same temperature range. The structural change was reversible in the temperature range below 45°C. However, the structural change was partially reversible when cooled down to the room temperature subsequent to heating to 65°C. On the other hand, the structural change in BSA at pH 2.3 was completely reversible in the temperature range of 2-65°C, probably because of the interactions between domains and subdomains, which disappeared due to the acid expansion. The secondary structure of disulphide bridges-cleaved BSA remained unchanged during the heat treatment up to 65°C at pH of 2.8 and 7.0.

BSA is a water-soluble protein that is widely used in the biochemical, pharmaceutical, and food industries. The molecule possesses two aminoacids, one between repeating units and the other in the ring, besides the many hydroxyl groups. Due to this interesting structure, BSA has the ability to form hydrogen bonds both within its own structure and with polar solvents, providing BSA with significant solubility or a tendency to form molecular associations [49].

Peters [50] mentioned that the solubility of albumins was related to their high total electric charge, with corresponding strong hydrophilicity and attraction for water molecules near neutrality, with albumins extremely soluble in water, 35%wt. in dilute salt solutions, and 50%wt. in water pure solutions. A similar analysis of protein solubility was proposed by Haworth [49], which highlights the hydrophobicity of the composition and the type of amino acids that compose it.

The fact that BSA exhibits such dense hydrogen bonding within its own structure determine show much it interacts with the solvents *via* hydrogen bonding, which is an important fact that should be kept in mind [51, 52].

The solubility parameter can be easily determined for liquids based on their heat of vaporisation; however, for polymers, this process is inapplicable due to their non-volatility. Polymer solubility parameter values may be evaluated by indirect methods, such as by finding the liquids that cause the maximum swelling network of the polymer or that yield a maximum limiting viscosity number from which the δ value of this solvent may be lower to that of the polymer. The alternative method is calculating δ from the group molar attraction constants [37].

MATERIALS AND METHODS

BSA (lyophilised and deionised powder, purity grade >98%) was obtained from Fedesa S.A.-UNSL; the BSA molecular weight is 66,500 g mol^{-1}, with the universal shape function, $\nu_{(a/b)}$=4.27 for physiological BSA [34, 35]. Measurements were taken from fresh BSA solutions of 0.2 and 1.0% wt. in volume with pH 6.5. Solutions and dissolutions were prepared with deionised water. Different temperatures were maintained using a HAAKE C thermostatic bath (±0.1°C). Analyses were done using an Ubbelohde "suspended level" viscometer (IVA 1), with a water draining time of 34.25 s. Even though this viscometer works independently of the volume of the solution, it was used here for performing at least three measurements for each concentration and was later washed until the draining time of the solvent was obtained [37]. The density of each solution was measured using an Anton Paar DMA35N densimeter.

RESULTS AND DISCUSSION

The viscosity of liquids is highly dependent on temperature and its complex relations. The change of viscosity at different temperatures is commonly calculated with an equation of the Arrhenius form:

$$\eta = A_{vf} \exp\left(\frac{E_{avf}}{RT}\right) \quad \textbf{(1)}$$

Where η is the viscosity (poise); E_{avf} is the energy of the viscous flow activation (cal/mol); R is the gas constant (1.98 cal/mol K) and T is the temperature (K). The pre-exponential factor A_{vf} is considered independent or approximately independent of the temperature. Fig. (**1**) is convenient for calculating E_{avf} in a discrete range of temperatures. For simplicity reasons, proteins and macromolecules may be treated as rigid molecules for a hydrodynamic study. It is worth noting that the size of proteins is much larger than that of solvent (water) molecules. The linear relation between logarithmic of viscosity and reverse temperature is represented with a slope E_{avf} of polymeric solution; relating this to water, results in the increased activation energy of viscous flow of a polymer solution in a given solvent, ΔE_{avf}. The increase in the activation energy with viscous flow was observed due to the higher resistance to the flow of the biopolymer with respect to solvent [53]. The increase in ΔE_{avf} was 680 cal/mol and A_{vf} 9.75×10^{-5} g/cm s, with $\sigma^2=0.9365$ for BSA in water solution.

The physical properties of a polymer solution depend on solvent, temperature, and concentration. At low concentrations, the polymer chains are separated from each other, where each chain occupies a spherical volume of radius R_g. In this solution, the polymer-polymer interactions are small and the polymer coil volume is determined by polymer-solvent thermodynamic interactions. The hydrodynamic volume occupied by a given polymer molecular weight it's related to intrinsic viscosity, [η], which is a parameter that can be determined by dilute solution viscosity measurements. Intrinsic viscosity probes the interaction of molecular structure with the solution.

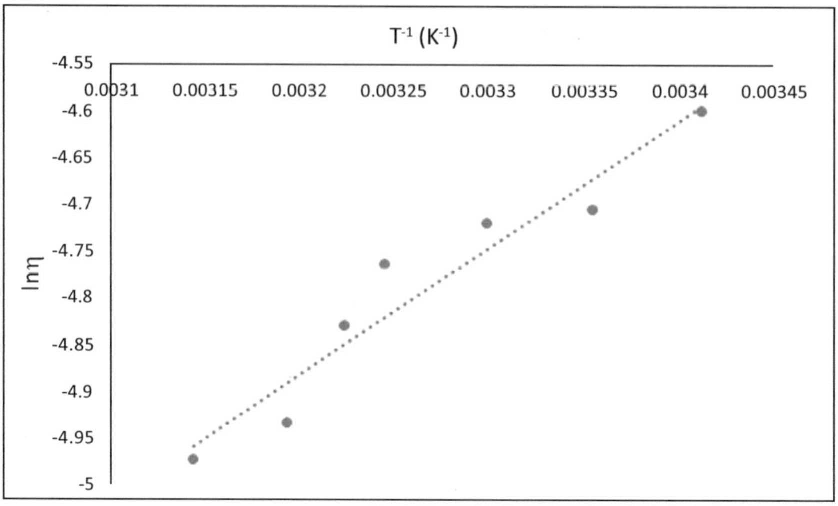

Fig. (1). Arrhenius plot for viscosity.

IUPAC recommends the term "increment of relative viscosity (η_i)", instead of "specific viscosity", because it has no attributions of specific quantity, meaning:

$$\eta_i = \eta_{sp} = \eta_r - 1 \quad \textbf{(2)}$$

It should also be remembered that:

$$\eta_r = \frac{\eta_s}{\eta_0} = \frac{\rho_s . t_s}{\rho_0 . t_0} \quad \textbf{(3)}$$

where the subindex "*s*" indicates "solution", and "*0*" indicates "solvent".

Since intrinsic viscosity is relevant for dilute solutions of BSA in the range of 0.2 to 1.0% wt., when high concentrations are used it is better to start with the first term of the Huggins equation "$\eta_{sp}/c = \eta_{red}$".

In Huggins' method [54], intrinsic viscosity [η] is defined as the ratio of the increase in relative viscosity (η_{sp}) to concentration (c in g/cm^3) when the latter tends towards zero (Fig. **2**).

$$\frac{\eta_{sp}}{c} = [\eta] + K_H [\eta]^2 c \quad \textbf{(4)}$$

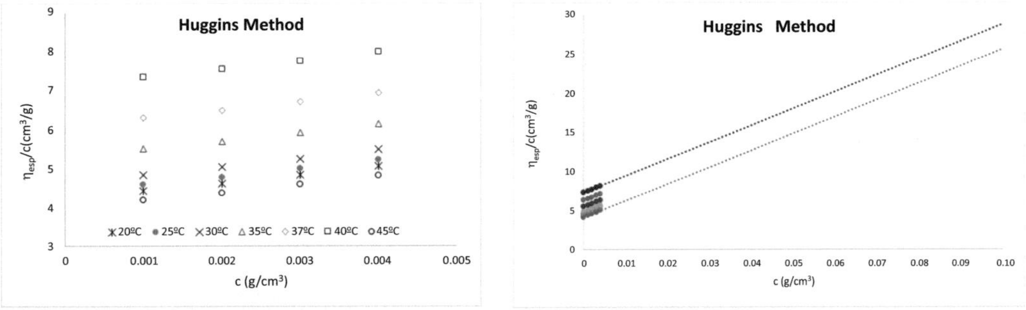

a- b-

Fig. (2). Effect of temperature in Huggins method, a- Normal and b- Extended.

The Huggins′ method shows slope of these curves do not cross, as they are parallel trend lines (see Fig. **2**).

In this work, the Martin′s method [55] is used:

$$\frac{\eta_{sp}}{c} = [\eta]e^{k_M[\eta]c}$$

$$\ln\frac{\eta_{sp}}{c} = \ln[\eta] + k_M[\eta]c \qquad \textbf{(5)}$$

While not the most popular method, this gives very good results, where its settings are superior to 0.96 and percentage relative error, respect to the Huggins method is below 4%. The intrinsic viscosity depends on intrinsic and extrinsic properties of the solution, such as molecular weight and structure of the dissolved polymer, ionic strength, solvent hydrophobicity, and temperature system [56].

In Fig. (**3**), ln(n_{sp}/c) is plotted as a function of concentration for different temperatures; data of calculated intrinsic viscosity can be seen in Table **3**.

Table 3. Data of intrinsic viscosity and constants of different methods.

	Temperature (°C)						
Methods	20	25	30	35	37	40	45
Huggins [η] k_H	4.4111 11.01	4.5741 10.24	4.8271 9.20	5.4831 7.54	6.2781 5.44	7.3191 4.01	4.1651 12.35

(Table 3) contd.....

	Temperature (ºC)						
Martin [η] k_M	4.4167 3.84	4.5790 5.15	4.8307 6.68	5.4849 8.52	6.2777 9.43	7.3162 10.10	4.1720 11.25
Kraemer [η] k_K	4.3985 10.02	4.5617 9.57	4.8149 8.76	5.4715 6.83	6.2672 5.23	7.3090 3.87	4.1523 10.78
Schulz-Blashke [η] $k_{S\text{-}B}$	4.4380 9.02	4.5997 8.45	4.8509 7.6740	5.5027 6.63	6.2936 4.75	7.3304 3.57	4.1942 9.99
Fuoss [η] k_F	4.0916 3.51	4.2500 3.64	4.4964 3.28	5.1387 2.97	5.9200 2.67	6.4992 2.35	3.8521 5.39
Fedors [η] c_{max}	4.4267 0.0250	4.5893 0.0259	4.8403 0.0270	5.4945 0.0305	6.2853 0.0347	7.3260 0.0404	4.1823 0.0238
Krasovskii-Baranov [η] α	4.6471 0.6532	4.7213 0.6512	4.8373 0.6484	5.1466 0.6423	5.5318 0.6364	6.0456 0.6303	4.5385 0.6563
Irzhak-Baranov [η]	5.4247	5.5834	5.8295	6.4661	7.2349	8.2371	5.1850
Kozicki-Kuang [η]	5.4941	5.6571	5.9101	6.5661	7.3611	8.4021	5.2481

Martin´s method, for determining the intrinsic viscosity, has the virtue of having an intersection for each of the temperatures by extrapolating the concentration of 0.03 g/cm^3 and the ln(η_{sp}/c) of 2.84 cm^3/g (Fig. **4**). A similar situation can be observed using Kraemer´s method, which also has an intersection of c=0.152 g/cm^3 for a value of (lnη_r)/c of 30.6 cm^3/g (Fig. **5**). This intersection phenomenon is due to the mathematical convergence of these equations.

Kraemer´s method [57] is determined with the following equation:

$$\frac{1}{c}\ln\eta_r = [\eta] + k_K[\eta]^2 c \tag{6}$$

Schulz-Blaschke´s method [58] is determined with the following equation:

$$\frac{\eta_{sp}}{c} = [\eta] + k_{S-B}[\eta]\eta_{sp} \tag{7}$$

In the Schulz-Blaschke´s method, the slopes of these curves converge towards zero (Fig. **6**).

The Fuoss method [59, 60] is determined with the following equation:

$$\frac{c}{\eta_{sp}} = \frac{1}{[\eta]} + k_{Fs}\frac{1}{[\eta]}c^{1/2} \tag{8}$$

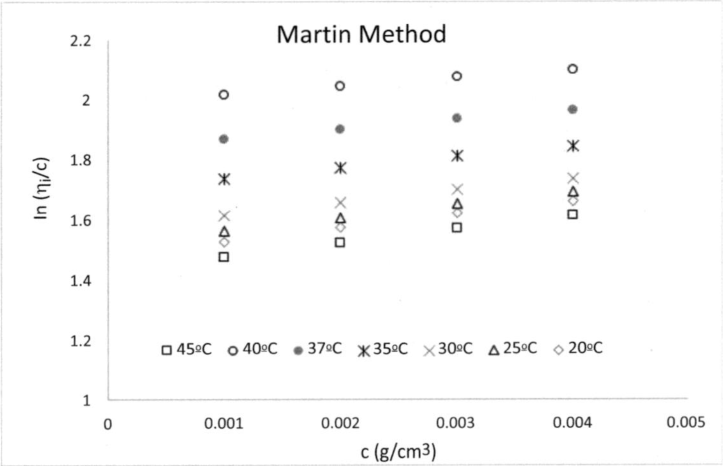

Fig. (3). Plot of $\ln(\eta_{sp}/c)$ *vs.* c for different temperatures.

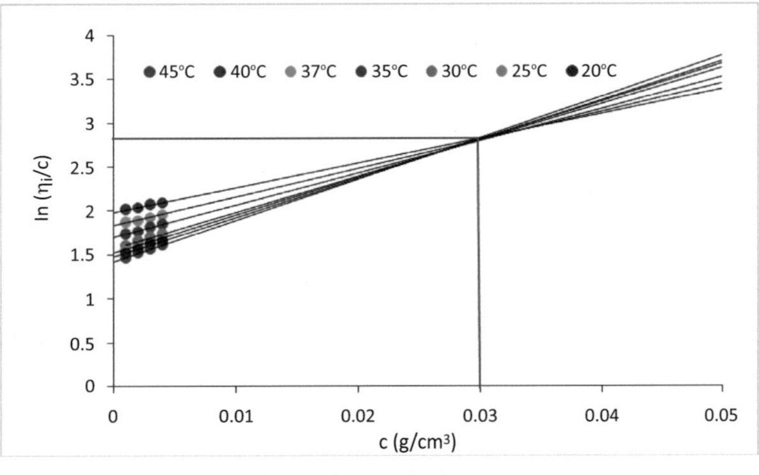

Fig. (4). Point of intersection temperature in Martin´s method.

In the Fuoss method [61, 62], a straight line is obtained when the reciprocal of reduced viscosity is plotted as a function of the square root of the polyelectrolyte concentration (Fig. **7**). It was usually assumed that this line could be extrapolated to zero polyelectrolyte concentration and that the intercept at zero concentration gives the reciprocal intrinsic viscosity.

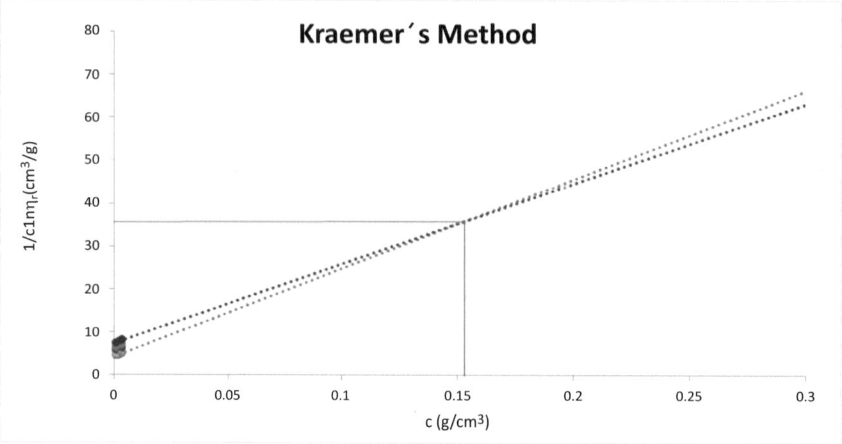

Fig. (5). Point of intersection temperature in Kraemer´s method.

However, careful investigations on the dilute solution behaviour revealed that the apparent unbounded rise in the reduced viscosity is always followed by a maximum, and the normal polymer behaviour is recovered as the polyelectrolyte concentration approaches→zero. Thus, the Fuoss method could not be employed to obtain the intrinsic viscosity and, in fact, this is now known to be one of the capital errors in polyelectrolyte solutions.

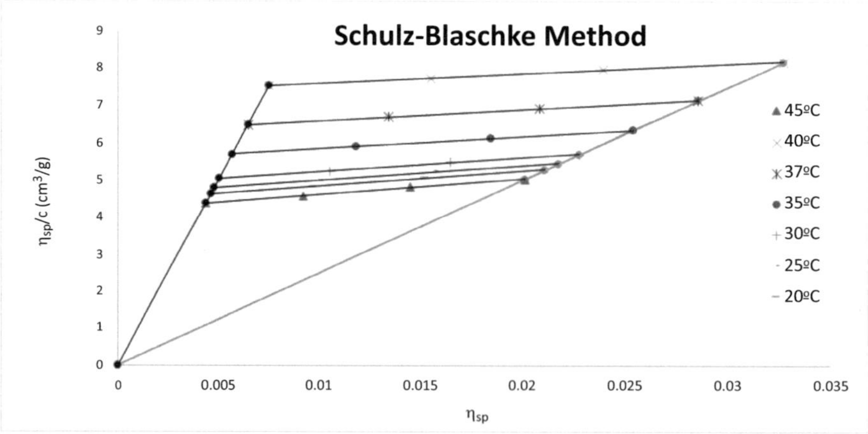

Fig. (6). Point of intersection of temperature in Schulz-Blaschke´s method.

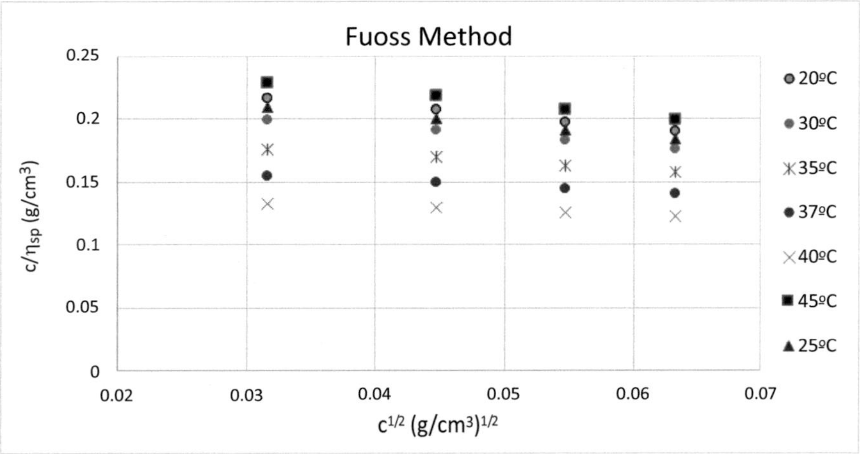

Fig. (7). The Fuoss method.

Fedors methods is [63] calculated for dilute and semi-dilute polymer solutions (Fig. **8**) with the following equation:

$$\frac{1}{\left[2\left(\eta_r^{1/2}-1\right)\right]}=\frac{1}{[\eta]c}-\frac{1}{[\eta]C_{\max}} \quad \textbf{(9)}$$

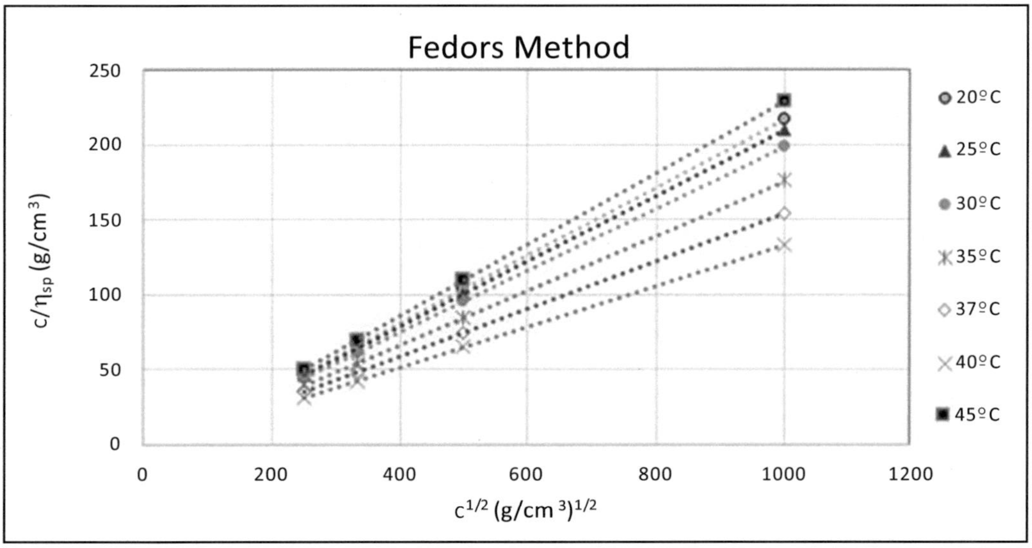

Fig. (8). Fedors' method.

C_{max} is a concentration parameter (in g/cm^3). Contrary to Eqs. (4) and (6), Fedors' equation does not contain any interaction constant (like k_H or k_M), but a concentration parameter, C_{max}, analogous to the volume fraction, corresponding to the maximum packing of solid particles [64]. The Fedors method converges to zero in c/n_{sp} and concentration.

$$\frac{1}{2}\left(\frac{1}{\eta_{sp}}+\frac{1}{\ln\eta_r}\right)=\frac{1}{[\eta]}-k_{He}c \quad \textbf{(10)}$$

The Heller equation contains an empirical constant, k_{He} [65] (see Fig. **9**).

Irzhak-Baranov proposed that $\partial\ln\eta_r/\partial c$ can be considered the current intrinsic viscosity in any point of the dependence $\ln\eta_r$ on c [66, 67]:

$$[\eta]\equiv\frac{\partial\ln\eta_r}{\partial c} \quad \textbf{(11)}$$

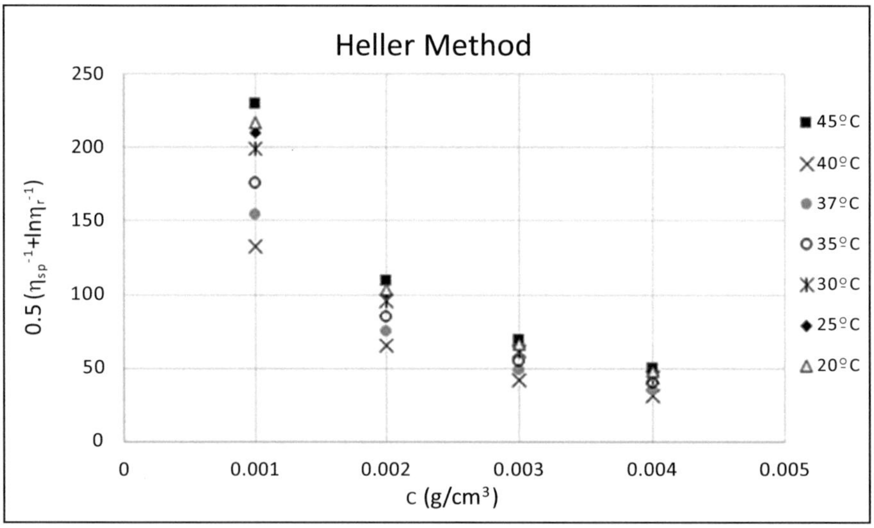

Fig. (9). Heller´s method.

It is clear that, at c → 0, [η] corresponds to the initial slope of the plot $\ln\eta_r$ *vs* c (see Fig. **11**), coinciding with [η] for the isolated macromolecule.

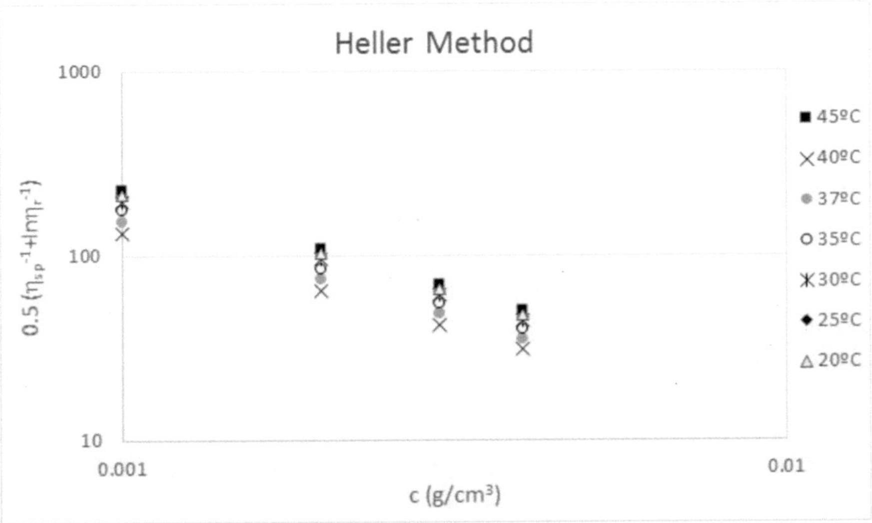

Fig. (10). Heller´s method with logarithmic scale.

At a finite polymer concentration, the intrinsic viscosity la viscosidad intrinseca characterises the specific volume occupied by the probe macromolecule placed into a solution with a uniform distribution of all other macromolecules. Such an approach can be considered an experimental realisation of the middle field concept. The [η] changes with variation of the polymer concentration.

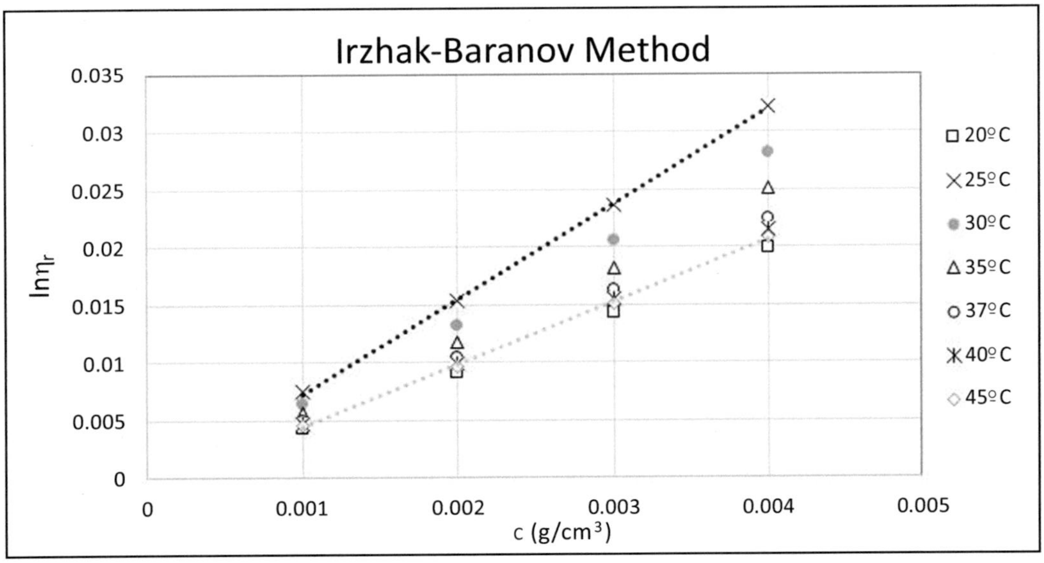

Fig. (11). Irzhak-Baranov method.

The method of determination of the intrinsic viscosity using Eq. (11) was employed in [66] when analysing the effect of compaction of the globules in flexible-chain polymers on passing to the moderate polymer concentration range in solvents of various aspects thermodynamic, and also when interpreting the hydrodynamic behaviour of rod-shaped macromolecules.

$$\ln \eta_r = ([\eta]c)^{\alpha} \quad \textbf{(12)}$$

The concentration dependence relative viscosity η_r for biopolymer in solution is quantitatively described by the relation of Krasovskii-Baranov [68, 69]:

where $\alpha = 1$ for the dilute solutions, and $\alpha = 0.7$-0.9 for the concentration range relating to moderately concentrated solutions. It should be noted that for biomacromolecules in solution, a transition from three-dimensional flow to planar flow occurs in the region of a critical concentration that corresponds to $[\eta]c \geq 2$. Double logarithmic of relative viscosity *vs.* concentration is applied, as shown in the following equation:

$$\log(\ln \eta_r) = \alpha \log[\eta] + \alpha \log c \quad \textbf{(13)}$$

It is possible to obtain the intrinsic viscosity of biopolymer solutions (see Fig. **12**).

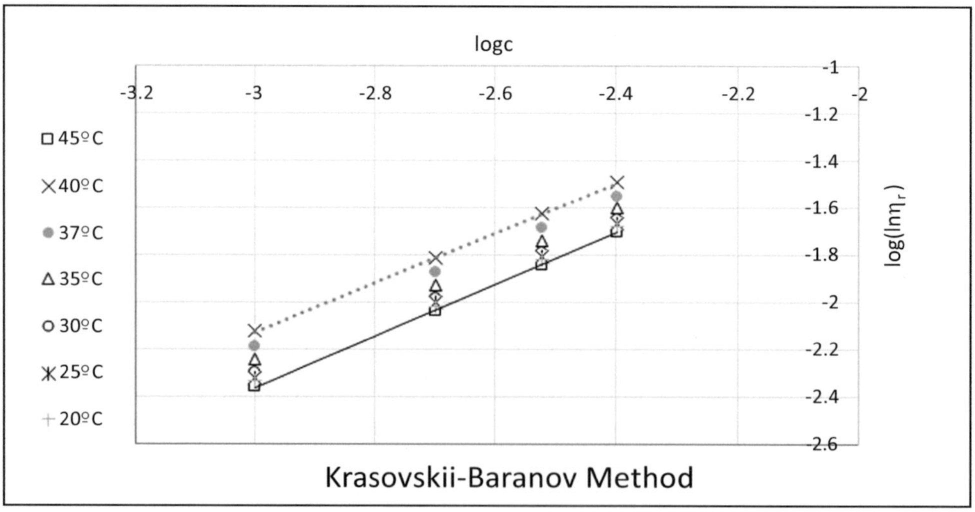

Fig. (12). Krashovskii-Baranov method.

The constant α, for this protein is equal to 0.64 ± 0.015, corresponding to moderate solution.

The intrinsic viscosity of a polymer solution is defined by:

$$\eta = \lim_{c \to 0}\left(\frac{\eta - \eta_0}{c\eta_0}\right) = \lim_{c \to 0}\frac{\eta_{sp}}{c} \quad \textbf{(14)}$$

This definition forms the basis of intrinsic viscosity determinations (Chou & Kokini [70]), which typically involve the extrapolation of experimental evaluations of the quotient η_{sp}/c *versus* the concentration "c" or specific viscosity "η_{sp}", to zero c or η_{sp}.

Application of L'Hôpital's rule to the defining expression yields of Kozicki-Kuang [71]:

$$[\eta] \equiv \left(\frac{\partial \eta_{sp}}{\partial c}\right)_{c=0} \quad \textbf{(15)}$$

which provides an alternative method for evaluation of the intrinsic viscosity as the derivative $d\eta_{sp}/dc$ at zero concentration (see Fig. **13**). Since η_{sp}, is known to be zero at zero concentration, the method does not involve the extrapolation data of η_{sp}*versus* c and is therefore not dependent on the availability of a theoretical relationship in the determination of [η], as required in an extrapolation method.

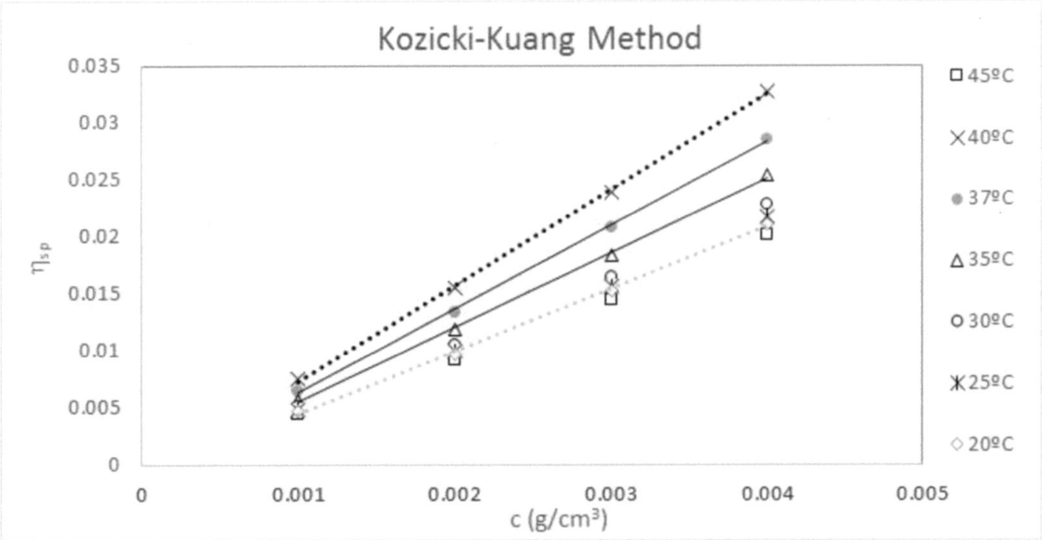

Fig. (13). Kozicki-Kuang method.

Any analytical expression for η_{sp} in terms of c which contains the origin, provides a good fit of experimental data in the low concentration range and is amenable to evaluation of the first derivative at c = 0 is sufficient for an independent evaluation of [η]. As can be seen, the problem associated with magnification of error in the quotient η_{sp}/c at low concentrations, made more pronounced when successive solutions are prepared by diluting a single solution, is avoided since η_{sp} and is used in place of the quotient η_{sp}/c [71].

$$\ln \eta_r = c[\eta]\left(\frac{1+Bc[\eta]^*}{1+Bc[\eta]}\right) \tag{16}$$

Recently, Wolf [72] has proposed a new approach to determine the intrinsic viscosity of polyelectrolytes based on phenomenological thermodynamic tools [73 - 77]. The Wolf equation is as follows:

where η_r and [η] are relative viscosity and intrinsic viscosity of the polymer, respectively. B is the viscometric interaction parameter and $[\eta]^*$ is only required for polyelectrolytes.

According to the Wolf equation, the intrinsic viscosity of polymer can be obtained from the initial slope of natural logarithm of relative viscosity *versus* concentration of the polymer. It is shown that Wolf's equation is applicable for charged and uncharged polymers [73]. When the neutral logarithm of relative viscosity *versus* concentration of polymer shows a linear relationship, intrinsic viscosity can be obtained from Huggins equation by setting B and $[\eta]^*$ as zero.

According to this relation, the intrinsic viscosities [η] of the polymers are given by the initial slope of ln η_r *vs* c; B represents a viscometric interaction parameter, analogous k_H. The additional parameter [η]c (as compared with the Huggins expression) is only required for polyelectrolytes in cases where the solvent does not contain sufficient amounts of extra salt.

According to experience, Equation (16) describes the experimental data accurately up to reduced polymer concentrations, [η]c, that are considerably larger than unity [75].

At finite polymer concentrations, the relation between the specific viscosity, η_{sp}, and the concentration of solute, c, can be expressed conveniently and most generally by the polynomial equation [78]:

$$\frac{\eta_{sp}}{c} = [\eta] + k_1[\eta]^2 c + k_2[\eta]^3 c^2 + k_3[\eta]^4 c^3 + \ldots, \tag{17}$$

Without the third term, this is the familiar Huggins equation [79], and the dimensionless parameter k_1 is called the Huggins slope coefficient or a hydrodynamic interaction constant [80], arising from hydrodynamic interactions between separate polymer chains [81].

Comparing the Taylor series of Equations (16) within the region of pair interaction (*i.e.* up to the square of c) yields the following interdependence of B and the Huggins coefficient k_H [74]:

$$B = \frac{0.5 - k_H}{1 - \left(\frac{[\eta]^*}{[\eta]} \right)} \qquad \textbf{(18)}$$

Electrolytes greatly affect the viscosity of solutions of hydropolymers and biopolymers. If a high concentration of counter ions is provided by the addition of salts, the electrostatic interactions are suppressed and chain segments can move freely as non-charged polymers; therefore, intrinsic viscosity can be treated by the Huggins equation [82 - 84]. By adding inorganic acid in biopolymeric solutions, the ionisation of ionisable groups was suppressed and the hydrodynamic volume of polyelectrolyte's chains was decreased; this leads to a reduction in intrinsic viscosity. For this reason, polyelectrolytes have the highest values of intrinsic viscosity in natural media [82 - 84].

Table **3** shows the intrinsic viscosity data and the respective constants calculated for each method (Huggins, Martin, Kraemer, Schulz-Blashke, Fuoss, Fedors, Krasovskii-Baranov, Irzhak-Baranov and Kozicki-Kuang).

The largest errors in the calculation of intrinsic viscosities are reported for the methods Irzhak-Baranov, Chou & Kikini and Kozicki-Kuang, with an error of a unit less value, *i.e.* 25%. Similar cases may be seen for Krasovskii-Baranov methods and Fuoss, but less so than the above (>10%). The rest of the methods have minor errors, of less than 1%, with respect to the Huggins method, which was chosen as a reference. The method of Young & Kokini is an abbreviated form of the equation of Huggins, but has the difficulty of strong dependence on the concentration of solute and its respective deviations in very dilute or concentrated areas. The Heller and Wolf methods are difficult to determine graphically and were therefore not taken into account in this work.

The intrinsic viscosity of the BSA solution is influenced by temperature, as shown in Fig. (**1**). The influence of temperature on the intrinsic viscosity is given by the parameter of chain flexibility (dln[η]/dT), which gives information about the conformation of the macromolecule chain in solution. The chain flexibility parameter (dln[η]/dT), whose value is 810.56K^{-1} with a $\sigma^2 = 0.8850$ (eliminating c-point), introduces error in the value of chain flexibility of BSA in water solution. Analysis of the relative stiffness parameter indicates that the molecular weight of BSA, 66,500 g/mol, is less flexible; however, three trends can be distinguished in Fig. (**14**), reflected in the stiffer chain curve with d(ln[η])/d(1/T) of -398.97 for A zone from 20-30°C (gel zone), -2759.1 for B zone from 35-40°C (active zone) and 5604.5 for C zone from 41-45°C (denatured protein). The point of intersection between zones A and B is 34.6°C, defining the lower limit of the active zone of the protein.

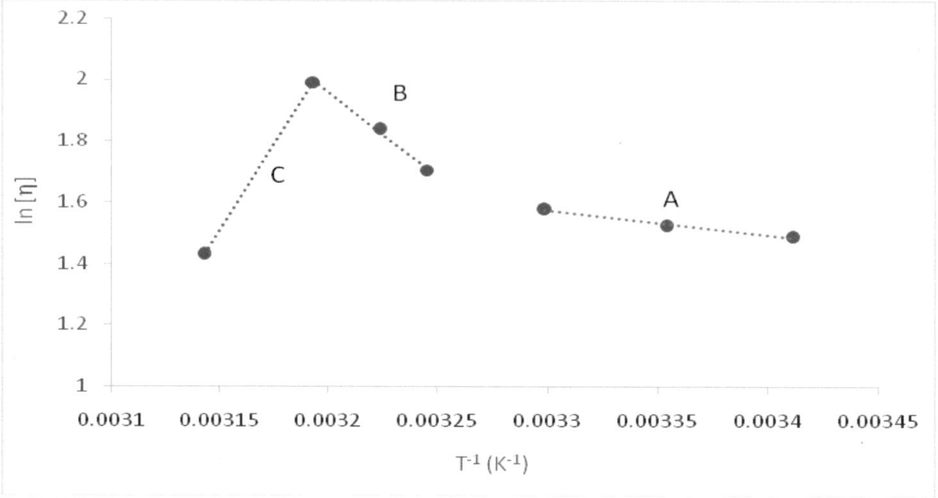

Fig. (14). Plot of ln[η] in function of T^{-1}, A corresponding to gel zone, B corresponding to active protein zone and C corresponding denatured protein zone.

An increasing intrinsic viscosity with increasing T can be caused by polymer-polymer interactions as well as by polymer-solvent interactions. The solvating envelope of the polymer chain can increase when the polymer-solvent interaction (solvent type) increases with temperature. The rising solvation of the chain leads to an expansion of the polymer coil and therefore to an increasing intrinsic viscosity. In this case, a rising temperature leads to decreasing polymer-solvent interactions, increasing intramolecular interactions, and therefore to a decreasing coil expansion [85].

The intrinsic viscosity measurements used to calculate the Mark-Houwink (M-H) parameters are generally performed for different molecular weights at a constant temperature, with the standard value of this temperature being 25°C or 37°C in the case of mammalian proteins, or under theta conditions for polymers and biopolymers. In the polymer industry, polysaccharides and proteins must circulate through pipes during transport processes, where pumps have a very high-energy expenditure and where temperatures must be greatly increased; at this point, calculation of the Mark-Houwink parameters becomes important. The M-H parameters are calculated at standardised temperatures; in many cases, these are not useful because of the errors they carry, and it then becomes difficult to calculate the molecular weight. It is therefore necessary to know the change in molecular weight as evidence of a change in the product obtained, as this may create a need to halt the production process, transport, or extrusion. The basic criterion is that the molecular weight does not change with temperature, or at least within one discrete range of temperatures, but that there is hydrodynamic change (intrinsic viscosity). The method is simple and requires iterative mathematical processing and the measurement of intrinsic viscosity at different temperatures.

Mark (1938) [86] and Houwink (1940) [87] independently correlated the intrinsic viscosity with molecular weight:

$$[\eta] = k\, M^{a} \qquad \textbf{(19)}$$

where k and a both are constants. The Mark-Houwink equation is applicable to many polymers and is extensively used to determine molecular weight. The constants k and a both vary with polymers and solvents [88 - 90].

Equation (18) describes the relationship between intrinsic viscosity and molecular weight. Since molecular weight is related to the size of the polymer chain, for proteins [56], the following equation is generally used:

$$[\eta] = k\left(\frac{M}{M_0}\right)^{a} \qquad \textbf{(20)}$$

where M_0 is the molecular weight of the amino acid repeating unit (monomer). The calculation of Mark-Houwink (M-H) parameters is carried out by the graphic representation of the following equation:

$$\ln[\eta] = \ln k + a \ln M_w \qquad \textbf{(21)}$$

where k and a are M-H constants, depending upon the type of polymer, the solvent, and the temperature of viscometric determinations. The exponent a is a function of polymer geometry and varies from 0.5 to 2.0. These constants can be determined experimentally by measuring the intrinsic viscosity of several polymer samples for which the molecular weight has been determined by an independent method (*e.g.*, osmotic pressure or light scattering).

Table 3. Mark-Houwink parameters for different temperatures.

T (K)	[η] (cm^3/g)	*k* (cm^3/g)	*a*
318.16	4.1770	0.00722	0.5727
313.16	7.3310	0.00725	0.6230
310.16	6.2900	0.00727	0.6090
308.16	5.4950	0.00729	0.5974
303.16	4.8390	0.00731	0.5849
298.16	4.5860	0.00736	0.5794
293.16	4.4230	0.00742	0.5754

Using the polymer standards, a plot of ln $[\eta]$ *versus* ln M_w usually gives a straight line. The slope of this line is the value of a and its intercept is equal to ln k. The M-H exponent bears the signature of a polymer chain's three-dimensional configuration in the solvent environment: a values from 0.0–0.5 reflect a rigid sphere in an ideal solvent; those from 0.5–0.8 a random coil in a good solvent; and from 0.8–2.0 a rigid or rod like configuration (stiff chain). The polymer molecule assumes a tighter configuration, and the solution has a lower intrinsic viscosity. The M-H "a" constant is close to 0.5 in "poor" solvents, whereas for a rigid or rod-like polymer molecule that is greatly extended in solution the "a" constant approaches a value of 2.0 [91].

Following this, the Mark-Houwink parameters are established by fixing molecular weight (M), and the intrinsic viscosity [η] was measured at each temperature to calculate the parameters. In the case of a protein, M_0 is the molecular weight representative of the amino acid monomers (calculated according to their percentage in the macromolecule). The a and k values are iterated and re-calculated, and a new molecular weight, M_1, is calculated. This procedure is repeated until the value of this molecular weight M_1 is very similar to M and with an error less than 5% or less than 1%. This procedure is then repeated for each temperature, and data are obtained for the Mark-Houwink parameters with respect to the standard molecular weight (M) [91]. The values obtained for k and a are plotted in Figs. (**15** and **16**), respectively.

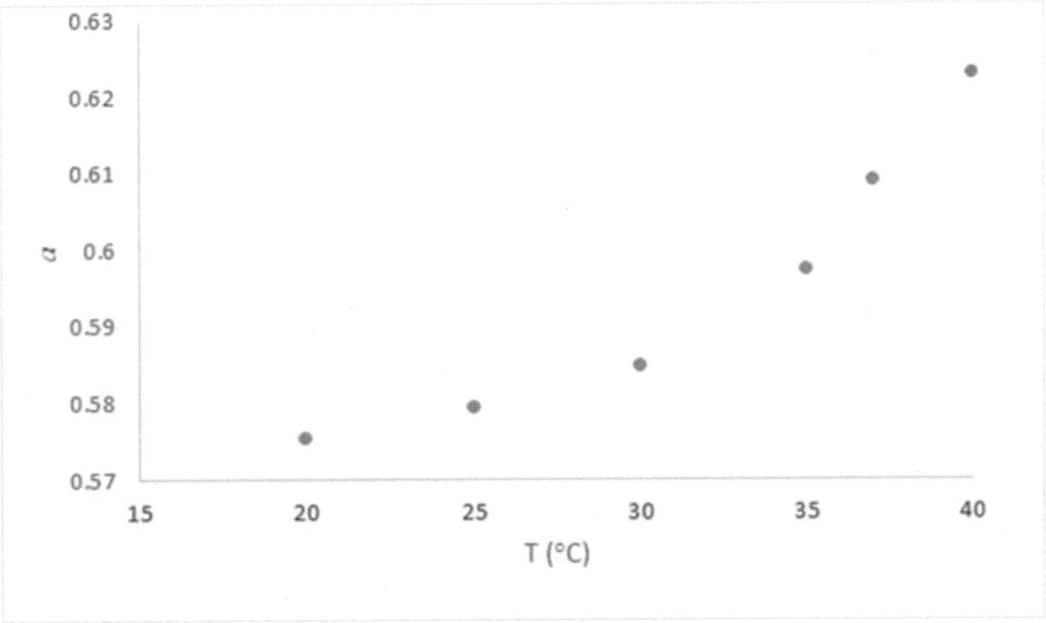

Fig. (15). Influence of temperature on the *a* Mark-Houwink parameter.

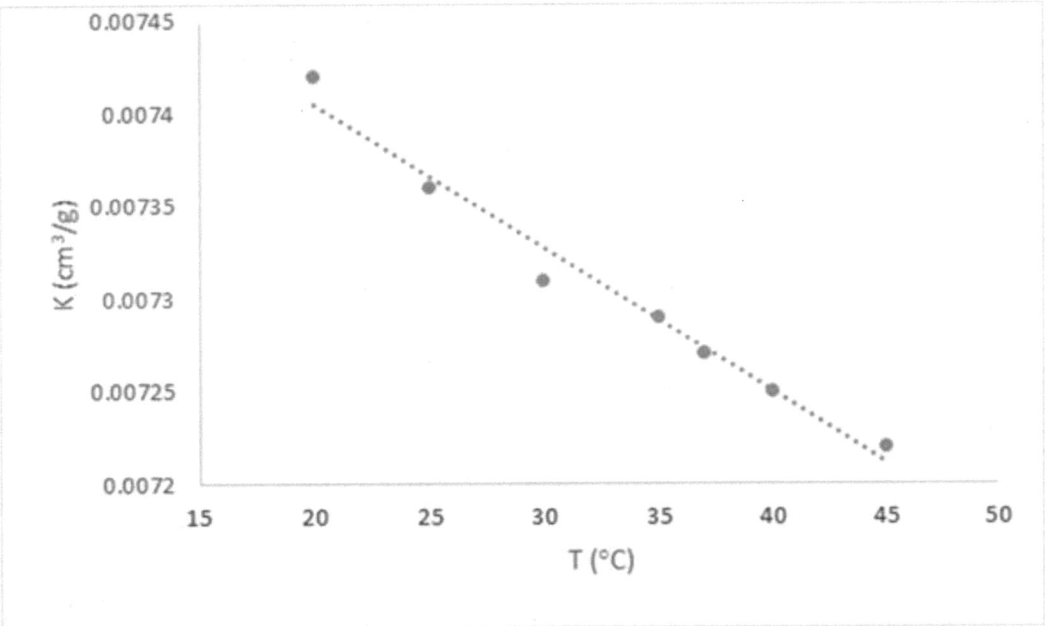

Fig. (16). Influence of temperature on the *k* Mark-Houwink parameter.

This iterative method for the calculation of Mark-Houwink parameters can be very useful for determining the molecular weight and hydrodynamic parameters of a macromolecule at different temperatures, for cases where these parameters are known only for a single temperature. The criterion for applicability is that the molecular weight does not change with temperature (within a discrete range); in other words, no bond breaking or bond formation occurs, although changes in hydrodynamics are permissible [92].

The hydrodynamic radius and intrinsic viscosity for proteins are higher at high molecular weights and decrease with increasing temperature. The classical values are used to calculate Mark-Houwink parameters, *a* and *k*, for each temperature. These studies of M-H parameters are usually carried out at a given temperature, obtaining a consistent result but in a very limited range of temperatures. The values at different temperatures show that BSA in aqueous solution behaves in a conformation that is predominantly confined to the random-coil conformation. The Mark-Houwink parameters for biopolymers may be varied with polymer solution and temperature. This is because the macromolecule changes its hydrodynamic radius with type of solution and temperature *via* changes in chain flexibility. In a good solvent, a temperature increase results in an intrinsic viscosity decrease and R_H<, because the entropy value increases with an increase in temperature and is unfavourable for an extended conformation (E_{avf} solution > E_{avf} solvent). Mark-Houwink values confirm that BSA behaves as a random-coil biopolymer for these conditions. Such classical equations relating the parameters of Mark-Houwink with temperature, which ultimately describe this type of thermodynamic parameter, are relations between properties of the solute with the solvent and temperature dependence.

CONCLUDING REMARKS

The intrinsic viscosity of BSA in water solutions at temperatures from 20 to 45°C is maintained in constant concentrations and pH; it is possible to determine the Mark-Houwink parameters by measuring intrinsic viscosity using classical methods (Huggins, Martin, Kraemer, Schulz-Blashke, Fuoss, Fedors, Krasovskii-Baranov, Irzhak-Baranov and Kozicki-Kuang). The largest errors in the calculations of the intrinsic viscosities are reported for the Irzhak-Baranov, Chou & Kikini and Kozicki-Kuang methods, with an error of a unit less value, *i.e.* 25%. Similar cases may be seen for Krasovskii-Baranov and Fuoss methods, but less so than above (>10%). The rest of the methods have minor errors, less than 1%, with respect to Huggins method, which was chosen as a reference. The temperature effect over classical functions is uneven. Martin´s method for determining the intrinsic viscosity has the virtue of having an intersection for each of the temperatures by extrapolating the concentration of 0.03 g/cm^3 and for the $\ln(\eta_i/c)$

of 2.84 cm^3/g. A similar situation can be observed using Kraemer's method, which also has an intersection of c=0.152 g/cm^3 for a value of $(\ln\eta_r)/c$ of 30.6 cm^3/g. This phenomenon of the intersection is due to the mathematical convergence of these equations. The Schulz-Blaschke's method has a slope of the curves that converge towards zero. In the Huggins method the slopes are the same and have a parallel behavior for each of the temperatures and different orders to the origin, ie different intrinsic viscosities.. The value given at different temperatures shows that BSA in aqueous solution predominantly behaves in a random-coil conformation. Thus, the intrinsic viscosity determinations are also made by different graphical methods such as Fuoss, Fedors, Heller, Irzhak-Baranov, Krasosvkii-Baranov, Kozicki-Kuang and others. The increase in activation energy of viscous flow occurs due to the higher resistance to the flow of the biopolymer with respect to solvent [53]. The increase of ΔE_{avf} is 680 cal/mol and A_{vf} 9.75×10^{-5} g/cm s, with σ^2=0.9365 for BSA in water solution. Analysis of the relative stiffness parameter indicates that the molecular weight of BSA, 66,500 g/mol, is less flexible, but three trends can be distinguished in Fig. (**3**); this is reflected in the stiffer chain curve, with $d(\ln[\eta])/d(1/T)$ of -398.97 for A region from 20-30ºC (gel zone), -2759.1 for B region from 35-40ºC (active zone) and 5604.5 for C region from 41-45ºC (denatured protein). The point of intersection between the zones A and B is 34.6ºC, defining the lower limit active zone of the protein.

CONFLICT OF INTEREST

The author (editor) declares no conflict of interest, financial or otherwise.

ACKNOWLEDGEMENTS

The authors thank Universidad Nacional de San Luis, Instituto de Física Aplicada (INFAP-CONICET). The UNSL projects 2-1712, 2-2414 and 2-1916 for their financial support. Dr. Rolando Curvale for their valuable contributions. The authors thank Dr. Sebastian Larregola for English corrections.

REFERENCES

[1] M.A. Rothschild, M. Oratz, and S.S. Schreiber, "Serum albumin", *Am. J. Dig. Dis.* , vol. 14, no. 10, pp. 711-744, 1969.
[http://dx.doi.org/10.1007/BF02233577] [PMID: 4898690]

[2] J.F. Foster, *Albumin Structure. Function and Uses.* Academic Press, 1977, pp. 53-84.
[http://dx.doi.org/10.1016/B978-0-08-019603-9.50010-7]

[3] M.J. Geisow, and G.H. Beaven, "Physical and binding properties of large fragments of human serum albumin", *Biochem. J.* , vol. 163, no. 3, pp. 477-484, 1977.
[http://dx.doi.org/10.1042/bj1630477] [PMID: 18136]

[4] J.F. Foster, *The Plasma Proteins,* F.W. Putnam, Ed., vol. 1. Academic Press: New York, 1960, pp. 179-233.

[http://dx.doi.org/10.1016/B978-1-4832-2810-5.50013-X]

[5] E.J. Williams, and J.F. Foster, "The aggregation of bovine plasma albumin at low pH", *J. Am. Chem. Soc.* , vol. 82, pp. 3741-3745, 1960. [http://dx.doi.org/10.1021/ja01499a068]

[6] W.F. Harrington, P. Johnson, and R.H. Ottewill, "Bovine serum albumin and its behaviour in acid solution", *Biochem. J.* , vol. 62, no. 4, pp. 569-582, 1956. [http://dx.doi.org/10.1042/bj0620569] [PMID: 13315216]

[7] J-R. Bae, "Ultrasonic relaxation study of proton transfer and conformational change in aqueous solution of bovine serum albumin", *Polymer (Korea).* , vol. 17, no. 3, pp. 308-315, 1993.

[8] "Albumin from Bovine Serum,", Product Information, Sigma Aldrich, 2000.

[9] E. L. Hess, and Cobure. Aspascia, "The intrinsic viscosity of mixed protein systems, including studies of plasma and serum", *J. Gen. Physiol.* , vol. 20, no. 33(5), pp. 511-523, 1950.

[10] V.L. Koenig, and J.D. Perrings, "Critical sedimentation and viscosity studies on crystalline bovine plasma albumin", *Arch. Biochem. Biophys.* , vol. 41, no. 2, pp. 367-377, 1952. [http://dx.doi.org/10.1016/0003-9861(52)90465-7] [PMID: 13008454]

[11] C. Tanford, and J.G. Buzzell, "Evaluation of Scheraga and Mandelkern Shape Factor for Bovine Serum Albumin", *J. Am. Chem. Soc.* , vol. 76, pp. 3356-3357, 1954. [http://dx.doi.org/10.1021/ja01641a091]

[12] J.T. Yang, and J.F. Foster, "Changes in the intrinsic viscosity and optical rotation of bovine plasma albumin associated with acid binding", *J. Am. Chem. Soc.* , vol. 76, pp. 1588-1595, 1954. [http://dx.doi.org/10.1021/ja01635a038]

[13] P.A. Charlwood, "Sedimentation and viscosity studies of bovine serum albumin in urea solution", *Can. J. Chem.* , vol. 33, no. 6, pp. 1043-1054, 1955. [http://dx.doi.org/10.1139/v55-120]

[14] C. Tanford, J.G. Buzzell, D.G. Rands, and S.A. Swanson, "The reversible expansion of bovine serum albumin in acid solutions", *J. Am. Chem. Soc.* , vol. 77, pp. 6421-6428, 1955. [http://dx.doi.org/10.1021/ja01629a003]

[15] J.F. Foster, and J.T. Yang, "Effect of temperature on the intrinsic viscosity and optical rotation of bovine plasma albumin in acid media", *J. Am. Chem. Soc.* , vol. 77, pp. 3895-3901, 1955. [http://dx.doi.org/10.1021/ja01619a064]

[16] G.I. Loeb, and H.A. Scheraga, "Hydrodynamic and thermodynamic properties of bovine serum albumin at low pH", *J. Phys. Chem.* , vol. 60, pp. 1633-1644, 1956. [http://dx.doi.org/10.1021/j150546a009]

[17] C. Tanford, and J.G. Buzzell, "The viscosity of aqueous solutions of bovine serum albumin between pH 4.3 and 10.5", *J. Phys. Chem.* , vol. 60, pp. 225-231, 1956. [http://dx.doi.org/10.1021/j150536a020]

[18] W.G. Martin, "Molecular weight of bovine plasma albumin, vitellenin, and their oxidation products in formic acid", *Can. J. Biochem. Physiol.* , vol. 39, pp. 1405-1417, 1961. [http://dx.doi.org/10.1139/o61-150] [PMID: 13767420]

[19] C. Tanford, K. Kawahara, and S. Lapanje, "Proteins in 6-M guanidine hydrochloride. Demonstration of random coil behavior", *J. Biol. Chem.* , vol. 241, no. 8, pp. 1921-1923, 1966. [PMID: 5947952]

[20] V. Bloomfield, "The structure of bovine serum albumin at low pH", *Biochemistry.* , vol. 5, no. 2, pp. 684-689, 1966. [http://dx.doi.org/10.1021/bi00866a039] [PMID: 5940950]

[21] C. Tanford, K. Kawahara, and S. Lapanje, "Proteins as random coils. I. Intrinsic viscosities and sedimentation coefficients in concentrated guanidine hydrochloride", *J. Am. Chem. Soc.* , vol. 89, pp.

729-736, 1967.
[http://dx.doi.org/10.1021/ja00980a001]

[22] P.G. Squire, P. Moser, and C.T. OKonski, "The hydrodynamic properties of bovine serum albumin monomer and dimer", *Biochemistry.*, vol. 7, no. 12, pp. 4261-4272, 1968.
[http://dx.doi.org/10.1021/bi00852a018] [PMID: 5750167]

[23] S.F. Sun, "The intrinsic viscosities of bovine serum albumin in n-propyl alcohol-water mixtures", *J. Phys. Chem.*, vol. 76, no. 1, pp. 128-129, 1972.
[http://dx.doi.org/10.1021/j100645a022] [PMID: 5007642]

[24] D.E. McMillan, "A comparison of five methods for obtaining the intrinsic viscosity of bovine serum albumin", *Biopolymers.*, vol. 13, no. 7, pp. 1367-1376, 1974.
[http://dx.doi.org/10.1002/bip.1974.360130708] [PMID: 4411599]

[25] T. Peters Jr, "Serum albumin", *Adv. Protein Chem.*, vol. 37, pp. 161-245, 1985.
[http://dx.doi.org/10.1016/S0065-3233(08)60065-0] [PMID: 3904348]

[26] M.Y. Khan, "Direct evidence for the involvement of domain III in the N-F transition of bovine serum albumin", *Biochem. J.*, vol. 236, no. 1, pp. 307-310, 1986.
[http://dx.doi.org/10.1042/bj2360307] [PMID: 3790080]

[27] T. Ohta. Nobutoshi Chikazumi, "Estimation of hydrodynamic volume of proteins using high-performance size-exclusion chromatography and intrinsic viscosity measurement: An attempt at universal calibration", *J. Liq. Chrom.*, vol. 14, no. 1, pp. 403-425, 1991.

[28] J.L. Richards, "Viscosity and the shapes of macromolecules a physical chemistry experiment using molecular-level models in the interpretation of macroscopic data obtained from simple measurements", *J. Chem. Educ.*, vol. 70, no. 8, pp. 685-689, 1993.
[http://dx.doi.org/10.1021/ed070p685]

[29] T. Saeed, and A. Salahuddin, "Preparation and characterization of fragments from the N-terminal end of bovine serum albumin under native conditions", *Biochem. J.*, vol. 296, no. Pt 2, pp. 347-350, 1993.
[http://dx.doi.org/10.1042/bj2960347] [PMID: 8257423]

[30] K. Monkos, "Viscosity of bovine serum albumin aqueous solutions as a function of temperature and concentration", *Int. J. Biol. Macromol.*, vol. 18, no. 1-2, pp. 61-68, 1996.
[http://dx.doi.org/10.1016/0141-8130(95)01057-2] [PMID: 8852754]

[31] S.E. Harding, "The intrinsic viscosity of biological macromolecules. Progress in measurement, interpretation and application to structure in dilute solution", *Prog. Biophys. Mol. Biol.*, vol. 68, no. 2-3, pp. 207-262, 1997.
[http://dx.doi.org/10.1016/S0079-6107(97)00027-8] [PMID: 9652172]

[32] M. Luisa Ferrer, R. Duchowicz, B. Carrasco, J. García de la Torre, and A.U. Acuña, "The conformation of serum albumin in solution: a combined phosphorescence depolarisation-hydrodynamic modelling study", *Biophys. J.*, vol. 80, pp. 2422-2430, 2001.
[http://dx.doi.org/10.1016/S0006-3495(01)76211-X] [PMID: 11325741]

[33] J. Lee, and A. Tripathi, "Intrinsic viscosity of polymers and biopolymers measured by microchip", *Anal. Chem.*, vol. 77, no. 22, pp. 7137-7147, 2005.
[http://dx.doi.org/10.1021/ac050932r] [PMID: 16285659]

[34] R. Curvale, M. Masuelli, and A.P. Padilla, "Intrinsic viscosity of bovine serum albumin conformers", *Int. J. Biol. Macromol.*, vol. 42, no. 2, pp. 133-137, 2008.
[http://dx.doi.org/10.1016/j.ijbiomac.2007.10.007] [PMID: 18022223]

[35] R.A. Curvale, and J.C. Cesco, "Intrinsic viscosity determination by “single-point” and “double-point” equations", *Appl. Rheol.*, vol. 19, no. 5, p. 53347, 2009.

[36] V. Sharma, A. Jaishankar, Y-C. Wang, and G.H. McKinley, "Rheology of globular proteins: apparent yield stress, high shear rate viscosity and interfacial viscoelasticity of bovine serum albumin solutions", *Soft Matter.*, vol. 7, no. 11, pp. 5150-5160, 2011.

[http://dx.doi.org/10.1039/c0sm01312a]

[37] M.A. Masuelli, "Study of bovine serum albumin solubility in aqueous solutions by intrinsic viscosity measurements", *Adv. Phys. Chem.* , pp. 1-8, 2013. [http://dx.doi.org/10.1155/2013/360239]

[38] P.S. Sarangapani, S.D. Hudson, K.B. Migler, and J.A. Pathak, "The limitations of an exclusively colloidal view of protein solution hydrodynamics and rheology", *Biophys. J.* , vol. 105, no. 10, pp. 2418-2426, 2013. [http://dx.doi.org/10.1016/j.bpj.2013.10.012] [PMID: 24268154]

[39] M. Djabourov, "Gelation—A review", *Polym. Int.* , vol. 25, no. 3, pp. 135-143, 1991. [http://dx.doi.org/10.1002/pi.4990250302]

[40] F. Franks, *Solution Properties of Proteins.* Elsevier, 1988. [http://dx.doi.org/10.1007/978-1-59259-437-5_3]

[41] E.J. Cohn, "The relation between the isoelectric point of globulin and its solubility and acid combining capacity in salt solution", *Proc. Natl. Acad. Sci. USA.* , vol. 6, no. 5, pp. 256-263, 1920. [http://dx.doi.org/10.1073/pnas.6.5.256] [PMID: 16586809]

[42] E.J. Cohn, "Studies in the physical chemistry of the proteins I. The solubility of certain proteins at their isoelectric points", *J. Gen. Physiol.* , vol. 4, no. 6, pp. 697-722, 1922. [http://dx.doi.org/10.1085/jgp.4.6.697] [PMID: 19871969]

[43] K. Monkos, "Concentration and temperature dependence of viscosity in lysozyme aqueous solutions", *Biochim. Biophys. Acta.* , vol. 1339, no. 2, pp. 304-310, 1997. [http://dx.doi.org/10.1016/S0167-4838(97)00013-7] [PMID: 9187251]

[44] K. Monkos, "Viscosity analysis of the temperature dependence of the solution conformation of ovalbumin", *Biophys. Chem.* , vol. 85, no. 1, pp. 7-16, 2000. [http://dx.doi.org/10.1016/S0301-4622(00)00127-7] [PMID: 10885394]

[45] K. Monkos, "On the hydrodynamics and temperature dependence of the solution conformation of human serum albumin from viscometry approach", *Biochim. Biophys. Acta.* , vol. 1700, no. 1, pp. 27-34, 2004. [http://dx.doi.org/10.1016/j.bbapap.2004.03.006] [PMID: 15210122]

[46] K-C. Fan, M. Song, and R-S. Lu, *Seventh International Symposium on Precision Engineering Measurements and Instrumentation.*, August 07, 2011, Conference Volume 8321Yunnan, China, 2011.

[47] R. Wetzel, M. Becker, J. Behlke, H. Billwitz, S. Böhm, B. Ebert, H. Hamann, J. Krumbiegel, and G. Lassmann, "Temperature behaviour of human serum albumin", *Eur. J. Biochem.* , vol. 104, no. 2, pp. 469-478, 1980. [http://dx.doi.org/10.1111/j.1432-1033.1980.tb04449.x] [PMID: 6244951]

[48] K. Takeda, A. Wada, K. Yamamoto, Y. Moriyama, and K. Aoki, "Conformational change of bovine serum albumin by heat treatment", *J. Protein Chem.* , vol. 8, no. 5, pp. 653-659, 1989. [http://dx.doi.org/10.1007/BF01025605] [PMID: 2610859]

[49] I.S. Haworth, *Protein Solubility. Chemistry: Foundations and Applications.* , 2004. http://www.encyclopedia.com/doc/1G2-3400900422.html

[50] J. Peters, *All Albumin.* Elsevier, 1993.

[51] K.H. Van Holde, *Physical Biochemistry.* Prentice Hall, 1971.

[52] C.N. Pace, S. Treviño, E. Prabhakaran, and J.M. Scholtz, "Protein structure, stability and solubility in water and other solvents", *Philos. Trans. R. Soc. Lond. B Biol. Sci.* , vol. 359, no. 1448, pp. 1225-1234, 2004. [http://dx.doi.org/10.1098/rstb.2004.1500] [PMID: 15306378]

[53] M.A. Masuelli, "Viscometric study of pectin. Effect of temperature on the hydrodynamic properties", *Int. J. Biol. Macromol.* , vol. 48, no. 2, pp. 286-291, 2011.

[http://dx.doi.org/10.1016/j.ijbiomac.2010.11.014] [PMID: 21134395]

[54] M.L. Huggins, "The viscosity of dilute solutions of long-chain molecules. IV. Dependence on concentration", *J. Am. Chem. Soc.* , vol. 64, no. 11, pp. 2716-2718, 1942. [http://dx.doi.org/10.1021/ja01263a056]

[55] A.F. Martin, "Abstr", In: *103rd Am. Chem. Soc. Meeting.* ACS, 1942, pp. 1-c.

[56] Alberto Masuelli Martin, and Gabriela Sansone Maria, " Products and Applications of Biopolymers", In: *Hydrodynamic properties of Gelatine. Studies from intrinsic viscosity measurements,* C. J. R. Verbeek, Ed., INTECH, 2012. ISBN 978-953-51-0226-7

[57] E.O. Kraemer, "Molecular weights of celluloses and cellulose derivates", *Ind. Eng. Chem.* , vol. 30, pp. 1200-1204, 1938. [http://dx.doi.org/10.1021/ie50346a023]

[58] G.V. Schulz, and F. Blaschke, "Eine Gleichung zur Berechnung der Viskositatszahl fur sehr kleine Konzentrationen", *J. Prakt. Chem.* , vol. 158, pp. 130-135, 1941. [http://dx.doi.org/10.1002/prac.19411580112]

[59] R.M. Fuoss, "Viscosity function for polyelectrolytes", *J. Polym. Sci., Polym. Phys. Ed.* , vol. 3, pp. 603-604, 1948.

[60] M. Raymond, "Fuoss. Errata: Viscosity function for polyelectrolytes. (J. Polymer Sci. 3 (1948) 603-604)", *J. Polym. Sci., Polym. Phys. Ed.* , vol. 4, pp. 96-96, 1949.

[61] B. Yan, and X. Cao, "Preparation of aqueous two-phase systems composed of two pH-response polymers and liquid-liquid extraction of demeclocycline", *J. Chromatogr. A.* , vol. 1245, pp. 39-45, 2012. [http://dx.doi.org/10.1016/j.chroma.2012.05.045] [PMID: 22658298]

[62] A. Bhattarai, and B. Das, "Solvodynamic properties of sodium polystyrenesulphonate in methanol-water mixed solvent media in absence and in the presence of a salt using viscometric method", *Macromol. Symp.* , vol. 315, pp. 52-59, 2012. [http://dx.doi.org/10.1002/masy.201250507]

[63] R.F. Fedors, "An equation suitable for describing the viscosity of dilute to moderately concentrated polymer solutions", *Polymer (Guildf.).* , vol. 20, pp. 225-228, 1979. [http://dx.doi.org/10.1016/0032-3861(79)90226-X]

[64] A. Durand, "Aqueous solutions of amphiphilic polysaccharides: Concentration and temperature effect on viscosity", *Eur. Polym. J.* , vol. 43, pp. 1744-1753, 2007. [http://dx.doi.org/10.1016/j.eurpolymj.2007.02.031]

[65] W. Heller, "Treatment of viscosity data on polymer solutions (an analysis of equations and procedures)", *J. Colloid Sci.* , vol. 9, pp. 547-573, 1954. [http://dx.doi.org/10.1016/0095-8522(54)90057-4]

[66] V.I. Irzhak, *"in Fizicheskie svoistva vyazkouprugikh polimerov (Physical Properties of Viscoelastic Polymers)", Sverdlovsk: Ural'sk.* Nauchn. Tsentr Akad. Nauk SSSR, 1981, pp. 76-81.

[67] V.G. Baranov, Y.V. Brestkin, S.A. Agranova, and V.N. Pinkevich, "Behavior of Polystyrene Macromolecules in Stiffened Good Solvent", *Vysokomolekulyarnye Soedineniya Seriya B.* , vol. 28, no. 11, pp. 841-843, 1986.

[68] V.G. Baranov, S.Y. Frenkel, S.A. Agranova, Y.V. Brestkin, V.N. Pinkevich, and B.M. Shabsels, "Vysokomol. Soedin", *Ser. B.* , vol. 29, no. 10, pp. 745-748, 1987.

[69] A.N. Krasovskii, S.S. Mnatsakanov, E.G. Guseva, A.I. Andreeva, E.V. Maryanina, and L.N. Ustinova, "Comparative-study of the structure and relative viscosity of aqueous-solutions of commercial photogelatin as a function of concentration", *Russ. J. Appl. Chem.* , vol. 66, no. 4, pp. 671-679, 1993.

[70] T.D. Chou, and J.L. Kokini, "Rheological properties and conformation of tomato paste pectins, citrus and apple pectins", *J. Food Sci.* , vol. 52, no. 6, pp. 1658-1664, 1987.

[http://dx.doi.org/10.1111/j.1365-2621.1987.tb05900.x]

[71] W. Kozicki, "P. Q. Kuang. “An Alternative Method for Evaluation of Intrinsic Viscosity", *Can. J. Chem. Eng.* , vol. 74, pp. 429-432, 1996.
[http://dx.doi.org/10.1002/cjce.5450740317]

[72] B.A. Wolf, "Polyelectrolytes Revisited: Reliable Determination of Intrinsic Viscosities”. Bernhard A. Wolf", *Macromol. Rapid Commun.* , vol. 28, pp. 164-170, 2007.
[http://dx.doi.org/10.1002/marc.200600650]

[73] M.V. Badiger, N.R. Gupta, J. Eckelt, and B.A. Wolf, "Intrinsic Viscosity of Aqueous Solutions of Carboxymethyl Guar in the Presence and in the Absence of Salt", *Macromol. Chem. Phys.* , vol. 209, pp. 2087-2093, 2008.
[http://dx.doi.org/10.1002/macp.200800238]

[74] A. Eich, and B.A. Wolf, "Intrinsic viscosities of polyelectrolytes: determination and modeling of the effects of extra salt", *ChemPhysChem.* , vol. 12, no. 15, pp. 2786-2790, 2011.
[http://dx.doi.org/10.1002/cphc.201100439] [PMID: 21919181]

[75] L. Ghimici, M. Nichifor, A. Eich, and B.A. Wolf, "Intrinsic viscosities of polyelectrolytes in the absence and in the presence of extra salt: Consequences of the stepwise conversion of dextran into a polycation", *Carbohydr. Polym.* , vol. 87, pp. 405-410, 2012.
[http://dx.doi.org/10.1016/j.carbpol.2011.07.067]

[76] M. Bercea, L.E. Nita, J. Eckelt, and B.A. Wolf, "Polyelectrolyte complexes: phase diagram and intrinsic viscosities of the system water/poly(2-vinylpyridinium-Br)/poly(styrene sulphonate-Na)", *Macromol. Chem. Phys.* , vol. 213, pp. 2504-2513, 2012.
[http://dx.doi.org/10.1002/macp.201200445]

[77] J. Eckelt, A. Knopf, and B.A. Wolf, "Polyelectrolytes: Intrinsic Viscosities in the Absence and in the Presence of Salt", *Macromol.* , vol. 41, pp. 912-918, 2008.
[http://dx.doi.org/10.1021/ma702054f]

[78] M. Bohdanecky, and J. Kovar, "Polymer Science Library", In: *Viscosity of Polymer Solutions,* A.D. Jenkins, Ed., vol. Chapter 2. Elsevier: Amsterdam, 1982.

[79] G. Challa, *Polymer Chemistry. An Introduction,* T.J. Kemp, J.F. Kennedy, Eds., Ellis Horwood: London, 1993, p. 134.

[80] W.R. Moore, "Viscosities of Dilute Polymer Solutions", *Prog. Polym. Sci.* , vol. 1, pp. 3-43, 1967.
[http://dx.doi.org/10.1016/0079-6700(67)90001-9]

[81] S. Kirincic, and C. Klofutar, "Viscosity of aqueous solutions of polyethylene glycols at 298.15 K", *Fluid Phase Eq.* , vol. 155, pp. 311-325, 1999.
[http://dx.doi.org/10.1016/S0378-3812(99)00005-9]

[82] A. Mehrdad, N. Samadiani, and L. Poormoosa, "Effect of temperature and hydrochloric acid on the intrinsic viscosity of poly(acrylic acid) in aqueous solutions", *J. Mol. Liq.* , vol. 187, pp. 177-182, 2013.
[http://dx.doi.org/10.1016/j.molliq.2013.06.018]

[83] S.E. Hill, and A. David, *Ledward, J.R. Mitchell. Functional Properties of Food Macromolecules.* Springer Science, 1998, pp. 63-80.

[84] M. Eli, *Pearce, Bob A. Howell, Richard A. Pethrick, Gennady E. Zaikov. Physical Chemistry Research for Engineering and Applied Sciences, Volume Three: High Performance Materials and Methods.* CRC Press, 2015, pp. 60-61.

[85] *Christian Clasen. Viscosimetry of polymers and polyelectrolytes.* Springer-Verlag Berlin Heidelberg: Germany, 2004.

[86] H. Mark, *Der feste Körper,* R. Sänger, Ed., Hirzel: Leipzig, 1938, pp. 65-104.

[87] R. Houwink, "Zusammenhang zwischen viscosimetrisch und osmotisch bestimm- ten

polymerisationsgraden bei hochpolymeren", *J. Prakt. Chem.* , vol. 157, p. 15, 1940. [http://dx.doi.org/10.1002/prac.19401570102]

[88] M.A. Masuelli, A. Takara, and A. Acosta, "Hydrodynamic properties of tragacanthin. Study of temperature influence", *J. Arg. Chem. Soc.* , vol. 100, pp. 25-34, 2013.

[89] D. Gomes, C.P. Borges, and J.C. Pinto, "Evaluation of parameter uncertainties during the determination of the intrinsic viscosity of polymer solutions", *Polymer (Guildf.).* , vol. 41, pp. 5531-5534, 2000. [http://dx.doi.org/10.1016/S0032-3861(99)00858-7]

[90] Y.H. Junpei, H. Matsuoka, H. Kitano, N. Ise, T. Yamaguchi, S. Saeki, and M. Tsubokawa, "Revisit to the intrinsic viscosity-molecular weight relationship of ionic polymers. 4. Viscosity behaviour of ethylene glycol/water solutions of sodium poly(styrenesu1phonate)", *Macromol.* , vol. 24, pp. 3206-3208, 1991. [http://dx.doi.org/10.1021/ma00011a025]

[91] M.A. Masuelli, "Hydrodynamic properties of whole Arabic gum", *Am. J. Food Sci. Tech.* , vol. 1, no. 3, pp. 60-66, 2013.

[92] M.A. Masuelli, "Mark-Houwink parameters for aqueous-soluble polymers and biopolymers at various temperatures", *J. Pol. Biopol. Phys. Chem.* , vol. 2, no. 2, pp. 37-43, 2014.

CHAPTER 3

Small Angle Scattering and *ab initio* Modeling

Denis Renard[a,*] and **C. Sanchez**[b]

[a] *INRA, UR1268 Biopolymères Interactions Assemblages, F-44300 Nantes Cedex, France*

[b] *UMR1208 Ingénierie des Agropolymères et Technologies Emergentes, INRA-Montpellier SupAgro-CIRAD-Université Montpellier, 2 Place Pierre Viala, F-34060 Montpellier, France*

Abstract: Small-angle scattering of X-rays (SAXS) or neutrons (SANS) is a well-established method to study the overall structure and structural transitions of biopolymers in solution. Determination of overall structural parameters such as the radius of gyration R_g, the molecular mass M_w or the hydrated volume V is now easily computed using appropriate free Softwares. In addition, increasingly sophisticated *ab initio* methods were developed over the last two decades for building structural models of biopolymers from x-ray and neutron solution scattering data. Even if most of the developments were focused on protein in solution, folded or intrinsically disordered, protein complexes or protein fibrils, *ab initio* methods were also used to calculate three-dimensional shape models of polysaccharides or nucleic acids. The models derived from experimental scattering pattern up to a resolution of 0.5 nm are classified as low resolution models compared to atomistic resolution structures using X-ray crystallography. The efficiency of the methods is illustrated in this chapter by focusing mainly on proteins or polysaccharides with unknown crystal structures and for which the *ab initio* reconstruction of three-dimensional models may help to define, in combination or not with other structural techniques, such as microscopy, overall dimensions and global shape. These methods improved the resolution and reliability of models derived from scattering data substantially and has made solution scattering, in combination with recent developments using size-exclusion chromatography, robotic sample changer or microfluidics, and a useful technique in high-throughput large-scale structural characterization of biopolymers.

Keywords: *Ab initio*, Complexes, Fibrils, Flexibility, Microfluidics, Modeling, Neutron, Pair distance distribution function, Polydispersity, Polysaccharide, Proteins, Radius of gyration, Sample changer, Size exclusion chromatography, Small-angle scattering, Three-dimensional structures, X-ray.

* **Corresponding author Denis Renard**: INRA, UR1268 Biopolymères Interactions Assemblages, F-44300 Nantes Cedex, France; Email: drenard@nantes.inra.fr

Martin Masuelli & Denis Renard (Eds.)

INTRODUCTION

Small-angle scattering (SAS) of X-rays (SAXS) or neutrons (SANS) is a fundamental tool in the structural characterization of biopolymers with molar masses ranging from a few kDa to several MDa [1]. Great progress has been made over the years in applying this technique to extract structural information from non-crystalline samples in the fields of physics, materials science and biology [2]. Over the last decade, major advances in instrumentation and computational methods have led to new and exciting developments in the application of SAS to structural biology [3]. In particular, the synergistic improvement in hardware as well as software resulted in an automated data collection and in-depth analysis of the scattering data [4 - 6]. The absence of radiation damage and possibility of contrast variation by solvent exchange (H_2O/D_2O) or specific deuteration makes SANS an extremely useful complementary tool for SAXS [7]. SAS experiments delivers structural information in the length scale ranging from 1 nm to 1 µm depending on the instrument and source used (neutron or X-ray), and the largest length scale being explored by using ultra small-angle scattering (USAS). SAS experiments can be fast in particular on a third generation synchrotron and require moderate amount of purified material (order of hundreds of micrograms to a few milligrams, depending on the SAS instrument and the source used). For moderate polydisperse samples, the coupling of size-exclusion chromatography (SEC) to on-line SAXS may help in the structural determination of each component present in the sample. Structural changes due to environmental conditions (pH, ionic strength, temperature,…) can be easily monitored and directly correlated with functional results (*i.e.* kinetics, spectroscopy or interactions studies in solution) [8]. SAS is also extremely useful in the quantitative characterization of mixtures and flexible systems, including time-resolved experiments to monitor assembly or folding processes. Weak protein-ligand interactions and structural modeling of transient complexes are also widely studied using SAXS [9].

Solution structure will be investigated in this chapter to characterize overall parameters of shape and dimensions of isolated or interacted biopolymers alone or in interactions but readers should keep in mind that due to the high penetration of X-ray or neutrons, (nano)-structures on solid or gelled samples may also be explored using this technique. The focus will be on *ab initio* modeling methods to probe into the three-dimensional structure of non-crystalline biopolymers and the characterization of intrinsic flexibility encountered particularly in proteins. Special care will be given to the analyses of the polydispersed samples where developments have been brought these last years in the on-line combination of size exclusion chromatography to SAXS measurements. Kinetics using rapid mixing techniques will be presented with a focus on microfluidics as an emerging tool to probe structural dynamics of biological samples. Finally, some developments dealing with the high throughput applications of SAS experiments, including combination of automated microfluidic platform, will be presented, generating a renewed interest in the wider applications of the technique in the structural biology community [3].

Basics of Small Angle Scattering

In a standard scattering experiment, the rotationally averaged scattering intensity of a biopolymer is obtained as a function of the scattering vector or momentum transfer, q:

$$q = \frac{4\pi sin\theta}{\lambda} \quad (1)$$

Where θ is half of the angle between the incident and scattered radiation and λ is the incident radiation wavelength. The SAS scattering intensity $I(q)$ can be expressed as a product of the form factor $P(q)$, which contains information on the shape and dimension of the scattering particles, and the structure factor $S(q)$ describing the interparticle interaction [10]:

$$I(q) = N\Delta\rho^2 P(q)S(q) \quad (2)$$

Where N is the number density of the particles, and $\Delta\rho = (\rho\text{-}\rho_0)$ is the so-called "contrast" (*i.e.* electron or neutron density, showing the difference between the scattering length density of the particle ρ and that of the solvent ρ_0). Using contrast variation by SANS, multiple data sets are recorded from the same object, which are different because of different contrasts of the components in the particle. The scattering length density of the particle ρ may be changed by selective deuteration or perdeuteration (*i.e.* deuteration in which all hydrogen atoms are replaced by deuterium) while ρ_0 is conveniently varied in a broad range by H_2O/D_2O exchange in solution [7]. Chemical method (for cellulose) [11] and perdeuterated algae culture (for phycocyanin) [12] were previously used for selective deuteration or perdeuteration to modulate $\Delta\rho$ until the developments of genetic engineering over the last fifteen years that now offers a large panel of cultured strains to produce partially or fully deuterated biopolymers such as polysaccharide [13], lipids [14], ribosomes [15], RNA [16], proteins [17] or complexes [7, 18]. In addition, modern neutron scattering facilities in the world (Institut Laue Langevin, Grenoble, France; Oak Ridge National Laboratory, Oak Ridge, USA; OPAL, Lucas Heights, Australia,...) offer the opportunity of deuterium labeling for structural biology studies. The other way to change $\Delta\rho$ and therefore the scattering intensity is to use the contrast variation method to highlight structural informations in particular from complexes from two or more chemical distinct components [7, 19, 20]. Utility of subunit-selective deuteration of protein complexes with SANS measurements is illustrated on (Fig. **1**) [18]. A ternary complex (A), consisting of protein A (blue), B (orange), and RNA (red) is assumed. In 0% D_2O solution without any deuteration (B), SANS does provide information about the overall shape of the complex, comparable to SAXS (but with lower sensitivity). In 42% D_20 solution (C), deuterated protein components have a positive contrast, while the protonated component match the contrast of the buffer and are "invisible" [18]. Under these conditions, the shape and center-o--mass distances between the deuterated component and the RNA can be determined. In 70% D_2O solution (D), the contrast of RNA is matched by the buffer, and the shape and center-of-mass distances between the deuterated components (positive contrast), along with the protonated components (negative contrast), can be determined.

In addition, to assist in planning SANS experiments with contrast variation on protein and nucleic acid complexes, users may now be helped by using the new software tool SASSIE Contrast Calculator (SCC) on the basis of the primary sequence and deuteration level of each protein or nucleic acid component [21].

Calculations of Structural Parameters from SAS Data

As a preliminary remark, all structural parameters described in this section are now easily computed from SAS data using the PRIMUS program [22] contained in the ATSAS program package [23] developed at EMBL by Svergun's group or the SasView (previously named SansView) analysis software developed by an international collaboration of facilities (http://www. sasview.org).

In the dilute region, the interparticle interaction can be neglected (*i.e.* $S(q) = 1$), and the analysis of scattering intensity $I(q)$ can thus furnish direct information on morphological features of the scattering particles. Among the first direct informations, Guinier analysis remains the most straightforward method to calculate the radius of gyration R_g and the forward scattering intensity $I(0)$ on the low q range [24] (Table **1**, equation (1)). Data-treatment programs such as GNOM for example are commonly used to determine R_g and $I(0)$ [25, 26]. The Guinier approximation is valid for reasonably small q values, typically for $qR_g < 1.3$. In practice, R_g and $I(0)$ are determined by the Guinier plot, *Ln* $I(q)$ *versus* q^2. The plot should be linear for pure monodisperse samples, whereby the slope of the line gives R_g, and its intersection with the ordinate provides the forward scattering intensity $I(0)$. It is important to note that the linearity of the Guinier plot is often a good indication of the sample quality. Nonlinearity from Guinier plot can reflect improper background subtraction, polydispersity or interparticle interactions (*i.e.*$S(q) \neq 1$) [5]. For example, the presence of aggregates in the sample leads to scattering curve and Guinier plot with a sharp increase in intensity at small q values, while a decrease in the intensity is observed at low q values when interparticle repulsions are present. Alternatively, Debye's approximation can be more precise than Guinier's as its validity extends to larger scattering vectors [27] (Table **1**, equation 2). Equation (1) in Table **1** is valid for arbitrary particle shapes. For very elongated particles, the radius of gyration of the cross-section R_c can be computed from $qI(q)$ *vs* q and for flattened particles the radius of gyration of the thickness R_t from $q^2I(q)$ *vs* q^2 [28].

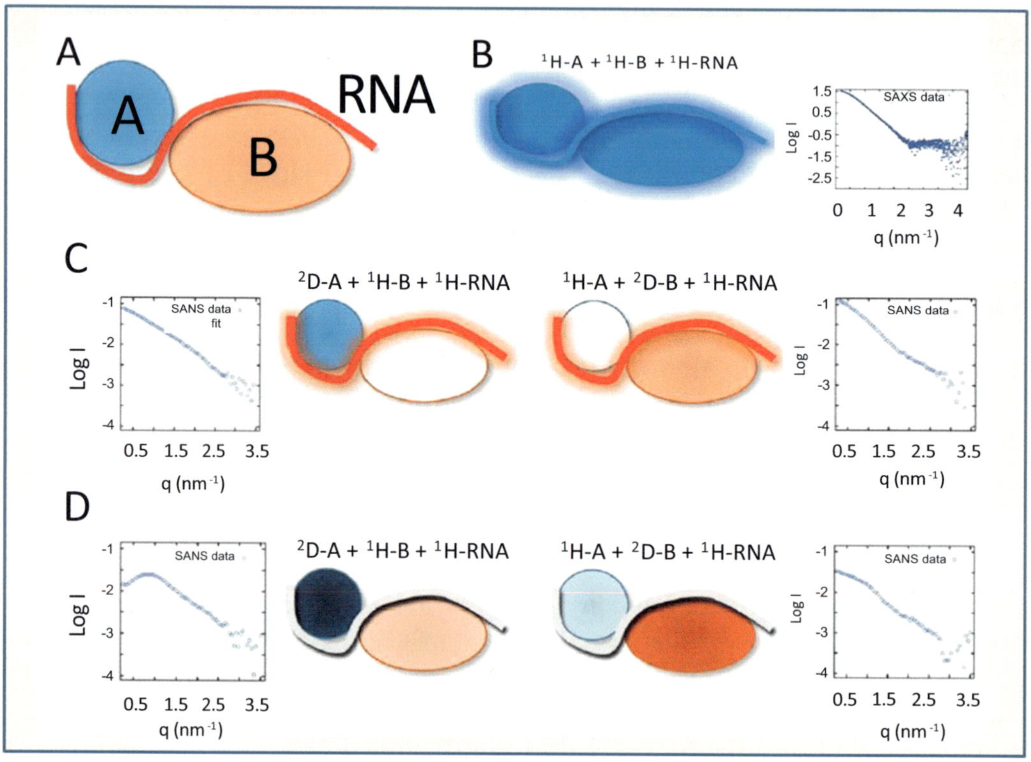

Fig. (1). Utility of subunit-selective deuteration of a ternary protein-protein-RNA complex with SANS measurements (see text for details) (adapted from reference 18).

From the Guinier analysis, the molecular weight M_w can be estimated based on knowledge of the forward scattering intensities and concentration of both the biopolymer of interest and a standard such as lyzozyme or bovine serum albumin (Table **1**, equation 3). Hence, the value of $I(0)$ divided by the biopolymer concentration is proportional to the biopolymer M_w. Significantly, in SAS experiments, using neutrons or X-rays, the absolute scattering intensity allowing the molar mass determination may also be obtained using the absolute value of water intensity in units of cross section (cm^{-1}) [29, 30]. An alternative method was recently proposed to determine M_w from SAXS data on a relative scale and without another scattering profile determined from a known standard protein [31]. The method is based on the determination of an analytical function relating the real volume of proteins in dilute solutions V and their apparent volume V' derived from the analysis of truncated SAXS data. Based on a systematic analysis of the known structures of 1148 proteins selected from the Protein Data Bank (PDB), simple linear relationships V *vs* V' were determined for different maximum q limits of SAXS data. In addition, this method allows for an unambiguous determination of the multimeric state of proteins with known molecular weights.

The Porod (excluded particle) volume of the particle can also be estimated by transforming the scattering data as $q^2I(q)$ *vs.* q (Kratky plot), capturing a defined area called the Porod invariant, Q (Table **1**, equations (4)) [32]. Q is directly related to $\Delta\rho$ that is scaled by the particle's volume V and concentration c. Thus, from the knowledge of Q and the forward scattering intensity $I(0)$, the particle volume can thus be determined in relative units (Table **1**, equation (6)). The determination of Q can in principle be used to determine the surface-to-volume ratio S/V (Table **1**, equation (7)) [10, 33]. The Porod-Debye law describes a fourth power law approximation between q and the scattering intensity $I(q)$, approximation that occurs within a limited range of scattering angles similar to the Guinier approximation. Application of the Porod-Debye law predicts a plateau within the low resolution region (*i.e.* high q values) of the SAS data when transformed by q^4 (Table **1**). The presence of the plateau validates the assumptions that the scattering particle exists with a sharp homogenous (electron or neutron) density contrast between the particle and solvent [32]. Limiting the determination of Q to the Porod-Debye region of the data (*i.e.* using the program PRIMUS) can now provide a reliable determination of the particle's macromolecular volume [32]. The Porod volume V allows one to make an alternative estimate of the M_w, by multiplying the volume V by the average mass density ρ_m ($g.cm^{-3}$), with the added advantage that this estimate is independent of errors in the sample concentration [3]. Typically, for a globular protein, with a mass density of 1.37 $g.cm^{-3}$, V (in nm^3) is 1.7 times the M_w (in kDa) [34]. These statements are however only valid for particles with homogeneous scattering length density and do not hold when inhomogeneity coming from flexible parts of the particle or assembly are present. In the case of flexible or unstructured particles, the volume of correlation, V_c, a SAS invariant derived from the scattered intensities, was recently introduced [35]. V_c is specific to the structural state of the particle, but independent of concentration and the requirements of a compact, folded particle.

Table 1. Graphical representations, structural parameters extracted from SAS experiments and equations associated. Adapted from references [13] and [22].

<table>
<tr><th>Graphical representation</th><th>Structural parameters</th><th>Equations</th></tr>
<tr><td>(a)
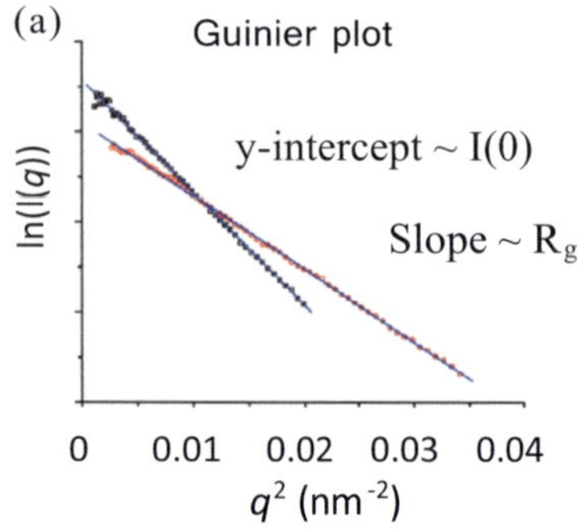
</td><td>*Radius of gyration* R_g
Forward scattering intensity $I(0)$</td><td>*Guinier analysis (Ln I(q) vs q²)*
$$I(q) = I(0)e^{\frac{-1}{3}R_g^2 q^2} \quad (1)$$
slope = $-^1/_3 R_g^2$; intercept = Ln I(0)
Debye analysis (Ln I(q) vs $q^2 R_g^2$)
$$\frac{I(q)}{I(0)} = \frac{2}{x^2}(x - 1 + e^{-x}) \quad (2)$$
with $x = q^2 R_g^2$</td></tr>
<tr><td></td><td>*Molecular weight* M_w</td><td>$$M_w = \frac{I(0)}{c} \frac{M_w^{stand}}{I_0^{stand}/c_{stand}} \quad (3)$$</td></tr>
<tr><td>(b)
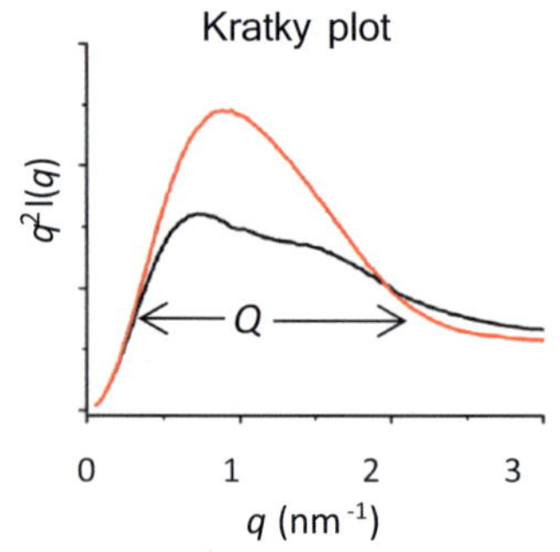
</td><td>*Porod volume* V</td><td>*Porod analysis ($q^2I(q)$ vs q)*
$$Q = \int_0^\infty q^2 I(q)dq = c2\pi^2(\Delta\rho)^2 V \quad (4)$$
Q: Porod invariant
and
$I(0) = N(\Delta\rho)^2 V^2$ (5)
N: number of particles per unit volume
then
$$V = \frac{2\pi^2 I(0)}{Q} \quad (6)$$</td></tr>
<tr><td>(c)
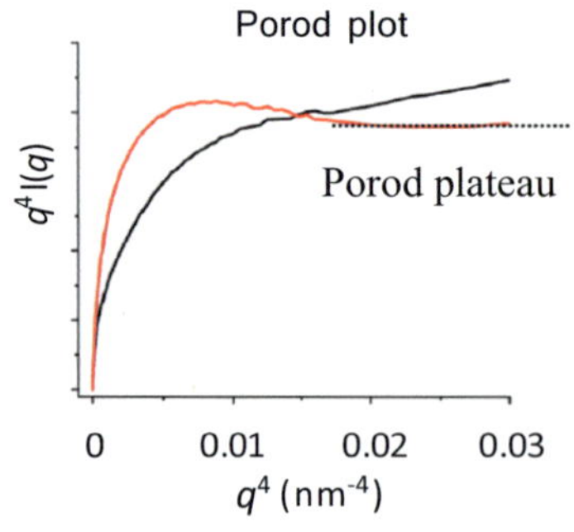
</td><td>*Surface-to-volume* S/V</td><td>*Porod-Debye analysis ($q^4I(q)$ vs q^4)*
$$\frac{S}{V} = \frac{\pi}{Q} \lim_{q \to \infty} q^4 I(q) \quad (7)$$
Porod plateau: I(q) ≈ $2\pi/q^4(\Delta\rho)^2 S$</td></tr>
</table>

(Table 1) contd.....

Graphical representation	Structural parameters	Equations
Indirect Fourier transformation (d) 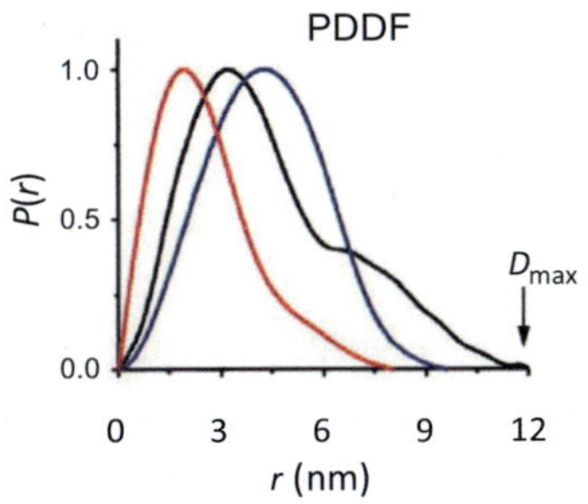	*Maximum distance D_{max}* *Radius of gyration R_g* *Forward scattering intensity I(0)*	*Indirect Fourier transform analysis* $p(r) = \frac{r^2}{2\pi^2}\int_0^\infty q^2 I(q)\frac{\sin(qr)}{qr}dq$ (8) p(r), pair distance distribution function

The scattering intensity profile can be converted into a pair distance distribution function, noted *p(r)*, a spherically averaged autocorrelation function of excess scattering density, using an indirect Fourier transformation [26, 36] (Table **1**, equation (7)). Globular particles display bell-shaped *p(r)* functions while two distances appear for two subunits assembly. The *p(r)* function is essentially a histogram of all pairwise distances, *r*, between electrons or neutrons in the sample. Because of the rotational averaging, no directional information is available about these distances [9].

The maximum distance in the particle D_{max} can be obtained directly from the *p(r)* function. D_{max} is a user defined variable that is optimized by the minimization of the χ^2 parameter corresponding to the goodness of fit between the experimental data and that calculated by the direct transform of the *p(r)* function. In the indirect transform program GNOM ([26]), the solution yielding the *p(r)* function is evaluated using perceptual criteria, providing the user with the means to easily identify reliable solutions and to obtain an optimal value of D_{max}. It is important to note that interparticle interference (*i.e.* $S(q) \neq 1$) will lead to an abrupt arrival of *p(r)* at zero suggesting that the chosen D_{max} is artificially short and for aggregated samples small peaks in *p(r)* at high *r* values or the *p(r)* function may never approach zero appropriately. The procedure to calculate D_{max} has also been automated in the program AUTOGNOM [37], where multiple GNOM runs are performed without user intervention across a range of D_{max} values [3].

The distance distribution function is a real space representation of the scattering data and allows one to graphically display the peculiarities of the particle shape. For example, globular particles yield bell-shaped profiles with a maximum at approximately $D_{max}/2$ and multi-domain particles often yield profiles with multiple shoulders and oscillations corresponding to intra and inter-subunit distances (See figure Table **1**) [3]. The parameters estimated from the indirect Fourier transform approach, *I(0)* and R_g, are more accurate than those obtained from a Guinier analysis as the entire scattering curve is considered for their estimation. From the *p(r)* function, an estimate of M_w can also be obtained, which is independent of the biopolymer shape.

Ab initio Modeling

In the absence of high resolution structural models, thc rcconstruction of low-resolution (about 1.5-2 nm) 3D models from SAS data alone is now a standard procedure and as such can also be considered a rapid characterization tool [3]. The basic principles behind shape determination were first developed in the 1960s where scattering profiles were computed from different geometrical shapes and compared with experimental data. These trial-and-error methods were refined in the 1970s by Stuhrmann [38, 39] where the shape of particles was described by an angular envelope function, the latter developed into a series of spherical harmonics [3]. This method was further developed into the first publicly available program SASHA [40 - 42]. The most popular *ab initio* bead-modeling program in current use is DAMMIN (Dummy Atom Model Minimisation) [43]. The algorithm represents a particle as a collection of densely packed beads inside a constrained (usually spherical) search volume, with a maximum diameter defined by the experimentally determined D_{max} (Fig. **2**) [3]. Each bead is randomly assigned to the solvent or solute, and the particle structure is described by a binary string X of length M. Shape reconstruction is conducted starting from a random initial approximation by simulated annealing, and by minimizing the goodness-o--the-fit χ^2 between the experimental and computed scattering profiles. At each step of the procedure, the assignment of a single bead is randomly changed leading to a new model. The solution is also constrained by the penalty term requiring that the beads must be connected and the model compact to ensure that physically feasible low-resolution structures are generated. Shape analysis is straightforward for monodisperse or almost monodisperse systems, where one of the components dominates the scattering profile [9]. A multiphase version of DAMMIN bead-modeling is implemented in the program MONSA [44], also widely used when contrast variation data from SANS and/or multiple SAXS curves from components of a complex are available [3]. Although DAMMIN is reasonably fast (taking several minutes to several hours), a new implementation called DAMMIF (where F refers to fast) was recently developed [45]. Information about particle symmetry, if available, can be explicitly used in the *ab initio* programs [7, 43, 46], further improving the resolution of the models. A program named DENFERT was recently developed for the structural modeling of biological macromolecules [47]. DENFERT is implementing a simulated annealing algorithm similar to DAMMIN program with the advantage that the hydration layer around bio-molecules is taken into account by introducing a second type of beads (hydration beads) in the model.

One must note that *ab initio* methods, seeking to reconstruct a 3D shape from a 1D scattering profile, cannot provide a unique solution and, when running the programs multiple times, somewhat different models are obtained. A comparison of these models ensures that the most persistent features are identified from a number of models that fit the data equally well but show variation in shape [3].

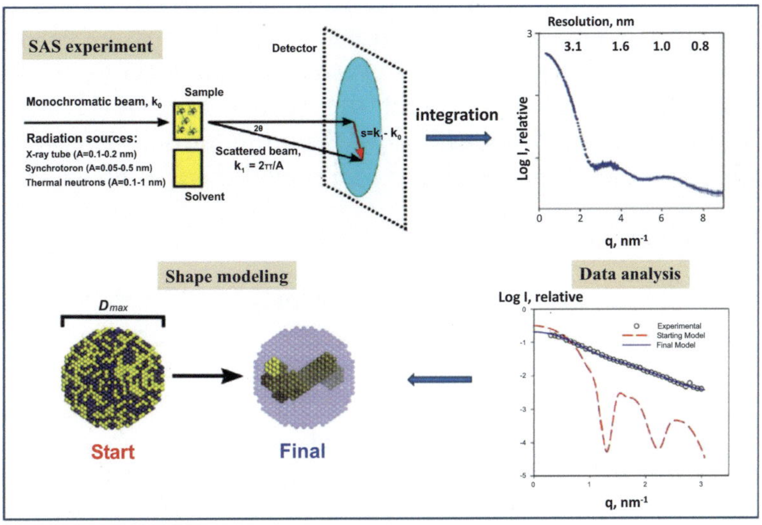

Fig. (2). Principle of small-angle scattering experiment (SAS), data analysis and *ab initio* shape modeling. The procedure starts from a spherical search volume with beads assigned to the particle in yellow and those assigned to solvent in blue. Fitting procedure is conducted until a final model is generated satisfying not only a fit to the experimental data, but also forming a compact and connected model of dummy atoms/beads. Adapted from references [3] and [5].

The programs SUPCOMB [48] and DAMAVER [49] have been developed for this purpose. SUPCOMB performs the rapid superposition of ensemble of models and calculates the degree of structural similarity. Further, SUPCOMB identifies the most probable ensemble member and DAMAVER averages the superposed models over the ensemble yielding a smoothed model containing the most persistent features in the reconstructions [3]. Low-resolution models obtained from SAS analyses are often validated with the calculations of the hydrodynamic parameters such as the sedimentation coefficient, diffusion coefficient, hydrodynamic radius, hydrated volume, maximum dimension and radius of gyration using the program HYDROPRO [50]. Some discrepancies may sometimes appear in the experimental and calculated hydrodynamic parameters when flexible and porous structures are studied because calculated hydrodynamic parameters are computed on a model that is first filled with beads based on the assumption that macromolecules are rigid [50], uncharged and impenetrable [51]. The low-resolution 3D model can provide useful information about the relationship between structure and function [1, 52, 53]. The *ab initio* analysis of biopolymers and assemblies using SAS can be usefully implemented by structural information coming from other methods such as NMR, electron microscopy, atomic force microscopy (AFM). In the case where high-resolution models coming from crystallography or NMR are available, SAXS comes into play to identify the biologically active conformations of protein crystal structures and/or to validate structural similarity between crystal and solution or to model multi-subunit proteins [3, 54, 55]. We direct the reader to several comprehensive reviews for additional details regarding SAS computations of concurrent high-resolution models and comparison with experimental data [3, 56 - 62].

Flexible Protein Systems

Flexibility can be encountered in the case of proteins when they are completely, partially or intrinsically disordered. An example of scattering curves (Kratky's representations) of folded, partially folded and completely unfolded particles is illustrated on (Fig. **3**). SAS were proven to be a powerful technique for the analysis of flexible systems [32, 63]. However, a recent study by Bernardó [64] clearly displayed the difficulties associated with meaningful interpretation of scattering profiles for highly flexible modular proteins. This study demonstrated that proteins may be wrongly identified as rigid from dynamically averaged SAXS profiles and that several indicators for flexibility should be monitored [3]. This is the reason why for proteins some preliminary calculations may be performed in order to identify the conformation of the protein and its possible flexibility. From the known number of residues N, the radius of gyration R_g, representing the z-average [2] over all accessible conformations in solution, can be calculated from Flory's equation $R_g = R_0 N^v$ with the constant R_0 which depends on the persistence length of the chain and $v = 0.6$ for an excluded-volume polymer (Table **2**) [65]. A compilation of R_g values for 26 chemically denatured proteins found a v value of 0.598 (~0.6) and a R_0 value of 1.927 [66]. These values are consistent with hydrodynamic equation from Wilkins *et al.* [67] and above all from those of Uversky [68] (Table **2**) giving some confidence to the random coil nature of the unfolded proteins. For comparison, equations for globular (folded) proteins are also given in Table **2** (Wyatt's equation extracted from http://wyatt.com based on HPSEC-MALLS calibration curve). The comparison of experimental R_g value from SAS and/or R_h value from dynamic light scattering (DLS) and those estimated from equations in Table **2** is a good starting strategy to assign flexibility (or disorder) to a protein. However, ambiguity still resides in protein conformations between chemically denatured state and naturally intrinsically disordered state that needs to be clarified using appropriate methodology. In the past, characterization of flexibility by means of SAS was done qualitatively, generally being restricted to the determination of a few simple geometric parameters [28]. Recent developments have allowed quantitative description of the conformational ensembles for flexible macromolecules in solution using an Ensemble Optimization Method (EOM) or BIBLIOMOD allowing the co-existence of multiple protein conformations in solution compatible with scattering data [63, 69 - 71].

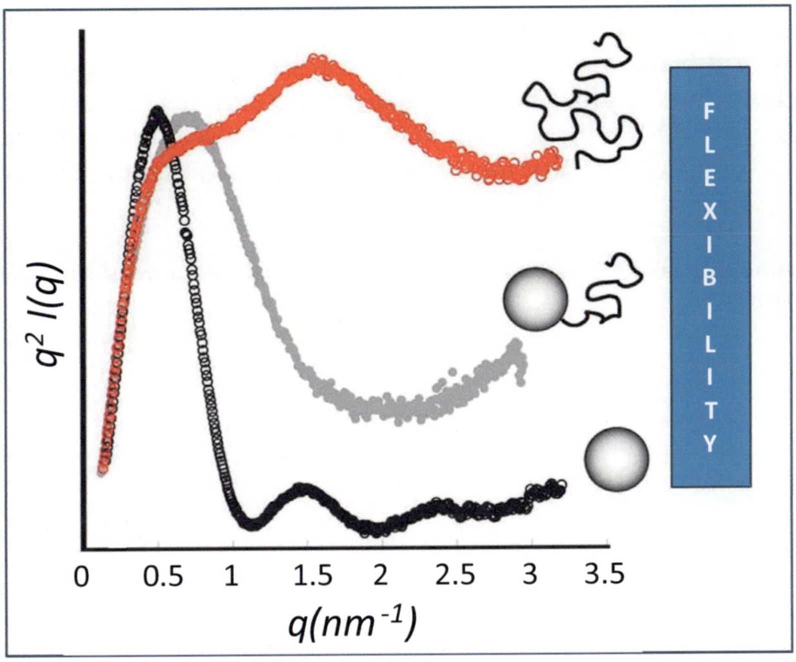

Fig. (3). Scattering data (Kratky plots) illustrating changes in the behavior of the scattering intensity for folded (sphere), partially folded (sphere-random coil) and completely unfolded (random coil) particles. Adapted from reference [32].

The efficiency of EOM was demonstrated in model and practical examples on completely or partially unfolded proteins and on multidomain proteins interconnected by linkers. In the latter case, EOM was able to distinguish between rigid and flexible proteins and to directly assess the interdomain contacts [70].

Glycosylated Proteins

The flexible and heterogeneous nature of carbohydrate chains often renders glycoproteins refractory to traditional structure determination methods. SAS can thus be used to obtain structural information where all-atom modeling of glycoproteins with flexible glycan chains can be applied to interpret the solution

SAS data [72]. Contrary to *ab initio* modeling, where low-resolution morphologies are calculated from SAS data but are unable to directly visualize the influence or position of the glycans, the all-atom ensemble modeling generates models where the spatial occupancy of the glycan chain in solution can be approximated. For simpler systems (single glycan, with a well-defined protein structure), all-atom modeling generates models in excellent agreement with the scattering pattern while for more complex systems (several glycan chains, or unknown protein substructure), the approach can still provide insightful models, though the orientations of glycans become poorly determined [72]. An alternative technique is to use SANS with contrast matching, as the scattering contrast of the carbohydrate is significantly different from that of the protein, to directly resolve glycans [73].

Table 2. Quantitative structural parameters interpretations of folded and chemically unfolded proteins from literature's equations.

	Static parameter		Hydrodynamic parameter	
	Unfolded protein		Random coil (unfolded) protein	Globular (folded) protein
Flory [65]	$R_g = R_0 N^v$	Wilkins et al. [67]	R_h (Å)= 2.21 $*N^{0.57}$	R_h (Å)= 4.75 $*N^{0.29}$
Kohn et al. [66]	$R_g = 1.927\ N^{0.60}$	Uversky [68]	R_h (Å)= 0.22 $M_w^{0.52}$(Urea) R_h (Å)= 0.19 $M_w^{0.54}$ (Guanidinium chloride)	R_h (Å)= 0.62 $M_w^{0.36}$ (g/mol)
		Wyatt (http://wyatt.com)		M_w (kDa)= 1.68 R_h (nm) $^{2.34}$
$\rho = R_g / R_h$	1.03		1.03	0.77

R_g: radius of gyration; R_0 = constant; n = exponential scaling factor; R_h: hydrodynamic radius; N: number of residues; M_w: molecular weight; ρ: structure factor

Protein Assemblies: Fibrils, Intermediate Filaments, Micelles, Coacervates

Amyloid fibrils are associated with critical diseases such as Alzheimer disease where a particular protein or polypeptide aggregates and forms insoluble fibrils [74]. Insulin fibrillation, as a common example of amyloid fibrillation where amyloid deposits have been observed in patients after subcutaneous insulin infusion [75], was recently studied by SAXS and *ab initio* modeling [76]. Authors clearly identified a helical oligomer unit composed of five to six insulin monomers that is likely to be a structural nucleus (Fig. **4**). The growth rate of the fibrils was found to be proportional to the amount of the helical oligomer present in solution, suggesting that these oligomers elongate the fibrils. This crucial oligomeric intermediate, serving both as the fibrillation nucleus, as well as the elongating species, would be an interesting target for small molecules in order to prevent or reduce amyloid fibril formation. The peptide hormone glucagon fibrillation was also studied using SAXS and different *ab initio* modeling tools [77]. The glucagon elongation pathway was found to be very similar to insulin fibrillation where oligomers of different sizes, depending on starting concentrations, evolve, after a lag time, to an intermediate species with rod-like shape and then to the final fibers.

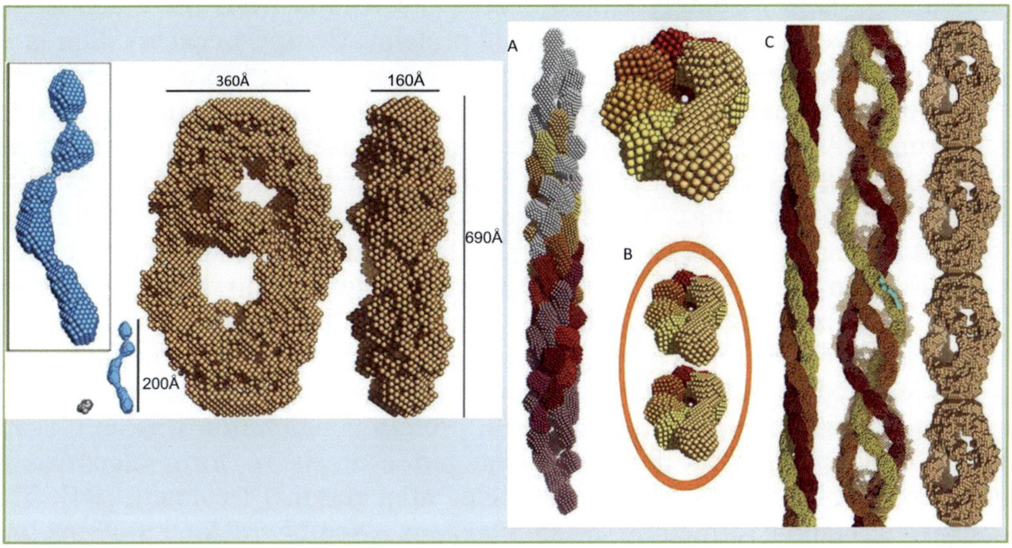

Fig. (4). (**left side**) The three components present during the insulin fibril elongation phase at relative scale: monomer in grey, oligomer in blue and repeating helical unit of the fibril in beige. Three intertwining protofibrils appear from the *ab initio* model of the repeating unit of the fibril. (**right side**) Model for assembly of the distinct insulin oligomers into protofilaments (A), protofibrils (B) and mature amyloid fibrils (C). Adapted from reference [66].

Mature fibers were also found to vary according to sample concentration. Intermediate filaments (IFs), along with microtubules and microfilaments, constitute the cytoskeleton of metazoan cells, were also studied by SAXS, the intermediates being likely to be very difficult to crystallize [78]. Authors studied the assembly of a human IF protein, vimentin, and identified the presence of tetramers laterally assembled in octamers with the increase of ionic strength. In addition, four octamers eventually associated into unit-length filaments (ULFs) that annealed longitudinally. Tetramers, octamers and ULFs represented distinct intermediates of vimentin assembly.

Food proteins and in particular caseins are well-known to form micelles in milk which are the transport vehicles of calcium phosphate required for bone growth of the neonate. SAS experiments revealed that a casein micelle is a homogeneous particle made of a protein matrix in which calcium phosphate clusters (2 nm radius) are dispersed and where the polydispersity in size is about 35% [79]. Indeed, the independence of the radius of gyration with contrast variation using SANS was found to indicate that the mass distribution within a casein micelle was homogeneous [80].

A particular protein assembly called coacervate results from the electrostatic interactions between two oppositely charged proteins. Complex coacervation is an universal process occurring between two oppositely charged (bio)polymers and has important application in encapsulation. Lactotransferrin, a strong cationic protein from milk, was mixed with β-casein or κ-casein, two anionic proteins forming soap-like micelles in solution. Stable protein-protein complexes were formed at low temperature while lactotransferrin was attached to β-casein micelles at high temperature [81]. From SAXS measurements, authors concluded that 1.2 β-casein bound 1 lactotransferrin when charge neutrality was reached while 5 κ-caseins were needed to neutralize the positive charges of lactotransferrin. Folded proteins such as lactotransferrin or lysozyme interacting with unstructured proteins such as caseins also formed coacervates [82, 83]. Combining electrostatic complexation with protein denaturation was an elegant way to self-assemble two oppositely charged proteins to form electrostatic complex micelles then charged nanoparticles after thermal treatment [84]. The diversity of structures formed by the spontaneous assembly of food proteins into microspheres, micelles or nanotubes [85] would benefit from more SAS studies (and in particular USAS to probe these large structures) in order to deeply understand the mechanism of formation and to better control the kinetics of the self-assembly process.

Beyond Proteins: What Else?

Lipid-Based Systems

Prominent example systems are multiple-component lipid-based systems even if *ab initio* modeling is not always applicable in this case due to large heterogeneity. X-ray or neutron scattering is a powerful tool to study the strong Bragg diffraction of lipid mesophases and associated bio-molecules interacting with them. Lipid systems can demonstrate the main advantages of X-ray or neutron scattering. The wide range of layer length-scales found in these systems is easily covered by X-ray or neutron measurements. For example, hydrocarbon tails ordering is on the sub-nanometer length-scale, while bilayer separation is typically few to tenths of nanometers [86]. Moreover, many of the functional characteristics of edible fat products depend on the colloidal fat crystal network present. Identifying the hierarchies of these colloidal networks and how they spontaneously self-assemble is important to understand their functionality and the oil binding capacity. USAXS, combined with modeling and computer simulations, has enabled significant advances to be made in understanding the nano- to micro-scale crystalline structures of edible oils in particular crystalline nanoplatelets [87]. A recent and elegant study used the contrast variation method on carrier discs made of lipids to determine the structure of membrane proteins [88]. Authors were thus able to prepare deuterated carrier discs from strain-based biosynthesis of lipids and membrane proteins that became invisible to neutrons; these 'stealth' carrier discs could be used as a general platform for low-resolution structural studies of membrane proteins [88]. Therefore, X-ray or neutron scattering is a key technique to study lipid structures and interactions, membrane elasticity and ion adsorption at various conditions [89 - 91]. Furthermore, in view of possible applications, such as drug delivery, non-intrusiveness is very important.

Drug Delivery Systems

X-ray scattering include numerous structural investigations on nanoparticles. Indirect transformation programs such as GNOM can compute the size distribution functions of spherical particles, spherical shells, and some other types of particles with simple shapes [5]. For example, nanoparticles made of copolymers and cholesterol were found to be a pearl necklace structure from *ab initio* modeling where ellipsoidal pearls mainly composed of cholesterol were covered by a copolymer shell; pearls were connected by bridges composed of hydrophilic copolymer chains [92]. The use of SAXS may allow the obtaining of the structure of nanoparticles and provide information on the mechanism of encapsulation such as in the case of dendrimer-based nanoparticles where SAXS clearly indicated that several dendrimers were required to stabilize the nanoparticles [93].

Polysaccharides

The relevance of small angle scattering and *ab initio* modeling to polysaccharide studies concerns both their practical as thickening and gelling agents and their fundamental aspects as complex and heterogeneous (hyper)-branched polymers. An example of branched, neutral or slightly acidic, complex polysaccharide is Acacia gum. Acacia gum is composed of 64 ramified 1,3-linked homo-galactan symmetrically arranged subunits, each of molecular mass 8000 $g.mol^{-1}$ [94]. Acacia gum is an hetero-polysaccharide since it contains ~2% of a polypeptide and is also described as hetero-polymolecular because it has, on the one hand, a variation in monomer composition and/or in the linking and branching of the monomer units and, on the other hand, a molecular mass distribution.

Fractionation of Acacia gum using hydrophobic interaction chromatography gives three molecular fractions, the bulk of the gum (88.4 wt. % of the total) being an arabinogalactan (AG)-peptide fraction with a weight-average molecular weight of $2.86.10^5$ $g.mol^{-1}$ [95]. From small angle neutron scattering experiments in charge screening conditions, the AG-peptide appeared to be a dispersion of two-dimensional structures with a radius of gyration of 6.5 nm and an inner dense branched structure [96].

Inverse Fourier transform of AG-peptide scattering form factor revealed a disk-like morphology with a diameter of 20 nm and a thickness below 2 nm. *Ab initio* calculations on the pair distance distribution function produced a porous oblate ellipsoid particle with a central intricated "network" (Fig. **5**). Both transmission electron microscopy and atomic force microscopy confirm the thin disk model and structural dimensions. The structural dimensions found would be in agreement with a phloem-mediated long-distance transport and could also explain the low viscosity of Acacia gum solutions, and its ability to self-assemble and to interact with proteins [96].

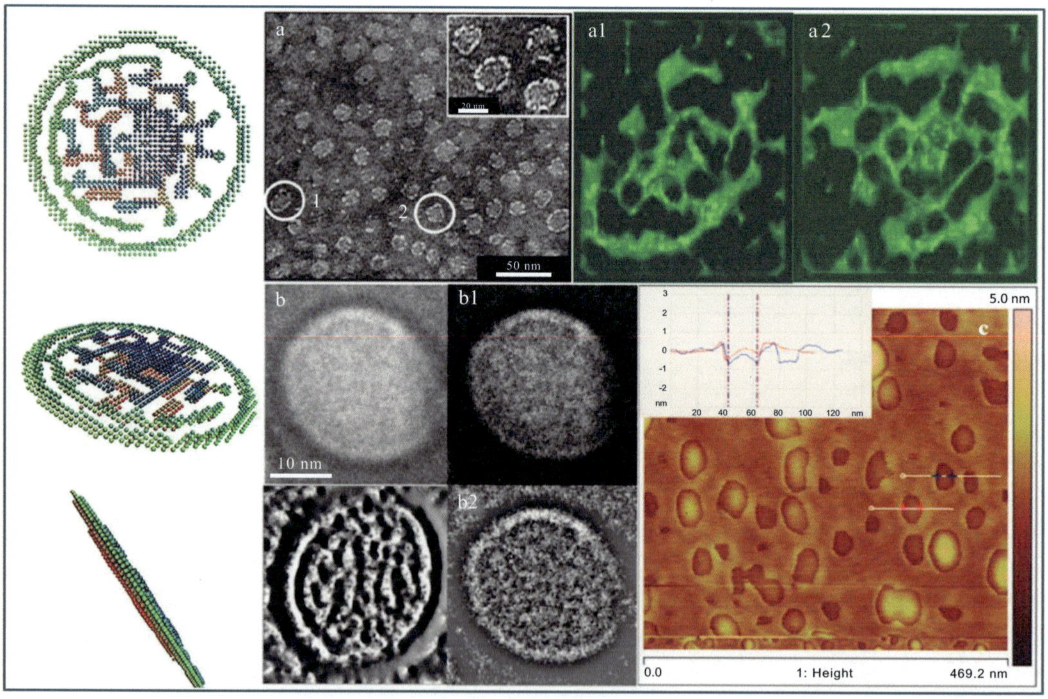

Fig. (5). (left side) Different views of the most probable DAM of arabinogalactan-peptide from Acacia gum using the DAM method as implemented in the program DAMMIN and (**right side**) microscopy images obtained using TEM (a, a1), cryo-TEM (b, b1, b2, b3) and AFM (c) (with inset corresponding to dimensions of the two macromolecules indicated in c). Adapted from reference [72].

An alternative method to *ab initio* modeling proposes to calculate SAXS data on polysaccharides using a new string of beads model in the 3D space [97].

The model enables one to simulate polysaccharide single helices with different pitch values, number of monomers per pitch, and cross-sectional radius, as well as different random coils with excluded volume taken into account. It was successfully applied to pullulan oligomers and gellan gum [97] and more recently on carboxymethyl cellulose [98]. The relevance of the semi-coarse-grained model on double and triple helices or non-linear polysaccharides needs however to be validated.

Small-angle scattering techniques can be used to probe the sol-gel transition of polysaccharides in which the mechanisms of gelation and the structure of the gels were investigated [99 - 106]. The kinetics of methylcellulose fibrils growth with temperature was for instance investigated using SANS and results showed that a small fibrils fraction began to form in the vicinity of 40-50 °C and that all the chains were incorporated into the fibrillar structure at $T \geq 70$ °C [107]. Moreover, USANS measurements revealed that heterogeneities in the gels persisted beyond the largest length scale accessed in scattering experiments (~ 20 μm), consistent with the observed optical turbidity.

SAS experiments can also be very useful to study the effect of advanced pretreatment technologies on the structure of polysaccharides such as those found in lignocellulosic biomass before their end transformation into fermentable sugars or conversion into renewable fuels, chemicals and materials. The complex interactions between plant cell walls and biomass pretreatment generate recalcitrance-related structural changes on multiple length scales for which SAXS and SANS, complementary to imaging techniques, offer several advantages like minimal sample preparation, versatile sample environment and *in situ* dynamic investigation of cell wall structures [108]. The supramolecular structure of cellulose, including the measurement of the cellulose microfibril angle and diameters and the alternating crystalline and amorphous domains in microfibrils, the supramolecular structure of lignin, the pores in cellulose fibres and plant cell walls, the enzymatic hydrolysis of cellulose, and finally the physical structural changes during biomass pretreatment were the main recent topics studied using SAS techniques [108].

Recent studies on the effect of ionic liquid (IL) on switchgrass by SANS demonstrated that IL dissolved the cellulose fibrils of switchgrass, although a supramolecular biopolymer network remained, indicating that IL was not sufficiently efficient to disrupt all of the physical entanglements and covalent linkages between cellulose, hemicellulose and lignin created during plant growth

[109, 110]. Despite all these promising results, the analysis of SAS data from lignocellulosic biomass samples is far from trivial, due to polydispersity in size and molecular structure, and more advances in data analysis methods are required to improve the current understanding of biomass pretreatment and recalcitrance such as the understanding of cellulases and solid substrate interactions.

Polydispersity: How to Solve the Problem

Model-free parameters previously defined are directly related to those of the single biopolymer. For polydispersed samples containing mixtures of different biopolymers or one biopolymer but present both in isolated and aggregated states, the interpretation of scattering data becomes more complicated. A novel approach, QUAFIT, has been recently suggested to analyze polydisperse mixtures of aggregates formed from the same initial monomer. The program fits the experimental scattering data on the assumption that any given aggregate is a result of optimum assembly of monomers according to a given point-symmetry group [111]. QUAFIT has been successfully applied to reconstruct the low resolution shapes of the decameric hemocyanin aggregates and to study the kinetics of their assembly and dissociation [112]. Analysis of self-associating proteins can also be performed by singular value decomposition in which the search identifies the association pathway and constants that provide the best agreement between reconstructed and observed data [113]. In addition, particular conditions (*e.g.*, temperature) can be varied to extract fundamental association parameters (ΔH_{ass}, ΔS_{ass}).

To overcome the limitation of polydispersed samples to SAS analyses, the coupling of a fractionation technique, in particular size exclusion chromatography, with visible light scattering methods such as dynamic light scattering (DLS) or multi-angle laser light scattering (MALLS) has been common in the field of structural biology for over two decades [114] (see also Chapter 5). The combination of SAXS with on-line chromatography (SEC-SAXS) was demonstrated much later by Mathew *et al.* [115]. In the case of a monodispersed solution, SEC-SAXS has the advantage of providing an in-built buffer blank and is also useful for demonstrating monodispersity by the invariance of particle size and shape across the elution peak since unlike DLS, SAXS has no intrinsic measurement of monodispersity [116]. Recent on-line SEC-SAXS combination was introduced on few beamlines to optimize data acquisition [6, 115, 117 - 119]. This combination is only possible at 3rd generation synchrotrons, where statistically meaningful data over a large q range (0.05-5 nm^{-1}) can be recorded in less than a second for concentrations below 0.1 wt. % [6]. A fine SAXS sampling of the SEC elution peak can then be performed, similarly to the SEC-MALLS combination (see Chapter 5). Difficult preparation conditions where the sample is unstable can thus be managed using this approach [120]. Complexes with low affinity can also be studied, provided the elution buffer contains one of the partners [121]. Challenging method to study high molecular weight biopolymers and elucidate their functions comes from the SEC-MALLS-SAXS combination to characterize both conformation and M_w. A proteoglycan from shark cartilage with a total M_w of 3.9 million and for the main component a M_w of 2 million was recently characterized [122]. In addition, the main component was found to be nearly equal to a chain with excluded volume and a persistence length ranging from 13.5 to 16.4 nm. The online coupling of the field-flow fractionation with SAXS has also been proposed for the detection of polyion complexes (PICs) [123], nanoparticles for toxicological studies [124] or polymers with broad M_w distribution [125, 126]. These coupling methods open great opportunities in a near future for the structural study of frequently encountered polydispersed biopolymers and large assemblies such as protein-polysaccharide or protein-polyphenol complexes.

Kinetics *via* Rapid Mixing Techniques

Beamlines specializing in rapid mixing methods are approaching the current limits of mixing times around 100 μs, which should fall further with continued development in microfluidics and chaotic mixers [116]. Recent examples of mixing studies include folding of cytochrome c during rapid dilution of guanidine hydrochloride, and unfolding dynamics of ubiquitin upon exposure to guanidine hydrochloride in a multichannel microfluidics mixer [127, 128]. Reconstructing three-dimensional shape envelopes from time-resolved small-angle X-ray scattering was also successfully performed [129].

Perspectives: "Fast and not Furious but Efficient"

An increasing demand of structure determination of biopolymers in solution comes today from more diverse research community that those of structural biology. High-throughput genomics and proteomics for instance are far outpacing macromolecular structure determinations, creating information gaps between the plethora of newly identified genes, known structures, and the structure-function relationship in the underlying biological networks [130]. This phenomenon encourages "scattering physicists community" to develop methods to go fast and get reliable structural data. High-brilliance synchrotron X-ray sources and parallel advances in hardware and computing first allow the reduction of data acquisition times to the millisecond range [131]. Second, the application of automated methods allows data processing and low resolution shape modeling to be completed within minutes [131 - 135]. Automated microfluidic sample-preparation platform for high throughput structural investigation of proteins, by screening in particular a wide set of conditions, was recently developed [136]. Moreover, *in situ* microfluidic dialysis for biological SAXS allowed the monitoring of structural changes in response to buffer exchange or protein concentration [137]. Graewert and Svergun provided an overview of the names, features and access details on many analysis packages developed over recent years [4] and Schneidman-Duhovny *et al.* provided a similar list for scattering profile calculation softwares [138]. The recent development of the BioSAXS sample changer installed at ESRF, PETRA and DLS third-generation synchrotrons is a perfect illustration of automated and high-throughput operation able to perform unattended measurements of up to several hundred samples per day [139]. All these developments pave the way for high-throughput studies that generate significant quantities of structural information over a short period of time for users community including novices and non-specialists [117, 133, 134, 140]. However, the absence of publication standards and the ease with which 3D models can be calculated against the inherently 1D scattering data means that an understanding of sample quality, data quality, and modeling assumptions is essential to have confidence in the results [141]. A repository for biological small-angle scattering data was recently proposed by the wwPDB small-angle scattering task force (SAStf) in order to improve reproducibility of the experiments by supporting transparency and accessibility of the data by rendering publicly available a comprehensive suite of original data, models and structural parameters (www.sasbdb.org) [142, 143].

CONFLICT OF INTEREST

The author (editor) declares no conflict of interest, financial or otherwise.

ACKNOWLEDGEMENTS

Declared none.

REFERENCES

[1] D. I. Svergun, and M. H. J. Koch, "Advances in structure analysis using small-angle scattering in solution", *Curr. Opin. Struct. Biol.* , vol. 12, pp. 654-660, 2002. [http://dx.doi.org/10.1016/S0959-440X(02)00363-9]

[2] L.A. Feigin, and D.I. Svergun, *Structure Analysis by Small-angle X-ray and Neutron Scattering.* Plenum Press: New York, 1987. [http://dx.doi.org/10.1007/978-1-4757-6624-0]

[3] H.D. Mertens, and D.I. Svergun, "Structural characterization of proteins and complexes using small-angle X-ray solution scattering", *J. Struct. Biol.* , vol. 172, no. 1, pp. 128-141, 2010. [http://dx.doi.org/10.1016/j.jsb.2010.06.012] [PMID: 20558299]

[4] M.A. Graewert, and D.I. Svergun, "Impact and progress in small and wide angle X-ray scattering (SAXS and WAXS)", *Curr. Opin. Struct. Biol.* , vol. 23, no. 5, pp. 748-754, 2013. [http://dx.doi.org/10.1016/j.sbi.2013.06.007] [PMID: 23835228]

[5] C.E. Blanchet, and D.I. Svergun, "Small-angle X-ray scattering on biological macromolecules and nanocomposites in solution", *Annu. Rev. Phys. Chem.*, vol. 64, pp. 37-54, 2013. [http://dx.doi.org/10.1146/annurev-physchem-040412-110132] [PMID: 23216378]

[6] J. Pérez, and Y. Nishino, "Advances in X-ray scattering: from solution SAXS to achievements with coherent beams", *Curr. Opin. Struct. Biol.*, vol. 22, pp. 670-678, 2012. [http://dx.doi.org/10.1016/j.sbi.2012.07.014]

[7] M.V. Petoukhov, and D.I. Svergun, "Joint use of small-angle X-ray and neutron scattering to study biological macromolecules in solution", *Eur. Biophys. J.*, vol. 35, no. 7, pp. 567-576, 2006. [http://dx.doi.org/10.1007/s00249-006-0063-9] [PMID: 16636827]

[8] M.V. Petoukhov, and D.I. Svergun, "Analysis of X-ray and neutron scattering from biomacromolecular solutions", *Curr. Opin. Struct. Biol.*, vol. 17, no. 5, pp. 562-571, 2007. [http://dx.doi.org/10.1016/j.sbi.2007.06.009] [PMID: 17714935]

[9] A.T. Tuukkanen, and D.I. Svergun, "Weak protein-ligand interactions studied by small-angle X-ray scattering", *FEBS J.*, vol. 281, no. 8, pp. 1974-1987, 2014. [http://dx.doi.org/10.1111/febs.12772] [PMID: 24588935]

[10] O. Glatter, and O. Kratky, *Small Angle X-Ray Scattering.* Academic Press: London, 1982.

[11] Y. Nishiyama, A. Isogai, T. Okano, M. Muller, and H. Chanzy, "Intracrystalline deuteration of native cellulose", *Macromolecules.*, vol. 32, pp. 2078-2081, 1999. [http://dx.doi.org/10.1021/ma981563m]

[12] M.C. Bellissent-Funel, J. Lal, K.F. Bradley, and S.H. Chen, "Neutron structure factors of in-vivo deuterated amorphous protein C-phycocyanin", *Biophys. J.*, vol. 64, no. 5, pp. 1542-1549, 1993. [http://dx.doi.org/10.1016/S0006-3495(93)81523-6] [PMID: 19431896]

[13] J.H. He, S.V. Pingali, S.P. Chundawat, A. Pack, A.D. Jones, and P. Langan, "Controlled incorporation of deuterium into bacterial cellulose", *Cellulose.*, vol. 21, pp. 927-936, 2014. [http://dx.doi.org/10.1007/s10570-013-0067-4]

[14] A. de Ghellinck, H. Schaller, V. Laux, M. Haertlein, M. Sferrazza, E. Maréchal, H. Wacklin, J. Jouhet, and G. Fragneto, "Production and analysis of perdeuterated lipids from Pichia pastoris cells", *PLoS One.*, vol. 9, no. 4, p. e92999, 2014. [http://dx.doi.org/10.1371/journal.pone.0092999] [PMID: 24747350]

[15] K. Vanatalu, T. Paalme, R. Vilu, N. Burkhardt, R. Jünemann, R. May, M. Rühl, J. Wadzack, and K.H. Nierhaus, "Large-scale preparation of fully deuterated cell components. Ribosomes from Escherichia coli with high biological activity", *Eur. J. Biochem.*, vol. 216, no. 1, pp. 315-321, 1993. [http://dx.doi.org/10.1111/j.1432-1033.1993.tb18148.x] [PMID: 8365413]

[16] X. Fang, J. R. Stagno, Y. R. Bhandari, X. Zuo, and Y.-X. Wang, "Small-angle X-ray scattering: a bridge between RNA secondary structures and three-dimensional topological structures", *Curr. Opin. Struct. Biol.*, vol. 30, no. 1, pp. 147-160, 2015. [http://dx.doi.org/10.1016/j.sbi.2015.02.010]

[17] X.J. Chen, K.L. Wilde, H. Wang, V. Lake, P.J. Holden, and A.P. Middelberg, "High yield expression and efficient purification of deuterated human protein galectin-2", *Food Bioprod. Process.*, vol. 90, pp. 563-572, 2012. [http://dx.doi.org/10.1016/j.fbp.2011.12.004]

[18] J. Hennig, and M. Sattler, "The dynamic duo: combining NMR and small angle scattering in structural biology", *Protein Sci.*, vol. 23, no. 6, pp. 669-682, 2014. [http://dx.doi.org/10.1002/pro.2467] [PMID: 24687405]

[19] W.T. Heller, "Small-angle neutron scattering and contrast variation: a powerful combination for studying biological structures", *Acta Crystallogr. D Biol. Crystallogr.*, vol. 66, no. Pt 11, pp. 1213-1217, 2010. [http://dx.doi.org/10.1107/S0907444910017658] [PMID: 21041939]

[20] W.T. Heller, V.S. Urban, G.W. Lynn, K.L. Weiss, H.M. O'Neill, and S.V. Pingali, "The Bio-SANS instrument at the High Flux Isotope Reactor of Oak Ridge National Laboratory", *J. Appl. Cryst.* , vol. 47, pp. 1238-1246, 2014. [http://dx.doi.org/10.1107/S1600576714011285]

[21] K.L. Sarachan, J.E. Curtis, and S. Krueger, "Small-angle scattering contrast calculator for protein and nucleic acid complexes in solution", *J. Appl. Cryst.* , vol. 46, pp. 1889-1893, 2013. [http://dx.doi.org/10.1107/S0021889813025727]

[22] P.V. Konarev, V.V. Volkov, A.V. Sokolova, M.H. Koch, and D.I. Svergun, "PRIMUS: a Windows PC-based system for small-angle scattering data analysis", *J. Appl. Cryst.* , vol. 36, pp. 1277-1282, 2003. [http://dx.doi.org/10.1107/S0021889803012779]

[23] P.V. Konarev, M.V. Petoukhov, V.V. Volkov, and D.I. Svergun, "ATSAS 2.1, a program package for small-angle scattering data analysis", *J. Appl. Cryst.* , vol. 39, pp. 277-286, 2006. [http://dx.doi.org/10.1107/S0021889806004699]

[24] A. Guinier, "La diffraction des rayons X aux très petits angles; application à l'étude de phénomènes ultramicroscopiques", *Ann. Phys. (Paris).* , vol. 12, pp. 161-237, 1939.

[25] D.I. Svergun, "Mathematical methods in small-angle scattering data-analysis", *J. Appl. Cryst.* , vol. 24, pp. 485-492, 1991. [http://dx.doi.org/10.1107/S0021889891001280]

[26] D.I. Svergun, "Determination of the regularization parameter in indirect-transform methods using perceptual criteria", *J. Appl. Cryst.* , vol. 25, pp. 495-503, 1992. [http://dx.doi.org/10.1107/S0021889892001663]

[27] P. Calmettes, D. Durand, M. Desmadril, P. Minard, V. Receveur, and J.C. Smith, "How random is a highly denatured protein?", *Biophys. Chem.* , vol. 53, no. 1-2, pp. 105-113, 1994. [http://dx.doi.org/10.1016/0301-4622(94)00081-6] [PMID: 17020841]

[28] D.I. Svergun, and M.H. Koch, "Small-angle scattering studies of biological macromolecules in solution", *Rep. Prog. Phys.* , vol. 66, pp. 1735-1782, 2003. [http://dx.doi.org/10.1088/0034-4885/66/10/R05]

[29] D. Orthaber, A. Bergmann, and O. Glatter, "SAXS experiments on absolute scale with Kratky systems using water as a secondary standard", *J. Appl. Cryst.* , vol. 33, pp. 218-225, 2000. [http://dx.doi.org/10.1107/S0021889899015216]

[30] G. Zaccai, and B. Jacrot, "SMALL-ANGLE NEUTRON-SCATTERING", *Ann. Rev. Biophys. Bioeng.* , vol. 1983, pp. 139-157, 1983. [http://dx.doi.org/10.1146/annurev.bb.12.060183.001035]

[31] H. Fischer, M. de Oliveira Neto, H.B. Napolitano, I. Polikarpov, and A.F. Craievich, "Determination of the molecular weight of proteins in solution from a single small-angle X-ray scattering measurement on a relative scale", *J. Appl. Cryst.* , vol. 43, pp. 101-109, 2010. [http://dx.doi.org/10.1107/S0021889809043076]

[32] R.P. Rambo, and J.A. Tainer, "Characterizing flexible and intrinsically unstructured biological macromolecules by SAS using the Porod-Debye law", *Biopolymers.* , vol. 95, no. 8, pp. 559-571, 2011. [http://dx.doi.org/10.1002/bip.21638] [PMID: 21509745]

[33] M.H. Koch, P. Vachette, and D.I. Svergun, "Small-angle scattering: a view on the properties, structures and structural changes of biological macromolecules in solution", *Q. Rev. Biophys.* , vol. 36, no. 2, pp. 147-227, 2003. [http://dx.doi.org/10.1017/S0033583503003871] [PMID: 14686102]

[34] M.V. Petoukhov, D. Franke, A.V. Shkumatov, G. Tria, A.G. Kikhney, M. Gajda, C. Gorba, H.D. Mertens, P.V. Konarev, and D.I. Svergun, "New developments in the ATSAS program package for

small-angle scattering data analysis", *J. Appl. Cryst.* , vol. 45, no. Pt 2, pp. 342-350, 2012. [http://dx.doi.org/10.1107/S0021889812007662] [PMID: 25484842]

[35] R. P. Rambo, and J. A. Tainer, "Accurate assessment of mass, models and resolution by small-angle scattering", *Nature.* , vol. 496, no. 7446, pp. 477-481, 2013. [http://dx.doi.org/10.1038/nature12070]

[36] O. Glatter, "A new method for evaluation of small-angle scattering data", *J. Appl. Crystallogr.* , vol. 10, no. 5, pp. 415-421, 1977. [http://dx.doi.org/10.1107/S0021889877013879]

[37] M.V. Petoukhov, P.V. Konarev, A.G. Kikhney, and D.I. Svergun, "ATSAS 2.1 - towards automated and web-supported small-angle scattering data analysis", *J. Appl. Cryst.* , vol. 40, pp. S223-S228, 2007. [http://dx.doi.org/10.1107/S0021889807002853]

[38] H.B. Stuhrmann, "Interpretation of small-angle scattering functions of dilute solutions and gases. A representation of the structures related to a one-particle-scattering function", *Acta Crystallogr.* , vol. A26, pp. 297-306, 1970. [http://dx.doi.org/10.1107/S0567739470000748]

[39] H.B. Stuhrmann, "New method for determination of surface form and internal structure of dissolved globular proteins from small-angle X-ray measurements", *Z. Phys. Chem.* , vol. 72, pp. 177-182, 1970. [http://dx.doi.org/10.1524/zpch.1970.72.4_6.177]

[40] D.I. Svergun, "Restoring three-dimensional structure of biopolymers from solution scattering", *J. Appl. Cryst.* , vol. 30, pp. 792-797, 1997. [http://dx.doi.org/10.1107/S0021889897001039]

[41] D.I. Svergun, V.V. Volkov, M.B. Kozin, and H.B. Stuhrmann, "New developments in direct shape determination from small-angle scattering. 2. Uniqueness", *Acta Crystallogr. A.* , vol. 52, pp. 419-426, 1996. [http://dx.doi.org/10.1107/S0108767396000177]

[42] D.I. Svergun, V.V. Volkov, M.B. Kozin, H.B. Stuhrmann, C. Barberato, and M.H. Koch, "Shape determination from solution scattering of biopolymers", *J. Appl. Cryst.* , vol. 30, pp. 798-802, 1997. [http://dx.doi.org/10.1107/S0021889897001714]

[43] D. I. Svergun, "Restoring Low Resolution Structure of Biological Macromolecules from Solution Scattering Using Simulated Annealing", *Biophys. J.* , vol. 76, pp. 2879-2886, 1999. [http://dx.doi.org/10.1016/S0006-3495(99)77443-6]

[44] D.I. Svergun, and K.H. Nierhaus, "A map of protein-rRNA distribution in the 70 S Escherichia coli ribosome", *J. Biol. Chem.* , vol. 275, no. 19, pp. 14432-14439, 2000. [http://dx.doi.org/10.1074/jbc.275.19.14432] [PMID: 10799526]

[45] D. Franke, and D.I. Svergun, "DAMMIF, a program for rapid ab-initio shape determination in small-angle scattering", *J. Appl. Cryst.* , vol. 42, no. Pt 2, pp. 342-346, 2009. [http://dx.doi.org/10.1107/S0021889809000338] [PMID: 27630371]

[46] D.I. Svergun, M.V. Petoukhov, and M.H. Koch, "Determination of domain structure of proteins from X-ray solution scattering", *Biophys. J.* , vol. 80, no. 6, pp. 2946-2953, 2001. [http://dx.doi.org/10.1016/S0006-3495(01)76260-1] [PMID: 11371467]

[47] A. Koutsioubas, and J. Perez, "Incorporation of a hydration layer in the 'dummy atom' ab initio structural modelling of biological macromolecules", *J. Appl. Cryst.* , vol. 46, pp. 1884-1888, 2013. [http://dx.doi.org/10.1107/S0021889813025387]

[48] M.B. Kozin, and D.I. Svergun, "Automated matching of high- and low-resolution structural models", *J. Appl. Cryst.* , vol. 34, pp. 33-41, 2001. [http://dx.doi.org/10.1107/S0021889800014126]

[49] V.V. Volkov, and D.I. Svergun, "Uniqueness of ab initio shape determination in small-angle

scattering", *J. Appl. Cryst.* , vol. 36, pp. 860-864, 2003. [http://dx.doi.org/10.1107/S0021889803000268]

[50] J. García De La Torre, M.L. Huertas, and B. Carrasco, "Calculation of hydrodynamic properties of globular proteins from their atomic-level structure", *Biophys. J.* , vol. 78, no. 2, pp. 719-730, 2000. [http://dx.doi.org/10.1016/S0006-3495(00)76630-6] [PMID: 10653785]

[51] D.J. Winzor, L.E. Carrington, and S.E. Harding, "Analysis of thermodynamic non-ideality in terms of protein solvation", *Biophys. Chem.* , vol. 93, no. 2-3, pp. 231-240, 2001. [http://dx.doi.org/10.1016/S0301-4622(01)00223-X] [PMID: 11804728]

[52] J.G. Grossmann, S.S. Hasnain, F.K. Yousafzai, and R.R. Eady, "Evidence for the selective population of FeMo cofactor sites in MoFe protein and its molecular recognition by the Fe protein in transition state complex analogues of nitrogenase", *J. Biol. Chem.* , vol. 276, no. 9, pp. 6582-6590, 2001. [http://dx.doi.org/10.1074/jbc.M005350200] [PMID: 11084023]

[53] D.I. Svergun, "Advanced solution scattering data analysis methods and their applications", *J. Appl. Cryst.* , vol. 33, pp. 530-534, 2000. [http://dx.doi.org/10.1107/S0021889800001370]

[54] M.V. Petoukhov, and D.I. Svergun, "Global rigid body modeling of macromolecular complexes against small-angle scattering data", *Biophys. J.* , vol. 89, no. 2, pp. 1237-1250, 2005. [http://dx.doi.org/10.1529/biophysj.105.064154] [PMID: 15923225]

[55] J. Wang, X. Zuo, P. Yu, I.J. Byeon, J. Jung, X. Wang, M. Dyba, S. Seifert, C.D. Schwieters, J. Qin, A.M. Gronenborn, and Y.X. Wang, "Determination of multicomponent protein structures in solution using global orientation and shape restraints", *J. Am. Chem. Soc.* , vol. 131, no. 30, pp. 10507-10515, 2009. [http://dx.doi.org/10.1021/ja902528f] [PMID: 19722627]

[56] M. V. Petoukhov, and D. I. Svergun, "Applications of small-angle X-ray scattering to biomacromolecular solutions", *Int. J. Biochem. Cell B.* , vol. 45, pp. 429-437, 2013. [http://dx.doi.org/10.1016/j.biocel.2012.10.017]

[57] F. Förster, B. Webb, K.A. Krukenberg, H. Tsuruta, D.A. Agard, and A. Sali, "Integration of small-angle X-ray scattering data into structural modeling of proteins and their assemblies", *J. Mol. Biol.* , vol. 382, no. 4, pp. 1089-1106, 2008. [http://dx.doi.org/10.1016/j.jmb.2008.07.074] [PMID: 18694757]

[58] J. J. P. Perry, and J. A. Tainer, "Developing advanced X-ray scattering methods combined with crystallography and computation", *Methods.* , vol. 59, pp. 363-371, 2013. [http://dx.doi.org/10.1016/j.ymeth.2013.01.005]

[59] T.D. Grant, J.R. Luft, J.R. Wolfley, H. Tsuruta, A. Martel, G.T. Montelione, and E.H. Snell, "Small angle X-ray scattering as a complementary tool for high-throughput structural studies", *Biopolymers.* , vol. 95, no. 8, pp. 517-530, 2011. [http://dx.doi.org/10.1002/bip.21630] [PMID: 21462184]

[60] D.I. Svergun, "Small-angle X-ray and neutron scattering as a tool for structural systems biology", *Biol. Chem.* , vol. 391, no. 7, pp. 737-743, 2010. [http://dx.doi.org/10.1515/bc.2010.093] [PMID: 20482320]

[61] C. Pons, M. D'Abramo, D. I. Svergun, M. Orozco, P. Bernadó, and J. Fernández-Recio, "Structural Characterization of Protein–Protein Complexes by Integrating Computational Docking with Small-angle Scattering Data", *J. Am. Chem. Soc.* , no. 7, pp. 4378-4386, 2010. [http://dx.doi.org/10.1016/j.jmb.2010.08.029]

[62] J. Blobel, P. Bernadó, D.I. Svergun, R. Tauler, and M. Pons, "Low-resolution structures of transient protein-protein complexes using small-angle X-ray scattering", *J. Am. Chem. Soc.* , vol. 131, no. 12, pp. 4378-4386, 2009. [http://dx.doi.org/10.1021/ja808490b] [PMID: 19275229]

[63] P. Bernadó, and D.I. Svergun, "Structural analysis of intrinsically disordered proteins by small-angle X-ray scattering", *Mol. Biosyst.* , vol. 8, no. 1, pp. 151-167, 2012. [http://dx.doi.org/10.1039/C1MB05275F] [PMID: 21947276]

[64] P. Bernadó, "Effect of interdomain dynamics on the structure determination of modular proteins by small-angle scattering", *Eur. Biophys. J.* , vol. 39, no. 5, pp. 769-780, 2010. [http://dx.doi.org/10.1007/s00249-009-0549-3] [PMID: 19844700]

[65] P. J. Flory, *Principles of polymer chemistry*. Ithaca and London: Cornell University Press, 1953.

[66] J.E. Kohn, I.S. Millett, J. Jacob, B. Zagrovic, T.M. Dillon, N. Cingel, R.S. Dothager, S. Seifert, P. Thiyagarajan, T.R. Sosnick, M.Z. Hasan, V.S. Pande, I. Ruczinski, S. Doniach, and K.W. Plaxco, "Random-coil behavior and the dimensions of chemically unfolded proteins", *Proc. Natl. Acad. Sci. USA.* , vol. 101, no. 34, pp. 12491-12496, 2004. [http://dx.doi.org/10.1073/pnas.0403643101] [PMID: 15314214]

[67] D.K. Wilkins, S.B. Grimshaw, V. Receveur, C.M. Dobson, J.A. Jones, and L.J. Smith, "Hydrodynamic radii of native and denatured proteins measured by pulse field gradient NMR techniques", *Biochemistry.* , vol. 38, no. 50, pp. 16424-16431, 1999. [http://dx.doi.org/10.1021/bi991765q] [PMID: 10600103]

[68] V.N. Uversky, "What does it mean to be natively unfolded?", *Eur. J. Biochem.* , vol. 269, no. 1, pp. 2-12, 2002. [http://dx.doi.org/10.1046/j.0014-2956.2001.02649.x] [PMID: 11784292]

[69] P. Bernadó, L. Blanchard, P. Timmins, D. Marion, R.W. Ruigrok, and M. Blackledge, "A structural model for unfolded proteins from residual dipolar couplings and small-angle x-ray scattering", *Proc. Natl. Acad. Sci. USA.* , vol. 102, no. 47, pp. 17002-17007, 2005. [http://dx.doi.org/10.1073/pnas.0506202102] [PMID: 16284250]

[70] P. Bernadó, E. Mylonas, M.V. Petoukhov, M. Blackledge, and D.I. Svergun, "Structural characterization of flexible proteins using small-angle X-ray scattering", *J. Am. Chem. Soc.* , vol. 129, no. 17, pp. 5656-5664, 2007. [http://dx.doi.org/10.1021/ja069124n] [PMID: 17411046]

[71] M. Pelikan, G.L. Hura, and M. Hammel, "Structure and flexibility within proteins as identified through small angle X-ray scattering", *Gen. Physiol. Biophys.* , vol. 28, no. 2, pp. 174-189, 2009. [http://dx.doi.org/10.4149/gpb_2009_02_174] [PMID: 19592714]

[72] M. Guttman, P. Weinkam, A. Sali, and K.K. Lee, "All-atom ensemble modeling to analyze small-angle x-ray scattering of glycosylated proteins", *Structure.* , vol. 21, no. 3, pp. 321-331, 2013. [http://dx.doi.org/10.1016/j.str.2013.02.004] [PMID: 23473666]

[73] A.E. Whitten, and J. Trewhella, "Small-angle scattering and neutron contrast variation for studying bio-molecular complexes", In: *Micro and Nano Technologies in Bioanalysis: Methods and Protocols,* J. W. Lee, R. S. Foote, Eds., Humana Press: New York , USA, 2009. [http://dx.doi.org/10.1007/978-1-59745-483-4_20]

[74] C.M. Dobson, "Protein folding and misfolding", *Nature.* , vol. 426, no. 6968, pp. 884-890, 2003. [http://dx.doi.org/10.1038/nature02261] [PMID: 14685248]

[75] F.E. Dische, C. Wernstedt, G.T. Westermark, P. Westermark, M.B. Pepys, J.A. Rennie, S.G. Gilbey, and P.J. Watkins, "Insulin as an amyloid-fibril protein at sites of repeated insulin injections in a diabetic patient", *Diabetologia.* , vol. 31, no. 3, pp. 158-161, 1988. [http://dx.doi.org/10.1007/BF00276849] [PMID: 3286343]

[76] B. Vestergaard, M. Groenning, M. Roessle, J.S. Kastrup, M. van de Weert, J.M. Flink, S. Frokjaer, M. Gajhede, and D.I. Svergun, "A helical structural nucleus is the primary elongating unit of insulin amyloid fibrils", *PLoS Biol.* , vol. 5, no. 5, p. e134, 2007. [http://dx.doi.org/10.1371/journal.pbio.0050134] [PMID: 17472440]

[77] C.L. Oliveira, M.A. Behrens, J.S. Pedersen, K. Erlacher, D. Otzen, and J.S. Pedersen, "A SAXS study

of glucagon fibrillation", *J. Mol. Biol.* , vol. 387, no. 1, pp. 147-161, 2009. [http://dx.doi.org/10.1016/j.jmb.2009.01.020] [PMID: 19385046]

[78] A.V. Sokolova, L. Kreplak, T. Wedig, N. Mücke, D.I. Svergun, H. Herrmann, U. Aebi, and S.V. Strelkov, "Monitoring intermediate filament assembly by small-angle x-ray scattering reveals the molecular architecture of assembly intermediates", *Proc. Natl. Acad. Sci. USA.* , vol. 103, no. 44, pp. 16206-16211, 2006. [http://dx.doi.org/10.1073/pnas.0603629103] [PMID: 17050693]

[79] C.G. De Kruif, "The structure of casein micelles: a review of small-angle scattering data", *J. Appl. Cryst.* , vol. 47, pp. 1479-1489, 2014. [http://dx.doi.org/10.1107/S1600576714014563]

[80] C.G. de Kruif, T. Huppertz, V.S. Urban, and A.V. Petukhov, "Casein micelles and their internal structure", *Adv. Colloid Interface Sci.* , vol. 171-172, pp. 36-52, 2012. [http://dx.doi.org/10.1016/j.cis.2012.01.002] [PMID: 22381008]

[81] C.G. de Kruif, J. Pedersen, T. Huppertz, and S.G. Anema, "Coacervates of lactotransferrin and β- or κ-casein: structure determined using SAXS", *Langmuir.* , vol. 29, no. 33, pp. 10483-10490, 2013. [http://dx.doi.org/10.1021/la402236f] [PMID: 23859400]

[82] S.G. Anema, and C.G. de Kruif, "Co-acervates of lactoferrin and caseins", *Soft Matter.* , vol. 8, pp. 4471-4478, 2012. [http://dx.doi.org/10.1039/c2sm00015f]

[83] S.G. Anema, and C.G. de Kruif, "Coacervates of lysozyme and β-casein", *J. Colloid Interface Sci.* , vol. 398, pp. 255-261, 2013. [http://dx.doi.org/10.1016/j.jcis.2013.02.013] [PMID: 23511012]

[84] X. Pan, S. Yu, P. Yao, and Z. Shao, "Self-assembly of beta-casein and lysozyme", *J. Colloid Interface Sci.* , vol. 316, no. 2, pp. 405-412, 2007. [http://dx.doi.org/10.1016/j.jcis.2007.09.007] [PMID: 17915243]

[85] S. Bouhallab, and T. Croguennec, "Spontaneous Assembly and Induced Aggregation of Food Proteins", In: *Polyelectrolyte Complexes in the Dispersed and Solid State II: Application Aspects,* M. Muller, Ed., vol. 256. springer publisher: New York, 2014, pp. 67-101. [http://dx.doi.org/10.1007/12_2012_201]

[86] M. Kornreich, R. Avinery, and R. Beck, "Modern X-ray scattering studies of complex biological systems", *Curr. Opin. Biotechnol.* , vol. 24, no. 4, pp. 716-723, 2013. [http://dx.doi.org/10.1016/j.copbio.2013.01.005] [PMID: 23369441]

[87] F. Peyronel, D.A. Pink, and A.G. Marangoni, "Triglyceride nanocrystal aggregation into polycrystalline colloidal networks: Ultra-small angle X-ray scattering, models and computer simulation", *Curr. Opin. Colloid Interface Sci.* , vol. 19, pp. 459-470, 2014. [http://dx.doi.org/10.1016/j.cocis.2014.07.001]

[88] S. Maric, N. Skar-Gislinge, S. Midtgaard, M.B. Thygesen, J. Schiller, H. Frielinghaus, M. Moulin, M. Haertlein, V.T. Forsyth, T.G. Pomorski, and L. Arleth, "Stealth carriers for low-resolution structure determination of membrane proteins in solution", *Acta Crystallogr. D Biol. Crystallogr.* , vol. 70, no. Pt 2, pp. 317-328, 2014. [http://dx.doi.org/10.1107/S1399004713027466] [PMID: 24531466]

[89] N.W. Schmidt, A. Mishra, G.H. Lai, M. Davis, L.K. Sanders, D. Tran, A. Garcia, K.P. Tai, P.B. McCray, A.J. Ouellette, M.E. Selsted, and G.C. Wong, "Criterion for amino acid composition of defensins and antimicrobial peptides based on geometry of membrane destabilization", *J. Am. Chem. Soc.* , vol. 133, no. 17, pp. 6720-6727, 2011. [http://dx.doi.org/10.1021/ja200079a] [PMID: 21473577]

[90] O. Szekely, A. Steiner, P. Szekely, E. Amit, R. Asor, C. Tamburu, and U. Raviv, "The structure of ions and zwitterionic lipids regulates the charge of dipolar membranes", *Langmuir.* , vol. 27, no. 12, pp. 7419-7438, 2011.

[http://dx.doi.org/10.1021/la200264s] [PMID: 21598965]

[91] A. Yaghmur, B. Sartori, and M. Rappolt, "The role of calcium in membrane condensation and spontaneous curvature variations in model lipidic systems", *Phys. Chem. Chem. Phys.* , vol. 13, no. 8, pp. 3115-3125, 2011.
[http://dx.doi.org/10.1039/C0CP01036G] [PMID: 21049109]

[92] S.K. Filippov, P. Chytil, P.V. Konarev, M. Dyakonova, C. Papadakis, A. Zhigunov, J. Plestil, P. Stepanek, T. Etrych, K. Ulbrich, and D.I. Svergun, "Macromolecular HPMA-based nanoparticles with cholesterol for solid-tumor targeting: detailed study of the inner structure of a highly efficient drug delivery system", *Biomacromolecules.* , vol. 13, no. 8, pp. 2594-2604, 2012.
[http://dx.doi.org/10.1021/bm3008555] [PMID: 22793269]

[93] E.V. Shtykoya, N.V. Kuchkina, Z.B. Shifrina, L.M. Bronstein, and D.I. Svergun, "Unusual Structural Morphology of Dendrimer/CdS Nanocomposites Revealed by Synchrotron X-ray Scattering", *J. Phys. Chem. C.* , vol. 116, pp. 8069-8078, 2012.
[http://dx.doi.org/10.1021/jp210998h]

[94] S.C. Churms, E.H. Merrifield, and A.M. Stephen, "SOME NEW ASPECTS OF THE MOLECULAR-STRUCTURE OF ACACIA-SENEGAL GUM (GUM ARABIC)", *Carbohydr. Res.* , vol. 123, pp. 267-279, 1983.
[http://dx.doi.org/10.1016/0008-6215(83)88483-3]

[95] D. Renard, L. Lavenant-Gourgeon, M.C. Ralet, and C. Sanchez, "Acacia senegal gum: continuum of molecular species differing by their protein to sugar ratio, molecular weight, and charges", *Biomacromolecules.* , vol. 7, no. 9, pp. 2637-2649, 2006.
[http://dx.doi.org/10.1021/bm060145j] [PMID: 16961328]

[96] C. Sanchez, C. Schmitt, E. Kolodziejczyk, A. Lapp, C. Gaillard, and D. Renard, "The acacia gum arabinogalactan fraction is a thin oblate ellipsoid: a new model based on small-angle neutron scattering and ab initio calculation", *Biophys. J.* , vol. 94, no. 2, pp. 629-639, 2008.
[http://dx.doi.org/10.1529/biophysj.107.109124] [PMID: 17526575]

[97] I. Dogsa, J. Strancar, P. Laggner, and D. Stopar, "Efficient modeling of polysaccharide conformations based on Small-Angle X-ray Scattering experimental data", *Polymer (Guildf.).* , vol. 49, pp. 1398-1406, 2008.
[http://dx.doi.org/10.1016/j.polymer.2008.01.043]

[98] I. Dogsa, M. Tomšič, J. Orehek, E. Benigar, A. Jamnik, and D. Stopar, "Amorphous supramolecular structure of carboxymethyl cellulose in aqueous solution at different pH values as determined by rheology, small angle X-ray and light scattering", *Carbohydr. Polym.* , vol. 111, pp. 492-504, 2014.
[http://dx.doi.org/10.1016/j.carbpol.2014.04.020] [PMID: 25037380]

[99] M. Rinaudo, "Non-covalent interactions in polysaccharide systems", *Macromol. Biosci.* , vol. 6, no. 8, pp. 590-610, 2006.
[http://dx.doi.org/10.1002/mabi.200600053] [PMID: 16881089]

[100] T. Turquois, C. Rochas, F.R. Taravel, J.L. Doublier, and M.A. Axelos, "Small-angle x-ray-scattering of kappa-carrageenan based systems - sols, gels, and blends with carob galactomannan", *Biopolymers.* , vol. 36, pp. 559-567, 1995.
[http://dx.doi.org/10.1002/bip.360360502]

[101] Y. Yuguchi, "Nano-Structure Analysis of Sugar Chains by Small Angle X-ray Scattering", *Trends Glycosci. Glycotechnol.* , vol. 21, pp. 1-12, 2009.
[http://dx.doi.org/10.4052/tigg.21.1]

[102] Y. Yuguchi, M. Mimura, S. Kitamura, H. Urakawa, and K. Kajiwara, "Structural characteristics of gellan in aqueous-solution", *Food Hydrocoll.* , vol. 7, pp. 373-385, 1993.
[http://dx.doi.org/10.1016/S0268-005X(09)80233-6]

[103] A. Gamini, S. Paoletti, R. Toffanin, F. Micali, L. Michielin, and C. Bevilacqua, "Structural investigations of cross-linked hyaluronan", *Biomaterials.* , vol. 23, no. 4, pp. 1161-1167, 2002.

[http://dx.doi.org/10.1016/S0142-9612(01)00231-9] [PMID: 11791920]

[104] Y. Shinohara, K. Kayashima, Y. Okumura, C.M. Zhao, K. Ito, and Y. Amemiya, "Small-angle X-ray scattering study of the pulley effect of slide-ring gels", *Macromolecules.* , vol. 39, pp. 7386-7391, 2006.
[http://dx.doi.org/10.1021/ma061037s]

[105] B.T. Stokke, K.I. Draget, O. Smidsrod, Y. Yuguchi, H. Urakawa, and K. Kajiwara, "Small-angle X-ray scattering and rheological characterization of alginate gels. 1. Ca-alginate gels", *Macromolecules.* , vol. 33, pp. 1853-1863, 2000.
[http://dx.doi.org/10.1021/ma991559q]

[106] T. Tada, T. Matsumoto, and T. Masuda, "Dynamic viscoelasticity and small-angle X-ray scattering studies on the gelation mechanism and network structure of curdlan gels", *Carbohydr. Polym.* , vol. 39, pp. 53-59, 1999.
[http://dx.doi.org/10.1016/S0144-8617(98)00157-X]

[107] J.R. Lott, J.W. McAllister, M. Wasbrough, R.L. Sammler, F.S. Bates, and T.P. Lodge, "Fibrillar Structure in Aqueous Methylcellulose Solutions and Gels", *Macromolecules.* , vol. 46, pp. 9760-9771, 2013.
[http://dx.doi.org/10.1021/ma4021642]

[108] G. Cheng, X. Zhang, B. Simmons, and S. Singh, "Theory, practice and prospects of X- ray and neutron scattering for lignocellulosic biomass characterization: towards understanding biomass pretreatment", *Energy Environ. Sci.* , vol. 8, pp. 436-455, 2015.
[http://dx.doi.org/10.1039/C4EE03147D]

[109] H. Wang, G. Gurau, S.V. Pingali, H.M. O'Neill, B.R. Evans, and V.S. Urban, "Physical Insight into Switchgrass Dissolution in Ionic Liquid 1-Ethyl-3-methylimidazolium Acetate", *ACS Sustain. Chem.& Eng.* , vol. 2, pp. 1264-1269, 2014.
[http://dx.doi.org/10.1021/sc500088w]

[110] H. Wang, G. Gurau, and R.D. Rogers, "Dissolution of Biomass Using Ionic Liquids", In: *Structures and Interactions of Ionic Liquids,* S. Zhang, J. Wang, X. Lu, Q. Zhou, Eds., vol. 151. springer publisher: New York, 2014, pp. 79-105.
[http://dx.doi.org/10.1007/978-3-642-38619-0_3]

[111] F. Spinozzi, and M. Beltramini, "QUAFIT: a novel method for the quaternary structure determination from small-angle scattering data", *Biophys. J.* , vol. 103, no. 3, pp. 511-521, 2012.
[http://dx.doi.org/10.1016/j.bpj.2012.06.037] [PMID: 22947867]

[112] F. Spinozzi, P. Mariani, I. Mičetić, C. Ferrero, D. Pontoni, and M. Beltramini, "Quaternary structure heterogeneity of oligomeric proteins: a SAXS and SANS study of the dissociation products of Octopus vulgaris hemocyanin", *PLoS One.* , vol. 7, no. 11, p. e49644, 2012.
[http://dx.doi.org/10.1371/journal.pone.0049644] [PMID: 23166737]

[113] T.E. Williamson, B.A. Craig, E. Kondrashkina, C. Bailey-Kellogg, and A.M. Friedman, "Analysis of self-associating proteins by singular value decomposition of solution scattering data", *Biophys. J.* , vol. 94, no. 12, pp. 4906-4923, 2008.
[http://dx.doi.org/10.1529/biophysj.107.113167] [PMID: 18212017]

[114] J. Wen, T. Arakawa, and J. S. Philo, "Size-Exclusion Chromatography with On-Line Light-Scattering, Absorbance, and Refractive Index Detectors for Studying Proteins and Their Interactions", *Anal. Biochem.* , no. 12, pp. 155-166, 1996.
[http://dx.doi.org/10.1006/abio.1996.0345]

[115] E. Mathew, A. Mirza, and N. Menhart, "Liquid-chromatography-coupled SAXS for accurate sizing of aggregating proteins", *J. Synchrotron Radiat.* , vol. 11, no. Pt 4, pp. 314-318, 2004.
[http://dx.doi.org/10.1107/S0909049504014086] [PMID: 15211037]

[116] N.M. Kirby, and N.P. Cowieson, "Time-resolved studies of dynamic biomolecules using small angle X-ray scattering", *Curr. Opin. Struct. Biol.* , vol. 28, pp. 41-46, 2014.

[http://dx.doi.org/10.1016/j.sbi.2014.07.007] [PMID: 25108308]

[117] G. David, and J. Perez, "Combined sampler robot and high-performance liquid chromatography: a fully automated system for biological small-angle X-ray scattering experiments at the Synchrotron SOLEIL SWING beamline", *J. Appl. Cryst.*, vol. 42, pp. 892-900, 2009. [http://dx.doi.org/10.1107/S0021889809029288]

[118] N.J. Gunn, M.A. Gorman, R.C. Dobson, M.W. Parker, and T.D. Mulhern, "Purification, crystallization, small-angle X-ray scattering and preliminary X-ray diffraction analysis of the SH2 domain of the Csk-homologous kinase", *Acta Crystallogr. Sect. F Struct. Biol. Cryst. Commun.*, vol. 67, no. Pt 3, pp. 336-339, 2011. [http://dx.doi.org/10.1107/S1744309110053728] [PMID: 21393838]

[119] Y. Watanabe, and Y. Inoko, "Size-exclusion chromatography combined with small-angle X-ray scattering optics", *J. Chromatogr. A.*, vol. 1216, no. 44, pp. 7461-7465, 2009. [http://dx.doi.org/10.1016/j.chroma.2009.02.053] [PMID: 19269643]

[120] G.S. Wright, S.S. Hasnain, and J.G. Grossmann, "The structural plasticity of the human copper chaperone for SOD1: insights from combined size-exclusion chromatographic and solution X-ray scattering studies", *Biochem. J.*, vol. 439, no. 1, pp. 39-44, 2011. [http://dx.doi.org/10.1042/BJ20110948] [PMID: 21722094]

[121] D. Didry, F-X. Cantrelle, C. Husson, P. Roblin, A.M. Moorthy, J. Perez, C. Le Clainche, M. Hertzog, E. Guittet, M.F. Carlier, C. van Heijenoort, and L. Renault, "How a single residue in individual β-thymosin/WH2 domains controls their functions in actin assembly", *EMBO J.*, vol. 31, no. 4, pp. 1000-1013, 2012. [http://dx.doi.org/10.1038/emboj.2011.461] [PMID: 22193718]

[122] Y. Watanabe, and Y. Inoko, "Characterization of a large glycoprotein proteoglycan by size-exclusion chromatography combined with light and X-ray scattering methods", *J. Chromatogr. A.*, vol. 1303, pp. 100-104, 2013. [http://dx.doi.org/10.1016/j.chroma.2013.06.048] [PMID: 23859798]

[123] S. Sakamoto, Y. Sanada, M. Sakashita, K. Nishina, K. Nakai, and S-i. Yusa, "Chain-length dependence of polyion complex architecture bearing phosphobetaine block explored using SAXS and FFF-MALS", *Polym. J.*, vol. 46, pp. 617-622, 2014. [http://dx.doi.org/10.1038/pj.2014.25]

[124] P. Knappe, L. Boehmert, R. Bienert, S. Kamutzki, B. Niemann, A. Lampen, and A.F. Thünemann, "Processing nanoparticles with A4F-SAXS for toxicological studies: Iron oxide in cell-based assays", *J. Chromatogr. A.*, vol. 1218, no. 27, pp. 4160-4166, 2011. [http://dx.doi.org/10.1016/j.chroma.2010.11.012] [PMID: 21129749]

[125] P. Knappe, R. Bienert, S. Weidner, and A.F. Thuenemann, "Characterization of poly(N-vinyl-2-pyrrolidone)s with broad size distributions", *Polymer (Guildf.).*, vol. 51, pp. 1723-1727, 2010. [http://dx.doi.org/10.1016/j.polymer.2010.02.039]

[126] P. Knappe, R. Bienert, S. Weidner, and A.F. Thuenemann, "Poly(acrylic acid): A Combined Analysis with Field-Flow Fractionation and SAXS", *Macromol. Chem. Phys.*, vol. 211, pp. 2148-2153, 2010. [http://dx.doi.org/10.1002/macp.201000163]

[127] R. Jain, M. Petri, S. Kirschbaum, H. Feindt, S. Steltenkamp, S. Sonnenkalb, S. Becker, C. Griesinger, A. Menzel, T.P. Burg, and S. Techert, "X-ray scattering experiments with high-flux X-ray source coupled rapid mixing microchannel device and their potential for high-flux neutron scattering investigations", *Eur Phys J E Soft Matter.*, vol. 36, no. 9, p. 109, 2013. [http://dx.doi.org/10.1140/epje/i2013-13109-9] [PMID: 24092048]

[128] S.V. Kathuria, C. Kayatekin, R. Barrea, E. Kondrashkina, R. Graceffa, L. Guo, R.P. Nobrega, S. Chakravarthy, C.R. Matthews, T.C. Irving, and O. Bilsel, "Microsecond barrier-limited chain collapse observed by time-resolved FRET and SAXS", *J. Mol. Biol.*, vol. 426, no. 9, pp. 1980-1994, 2014. [http://dx.doi.org/10.1016/j.jmb.2014.02.020] [PMID: 24607691]

[129] J. Lamb, L. Kwok, X. Qiu, K. Andresen, H.Y. Park, and L. Pollack, "Reconstructing three-dimensional shape envelopes from time-resolved small-angle X-ray scattering data", *J. Appl. Cryst.* , vol. 41, no. Pt 6, pp. 1046-1052, 2008. [http://dx.doi.org/10.1107/S0021889808028264] [PMID: 19529835]

[130] R.P. Rambo, and J.A. Tainer, "Super-resolution in solution X-ray scattering and its applications to structural systems biology", *Annu. Rev. Biophys.* , vol. 42, pp. 415-441, 2013. [http://dx.doi.org/10.1146/annurev-biophys-083012-130301] [PMID: 23495971]

[131] C.M. Jeffries, and D.I. Svergun, "High-throughput studies of protein shapes and interactions by synchrotron small-angle X-ray scattering", In: *Structural Proteomics: High-Throughput Methods,* R.J. Owens, Ed., Springer Science+Business Media: New York, 2015, pp. 277-301. [http://dx.doi.org/10.1007/978-1-4939-2230-7_15]

[132] G. Benecke, W. Wagermaier, C. Li, M. Schwartzkopf, G. Flucke, R. Hoerth, I. Zizak, M. Burghammer, E. Metwalli, P. Müller-Buschbaum, M. Trebbin, S. Förster, O. Paris, S.V. Roth, and P. Fratzl, "A customizable software for fast reduction and analysis of large X-ray scattering data sets: applications of the new DPDAK package to small-angle X-ray scattering and grazing-incidence small-angle X-ray scattering", *J. Appl. Cryst.* , vol. 47, no. Pt 5, pp. 1797-1803, 2014. [http://dx.doi.org/10.1107/S1600576714019773] [PMID: 25294982]

[133] S. Classen, I. Rodic, J. Holton, G.L. Hura, M. Hammel, and J.A. Tainer, "Software for the high-throughput collection of SAXS data using an enhanced Blu-Ice/DCS control system", *J. Synchrotron Radiat.* , vol. 17, no. 6, pp. 774-781, 2010. [http://dx.doi.org/10.1107/S0909049510028566] [PMID: 20975223]

[134] G.L. Hura, A.L. Menon, M. Hammel, R.P. Rambo, F.L. Poole II, S.E. Tsutakawa, F.E. Jenney Jr, S. Classen, K.A. Frankel, R.C. Hopkins, S.J. Yang, J.W. Scott, B.D. Dillard, M.W. Adams, and J.A. Tainer, "Robust, high-throughput solution structural analyses by small angle X-ray scattering (SAXS)", *Nat. Methods.* , vol. 6, no. 8, pp. 606-612, 2009. [http://dx.doi.org/10.1038/nmeth.1353] [PMID: 19620974]

[135] T.D. Grant, J.R. Luft, L.G. Carter, T. Matsui, T.M. Weiss, A. Martel, and E.H. Snell, "The accurate assessment of small-angle X-ray scattering data", *Acta Crystallogr. D Biol. Crystallogr.* , vol. 71, no. Pt 1, pp. 45-56, 2015. [http://dx.doi.org/10.1107/S1399004714010876] [PMID: 25615859]

[136] J.P. Lafleur, D. Snakenborg, S.S. Nielsen, M. Moller, K.N. Toft, and A. Menzel, "Automated microfluidic sample-preparation platform for high-throughput structural investigation of proteins by small-angle X-ray scattering", *J. Appl. Cryst.* , vol. 44, pp. 1090-1099, 2011. [http://dx.doi.org/10.1107/S0021889811030068]

[137] M. Skou, S. Skou, T.G. Jensen, B. Vestergaard, and R.E. Gillilan, "In situ microfluidic dialysis for biological small-angle X-ray scattering", *J. Appl. Cryst.* , vol. 47, no. Pt 4, pp. 1355-1366, 2014. [http://dx.doi.org/10.1107/S1600576714012618] [PMID: 25242913]

[138] D. Schneidman-Duhovny, S.J. Kim, and A. Sali, "Integrative structural modeling with small angle X-ray scattering profiles", *BMC Struct. Biol.* , vol. 12, p. 17, 2012. [http://dx.doi.org/10.1186/1472-6807-12-17] [PMID: 22800408]

[139] A. Round, F. Felisaz, L. Fodinger, A. Gobbo, J. Huet, C. Villard, C.E. Blanchet, P. Pernot, S. McSweeney, M. Roessle, D.I. Svergun, and F. Cipriani, "BioSAXS Sample Changer: a robotic sample changer for rapid and reliable high-throughput X-ray solution scattering experiments", *Acta Crystallogr. D Biol. Crystallogr.* , vol. 71, no. Pt 1, pp. 67-75, 2015. [http://dx.doi.org/10.1107/S1399004714026959] [PMID: 25615861]

[140] S. Classen, G.L. Hura, J.M. Holton, R.P. Rambo, I. Rodic, P.J. McGuire, K. Dyer, M. Hammel, G. Meigs, K.A. Frankel, and J.A. Tainer, "Implementation and performance of SIBYLS: a dual endstation small-angle X-ray scattering and macromolecular crystallography beamline at the Advanced Light Source", *J. Appl. Cryst.* , vol. 46, no. Pt 1, pp. 1-13, 2013.

[http://dx.doi.org/10.1107/S0021889812048698] [PMID: 23396808]

[141] D.A. Jacques, and J. Trewhella, "Small-angle scattering for structural biologyexpanding the frontier while avoiding the pitfalls", *Protein Sci.* , vol. 19, no. 4, pp. 642-657, 2010. [http://dx.doi.org/10.1002/pro.351] [PMID: 20120026]

[142] E. Valentini, A. G. Kikhney, G. Previtali, C. M. Jeffries, and D. I. Svergun, "SASBDB, a repository for biological small-angle scattering data", *Nucleic Acids Res.* , vol. 43, pp. D357-63, 2015. [http://dx.doi.org/10.1093/nar/gku1047]

[143] J. Trewhella, W.A. Hendrickson, G.J. Kleywegt, A. Sali, M. Sato, T. Schwede, D.I. Svergun, J.A. Tainer, J. Westbrook, and H.M. Berman, "Report of the wwPDB Small-Angle Scattering Task Force: data requirements for biomolecular modeling and the PDB", *Structure.* , vol. 21, no. 6, pp. 875-881, 2013. [http://dx.doi.org/10.1016/j.str.2013.04.020] [PMID: 23747111]

CHAPTER 4

High-Performance Size-Exclusion Chromatography coupled with on-line Multi-angle Laser Light Scattering (HPSEC-MALLS)

Agnès Rolland-Sabaté*

UMR408 Sécurité et Qualité des Produits d'Origine Végétale, INRA, Université Avignon, F-84000 Avignon, France; UR1268 Biopolymères Interactions Assemblages, INRA, F-44316 Nantes, France

Abstract: High-Performance Size-Exclusion Chromatography (HPSEC) is widely used for the determination of the molar mass and size distribution of biopolymers in aqueous or organic solvents. Elements of the theory of fractionation by HPSEC, column calibration and online light scattering detection are given, showing that coupling HPSEC with multi-angle laser light scattering detection makes it easier to obtain molar mass distributions since light (MALLS) scattering gives the weight-average molar mass at each elution volume of the chromatogram. Some applications of HPSEC-MALLS for the macromolecular characterization of starches, glycogens, dextrans, celluloses, hemicelluloses, β-glucans, pectins, gums, alginates, carrageenans, chitosans, lignins, proteins and peptides are also presented.

Keywords: Cellulose, Chitosan, Gum, High-Performance Size-Exclusion Chromatography, Lignin, Multi-angle laser light scattering, Polysaccharide, Protein, Starch, Static and dynamic light scattering.

INTRODUCTION

Biopolymers found in nature are generally highly polydisperse. Their size and molar mass distributions are directly linked to their synthesis conditions and have an impact on their properties and their behavior in terms of degradation and other modifying processes. To understand their behavior, it is therefore important to determine these distributions. Size-exclusion chromatography (SEC, or gel permeation chromatography, GPC) is a very popular technique used for the determination of the molar mass distribution of polymers. It is a liquid chromatographic technique that allows the fractionation of molecules as a

* **Corresponding author Agnès Rolland-Sabaté:** UMR408 Sécurité et Qualité des Produits d'Origine Végétale, INRA, Université Avignon, F-84000 Avignon, France; Tel: +33 (0)4 32 72 25 22; Fax: +33 (0)4 32 72 24 92; E-mail: agnes.rolland-sabate@inra.fr

Martin Masuelli & Denis Renard (Eds.)

function of their hydrodynamic volume, *i.e.*, their size (which is proportional to their molar mass), by eluting across one or several columns filled with a porous stationary phase. The development of high-performance liquid chromatography leads to the appearance of high-performance SEC (HPSEC), which is highly precise (repeatable), robust and automatic. Since this is a size fractionation technique, the molar mass distribution is obtained by calibration of the system with standards. However, the relationship between size and molar mass is not that simple, depending on the macromolecules (in particular, in the case of branched ones), and is affected by operating conditions. That is why the development of HPSEC coupled with multi-angle laser light scattering (MALLS) detection is a real added value for the determination of molar mass distributions since light scattering gives the weight-average molar mass ($\overline{M}_w$) at each elution volume of the chromatogram. In this chapter, a brief overview of the theory of fractionation by SEC and online light scattering detection is first given, followed by some elements on column choice and applications of HPSEC-MALLS for the characterization of some biopolymers such as polysaccharides and proteins.

SEC THEORY

Fractionation Mechanism

During elution, molecules that are too large to enter the pores of the stationary phase, are eluted in the interparticular volume of the system (or total exclusion volume, V_0), whereas molecules that are sufficiently small to freely diffuse in and out of the pores are eluted in the interparticular volume of the system (or total column volume, V_t), equal to $V_0 + V_p$, where V_p is the pore volume. Intermediate-size molecules enter the pores and are separated by decreasing their size; the smaller they are, the more molecules enter the pores. Elution volume is expressed as follows [1]:

$$V_i = V_0 + K_{SEC}\, V_p \quad \textbf{(1)}$$

$$K_{SEC} = \frac{c_p}{c_0} \quad \textbf{(2)}$$

V_p is the accessible volume, K_{SEC} is the distribution coefficient, c_p is the mean concentration of the polymer in the pore volume, and c_0 is the interstitial volume. In the case of a pure SEC mechanism, *i.e.*, based on size only, $0 < K_{SEC} < 1$. A K_{SEC}=0 corresponds to an elution in the total exclusion volume, and when K_{SEC}=1, the solute is eluted in the total volume. If $K_{SEC} > 1$, the separation is controlled by enthalpic interactions that depend on the chemical composition of the polymer.

SEC Calibration

The aim of calibration is to obtain a relationship that links the elution volume to the molar mass of a sample. This technique is useful when only the concentration is determined online. The column calibration must be done in the same chromatographic conditions: eluent, temperature, elution flow, injected concentration and volume.

The classical calibration technique is based on the analysis of a series of monodisperse standards with known molar mass and the same chemical nature as the sample to be analyzed. The calibration curve shows the plot of the logarithm of the molar mass (*logM*) of the standards *versus* the elution volume (V_i), and must cover the entire separation range, from the excluded volume to the total volume [1, 2]. An alternative method is the use of polydisperse standards [1, 3, 4]. These two methods require standards of the same type as the sample because the size and the hydrodynamic volume depend on the length of the polymeric chains, their structure (linear or branched), their conformation (sphere or rod) and the solvent. That is why these methods are generally difficult to use, in particular, for branched samples since the standard structure must be exactly the same as the sample.

The universal calibration method was introduced by Benoit [5, 6]. It uses the hydrodynamic volume, V_η:

$$V_\eta \propto [\eta]\, M \qquad \textbf{(3)}$$

where $[\eta]$ is the intrinsic viscosity and M is the molar mass. The calibration curve, $\log[\eta]M$ *versus* V_i, is obtained with a series of monodisperse standards with known molar mass and intrinsic viscosity. The molar mass is then calculated with the Mark-Houwink-Sakurada (MHS) equation:

$$M = \left(\frac{V_\eta}{K_a} \right)^{1/(a+1)} \qquad \textbf{(4)}$$

where a and K_a are constants, also referred as the MHS parameters. The advantage of this technique is the possibility to use standards with different chemical natures. However, it is valid only in the case of a pure exclusion effect, and the MHS parameters of the standards and the sample have to be known. More information can be found in the review of Kostanski *et al.* [7] where the different SEC calibration methodologies for simple and complex polymers were compared.

Secondary Effects in HPSEC

The secondary effects described here are those that induce a separation of the sample with a mechanism that differs from pure exclusion and that would produce a bias in the interpretations.

When extremely high molar mass polymers are fractionated, degradation of the chains by shear scission could occur. This happens when chains go through the pores or frits and, in particular, when high flows and/or viscous solvents are used, inducing high pressure and high shear rates in the system [8]. Cave *et al.* [9] clearly observed such shear degradations on the amylopectin fraction of starch macromolecules. These degradations could be limited by using low flow rates [10].

When the macromolecules are particularly large compared to the pore size, a separation according to the hydrodynamic effect is possible [11]. The macromolecules are then excluded from the gel and fractionated in the interparticular volume in a laminar flow. The large molecules are eluted at a higher speed than the smaller ones that are closer to the walls (where the speed is quasi-null). This separation can be combined with SEC for the fractionation of very large and highly polydisperse biopolymers such as starch macromolecules [12].

Interactions take place with the stationary phase when it is not inert to the sample. This could produce adsorption due to hydrophobic interactions that produce a shoulder following the sample peak. These interactions could be reduced by increasing the temperature or by adding an organic compound to the eluent [13]. Ionic interactions may occur when the stationary phase has charges or a polyelectrolyte. This is typically the case when working with proteins. This induces a peak at the total excluded volume. Using eluents with sufficient ionic strength, such as buffers, makes it possible to avoid this effect [14].

When a sample is injected into an HPSEC system at an excessively high concentration, the equilibrium between the polymer concentration in the pores and in the interstitial volume cannot be established because of the high viscosity. Such overloading produces a peak at low elution volumes. This effect could be reduced by decreasing the sample concentration and the flow rates [15].

Performances of an HPSEC System

The performances of an HPSEC system are directly linked to the band broadening [1], which is a common experimental effect for every fractionating system based on size, and depend on the connections between the system and the detectors. The band broadening could produce a peak broadening and/or distortion because the content of the detector cell at a given elution volume V_i does not correspond to the fraction of the hydrodynamic volume V_{Hi}, but instead corresponds to a mixture of contiguous fractions.

A chromatographic system can be theoretically divided into *N* theoretical plates corresponding to adjoining separation zones, where each zone has a length such that there can be complete equilibrium of the solute between the mobile and stationary phases within the zone. The length of a theoretical plate in the column is referred to as the height equivalent to a theoretical plate or the plate height [1]. The performances of an HPSEC system are then determined using the following values: the number of theoretical plates, *N*, the plate height, *H*, the asymmetry factor, *AF*, and the resolution of the system, R_s.

The number of theoretical plates characterizes the efficacy of the separation system in terms of signal broadening and is expressed, if peaks have a Gaussian shape [4], by the equations:

$$N = 5{,}54 \quad x \quad \left(\frac{V_i}{W_{1/2}}\right)^2 \tag{5}$$

$$H = \frac{L}{N} \tag{6}$$

where $W_{1/2}$ is the peak width at one-half the peak height and *L* is the column length.

The asymmetry factor is:

$$AF = \frac{b}{a} \tag{7}$$

where *a* is the difference of elution volumes at 1/10 the peak height, between the beginning and the maximum of the elution peak, and *b* is the difference of elution volumes at 1/10 the peak height, between the maximum and the end of the elution peak.

$$R_S = \frac{2\ (V_{i2} - V_{i1})}{(W_{b2} + W_{b1})} \tag{8}$$

The resolution of a fractionating system, R_S, between two solutes 1 and 2, is obtained from the chromatogram of a mixture of these two solutes using the following equation:

Where W_b is the chromatogram peak width formed by the intersection of the tangents to the curve inflexion points with the baseline [1]. The separation between the two solutes is complete when this ratio is equal to unity. The resolution is the best when the peak has a low dispersity, *i.e.*, for low $W_{1/2}$ values. The efficacy of a SEC system decreases, thus increasing elution rates and the amount of sample injected.

On the basis of these theoretical considerations, it can be concluded that the choice of the column, the mobile phase, elution rates and the amount of injected sample depend on the biopolymer to be analyzed and are crucial to obtain a reliable fractionation. Recent reviews provide some aspects for for the analysis of proteins [16, 17] and polysaccharides [1, 18] by SEC.

COLUMNS

A SEC column is defined by its dimensions (length and internal diameter), by the nature of its stationary phase (semi-rigid gel, porous particles) and by its physical properties (exclusion limits, pore size, particle diameter). The range of porosity of the stationary phase must allow the separation in the range of the expected sample size. The sample must be completely soluble in the eluent. The stationary phase must be stable in the eluent to have a mechanism of fractionation based on size only. In terms of columns, it is necessary to differentiate between aqueous and organic HPSEC. For aqueous HPSEC, column-packing materials must have highly hydrophilic surfaces with minimal ionic or hydrophobic sites, in addition to a good mechanical and chemical stability. Polymeric and silica phases are available for aqueous HPSEC [19, 20]. Silica-based columns are available in different porosities (covering a wide range of molar mass), with different particle sizes and types of bonding. The most common silica packings consist of chemically-bonded 1,2-propanediol functional groups that provide a hydrophilic surface. This stationary phase blocks or reacts with many of the acidic silanol groups, neutralizing the surface. It is therefore ideal for HPSEC separation of biopolymers [20]. Grafting polymers onto silica makes it possible to mask surface silanol groups and to design the porosity and the stability of the packing material. These columns can be used at very high mobile phase velocities (to decrease the analysis time) and with up to 50% polar organic solvent, but their pH resistance is not wide (since a pH higher than 7.5 would solubilize the silica) and their maximum operating temperature is 30°C. They are designed for protein analysis. Methacrylate-based columns are also used in HPSEC for biopolymers. Their advantage is the high chemical resistance of the polyhydroxymethacrylate particles to extreme pH, high temperatures and organic solvents. Polyvinyl alcohol-based columns may also be used and are suitable for aqueous and for organic HPSEC. Like the methacrylate-based columns, they have a resistance to extreme pH, high temperatures and organic solvents. They are generally used for peptide and polysaccharide analysis. Styrene divinyl benzene copolymer-based columns are used for the analysis of biopolymers in organic solvents, for example, xylan in DMSO or cellulose in DMAc. They have a high resistance to pH, very high temperatures and organic solvents. Interested readers can find more information in the recent review of Barth *et al.* [20].

SEC COUPLING WITH DETECTORS

Concentration and Chemical Detectors

The most commonly used online detectors are concentration detectors. Among them, differential refractometric detectors (DRI) are widely used because of their universality. They measure the difference of the refractive index of a solvent with a solute and are used for biopolymers in solution when their refractive index sufficiently differs from that of the solvent. This is the case for proteins and polysaccharides in aqueous solvents, for example. UV-Vis and IR detectors are concentration detectors that are also sensitive to chemical groups that absorb the light at a chosen wavelength and are mostly used for proteins and aromatics (UV) and polyolefins (IR) [1]. Other online concentration detectors are available but are less commonly used: evaporative light scattering detectors, fluorescence detectors, dielectric detectors and ionization detectors [1].

Chemical composition distribution could be obtained by coupling HPSEC to chemical detectors (UV/vis, IR, fluorescence). HPSEC may also be coupled with mass spectrometry detectors (electrospray, matrix-assisted laser desorption/ionization time-of-flight, MALDI-TOF and induced coupled plasma (ICP)) [1], and a recent paper reports on the feasibility of determining polymer chemical heterogeneity by coupling SEC with continuous off-line RAMAN detection [21].

Light Scattering (LS) Detectors

Online static light scattering (multi-angle laser light scattering, MALLS) allows the determination of the weight-average molar mass $\overline{M}_w$, and the z- average radius of gyration $\overline{R}_G$. In dynamic mode (DLS, dynamic light scattering, QELS, quasi-elastic light scattering), it allows the determination of the translational diffusion coefficient D_t, and the hydrodynamic radius $\overline{R}_h$ of the macromolecule. A light scattering experiment consists of measuring the light scattered at different angles and concentrations around a diluted solution of macromolecules or particles submitted to a laser with a wavelength of between 400 nm and 700 nm. The Rayleigh ratio ΔR_θ is used instead of the scattered intensity value to overcome the observation distance R between the observation and the scattering points:

$$\Delta R_\theta = \frac{(I_0 - I_s)R^2}{I_0} \tag{9}$$

Where I_θ is the scattering intensity at the angle θ, I_s the scattering intensity of the solvent, and I_0 is the incident light intensity. The light scattered at an angle θ relative to the incident light corresponds to the scattering vector with the following module:

$$q = \frac{4\pi n_0}{\lambda_0} \sin\left(\frac{\theta}{2}\right) \tag{10}$$

Where n_0 is the refractive index of the pure solvent and λ_0 the wavelength of the incident beam.

In static light scattering (SLS), the scattered intensity is averaged over a certain period of time that is sufficiently long to smooth any internal mobility off the macromolecule (1-2s). For a polymer solution at a concentration c, the fundamental light scattering equation is [22]:

$$\frac{Kc}{\Delta R_\theta} = \left(\frac{1}{\overline{M_w}} + 2A_2c\right)\left(1 + \frac{1}{3}q^2\overline{R_G^2} + \ldots\ldots\right) \tag{11}$$

Where A_2 is the second Virial coefficient. The optical constant is:

$$K = \frac{4\pi^2}{\lambda_0^4 N_a} n^2 \left(\frac{dn}{dc}\right)^2 \tag{12}$$

Where dn/dc is the refractive index increment of the solution and n is the refractive index of the solution.

If the macromolecules in solution are isotropic and small compared to the wavelength (< $\lambda/20$), the scattering intensity is not dependent on the angle θ and eq. (11) becomes:

$$\frac{Kc}{\Delta R_{90^0}} = \left(\frac{1}{\overline{M}_w} + 2A_2c\right) \quad \textbf{(13)}$$

$\overline{R}_G$ is then not accesible with LS. When the macromolecules have at least one dimension > $\lambda/20$, there is an angular dependence. This is the most common case, except for very small globular proteins. The function *P(θ)* quantifies this dependence:

$$P(\theta) = 1 - \frac{1}{3} q^2 \overline{R}_G^2 + \ldots \quad \textbf{(14)}$$

When a light scattering detector is coupled with HPSEC, the general equation (11) is, for each fraction *i* of concentration c_i:

$$\frac{Kc_i}{\Delta R_{\theta i}} = \left(\frac{1}{M_i\ P_i\ (\theta)} + 2A_2c_i\right) \quad \textbf{(15)}$$

Because of the dilution in the separation system, $2A2c_i$ becomes negligible. Then:

$$\frac{Kc_i}{\Delta R_{\theta i}} = \frac{1}{M_i\ P_i\ (\theta)} \quad \textbf{(16)}$$

$\overline{M}_w$, and $\overline{R}_G$ are then determined by extrapolation of the data at a null angle by using different diagrams depending on the nature of the polymer to be characterized, and A_2 is not accessible with online LS measurement.

In the Zimm diagram [22], $Kc/\Delta R_\theta$ is plotted *versus* $sin^2(\theta/2)$. The slope of the curve gives $\overline{R}_G^2$, whereas the intersection of this curve with the *y*-axis gives $1/\overline{M}_w$. In the Berry diagram [23], the square root of $Kc/\Delta R_\theta$ is plotted *versus* $sin^2(\theta/2)$, and the intersection of this curve with the *y*-axis gives the square root of $1/\overline{M}_w$. In the Debye diagram [24], $\Delta R_\theta/Kc$ is plotted *versus* $sin^2(\theta/2)$, and the intersection of this curve with the *y*-axis gives $\overline{M}_w$. The Zimm diagram is the most commonly used whereas the Berry diagram is better for high molar mass macromolecules and the Debye diagram is better for the characterization of small globular proteins.

c_i is calculated from the concentration detector response. It is important to know the *dn/dc* and/or the absorption coefficient (if a UV detector is used to determine the concentration) because the concentration value impacts the molar mass calculation. In particular, the *dn/dc* value must be known because it is used in any case in the LS equation with a square ponderation (eq. (11)). A *dn/dc* value of 0.146 $mL.g^{-1}$ is generally used for glucans and a value of 0.168 $mL.g^{-1}$ for proteins.

Thus, by combining a MALLS and a concentration detector, the quantities c_i, M_i, and R_{Gi} (respectively, mass concentration, molar mass, and radius of gyration of the fraction i of the chromatogram) can be obtained, and from these data, the size and molar mass distributions and the dispersity of the sample can be obtained. Since each fraction is considered as quasi-monodisperse, the weight, number and z-average molar masses of the fraction i are equal ($M_{wi} = M_{ni} = M_{zi}$) for linear polymers, but $M_{wi} \neq M_{ni}$ when the number of long chain branching is significant. Number-average molar mass $\bar{M}_n$, $\bar{M}_w$, the dispersity index $\bar{M}_w/\bar{M}_n$, and the root-mean-square radius $\bar{R}_G$, can then be calculated using the summations taken over one peak:

$$\bar{M}_n = \frac{\sum_i c_i}{\sum_i c_i/M_i} \quad (17)$$

$$\bar{M}_w = \frac{\sum_i c_i M_i}{\sum_i c_i} \quad (18)$$

$$\bar{R}_G^2 = \frac{\sum_i c_i M_i < R_G^2 >_i}{\sum_i c_i M_i} \quad (19)$$

Quasi-elastic light scattering (QELS) is based on the study of temporary concentration fluctuation linked to the movement of macromolecules or particles in solution. These fluctuations of concentration induce variations in the light signal and are quantified by an autocorrelation function $G(\tau)$ such that:

$$G\,\tau = \lim_{T\to\infty}\frac{1}{2T}\int_{-T}^{T} I(\tau)J(\tau+t)dt = <I(\tau)J(\tau+t)> \qquad \textbf{(20)}$$

where T is the integration period, t is a time displacement, and $I(\tau)$ and $J(\tau)$ are signal intensities depending on time. In QELS, a correlator allows the experimental measurement of the autocorrelation function of the scattered intensity I_d:

$$G^{(2)}(\tau) = <I_d(0)I_d(\tau)> \qquad \textbf{(21)}$$

Thus, the scattered intensity at time zero is compared to the one scattered at 10^{-6}-10^{-3} s. This autocorrelation function, which is difficult to interpret, is converted to an autocorrelation function of the diffused electric field $g_1(\tau)$ linked to the dynamics of the macromolecules in solution:

$$G_2(\tau) = B[1+\beta\,|g_1(\tau)|^2] \qquad \textbf{(22)}$$

B is the baseline and β the consistency factor. $g_1(\tau)$ can then be broken down into a distribution of exponentials:

$$g_1(\tau) = \int_0^{\infty} w(\Gamma)\,e^{-\Gamma t}\,d\Gamma \qquad \textbf{(23)}$$

$w(\Gamma)$ is the distribution function in frequency of relaxation Γ. $g_1(\tau)$ could also be expressed as a function of relaxation time ($\tau=1/\Gamma$):

$$g_1(\tau) = \int_0^{\infty} \tau\,A(\tau)\,e^{-t/\tau}\,d\log\tau \qquad \textbf{(24)}$$

$A(\tau)$ is the distribution function in relaxation time τ. Various methods based on the use of different mathematical algorithms are available to approach $A(\tau)$ and τ. The most frequently used are the cumulant [25] and the methods based on the Laplace inversion of $g_1(\tau)$ such as CONTIN [26].

In diluted solution, each molecule has a Brownian motion that does not depend on the other molecules. For $q\,\overline{R}_G << 1$, the macromolecule has small dimensions compared to the observation window q^{-1}, quasi-elastic scattered light then comes from the Brownian motion only and the autocorrelation function $g_1(t)$ is a monoexponential characteristic of a monodisperse polymer:

$$g_1(t) = e^{-\Gamma t} \quad \textbf{(25)}$$

The translational diffusion coefficient of the molecule $\overline{D}$ is:

$$\overline{D} = \left(\Gamma/q^2\right)_{q\to 0} \quad \textbf{(26)}$$

The Stokes-Einstein equation then makes it possible to calculate the hydrodynamic radius of an equivalent sphere (or Stokes radius) $\overline{R}_h$:

$$\overline{R}_h = \frac{k_B T}{6\pi\pi_0 \overline{D}} \quad \textbf{(27)}$$

where k_B is the Boltzman constant (1.38 x 10^{-23} J.K^{-1}), T the temperature and η_0 the solvent viscosity.

When the macromolecules are polydisperse, each fraction i has its own autocorrelation function, and the previously cited algorithms are useful to solve the system.

In the case of online QELS, each elugram fraction has its own autocorrelation function (eq. (22)), and because of a very low dispersion inside each fraction, the relaxation time can be obtained by adjusting the curve with a monoexponential function (eq. (25)).

For reliable measurement with online QELS, it is necessary to check that the sample will stay long enough in the measurement cell (> 50τ) in the measurement volume: this implies the use of relatively small elution flow rates (< 1 mL.min^{-1}) and increases the analysis time. Moreover, the acquisition time must be long enough ($\geq$10000τ) in order to have a good average for the whole measurement. Finally, the observation angle influences the relaxation time and the measurement volume, so it has to be carefully chosen: it has to be sufficiently high and adapted to the size of the sample to be analyzed. This measurement at one angle only should be limited to medium-sized macromolecules (< 70 nm) [12, 27].

The principal advantage of this coupling with light scattering is that the molar mass and the size are obtained without any calibration. Nevertheless, the band broadening previously described can induce errors in molar mass and molar mass distribution determinations. A mathematical correction could be applied to the data, specific to the studied system [7, 28 - 30]. In HPSEC, the band broadening produces errors that are higher when the macromolecules are monodisperse and exhibit a relatively low molar mass (< 30000 g.mol^{-1}) [31]. Few software programs for data acquisition and treatment have an automatic correction of the band broadening, so it is difficult to know how they do it.

Viscometric Detectors and Multi Detection

The viscosity of a solution significantly increases when macromolecules are solubilized in a solvent. The simplest method to characterize a polymer in solution is the measurement of this increase in viscosity. An online viscometer (VS) measures the pressure in a capillary caused by a liquid. This pressure is proportional to the eluent viscosity. The relative viscosity ηr, is:

$$\eta_{ir} = \left(\frac{\eta_1}{\eta_0}\right) = \left(\frac{\Delta P_1}{\Delta P_0}\right) \tag{28}$$

where η_1 and η_0 are the solute and the solvent viscosities, and ΔP_1 and ΔP_0 are the solute and the pure mobile phase pressures. By coupling a differential refractometer and a viscometer to a HPSEC system, the intrinsic viscosity can be determined considering a negligible concentration in the eluted fraction:

$$[\eta] = \left(\frac{\ln \, {}^{\Delta P_1}\!/_{\Delta P_0}}{c}\right) \tag{29}$$

The intrinsic viscosity $[\eta]_i$ can be determined for each elution volume, making it possible to determine the intrinsic viscosity distributions. The molar mass of each fraction can then be determined *via* the universal calibration, and given the MHS parameters (see eq. (3) and eq. (4)):

$$M_i = \left(\frac{V_{\eta i}}{[\eta]_i}\right) \tag{30}$$

where $V_{\eta i}$ is the hydrodynamic volume of the fraction *i*. This molar mass is considered to be equal to the number-average molar mass of the fraction M_{ni} by numerous authors [7, 32, 33].

By coupling a concentration detector, a viscometric detector and a light scattering detector (triple detection), it is possible to obtain c_i, M_i, R_i and $[\eta]_i$ without any calibration. The viscometric hydrodynamic radius $R_{\eta i}$ can then be obtained by the following equation:

$$[\eta]_i M_i = \frac{10\pi}{3} N_A R_{\eta i}^{3} \quad \textbf{(31)}$$

where N_A is the Avogadro number. The weight distributions in R_η can then be established. Since HPSEC fractionate according to size and not to molar mass, studied fractions of branched polymers have a certain dispersity in structure and molar mass [32, 33]. That is why the actual tendency for the description of these polymers is based on the analysis of their structure along their size distribution. Gaborieau *et al.* [34] showed that number distributions in hydrodynamic radii can also be established by coupling a DRI, a VS and universal calibration. By doing this, it is possible to access the dispersity of each fraction by dividing the weight-average molar mass obtained by MALLS and the number-average molar mass obtained by coupling a DRI, a VS and universal calibration [34, 35]. Nevertheless, for branched polymers, the validity of the universal calibration is controversial [35] and must be checked in order to obtain the proper number of hydrodynamic radius distributions.

A UV-Vis detector is often coupled with a triple detection system as another concentration detector. This quadruple detector system allows the determination of copolymer and conjugated proteins molar masses, in addition to the amount of comonomers in the distribution.

Other Accessible Structural Parameters

By plotting the molar mass *versus* the radii or the intrinsic viscosity obtained from HPSEC-MALLS coupled with QELS or VS, conformational data can be determined by using the following equations:

$$R_{Gi} = K_G M_{wi}^{\nu G} \quad \textbf{(32)}$$

$$R_{hi} = K_H M_{wi}^{\nu H} \quad \textbf{(33)}$$

$$[\eta]_i = K_a M_{\eta i}^{a} \quad \textbf{(34)}$$

where R_{hi} is the hydrodynamic radius for each slice i of the chromatogram, v_G and v_H are the corresponding scaling law coefficients, and K_G and K_H are the corresponding constants. The values of v_G, v_H and a are dependent on polymer shape, temperature and polymer-solvent interactions, and for a monodisperse polymer with a sufficiently high molar mass: $v_G = v_H = 0.33$, $a=0$ for a sphere; $v_G = v_H = 0.5\text{-}0.6$, $a=0.5\text{-}0.8$ for a linear random coil (according to Flory's theory [36]); and $v_G = v_H = 1$, $a=1.8$ for a rod. The prefactor is a non-universal constant: it depends on the detailed monomer structure and on the solvent chosen [37]. These scaling laws relative to molar masses and radii or $[\eta]$ are widely used to access the behavior of biopolymers in solution [27, 38 - 42]. If the log-log plots of radii or $[\eta]$ *vs.* molar mass are non-linear, this is an indication that the macromolecular system adopts more than one conformation. For example, amylose [39] and pectin [43, 44] adopt more than one conformation. For a non-homologous series of polysaccharides, any attempt to extract the molar mass dependency of $[\eta]$, for example, will lead to misinterpretation [45].

In the case of objects that have a stable structure along the size distribution (linear relation between log R_G and log M_w), a global fractal dimension can also be obtained by coupling MALLS to HPSEC according to the following equation [46, 47]:

$$M_i = K'_G R_{Gi}^{\ 1/vG} = K'_G R_{Gi}^{\ dfg} \quad \textbf{(35)}$$

where d_{fg} is the global fractal dimension. For a self-similar macromolecule, the fractal dimension provides information about the macromolecule architecture. d_f is 3 for a sphere, 2 for a flat disc and 1 for thin rods, whereas a d_f of 1.7 was found for random coil macromolecules in good solvent (chains with excluded volume), and a d_f between 2.0 and 2.5 was obtained by simulation using percolation theory for self-similar branched molecules highly swollen in a good solvent and with low swelling of the solvent, respectively.

By using MALLS detection coupled with SEC, it is possible to obtain the molar mass and the radius of gyration distributions of biopolymers. Adding online QELS makes it possible to access the hydrodynamic radius distribution and, consequently, the conformational and structural heterogeneity of the sample along its size and molar mass distributions *via* the ratio of R_{Gi} to R_{hi}, also referred to as the structure factor ρ_i [27]. The $\overline{R}_G/\overline{R}_h$ ratio ($=\rho$) is 0.778 for a compact sphere and 2 for rigid rods [47 - 49]. For macromolecules, ρ values are between 1 and 2, the latter value corresponding to elongated conformations. ρ decreases when branching increases and increases with polydispersity [47 - 49].

The apparent particle density, which describes the apparent volume of the molecules, is accessible for each chromatogram slice from R_{Gi} and M_{wi} *via* the following equation:

$$d_{Gappi} = M_i / [N_a (4\pi/3) R_{Gi}^3] \quad \textbf{(36)}$$

The apparent particle density is linked to branching density since highly branched macromolecules are expected to be denser than less branched ones [27, 50], and it has also been used to understand variations in ρ_i values [25]. Kratky plots are another means to obtain structural information from light scattering data processed on angular dependence [51, 52]. To detect branches and obtain information on the branching pattern of a molecular structure [49], Kratky plots of macromolecules with $\overline{R}_G > 100$ nm are compared to theoretically calculated plots for differently shaped and structured macromolecules. Kratky plots were constructed from HPSEC-MALLS data to determine amylopectin branching characteristics [53, 54].

Shrinking factors (or branching parameters) also provide information about the branching structure of polymers. They have been widely used to characterize branched polymers in solution using the following equations [49, 55]:

$$g_M = \overline{R}^2_{Gw(br)} / \overline{R}^2_{Gw(lin)} \quad \textbf{(37)}$$

$$g' = [\eta]_{(br)} / [\eta]_{(lin)} \quad \textbf{(38)}$$

$$h = \overline{R}_{h(br)} / \overline{R}_{h(lin)} \quad \textbf{(39)}$$

where g_M, g' and h are the branching parameter, the viscosimetric and the dynamic shrinking factors, respectively; $\overline{R}_{Gw}$ is the weight-average radius of gyration, and the subscripts *br* and *lin* denote branched polymers and the corresponding linear polymers at the same molar mass, respectively. The shrinking factors could be determined using HPSEC-MALLS-VS (or HPSEC-MALLS-QELS) data. Nevertheless, to obtain these shrinking factors, it is necessary to have the linear equivalent with the same molar mass, which is not that simple, especially in the case of biopolymers.

Zimm & Stockmayer [56] derived the g_M expression on the basis of a random branching model with a linear growth of branches. Thus, for a monodisperse, highly branched polymer with a functionality of 3:

$$g_M = \left[\left(\frac{1+\overline{B}}{7} \right)^{1/2} + \left(\frac{4\,\overline{B}}{9\pi} \right) \right]^{-1/2} \quad \textbf{(40)}$$

where $\overline{B}$ is the number of branching points in the macromolecule. For a polydisperse polymer, highly branched with a functionality of 3:

$$g_M = \frac{6}{\overline{B}} \left[\frac{1}{2} \left(\frac{2+\overline{B}}{7} \right)^{1/2} \ln \left(\frac{\left(2+\overline{B}\right)^{1/2} + \overline{B}^{1/2}}{\left(2+\overline{B}\right)^{1/2} - \overline{B}^{1/2}} \right) - 1 \right] \quad \textbf{(41)}$$

On the basis of these equations, the branching degree (BD) can be obtained *via* $\overline{B}/\overline{DP}_w$, followed by the average chain length $\overline{DP}_w/\overline{B}$.

The flexibility of a linear chain may also be approached *via* the determination of the persistence length (L_p) of equivalent worm-like chains [49]. The persistence length is defined as the average projection length along the initial direction of the polymer chain. For a theoretical random coil, L_p tends to 0 and to the infinite for the equivalent rod [55], although in practice, the limits are 1 nm for random coils (*e.g.*, pullulan) and 200 nm for a rod (*e.g.*, schizophyllan). The L_p of different polysaccharides were extracted from $\overline{R}_G$ and $\overline{M}_w$ data obtained by HPSEC-MALLS [39, 57] or from $\overline{R}_G$, $\overline{M}_w$ and *[η]* data obtained by HPSEC-MALLS-VS [45, 57]. They were estimated using different models [58, 59]. The choice of the model is important as it impacts the L_p value [60].

Relationships between radii or *[η]* and molar masses, shrinking factors and the ρ parameter have been determined to date using average values, thus, without prior fractionation of the samples. Since these entities are strongly influenced by the dispersity of samples, using a separation technique like HPSEC prior to characterization makes it possible to access these parameters without the bias introduced by the dispersity and to thus approach the structural heterogeneity of the sample. In addition, coupling HPSEC to MALLS increases the number of points in the plots, thus increasing the accuracy of the determination of the structural parameters.

APPLICATIONS FOR POLYSACCHARIDES

Starch, Glycogen and other α-Glucans

Starch in plants and glycogen in animals constitute the major energy source in living organisms. They are both mainly made up of α-(1,4)-linked α-D-glucosyl units. Nevertheless, glycogen has a highly quasi-randomly branched structure with 7-10% of α-(1,6) branching points and is soluble in cold water, whereas starch is composed of ~25% amylose, which has very few α-(1,6) branches and ~75% amylopectin with 5-6% of α-(1,6) branching points. The branching structure of starch induces its insolubility in cold water. Numerous studies have dealt with HPSEC-MALLS for the macromolecular characterization of starches and glycogens. They involve silica-based or polymer-based stationary phases and are carried out either in aqueous solvents with added salts (NaCl, KCl, $NaNO_3$, NaN_3) [12, 27, 39, 40, 50, 53, 54, 61 - 70], NaOH or KOH [71 - 73], or organic solvents, *e.g.*, dimethylsulfoxide (DMSO) with LiBr [68, 74], DMSO/water [75] or dimethylacetamide (DMAc) with LiBr [76, 77].

Prior to analysis, the sample must be solubilized in the eluent. DMSO is considered as the best solvent for starch. However, it facilitates shear degradation in the column [9] and the dn/dc of starch polysaccharides in DMSO (0.07 $mL.g^{-1}$) is lower than in aqueous solvents (0.146 $mL.g^{-1}$), thus producing large errors in molar mass determinations. For NaOH or KOH (0.1 N), the limiting factor is the system's compatibility. Native starch macromolecules exhibit an extremely high dispersity (10^4 to 10^8 $g.mol^{-1}$), and no SEC matrix is adapted for the extremely high molar mass of amylopectin, which is then often retained in the column, degraded by shearing or totally excluded from the gel [78]. However, the analyses on silica-based columns in water with NaN_3 [12, 40, 50, 54, 61] and on polymeric columns in DMAc/LiBr [76] are quantitative, but even if they make it possible to differentiate the amylose and amylopectin fractions, they do not allow a complete separation between these two entities (Fig. **1**).

Analysis of synthetic amyloses by HPSEC-MALLS in aqueous media [39] and DMSO [75] allows the determination of molar mass (10^4-10^5 $g.mol^{-1}$), the radius of gyration (30-52 nm), the scaling law coefficient ν_G (0.5-0.6), and the persistence length (1.5) [39], thus confirming the random coil and high flexibility behavior of amyloses, and showing that DMSO and KOH (0.1 N) are good solvents for amylose, whereas water is a θ solvent. Pullulans, as standard substances, are often used to calibrate HPSEC columns. Roger *et al.* [39] found that they can be used to replace the linear amylose standards for chromatographic calibration purposes in 0.1 M KOH, but not in water. It has also been shown using HPSEC-MALLS on natural amyloses to determine their molar mass distributions and conformational properties in 0.1 N KOH, that they are slightly branched with a probable comb-like structure [71, 72]. It is also possible to extract the following values from HPSEC-MALLS data on total starches: the molar mass (~10^8 $g.mol^{-1}$), the radius of gyration (127-250 nm) and the scaling law coefficient ν_G (0.36-0.49) of the amylopectin population. They are a function of botanical origin and correspond to a compact sphere behavior in relation to the branched structure of amylopectin [12, 40, 65]. Amylopectin branching information has also been extracted from Kratky plots using HPSEC-MALLS data [53, 54]. However, molar mass, $\overline{R}_G$, and conformational and branching values of amylopectin extracted from HPSEC-MALLS data are biased because of the poor fractionation of amylopectin by HPSEC (which probably contains aggregate molecules) and of the high angular dependency of the scattered light that arises from the poor fractionation, making the extrapolation to the zero angle difficult.

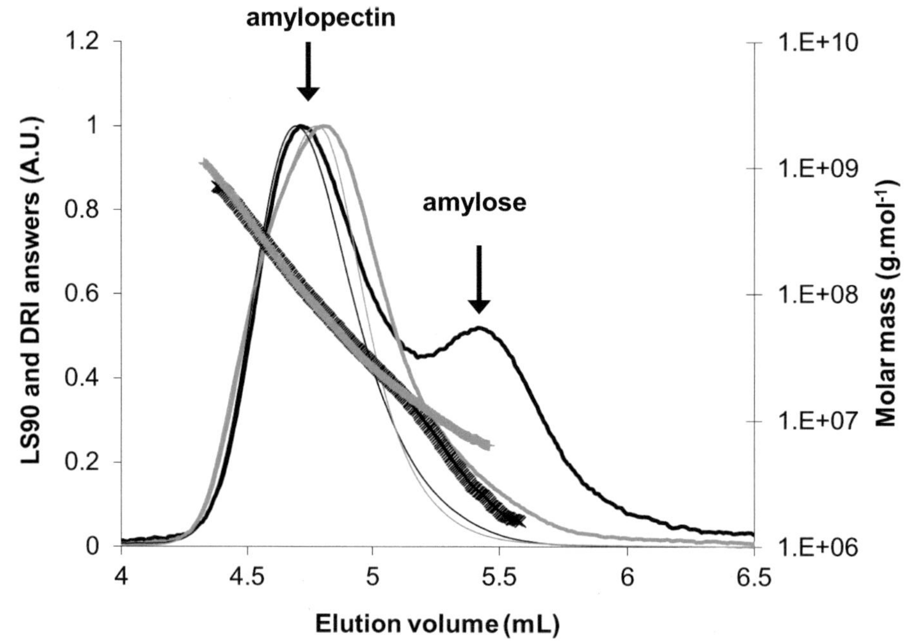

Fig. (1). Chromatograms and molar mass distribution of amylose free potato starch (0% amylose, gray) and wild type potato starch (22% amylose, black), obtained by HPSEC-MALLS-DRI in water with 0.02% NaN_3 with a Shodex KW802.5 silica-based column.

HPSEC-MALLS is performed for the analysis of natural [27, 62, 66, 73] and synthetic [68 - 70] glycogens. By coupling HPSEC–MALLS to QELS [27] or VS [68] detection, it is possible to determine average molar masses and molar mass and size distributions, and to show their dependence according to the glycogen origin [27] and synthesis conditions [68]. Conformation and branching information is also extracted from scaling relationships, ρ parameters, apparent densities and shrinking factors [27]. The HPSEC-MALLS-QELS analysis thus makes it possible to reveal the various conformations of natural glycogen functions as of their origin and along their size distribution [27]. By comparing the scaling laws, ρ parameters and apparent densities of glycogens to dextrans, amyloses and pullulans extracted from HPSEC-MALLS-QELS data, it was possible to distinguish polysaccharides with different branching patterns (Fig. **2**) [27]. However, deviations from the scaling law behavior were observed (Fig. **2**) [27]. The universality of scaling behavior no longer seems to be possible if structural heterogeneity governs the system.

Some studies focus on the HPSEC-MALLS characterization of dextrans [27, 79 - 82], which have a backbone of α-(1,6)-linked D-glucosyl units with randomly distributed side chains connected *via* α-(1,3), α-(1,2) or α-(1,4) linkages and are produced by lactic acid bacteria. In particular, the scaling law exponents, ρ values, apparent densities and shrinking factors for a series of commercial dextrans in aqueous solvents can be extracted [27, 80, 81]. By performing HPSEC-MALLS analyses on dextrans in DMSO and aqueous solvents, higher molar masses were found in aqueous solution, which are attributed to the presence of aggregates in aqueous solution that are less prevalent in DMSO [82]. HPSEC-MALLS is also used to characterize the molar mass, size and conformation of CM-arabinan, dextran, and mixed dextran 270K and CM-arabinan [83], Deae-Dextran [84], chemically-modified exopolysaccharide pullulans [85], or to reveal the degradation of alternan in HPSEC columns with DMSO [86].

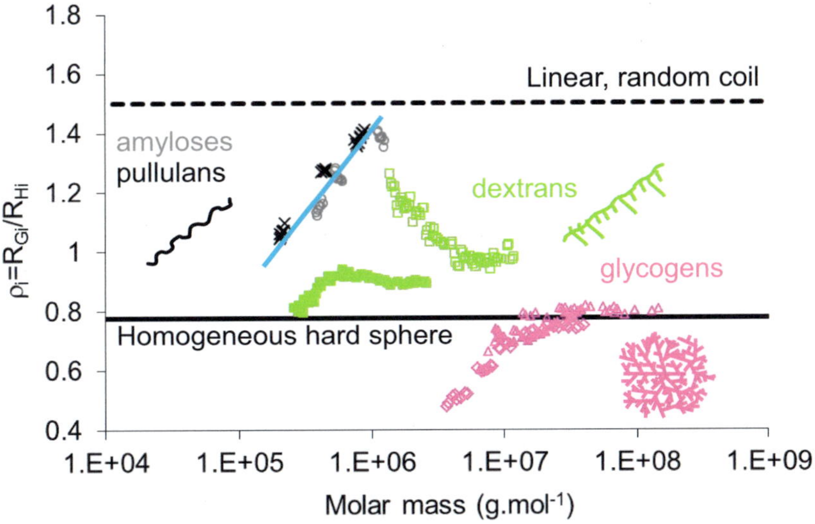

Fig. (2). Plot of ρ_i *versus* molar mass (adapted from [27]).
Linear random coils (pullulans, X; amyloses, ○); comb-like branched dextrans (■, □); highly branched glycogens (rabbit liver, △; oyster, ◇; phytoglycogen, ▲). Theoretical curves for the homogeneous hard sphere (σ=0.778) and for the random coil, linear chain, monodisperse (σ=1.504). The blue line is shown to guide the eyes.

Cellulose, Hemicelluloses and some other β-Glycans

HPSEC is the prevailing technique for molar mass determination of β-glycans. Cellulose is the most abundant natural resource worldwide. It consists of a linear chain of β-(1,4)-linked glucose units that is insoluble in water. It is used for many applications in industry (paper, textiles, *etc.*) in a derivatized form, and its variability in chain length impacts its properties and chemical behavior. In particular, molar mass determination is often used to evaluate paper condition and its degradation kinetics [87, 88]. The standard method of determining the molar mass distribution of celluloses is HPSEC-MALLS, and the most advanced methods have been well established in the literature in recent years [57, 89 - 95]. Two particular methods are used: indirect dissolution of cellulose in DMAc/LiCl [57, 92, 96 - 101] and a derivatization procedure with phenyl isocynate in pyridine to cellulose tricarbanilate (CTC), with further dissolution in tetrahydrofuran (THF) then used as an eluent [91, 95, 102, 103]. The latter method is less utilized. Nevertheless, it has been improved by determining the *dn/dc* and A_2 using an "online" method to avoid errors resulting from cellulose derivatization by-products present in the CTC solution [42]. The A_2 influence on cellulose molar mass calculations and cellulose molecular dimensions have thus been determined, providing better quality parameters for the calculation of the molar mass of cellulose from HPSEC-MALLS measurements. Moreover, a round robin test was performed on HPSEC for a wide range of different pulp samples in order to survey the status quo of the methods available to date [104]. Current state-of-the-art HPSEC-DRI, HPSEC-MALLS-DRI with viscosity detectors, and various dissolution protocols were applied. A variation of 36% of the $\overline{M}_w$ was obtained across all methods. The two major influencing parameters are sample preparation, *i.e.*, derivatization and dissolution methods, and the type of molar mass evaluation in HPSEC, *i.e.*, through calibration or absolute measurement techniques. Both dissolution approaches (CTC derivatization and DMAc/LiCl) give acceptable results, but CTC derivatization is recommended when samples are difficult to dissolve and the available sample amount is very low (< 2 mg of cellulose), with DMAc/LiCl as an eluent, four columns in series and LS techniques to determine absolute molar mass. HPSEC-MALLS is performed in aqueous solvents for water-soluble cellulose derivatives to obtain molar masses and conformational information on methylhydroxyethyl cellulose [105], hydroxypropylmethyl cellulose [106], hydroxypropyl cellulose [107, 108] and carboxymethyl cellulose [109, 110]. Some authors have also revealed the persistence length of cellulose or cellulose derivatives [57, 109] *via* the analysis of HPSEC-MALLS-VS data.

Arabinoxylan is a hemicellulose that consists of a backbone of β-(1,4)-linked D-xylopyranosyl units with α-L-arabinofuranosyl units attached to the O-2 or O-3 positions, or both. HPSEC is commonly applied to monitor processes of enzymatic degradation [111, 112] or to assess extractions with respect to their chain-degrading properties [113] and to thus determine the structure of the extracted fractions [114]. When comparing HPSEC-MALLS analysis in DMSO and in aqueous solvent, it was shown that cereal arabinoxylans have a greater tendency to aggregate in aqueous solutions. Thus, using DMSO as the eluent provides a better, albeit not absolutely accurate, means of molecular size determination of these molecules [115]. Moreover, Pitkanen *et al.* [116] pointed out that analyzing arabinoxylans in aqueous HPSEC-MALLS produces an overestimation of molar mass and size, even if the aggregate concentrations are low.

Cereal β-glucans are made up of alternating blocks of β-(1,3) and β-(1,4)-linked glucose units. They are associated with beneficial health effects probably linked to their solution behavior and viscosity. Round Robin tests were done on β-D-glucan standards isolated from barley by notably using HPSEC-MALLS-VS in order to allow access to well-characterized β-glucan standards for molar mass determinations [117]. Molar mass, molecular size, conformation, shape and chain flexibility (L_p) of oat β-glucan can be obtained by HPSEC-MALLS-VS in aqueous solvent to monitor its hydrolysis by different enzymes [118, 119]. Moreover, HPSEC-MALLS-VS in LiBr–DMSO has been shown to be useful for monitoring the effect of oxidative degradation of barley β-glucan, even if some aggregates may bias the molar mass determination [120].

Since each β-glucan has different aggregation and solubilization behavior, setting the best analysis conditions is very important. For this purpose, comprehensive work was done on the analysis of water-soluble β-(1,3)-D-glucans with β-(1,6--linked branches (SBG) isolated from the cell walls of *Saccharomyces cerevisiae* using HPSEC-MALLS in various solvents and temperatures [121, 122]. This investigation resulted in a simple and reliable protocol for preparing unaggregated, fully dissolved and undegraded SBG in DMAc/LiCl, which is well suited as a standard analysis of the molar mass distribution of SBG-like molecules without chemical derivatization. The electron irradiation effect on scleroglucans can be investigated by means of HPSEC-MALLS [124]. It shows a decrease of the molar mass function of the energy dose, but HPSEC does not make it possible to highlight the presence of low and high molar mass fractions, which are detectable by means of flow field-flow fractionation (F4) coupled with MALLS (see Chapter 5). HPSEC-MALLS-VS also makes it possible to study the molecular properties of linear and nanosized cyclic structures of scleroglucan. Using a series of scleroglucans containing different known fractions of linear and circular molecules, it was possible to extrapolate the molar mass and the radii to a situation of only circular species, thus allowing the determination of the linear mass per unit length (M_L) of the linear and the circular morphologies [123]. This made it possible to determine the pattern of reassociation of the individual chains. Moreover, the conformational properties (scaling law relationships and L_p) of a β-D-glucan from Lentinus, determined from HPSEC-MALLS-VS data [125], made it possible to establish that the predominant species of the glucan in 0.5 M NaCl aqueous solution exist as triple-helical chains with high rigidity, and in DMSO as single-flexible chains. HPSEC-MALLS is also performed to characterize the molar mass, size and conformation of various other β-glycans, including a water-insoluble β-(1,3), a further carboxymethylated D-glucan [126], an exopolysaccharide [127], xyloglucan, konjac glucomannan, larch arabinogalactan, CM-arabinan mixed xyloglucan [83], galactomannan [128, 129] and heparin [130]. However, the polyelectrolyte characteristics and the high molar mass of hyaluronan make its structural characterization difficult to achieve with HPSEC-MALLS because of the lack of relevant HPSEC packing materials with sufficient pore sizes and the possible shear degradation of its chains [131, 132].

Pectins and Gums

Pectins are a complex family of heteropolysaccharides extracted from the cell walls of dicotyledons. They consist of two main structural elements: the homogalacturonan (HG) and type I rhamnogalacturonan (RG-I) regions. The flexibility of pectins plays an important role in their structure–function relationships. Several studies have been done on the analysis of pectin macromolecular structure using HPSEC-MALLS-VS. On the basis of the data, molar mass and size distributions are extracted, as well as scaling law parameters and L_p. Morris *et al.* have thus determined the molecular flexibility [133] but also demonstrated that fractions of sugar beet pectin rich in neutral sugar side chains (RG-I regions) are more compact than those rich in galacturonic acid (HG regions) [45]. Therefore, it should be noted that any measurements on a non-fractionated material will only give the $\overline{M}_w$, *[η]* or conformation and may not necessarily be representative. Using HPSEC-MALLS-VS to separately characterize HG and RG-I fractions, Morris *et al.* [134] also showed that the RG-I fraction has a high $\overline{M}_w$ (188,000 g.mol^{-1}) but a low *[η]*, in line with a random coil conformation (L_p=1.4 nm), and that the HG fraction has a relatively low $\overline{M}_w$ (20,000 g.mol^{-1}) but a rather high *[η]*, in line with a rigid conformation in solution (L_p = 9.8 nm). The molar mass and size distribution of pectins extracted by microwave from orange albedo [135] and flash-extracted from fresh sugar beet pulp [136] have also been determined by HPSEC-MALLS-VS. Chromatography revealed bimodal distributions and that the low-molar-mass fraction increases with increasing extraction time [135, 136]. Scaling law exponents revealed a spherical and compact high molar mass fraction and a coiled low molar mass fraction that is more asymmetric in shape. The efficiency of LiAc/HAc buffer against $NaNO_3$ solution as mobile phases for measuring the molecular properties of commercial pectins (more soluble in $NaNO_3$ solution) and microwave-extracted pectins (more soluble in LiAc/HAc buffer) was compared using HPSEC-MALL--VS [44]. This study led to the conclusion that pectin is more associated in $NaNO_3$ than in LiAc/HAc and that aggregated pectin tends to be more soluble than less aggregated and/or associated pectin. More recently, HPSEC-MALLS-QELS-VS-UV-DRI was combined to study the macromolecular distribution of polysaccharide fractions extracted from sugar beet pulp [137]. By using this set up, it is possible to evaluate not only the molar mass and size distributions but also the amount of proteins linked to pectins in each fraction and, from the analysis of the ρ parameters along the distribution, to identify different conformations, depending on the richness of linked proteins.

Gums are natural surface-active polysaccharides with emulsifying properties that are used as thickeners in different types of applications. Gum arabic (Acacia gum) is a natural arabinogalactan-protein type polysaccharide. Around 70-80% of gum arabic is composed of a branched, complex polysaccharide in which the main chain consists of β-D-galactopyranosyl units substituted in position 6 by side chains (arabinogalactan AG) containing very little protein; the remainder consists of an arabinogalactan-protein complex (AGP), and a minor fraction of a glycoprotein (GP). HPSEC-MALLS-DRI-UV in aqueous solvent is now the standard technique for the characterization of gum arabic to determine the molar mass distribution of the whole gum and the individual AGP, AG and GP fractions [138 - 141]. The structures of the AGP fraction [142] and of the GP fraction [143] of Acacia senegal gum isolated from hydrophobic interaction chromatography were determined using HPSEC-MALLS. AGP in solution has a $\overline{M}_w$ of 1.86×10^6 g mol^{-1} and a $\overline{R}_G$ of 30 nm. The study of the scaling relations from molar mass and radii made it possible to highlight two types of conformations depending on the molar mass: a compact conformation for the low molar mass population with long-chain branching, and an elongated conformation for the high molar mass population with short-chain branching. On the basis of the two exponent values found in the R_G *vs.* molar mass relationship extracted from HPSEC-MALLS data, it was suggested that GP in solution is a mixture of spheroidal monomers and more anisotropic oligomers [143]. The analysis of HPSEC-MALLS data from two varieties of Acacia gum, A. senegal and A. seyal, polysaccharides that presented different branching degrees and branching patterns [144] suggested that macromolecules of both acacia gums adopted an ellipsoid-like conformation in solution even if the molar mass of A. seyal macromolecules was larger and appeared to be more compact and less viscous. For both Acacia gums, the anisotropy of macromolecules increased with the molar mass, although A. senegal macromolecules were the most anisotropic ones. Analyzing various Acacia gums from Nigeria and Sudan by HPSEC-MALLS-DRI-UV revealed the presence of an additional small proportion of very low molar mass proteinaceous material in addition to the AG, AGP and GP fractions [145]. Carboxymethyl hydroxypropyl guar gum has also been characterized in aqueous solution using HPSEC-MALL--DRI and viscometry [146]. The chain conformation and flexibility extracted *via* scaling laws and L_p are indicative of a more rigid chain for functionalized guar gum compared to native guar gum. The fractionation of gum Arabic by HPSEC and F4 [147, 148] are reliable, particularly when agarose-based HPSEC columns are used at low system pressure (which avoids shear degradation and the lost of the high molar mass fractions) [148]. Nevertheless, F4 seems to produce a better resolution between the protein-rich fraction and the protein-poor fraction [147].

Marine Biopolymers

Alginates (alginic acid) are linear copolymers of β-(1,4)-linked D-mannuronate and α-(1,4)-linked L-guluronate, which are difficult to analyze because of their high viscosity and their polyelectrolyte character. However, they are widely studied by HPSEC-MALLS-VS. The solution behavior of polyelectrolytes depends on the ionic strength. Working at a high ionic strength for HPSEC measurements allows the suppression of ionic effects. Vold *et al.* [149] showed that working in 0.33 M Na_2SO_4 represents a practical limit of ionic strength to HPSEC studies of alginates and that 0.1 M is the best ionic strength. They also showed that naturally occurring alginates and epimerized mannuronans have the same conformations (same size-mass and $[\eta]$-mass relationships), and the same flexibility, applying the worm-like chain model (L_p=15 nm). In contrast, oxidation causes depolymerization and a pronounced decrease in chain flexibility (L_p=4 nm at 44% oxidation). The mechanical degradation of alginate by dynamic high and ultra-high pressure homogenization (HPH) was studied compared to other polysaccharides by HPSEC-MALLS-VS in 0.1 M NaCl [150]. This revealed that the decrease of molar mass, radii and $[\eta]$ during HPH, and that the differences of polysaccharide structures and conformation (branching, flexibility) have a stronger impact on HPH treatments than polymer charge or molar mass. The separation of ultra-high viscosity (UHV) alginate components in the columns is not the result of a simple equilibrium between exclusion and permeation of the macromolecules in the stationary phase [151] but, instead, of complex processes such as flow suppression in pores, shear degradation, hydrodynamically-induced diffusion and viscous fingering [152]. Indeed, an inhomogeneity in molar mass, which is the result of a coelution of UHV fractions with LV fractions, was observed [151].

Carrageenans consist of galactose and 3,6-anhydrogalactose, which can be both sulfated and non-sulfated, and alternatively linked by β-(1,4) and α-(1,3). Carrageenans have been studied by HPSEC-MALLS so far [153, 154] to describe the relationships between the mechanical properties of the carrageenan gels and the molar mass together with hydrolysis behavior. More recently, the solution behavior (*e.g.*, varying temperature or ionic strength conditions) of many carrageenans was studied using HPSEC-MALLS [155 - 157]. The average molar masses, molar mass distributions and scaling law coefficients were determined for κ, ι and λ-carrageenans in various eluents containing a 0.1 M concentration of different ionic salts, namely LiCl, NaCl, KCl and NaI [155]. Variations in molar mass function of the salt used and the carrageenan type were revealed, *e.g.*, κ-carrageenan, which is not soluble in KCl, has average molar mass and radius in NaI twice those obtained in NaCl and LiCl, whereas λ-carrageenan yields very similar molar masses and radii of gyration in the four eluents. A comprehensive study for the characterization of carrageenans was conducted by Berth *et al.* [156]. They found that suitable conditions for the macromolecular characterization of carrageenan by HPSEC-MALLS are at 60 °C and added salt (*e.g.*, $NaNO_3$) equivalent to an ionic strength of 0.1 M, which is enough to suppress intramolecular charge effects on the conformation due the polyelectrolyte character of carrageenan, as well as electrostatic interactions on HPSEC. Moreover, the molar mass of carrageenan considerably increases when temperature decreases, thus confirming LS batch measurements in 0.2 M LiCl or LiI, suggesting the formation of double-stranded chains in the ordered state at lower temperatures [158, 159]. HPSEC-MALLS-VS coupling in 0.1 M $LiNO_3$ also makes it possible to characterize the size and molar mass distributions of sulfated polysaccharides and to highlight that chemical composition, molar mass and structure vary depending on the origin, the harvest period and the extraction methods [160].

Chitosan is a soluble polycationic derivative of insoluble chitin based on β-(1,4--linked poly (N-glucosamine). It has been widely considered for use in the food, cosmetics and pharmaceutical industries. HPSEC-MALLS is commonly used to investigate the structure and conformation of chitosan. Morris *et al.* [161] measured the $\overline{M}_w$ and $[\eta]$ of chitosans of different molar masses stored at 4, 25 or 40 °C using HPSEC-MALLS and extracted their solution conformation from the MHKS power law exponent (a=0.95), which is consistent with a stiff coil conformation for all of the chitosans studied. Branched chitosans have also been analyzed by HPSEC-MALLS-VS [162]. The extraction of macromolecular characteristics revealed an increase of the $\overline{M}_w$ and the dispersity with the degree of branching of 1.2-2.1. Moreover, the compactness of the structure in solution was higher than linear chitosan of the same molar mass (decrease of $\overline{R}_G$ and $[\eta]$), with a corresponding change in the calculated viscosity (g') and the geometric (g_M) shrinking factors. However, HPSEC separation of acetylated chitosans does not make it possible to characterize the larger fractions corresponding to a part of the aggregated polymer [163]. This was demonstrated by the discrepancy of the average molar mass obtained by HPSEC-MALLS and batch MALLS measurements for a series of acetylated chitosans in 0.2 M acetic acid/0.15 M ammonium acetate buffer (pH 4.5) [163]. Moreover, the increasing tendency to self-association with the degree of acetylation (DA) induced a lower efficiency of the chromatographic separation (increase of the retention on the column). On the basis of the scaling equations and the L_p, determined from HPSEC-MALLS data, it was possible to show a more or less flexible worm-like chain, although no differences according to DA were revealed. More recently, Alumairi *et al.* [164] showed that the polydisperse distributions, the molar mass and the semi-flexible conformation suggested by the analysis of the scaling laws obtained by HPSEC-MALLS-VS in 0.2 M acetate buffer (pH 4.3) are in total agreement with those obtained from analytical ultracentrifugation for chitosans of different molar masses and DA.

Lignin

Lignin is, after cellulose, the second most abundant biopolymer worldwide, and the main source of aromatic structures on earth. It is a biopolymer dominated by cross-linked phenylpropanoid subunits, primarily *p*-coumaryl, coniferyl and sinapyl alcohols. Its complex structure considerably varies depending on the plant species and the isolation process. Many studies are devoted to the determination of molar mass distributions of lignins by HPSEC, including interesting round robin tests [165, 166], yet the large majority of these studies are based on standard calibration, using polystyrene standards, which leads to the under-estimation of the molar mass of the branched lignin macromolecules. The use of HPSEC-MALLS [165, 167 - 171] clearly improves these determinations. It should be noted that the light scattering detector must be equipped with narrow band-pass filters to eliminate the fluorescence of lignins that could bias the results. Prior to analysis, lignins are solubilized in alkali or in organic solvents (DMSO/H2O/LiBr, DMAc/LiCl, THF). In addition, in order to improve the solubility of lignins in organic HPSEC eluents such as THF, derivatization is carried out through methylation, acetylation or silylation. However, some of them, like the widely used acetobromination, may possibly cleave some bonds within the lignin macromolecule, inducing further bias in the molar mass determination. A method using HPSEC-MALLS-QELS-DRI-UV was developed to characterize and track changes in molar mass, size and molecular conformation of lignin molecules [170, 171]. After studying the effects of pH, flow conditions and mobile phase compositions (DMSO, water, NaOH, and LiBr), the authors concluded that working in 0.1 M NaOH gave the best separation and the most reliable molar mass distributions without aggregates [171]. In addition, thanks to MALLS, they also revealed that the lignin aggregated at pH values below 8.5

[171] and that the aggregate formation depends to a great extent on the lyophilization of the lignin [170]. The aggregation of spruce milled wood lignin and guaiacyl lignin polymer models has also been demonstrated by analyzing them with HPSEC-MALLS in *N,N*-dimethylformamide (DMF), and thus revealing two eluting populations, one of which is in the range of 10^5-10^8 g.mol^{-1} [168]. By investigating the molar mass and radius distributions of this bimodal elution profile, intense associations and high density of the insoluble fractions were revealed. Moreover the soluble fractions exhibited a higher $\overline{R}_G$, which was the result of retarded elution due to interaction with the column material. The authors then characterized these fractions by HPSEC-MALLS in THF after acetylation to determine molar masses and conformations in molecular dispersed solutions. A method to analyze lignosulfonate fractions by HPSEC-MALLS in aqueous phosphate buffer containing DMSO and SDS was developed [167]. The molar mass distributions and conformational properties of lignosulfonates of various origins thus obtained [169] made it possible to reveal that molar mass depends on the origin of the lignosufonates. Based on the HPSEC-MALLS results, an HPSEC-UV method where lignosulfonate fractions were used as broad molar mass calibration standards has also been developed.

APPLICATIONS FOR PROTEINS AND PEPTIDES

The molar mass distribution of peptides and proteins is vital information for understanding their physical/chemical properties. HPSEC has been used to date not only for molar mass determination but also for studying the interaction between polypeptides and proteins, aggregation phenomena and as the standard separation technique for the quantification of protein dimers, trimers and oligomers. The main advantage of HPSEC is the mild elution conditions that allow the characterization of the protein, with low impact on the conformational and the local environment.

Many problems in protein and peptide fractionation by HPSEC are linked to their electrolyte character, which leads to electrostatic and hydrophobic interactions with the stationary phase [13, 16, 17, 172 - 178]. This can result in adsorption of the protein [179], shifts in retention time [180], peak tailing or peak asymmetry [181], or to changes in the three-dimensional conformation of the protein [172, 178]. Yet, the interactions between the proteins and the stationary phase can be minimized by proper adjustment of the mobile phase composition. A common approach is to increase the ionic strength using a buffered solution with an ionic strength of 50-200 mM. The pH of the mobile phase also has an effect on the protein/stationary phase interactions, especially when working at low ionic strengths [173]. It should be close to the isoelectric point of the protein in order to minimize the interactions. The most commonly used buffer is phosphate

(10-200 mM) and pH 6-7.2. The peak shape and the recovery of proteins and aggregates can also be increased by the addition of salts (NaCl, KCl or Na_2SO_4 at 10-250 mM) [175, 178] and the use of some additives in the mobile phase such as arginine [16, 182, 183]. Some organic modifiers (*e.g.*, methanol and ethanol) can also be added (5-10%) to reduce the strength of hydrophobic interactions [1].

By far, the most commonly applied technique for the detection of polypeptides and proteins is UV, although HPSEC-MALLS is increasingly applied to determine the molar mass, size distribution and conformation of polypeptides [184], proteins and their aggregates [41, 185 - 202] and complexes with biomolecules [203]. The precision and accuracy of the determination of the molar mass of proteins by HPSEC-MALLS-UV and HPSEC-MALLS-DRI were compared [190 - 199]. These two systems give similar results even if HPSEC-MALLS-UV has been shown to have slightly greater coefficients of variation. For peptides, Oliva *et al.* [193] showed that HPSEC-MALLS produce a higher bias in the molar mass determination than MALDI-TOF and that the *dn/dc* provided the most significant contribution to the combined uncertainty. In any case, since the intensity of MALLS is proportional to the molar mass, its sensitivity is very low for small peptides [176, 193].

HPSEC-MALLS-DRI-UV is valuable for determining the molar mass and protein fraction distributions of polymer bioconjugates and for characterizing the reaction efficiency, degree of substitution and valency of the products of conjugation of either peptides or proteins to a biopolymer scaffold [203]. The amount of material needed for this method depends upon instrument sensitivity, molar mass and reagent physiochemical parameters, but is significantly less than standard methods. Accurate measurement of protein fractions requires that the bioconjugate components have different *dn/dc* and UV absorbance coefficients.

HPSEC-MALLS provides information not only on proteins and their dimers but also on high-order aggregates [16, 17, 192, 194 - 196]. For human IgG aggregates containing trace amounts of higher order aggregates, HPSEC-MALLS is significantly more sensitive to the high molar mass aggregates than HPSEC-UV and enables their confirmation and analysis [191]. HPSEC-MALLS is also valuable for the investigation of aggregation and degradation of glycosylated and nonglycosylated proteins, and antibodies under various conditions (addition of detergent, pH, protein concentration and heat stress). The determination of molar mass is absolute, which avoids misleading results caused by molecular shape or interactions with the column matrix [196]. Several glutenin polymers purified from wheat flours of different cultivars were accurately characterized by HPSEC-MALLS [41] using phase with a very high exclusion limit [188]. The size and molar mass distributions of glutenin polymers was found to be a complex mixture of high- and low-molar mass fractions, and the relative percentage can be used to differentiate the different wheat genotypes [41, 188] and is highly correlated with the percentage of high-molecular subunits in glutenins [188]. The dependence of size upon mass also revealed significant differences in the molecular conformation function of the genotype [188]. However, these determinations are difficult and the results depend on the buffer used and the SEC columns [41]. HPSEC-MALLS has also been used to monitor the changes in the aggregate size and molar mass distribution during the hydrolysis phenomenon of bovine casein [197, 198], which was resolved by HPSEC into two major peaks corresponding to aggregates and monomers.

HPSEC-MALLS-UV-DRI-VS is also used for extensive characterization of protein samples [200] by providing access to the molecular density, size of the molecules and structural changes, in addition to molar mass.

CONCLUDING REMARKS

The HPSEC-MALLS technique is widely used for biopolymer characterization because it is a very powerful technique to determine molar mass and size distributions together with the conformation, shape and flexibility of the polymers, in just one step. It is a routine and well-established analytical separation method but it has many limitations and pitfalls [204]. It has a large surface area, which facilitates the interaction of the sample with the column packing, thus inducing sample lost, anchoring of branched polymers or abnormal elution when the elution must follow a pure SEC mechanism for the accurate determination of the conformation parameters. In addition, the column packing may cause shear degradation, especially in the case of branched polymers, and when aggregates are present in the sample, the macromolecular characteristics obtained are biased.

CONFLICT OF INTEREST

The author (editor) declares no conflict of interest, financial or otherwise.

ACKNOWLEDGEMENTS

The author thanks Sophie Guilois for her help in formatting the manuscript.

REFERENCES

[1] A.M. Striegel, W.W. Yau, J.J. Kirkland, and D.D. Bly, *Modern Size-Exclusion Liquid Chromatography. Practice of Gel Permeation and Gel Filtration Chromatography.* 2nd ed Wiley: Hoboken, 2009.

[2] J. Cazes, "XXIX. Gel Permeation Chromatography", *J. Chem. Educ.*, vol. 43, pp. A657-A625, 1966.

[3] S.T. Balke, A.E. Hamielec, B.P. Leclair, and S.L. Pearce, "Gel Permeation Chromatography", *Ind. Eng. Chem. Prod. Res. Dev.*, vol. 8, pp. 54-57, 1969.

[http://dx.doi.org/10.1021/i360029a008]

[4] W.W. Yau, J.J. Kirkland, and D.D. Bly, *Modern size exclusion liquid chromatography.* ASA: Wiley-Interscience: New York, 1979.

[5] H. Benoit, and C. Picot, "Etude par diffusion de la lumière des solutions macromoléculaires moyennement concentrées", *Pure Appl. Chem.* , vol. 12, pp. 545-561, 1966. [http://dx.doi.org/10.1351/pac196612010545]

[6] Z.R. Grubisic, "P.; Benoit, H., "A universal calibration for gel permeation chromatography", *Polym. Letters.* , vol. 5, pp. 753-759, 1967. [http://dx.doi.org/10.1002/pol.1967.110050903]

[7] L.K. Kostanski, D.M. Keller, and A.E. Hamielec, "Size-exclusion chromatography-a review of calibration methodologies", *J. Biochem. Biophys. Methods.* , vol. 58, no. 2, pp. 159-186, 2004. [http://dx.doi.org/10.1016/j.jbbm.2003.10.001] [PMID: 14980789]

[8] H.G. Barth, and F.J. Carlin, "A review of polymer shear degradation in size-exclusion chromatography", *J. Liq. Chromatogr.* , vol. 7, pp. 1717-1738, 1984. [http://dx.doi.org/10.1080/01483918408068832]

[9] R.A. Cave, S.A. Seabrook, M.J. Gidley, and R.G. Gilbert, "Characterization of starch by size-exclusion chromatography: the limitations imposed by shear scission", *Biomacromolecules.* , vol. 10, no. 8, pp. 2245-2253, 2009. [http://dx.doi.org/10.1021/bm900426n] [PMID: 19627139]

[10] J.C. Giddings, "Analysis of fundamental obstacles to the size exclusion chromatography of polymers of ultra high molecular weight", In: *Adv. Chromatogr.* vol. 20. 1982, pp. 217-258.

[11] G. Stegeman, J.C. Kraak, and H. Poppe, "Hydrodynamic and size-exclusion chromatography of polymers on porous particles", *J. Chromatogr. A.* , vol. 550, pp. 721-739, 1991. [http://dx.doi.org/10.1016/S0021-9673(01)88577-0]

[12] A. Rolland-Sabaté, S. Guilois, B. Jaillais, and P. Colonna, "Molecular size and mass distributions of native starches using complementary separation methods: asymmetrical flow field flow fractionation (A4F) and hydrodynamic and size exclusion chromatography (HDC-SEC)", *Anal. Bioanal. Chem.* , vol. 399, no. 4, pp. 1493-1505, 2011. [http://dx.doi.org/10.1007/s00216-010-4208-4] [PMID: 20878394]

[13] C.T. Mant, J.M. Parker, and R.S. Hodges, "Size-exclusion high-performance liquid chromatography of peptides. Requirement for peptide standards to monitor column performance and non-ideal behaviour", *J. Chromatogr. A.* , vol. 397, pp. 99-112, 1987. [http://dx.doi.org/10.1016/S0021-9673(01)84993-1] [PMID: 3654835]

[14] E. Pfannkoch, K.C. Lu, F.E. Regnier, and H.G. Barth, "Characterization of Some Commercial High-Performance Size- Exclusion Chromatography Columns for Water-Soluble Polymers", *J. Chromatogr. Sci.* , vol. 18, pp. 430-441, 1980. [http://dx.doi.org/10.1093/chromsci/18.9.430]

[15] Y. Kato, S. Kido, and T. Hashimoto, "High-resolution gel-permeation chromatography", *J. Polym. Sci., Polym. Phys. Ed.* , vol. 11, pp. 2329-2337, 1973.

[16] P. Hong, S. Koza, and E.S. Bouvier, "Size-Exclusion Chromatography for the Analysis of Protein Biotherapeutics and their Aggregates", *J. Liquid Chromatogr. Relat. Technol.* , vol. 35, no. 20, pp. 2923-2950, 2012. [PMID: 23378719]

[17] S. Fekete, A. Beck, J.-L. Veuthey, and D. Guillarme, "Theory and practice of size exclusion chromatography for the analysis of protein aggregates", In: *JPBA Reviews 2014.* vol. 101. 2014, pp. 161-173. [http://dx.doi.org/10.1016/j.jpba.2014.04.011]

[18] M. Gaborieau, and P. Castignolles, "Size-exclusion chromatography (SEC) of branched polymers and

polysaccharides", *Anal. Bioanal. Chem.* , vol. 399, no. 4, pp. 1413-1423, 2011. [http://dx.doi.org/10.1007/s00216-010-4221-7] [PMID: 20967430]

[19] C. Wu, *Handbook of Size Exclusion Chromatography.* vol. 69. Chromatographic science series: New York, USA, 1995.

[20] H.G. Barth, and G.D. Saunders, "The state of the art and future trends of size-exclusion chromatography packing and columns", *LC GC N. Am.* , vol. 30, pp. 544-563, 2012.

[21] L. Pitkanen, A.A. Urbas, and A.M. Striegel, "On the feasibility of determining polymer chemical heterogeneity by SEC with continuous off-line Raman detection", *Polym. Chem.* , vol. 6, pp. 4864-4874, 2015.
[http://dx.doi.org/10.1039/C5PY00189G]

[22] B.H. Zimm, "Apparatus and methods for measurement and interpretation of the angular variation of light scattering; Preliminary results on polystyrene solutions", *J. Chem. Phys.* , vol. 16, pp. 1099-1116, 1948.
[http://dx.doi.org/10.1063/1.1746740]

[23] G.C. Berry, "Thermodynamic and conformational properties of polystyrene. I. Light-Scattering studies on dilute solutions of finear polystyrene", *J. Chem. Phys.* , vol. 44, pp. 4550-4564, 1966.
[http://dx.doi.org/10.1063/1.1726673]

[24] P. Debye, "Molecular-weight determination by light scattering", *20th National Colloid Symposium.* , 1946 Ithaca, New York

[25] D.E. Koppel, "Analysis of macromolecular polydispersity in intensity correlation spectroscopy: the method of cumulants", *J. Chem. Phys.* , vol. 57, pp. 4814-4820, 1972.
[http://dx.doi.org/10.1063/1.1678153]

[26] S.W. Provencher, "Low-bias macroscopic analysis of polydispersity", *Biochem. Soc. Trans.* , vol. 19, no. 2, pp. 479-480, 1991.
[http://dx.doi.org/10.1042/bst0190479] [PMID: 1889638]

[27] A. Rolland-Sabaté, M.G. Mendez-Montealvo, P. Colonna, and V. Planchot, "Online determination of structural properties and observation of deviations from power law behavior", *Biomacromolecules.* , vol. 9, no. 7, pp. 1719-1730, 2008.
[http://dx.doi.org/10.1021/bm7013119] [PMID: 18547102]

[28] J.V. Castro, K.Y. van Berkel, G.T. Russell, and R.G. Gilbert, "General solution to the band-broadening problem in polymer molecular weight distributions", *Aust. J. Chem.* , vol. 58, pp. 178-181, 2005.
[http://dx.doi.org/10.1071/CH05002]

[29] D. Konkolewicz, J.W. Taylor, P. Castignolles, A. Gray-Weale, and R.G. Gilbert, "Toward a more general solution to the band-broadening problem in size separation of polymers", *Macromolecules.* , vol. 40, pp. 3477-3487, 2007.
[http://dx.doi.org/10.1021/ma062973a]

[30] G. Meira, M. Netopilik, M. Potschka, I. Schnoll-Bitai, and J. Vega, "Band broadening function in size exclusion chromatography of polymers: Review of some recent developments", *Macromol. Symp.* , vol. 258, pp. 186-197, 2007.
[http://dx.doi.org/10.1002/masy.200751221]

[31] A.M. Striegel, "Longitudinal diffusion in size-exclusion chromatography: a stop-flow size-exclusion chromatography study", *J. Chromatogr. A.* , vol. 932, no. 1-2, pp. 21-31, 2001.
[http://dx.doi.org/10.1016/S0021-9673(01)01214-6] [PMID: 11695865]

[32] M. Gaborieau, R.G. Gilbert, A. Gray-Weale, J.M. Hernandez, and P. Castignolles, "Theory of multiple-detection size-exclusion chromatography of complex branched polymers", *Macromol. Theor. Simul.* , vol. 16, pp. 13-28, 2007.
[http://dx.doi.org/10.1002/mats.200600046]

[33] A.E. Hamielec, and A.C. Ouano, "Generalized universal molecular weight parameter in GPC", *J. Liq. Chromatogr.* , vol. 1, pp. 111-120, 1978.
[http://dx.doi.org/10.1080/01483917808068382]

[34] M. Gaborieau, J. Nicolas, M. Save, B. Charleux, J.P. Vairon, R.G. Gilbert, and P. Castignolles, "Separation of complex branched polymers by size-exclusion chromatography probed with multiple detection", *J. Chromatogr. A.* , vol. 1190, no. 1-2, pp. 215-223, 2008.
[http://dx.doi.org/10.1016/j.chroma.2008.03.031] [PMID: 18378255]

[35] M. Netopilik, J. Podesva, J. Lokaj, and P. Kratochvil, "Number-average molecular weight of branched polymers from SEC with viscosity detection and universal calibration", *Polym. Int.* , vol. 57, pp. 1152-1158, 2008.
[http://dx.doi.org/10.1002/pi.2458]

[36] P.J. Flory, *Principles in Polymer Chemistry.* Cornell University Press: Ithaca, New York, 1953.

[37] P.G. De Gennes, *Scaling Concepts in Polymer Physics.* Cornwell University Press: Ithaca, New York, 1979.

[38] J.E. Rollings, "Use of on-line laser light scattering coupled to chromatographic separations for the determination of molecular weight, branching, size and shape distributions of polysaccharides", In: *Laser light scattering in biochemistry,* S.E. Harding, D.B. Satelle, V.A. Bloomfield, Eds., 1992, pp. 275-293.

[39] P. Roger, M.A. Axelos, and P. Colonna, "SEC-MALLS and SANS studies applied to solution behavior of linear alpha-glucans", *Macromolecules.* , vol. 33, pp. 2446-2455, 2000.
[http://dx.doi.org/10.1021/ma991302s]

[40] A. Rolland-Sabaté, N.G. Amani, D. Dufour, S. Guilois, and P. Colonna, "Macromolecular characteristics of ten yam (Dioscorea spp) starches", *J. Sci. Food Agric.* , vol. 83, pp. 927-936, 2003.
[http://dx.doi.org/10.1002/jsfa.1410]

[41] R. Mendichi, S. Fisichella, and A. Savarino, "Molecular weight, size distribution and conformation of Glutenin from different wheat cultivars by SEC-MALLS", *J. Cereal Sci.* , vol. 48, pp. 486-493, 2008.
[http://dx.doi.org/10.1016/j.jcs.2007.11.006]

[42] D. Pawcenis, J. L. Thomas, T. Lojewski, J. M. Milczarek, and J. Lojewska, "Towards determination of absolute molar mass of cellulose polymer by size exclusion chromatography with mulitple angle laser light scattering detection", *J. Chromatogr. A.* , vol. 1409, pp. 53-9, Aug 28, 2015.
[http://dx.doi.org/10.1016/j.chroma.2015.06.042]

[43] H. Anger, and G. Berth, "Gel-permeation chromatography of sunflower pectin", *Carbohydr. Polym.* , vol. 5, pp. 241-250, 1985.
[http://dx.doi.org/10.1016/0144-8617(85)90033-5]

[44] M.L. Fishman, H.K. Chau, F. Kolpak, and J. Brady, "Solvent effects on the molecular properties of pectins", *J. Agric. Food Chem.* , vol. 49, no. 9, pp. 4494-4501, 2001.
[http://dx.doi.org/10.1021/jf001317l] [PMID: 11559160]

[45] G.A. Morris, and M.C. Ralet, "The effect of neutral sugar distribution on the dilute solution conformation of sugar beet pectin", *Carbohydr. Polym.* , vol. 88, pp. 1488-1491, 2012.
[http://dx.doi.org/10.1016/j.carbpol.2012.02.020]

[46] R. Hanselmann, W. Burchard, M. Ehrat, and H.M. Widmer, "Structural properties of fractionated starch polymers and their dependence on the dissolution process", *Macromolecules.* , vol. 29, pp. 3277-3282, 1996.
[http://dx.doi.org/10.1021/ma951452c]

[47] W. Burchard, "Solution properties of branched macromolecules", In: *Adv. Polym. Sci.* vol. 143. SPRINGER-VERLAG BERLIN: Berlin, 1999, pp. 113-194.

[48] W. Burchard, "Particle scattering factors of some branched polymers", *Macromolecules.* , vol. 10, pp.

919-927, 1977.
[http://dx.doi.org/10.1021/ma60059a008]

[49] W. Burchard, "Static and dynamic Light scattering from branched polymes and biopolymers", *Adv. Polym. Sci.* , vol. 48, pp. 1-124, 1983.
[http://dx.doi.org/10.1007/3-540-12030-0_1]

[50] A. Rolland-Sabaté, T. Sanchez, A. Buleon, P. Colonna, B. Jaillais, H. Ceballos, and D. Dufour, "Structural characterization of novel cassava starches with low and high-amylose contents in comparison with other commercial sources", *Food Hydrocoll.* , vol. 27, pp. 161-174, 2012.
[http://dx.doi.org/10.1016/j.foodhyd.2011.07.008]

[51] W. Burchard, and H.J. Vogel, "Particle scattering form factors of chemical and conformational triblock copolymers", *Comput. Theor. Polym. Sci.* , vol. 10, pp. 133-148, 2000.
[http://dx.doi.org/10.1016/S1089-3156(99)00066-5]

[52] O. Kratky, and G. Porod, "Diffuse small-angle scattering of X-rays in colloid systems", *J. Colloid Sci.* , vol. 4, no. 1, pp. 35-70, 1949.
[http://dx.doi.org/10.1016/0095-8522(49)90032-X] [PMID: 18110601]

[53] T. Aberle, W. Burchard, W. Vorwerg, and S. Radosta, "Conformational Contributions of Amylose and Amylopectin to the Structural-Properties of Starches from Various Sources", *Starke.* , vol. 46, pp. 329-335, 1994.
[http://dx.doi.org/10.1002/star.19940460903]

[54] L.A. Bello-Perez, P. Colonna, P. Roger, and O. Paredes-Lopez, "Structural properties of starches dissolved by microwave heating", *Starke.* , vol. 50, pp. 137-141, 1998.
[http://dx.doi.org/10.1002/(SICI)1521-379X(199804)50:4 < 137::AID-STAR137 > 3.0.CO;2-J]

[55] S.E. Harding, K. Day, R. Dhami, and P.M. Lowe, "Further observations on the size, shape and hydration of kappa-Carrageenan in dilute solution", *Carbohydr. Polym.* , vol. 32, pp. 81-87, 1997.
[http://dx.doi.org/10.1016/S0144-8617(96)00167-1]

[56] B.H. Zimm, and W.H. Stockmayer, "The dimensions of chain molecules containing branches and rings", *J. Chem. Phys.* , vol. 17, pp. 1301-1314, 1949.
[http://dx.doi.org/10.1063/1.1747157]

[57] M. Kes, and B.E. Christensen, "A re-investigation of the Mark-Houwink-Sakurada parameters for cellulose in Cuen: a study based on size-exclusion chromatography combined with multi-angle light scattering and viscometry", *J. Chromatogr. A.* , vol. 1281, pp. 32-37, 2013.
[http://dx.doi.org/10.1016/j.chroma.2013.01.038] [PMID: 23398992]

[58] P. Roger, and P. Colonna, "The Influence of Chain-Length on the Hydrodynamic Behavior of Amylose", *Carbohydr. Res.* , vol. 227, pp. 73-83, 1992.
[http://dx.doi.org/10.1016/0008-6215(92)85061-4]

[59] G.A. Morris, G.G. Adams, and S.E. Harding, *On hydrodynamic methods for the analysis of the sizes and shapes of polysaccharides in dilute solution: a short review.* vol. 42. Food Hydrocolloid, 2014, pp. 318-334.

[60] M.S. Kök, A.S. Abdelhameed, S. Ang, G.A. Morris, and S.E. Harding, "A novel global hydrodynamic analysis of the molecular flexibility of the dietary fibre polysaccharide konjac glucomannan", *Food Hydrocoll.* , vol. 23, pp. 1910-1917, 2009.
[http://dx.doi.org/10.1016/j.foodhyd.2009.02.002]

[61] L.A. Bello-Perez, P. Roger, B. Baud, and P. Colonna, "Macromolecular features of starches determined by aqueous high-performance size exclusion chromatography", *J. Cereal Sci.* , vol. 27, pp. 267-278, 1998.
[http://dx.doi.org/10.1006/jcrs.1998.0186]

[62] C.E. Ioan, T. Aberle, and W. Burchard, "Solution properties of glycogen. 1. Dilute solutions", *Macromolecules.* , vol. 32, pp. 7444-7453, 1999.

[http://dx.doi.org/10.1021/ma990600m]

[63] S.G. You, and M.S. Izydorczyk, "Molecular characteristics of barley starches with variable amylose content", *Carbohydr. Polym.* , vol. 49, pp. 33-42, 2002. [http://dx.doi.org/10.1016/S0144-8617(01)00300-9]

[64] S. Radosta, W. Vorwerg, A. Ebert, A.H. Begli, D. Grulc, and M. Wastyn, "Properties of low-substituted cationic starch derivatives prepared by different derivatisation processes", *Starke.* , vol. 56, pp. 277-287, 2004. [http://dx.doi.org/10.1002/star.200300210]

[65] F.A. Tetchi, A. Rolland-Sabate, G.N. Amani, and P. Colonna, "Molecular and physicochemical where science meets business characterisation of starches from yam, cocoyam, cassava, sweet potato and ginger produced in the Ivory Coast", *J. Sci. Food Agric.* , vol. 87, pp. 1906-1916, 2007. [http://dx.doi.org/10.1002/jsfa.2928]

[66] G.A. Morris, S. Ang, S.E. Hill, S. Lewis, B. Schafer, U. Nobbmann, and S.E. Harding, "Molar mass and solution conformation of branched alpha(1 - > 4), alpha(1 - > 6) Glucans. Part I: Glycogens in water", *Carbohydr. Polym.* , vol. 71, pp. 101-108, 2008. [http://dx.doi.org/10.1016/j.carbpol.2007.05.029]

[67] S.H. Yoo, C. Perera, J. Shen, L. Ye, D.S. Suh, J.L. Jane, C. Perera, J. Shen, L. Ye, D-S. Suh, and J-L. Jane, "Molecular structure of selected tuber and root starches and effect of amylopectin structure on their physical properties", *J. Agric. Food Chem.* , vol. 57, no. 4, pp. 1556-1564, 2009. [http://dx.doi.org/10.1021/jf802960f] [PMID: 19199605]

[68] J. Ciric, J. Oostland, J.W. de Vries, A.J. Woortman, and K. Loos, "Size exclusion chromatography with multi detection in combination with matrix-assisted laser desorption ionization-time-of-flight mass spectrometry as a tool for unraveling the mechanism of the enzymatic polymerization of polysaccharides", *Anal. Chem.* , vol. 84, no. 23, pp. 10463-10470, 2012. [http://dx.doi.org/10.1021/ac302704q] [PMID: 23121513]

[69] M. Ulbrich, S. Radosta, B. Kiessler, and W. Vorwerg, "Interaction of cationic starch derivatives and cellulose fibres in the wet end and its correlation to paper strength with a statistical evaluation", *Starke.* , vol. 64, pp. 972-983, 2012. [http://dx.doi.org/10.1002/star.201200033]

[70] F. Grimaud, C. Lancelon-Pin, A. Rolland-Sabaté, X. Roussel, S. Laguerre, A. Viksø-Nielsen, J.L. Putaux, S. Guilois, A. Buléon, C. DHulst, and G. Potocki-Véronèse, "In vitro synthesis of hyperbranched α-glucans using a biomimetic enzymatic toolbox", *Biomacromolecules.* , vol. 14, no. 2, pp. 438-447, 2013. [http://dx.doi.org/10.1021/bm301676c] [PMID: 23311582]

[71] P. Roger, and P. Colonna, "Molecular weight distribution of amylose fractions obtained by aqueous leaching of corn starch", *Int. J. Biol. Macromol.* , vol. 19, no. 1, pp. 51-61, 1996. [http://dx.doi.org/10.1016/0141-8130(96)01101-4] [PMID: 8782720]

[72] P. Roger, V. Tran, J. Lesec, and P. Colonna, "Isolation and characterisation of single chain amylose", *J. Cereal Sci.* , vol. 24, pp. 247-262, 1996. [http://dx.doi.org/10.1006/jcrs.1996.0057]

[73] A. Rolland-Sabate, P. Colonna, G. Potocki-Veronese, P. Monsan, and V. Planchot, "Elongation and insolubilisation of [alpha]-glucans by the action of Neisseria polysaccharea amylosucrase", *J. Cereal Sci.* , vol. 40, pp. 17-30, 2004. [http://dx.doi.org/10.1016/j.jcs.2004.04.001]

[74] W. Yokoyama, J.J. Renner-Nantz, and C.F. Shoemaker, "Starch molecular mass and size by size-exclusion chromatography in DMSO-LiBr coupled with multiple angle laser light scattering", *Cereal Chem.* , vol. 75, pp. 530-535, 1998. [http://dx.doi.org/10.1094/CCHEM.1998.75.4.530]

[75] S. Radosta, M. Haberer, and W. Vorwerg, "Molecular characteristics of amylose and starch in

dimethyl sulfoxide", In: *Biomacromolecules.* vol. 2. 2001, pp. 970-978.

[76] A.M. Striegel, and J.D. Timpa, "Molecular Characterization of Polysaccharides Dissolved in Me(2)Nac-Licl by Gel-Permeation Chromatography", *Carbohydr. Res.* , vol. 267, pp. 271-290, 1995. [http://dx.doi.org/10.1016/0008-6215(94)00307-2]

[77] B.P. Wasserman, and J.D. Timpa, "Rapid quantitative measurement of extrusion-induced starch fragmentation by automated gel permeation chromatography", *Starke.* , vol. 43, pp. 389-392, 1991. [http://dx.doi.org/10.1002/star.19910431005]

[78] M.J. Gidley, I. Hanashiro, N.M. Hani, S.E. Hill, A. Huber, J-L. Jane, Q. Liu, G.A. Morris, A. Rolland-Sabaté, A.M. Striegel, and R.G. Gilbert, "Reliable measurements of the size distributions of starch molecules in solution: Current dilemmas and recommendations", *Carbohydr. Polym.* , vol. 79, pp. 255-261, 2010.
[http://dx.doi.org/10.1016/j.carbpol.2009.07.056]

[79] W.S. Bahary, M.P. Hogan, M. Jilani, and M.P. Aronson, "Size-Exclusion Chromatography of Dextrans and Pullulans with Light-Scattering, Viscosity, and Refractive-Index Detection", In: *Chromatographic Characterization of Polymers.* vol. 247. AMER CHEMICAL SOC: Washington, 1995, pp. 151-166.
[http://dx.doi.org/10.1021/ba-1995-0247.ch012]

[80] C.E. Ioan, T. Aberle, and W. Burchard, "Structure properties of dextran. 2. Dilute solution", *Macromolecules.* , vol. 33, pp. 5730-5739, 2000.
[http://dx.doi.org/10.1021/ma000282n]

[81] C.E. Ioan, T. Aberle, and W. Burchard, "Structure properties of dextran. 3. Shrinking factors of individual clusters", *Macromolecules.* , vol. 34, pp. 3765-3771, 2001.
[http://dx.doi.org/10.1021/ma000479f]

[82] N.H. Maina, L. Pitkänen, S. Heikkinen, P. Tuomainen, L. Virkki, and M. Tenkanen, "Challenges in analysis of high-molar mass dextrans: comparison of HPSEC, AsFlFFF and DOSY NMR spectroscopy", *Carbohydr. Polym.* , vol. 99, pp. 199-207, 2014.
[http://dx.doi.org/10.1016/j.carbpol.2013.08.021] [PMID: 24274497]

[83] K.-L. Cheong, D.-T. Wu, J. Zhao, and S.-P. Li, "A rapid and accurate method for the quantitative estimation of natural polysaccharides and their fractions using high performance size exclusion chromatography coupled with multi-angle laser light scattering and refractive index detector", *J. Chromatogr. A.* , vol. 1400, pp. 98-106, Jun 26, 2015. (Epub 2015 May 2015)
[http://dx.doi.org/10.1016/j.chroma.2015.04.054]

[84] R. Hanselmann, W. Burchard, R. Lemmes, and D. Schwengers, "Characterization of Deae-Dextran by Means of Light-Scattering and Combined Size-Exclusion Chromatography Low-Angle Laser-Light Scattering Viscometry", *Macromol. Chem. Phys.* , vol. 196, pp. 2259-2275, 1995.
[http://dx.doi.org/10.1002/macp.1995.021960715]

[85] L. Picton, G. Mocanu, D. Mihai, A. Carpov, and G. Muller, "Chemically modified exopolysaccharide pullulans: Physico- chemical characteristics of ionic derivatives", *Carbohydr. Polym.* , vol. 28, pp. 131-136, 1995.
[http://dx.doi.org/10.1016/0144-8617(95)00112-3]

[86] A.M. Striegel, S.L. Isenberg, and G.L. Côté, "An SEC/MALS study of alternan degradation during size-exclusion chromatographic analysis", *Anal. Bioanal. Chem.* , vol. 394, no. 7, pp. 1887-1893, 2009.
[http://dx.doi.org/10.1007/s00216-009-2895-5] [PMID: 19554314]

[87] A. Potthast, T. Rosenau, H. Sixta, and P. Kosma, "Degradation of cellulosic materials by heating in DMAc/LiCl", *Tetrahedron Lett.* , vol. 43, pp. 7757-7759, 2002.
[http://dx.doi.org/10.1016/S0040-4039(02)01767-7]

[88] A.M. Emsley, and G.C. Stevens, "Kinetics and mechanisms of the low-temperature degradation of cellulose", *Cellulose.* , vol. 1, pp. 26-56, 1994.

[http://dx.doi.org/10.1007/BF00818797]

[89] M. Strlic, J. Kolar, M. Zigon, and B. Pihlar, "Evaluation of size-exclusion chromatography and viscometry for the determination of molecular masses of oxidised cellulose", *J. Chromatogr. A.* , vol. 805, pp. 93-99, 1998. [http://dx.doi.org/10.1016/S0021-9673(98)00008-9]

[90] A.L. Dupont, "Study of the degradation of gelatin in paper upon aging using aqueous size-exclusion chromatography", *J. Chromatogr. A.* , vol. 950, no. 1-2, pp. 113-124, 2002. [http://dx.doi.org/10.1016/S0021-9673(02)00010-9] [PMID: 11990984]

[91] R. Stol, J.L. Pedersoli Jr, H. Poppe, and W.T. Kok, "Application of size exclusion electrochromatography to the microanalytical determination of the molecular mass distribution of celluloses from objects of cultural and historical value", *Anal. Chem.* , vol. 74, no. 10, pp. 2314-2320, 2002. [http://dx.doi.org/10.1021/ac0111309] [PMID: 12038756]

[92] A.L. Dupont, "Cellulose in lithium chloride/N,N-dimethylacetamide, optimisation of a dissolution method using paper substrates and stability of the solutions", *Polymer (Guildf.).* , vol. 44, pp. 4117-4126, 2003. [http://dx.doi.org/10.1016/S0032-3861(03)00398-7]

[93] A.L. Dupont, and G. Harrison, "Conformation and dn/dc determination of cellulose in N,N-dimethylacetamide containing lithium chloride", *Carbohydr. Polym.* , vol. 58, pp. 233-243, 2004. [http://dx.doi.org/10.1016/j.carbpol.2004.07.016]

[94] P. Rudolph, F. J. Ligterink, J. L. Pedersoli, M. Van Bommel, J. Bos, H. A. Aziz, J. Havermans, H. Scholten, D. Schipper, and W. Kautek, "Characterization of laser-treated paper", *Appl. Phys. a-Mater.* , vol. 79, pp. 181-186, Jul 2004. [http://dx.doi.org/10.1007/s00339-004-2650-x]

[95] T. Łojewski, K. Zieba, and J. Lojewska, "Size exclusion chromatography and viscometry in paper degradation studies. New Mark-Houwink coefficients for cellulose in cupri-ethylenediamine", *J. Chromatogr. A.* , vol. 1217, no. 42, pp. 6462-6468, 2010. [http://dx.doi.org/10.1016/j.chroma.2010.07.071] [PMID: 20833398]

[96] A.M. Striegel, "Theory and applications of DMAC/LICL in the analysis of polysaccharides", *Carbohydr. Polym.* , vol. 34, pp. 267-274, 1997. [http://dx.doi.org/10.1016/S0144-8617(97)00101-X]

[97] T. Schult, S.T. Moe, T. Hjerde, and B.E. Christensen, "Size exclusion chromatography of cellulose dissolved in LiCl/DMAC using macroporous monodisperse poly(styrene-co-divinylbenzene) particles", *J. Liquid Chromatogr. Relat. Technol.* , vol. 23, pp. 2277-2288, 2000. [http://dx.doi.org/10.1081/JLC-100100487]

[98] M. Yamamoto, R. Kuramae, M. Yanagisawa, D. Ishii, and A. Isogai, "Light-scattering analysis of native wood holocelluloses totally dissolved in LiCl-DMI solutions: high probability of branched structures in inherent cellulose", *Biomacromolecules.* , vol. 12, no. 11, pp. 3982-3988, 2011. [http://dx.doi.org/10.1021/bm201211z] [PMID: 21928815]

[99] L. Hardelin, E. Perzon, B. Hagstrom, P. Walkenstrom, and P. Gatenholm, "Influence of molecular weight and rheological behavior on electrospinning cellulose nanofibers from ionic liquids", *J. Appl. Polym. Sci.* , vol. 130, pp. 2303-2310, 2013. [http://dx.doi.org/10.1002/app.39449]

[100] M. Kes, and B.E. Christensen, "A re-investigation of the Mark-Houwink-Sakurada parameters for cellulose in Cuen: a study based on size-exclusion chromatography combined with multi-angle light scattering and viscometry", *J. Chromatogr. A.* , vol. 1281, pp. 32-37, 2013. [http://dx.doi.org/10.1016/j.chroma.2013.01.038] [PMID: 23398992]

[101] W. Ahmad, S. Kuitunen, H. Sixta, and V. Alopaeus, "Population balance based modeling of changes in cellulose molecular weight distribution during ageing", *Cellulose.* , vol. 22, pp. 151-163, 2015.

[http://dx.doi.org/10.1007/s10570-014-0529-3]

[102] J.M. Lauriol, J. Comtat, P. Froment, F. Pla, and A. Robert, "Molecular weight distribution of cellulose by on-line size exclusion chromatography – low angle laser lightscattering; part I: basic experiments and treatment of data", *Holzforschung.*, vol. 41, pp. 165-169, 1987. [http://dx.doi.org/10.1515/hfsg.1987.41.2.109]

[103] F. Kačík, P. Šmíra, D. Kačíková, V. Veľková, A. Nasswettrová, and V. Vacek, "Chemical alterations of pine wood saccharides during heat sterilisation", *Carbohydr. Polym.*, vol. 117, pp. 681-686, 2015. [http://dx.doi.org/10.1016/j.carbpol.2014.10.065] [PMID: 25498688]

[104] A. Potthast, S. Radosta, B. Saake, S. Lebioda, T. Heinze, U. Henniges, A. Isogai, A. Koschella, P. Kosma, T. Rosenau, S. Schiehser, H. Sixta, M. Strlic, G. Strobin, W. Vorwerg, and H. Wetzel, "Comparison testing of methods for gel permeation chromatography of cellulose: coming closer to a standard protocol", *Cellulose.*, vol. 22, pp. 1591-1613, 2015. [http://dx.doi.org/10.1007/s10570-015-0586-2]

[105] P. Pfefferkorn, J. Beister, A. Hild, H. Thielking, and W.M. Kulicke, "Determination of the molar mass and the radius of gyration, together with their distributions for methylhydroxyethylcelluloses", *Cellulose.*, vol. 10, pp. 27-36, 2003. [http://dx.doi.org/10.1023/A:1023046306892]

[106] K. Jumel, S. E. Harding, J. R. Mitchell, K. M. To, I. Hayter, and J. E. Omullane, *Carbohydr. Polym.*, vol. 29, pp. 105-109, 1996. [http://dx.doi.org/10.1016/0144-8617(96)00009-4]

[107] B. Wittgren, and B. Porsch, "Molar mass distribution of hydroxypropyl cellulose by size exclusion chromatography with dual light scattering and refractometric detection", *Carbohydr. Polym.*, vol. 49, pp. 457-469, 2002. [http://dx.doi.org/10.1016/S0144-8617(01)00352-6]

[108] B. Wittgren, M. Stefansson, and B. Porsch, "Interactions between sodium dodecyl sulphate and non-ionic cellulose derivatives studied by size exclusion chromatography with online multi-angle light scattering and refractometric detection", *J. Chromatogr. A.*, vol. 1082, no. 2, pp. 166-175, 2005. [http://dx.doi.org/10.1016/j.chroma.2005.05.047] [PMID: 16035358]

[109] C.W. Hoogendam, A. de Keizer, M.A. Stuart, B.H. Bijsterbosch, J.A. Smit, J. van Dijk, P.M. van der Horst, and J.C. Batelaan, "Persistence length of carboxymethyl cellulose as evaluated from size exclusion chromatography and potentiometric titrations", *Macromolecules.*, vol. 31, pp. 6297-6309, 1998. [http://dx.doi.org/10.1021/ma971032i]

[110] W.M. Kulicke, C. Clasen, and C. Lohman, "Characterization of water-soluble cellulose derivatives in terms of the molar mass and particle size as well as their distribution", *Macromol. Symp.*, vol. 223, pp. 151-174, 2005. [http://dx.doi.org/10.1002/masy.200550511]

[111] P. Katapodis, M. Vardakou, E. Kalogeris, D. Kekos, B.J. Macris, and P. Christakopoulos, "Enzymic production of a feruloylated oligosaccharide with antioxidant activity from wheat flour arabinoxylan", *Eur. J. Nutr.*, vol. 42, no. 1, pp. 55-60, 2003. [http://dx.doi.org/10.1007/s00394-003-0400-z] [PMID: 12594542]

[112] L.E. Rasmussen, and A.S. Meyer, "Size exclusion chromatography for the quantitative profiling of the enzyme-catalyzed hydrolysis of Xylo-oligosaccharides", *J. Agric. Food Chem.*, vol. 58, no. 2, pp. 762-769, 2010. [http://dx.doi.org/10.1021/jf903200h] [PMID: 19994888]

[113] D.J. Rose, and G.E. Inglett, "Two-stage hydrothermal processing of wheat (Triticum aestivum) bran for the production of feruloylated Arabinoxylooligosaccharides", *J. Agric. Food Chem.*, vol. 58, no. 10, pp. 6427-6432, 2010. [http://dx.doi.org/10.1021/jf100058v] [PMID: 20408584]

[114] M.L. Fishman, L.W. Doner, H.K. Chau, and P.D. Hoagland, "Characterization of hemicellulose B from corn fiber by high-performance size exclusion chromatography with on-line molar mass and viscometric detection", *Inter. J. Polym. Anal. Ch.* , vol. 5, pp. 359-379, 2000. [http://dx.doi.org/10.1080/10236660008034633]

[115] L. Pitkänen, L. Virkki, M. Tenkanen, and P. Tuomainen, "Comprehensive multidetector HPSEC study on solution properties of cereal arabinoxylans in aqueous and DMSO solutions", *Biomacromolecules.* , vol. 10, no. 7, pp. 1962-1969, 2009. [http://dx.doi.org/10.1021/bm9003767] [PMID: 19499890]

[116] L. Pitkänen, M. Tenkanen, and P. Tuomainen, "Behavior of polysaccharide assemblies in field-flow fractionation and size-exclusion chromatography", *Anal. Bioanal. Chem.* , vol. 399, no. 4, pp. 1467-1472, 2011. [http://dx.doi.org/10.1007/s00216-010-4160-3] [PMID: 20820993]

[117] B.E. Christensen, A.S. Ulset, M.U. Beer, B.E. Knuckles, D.L. Williams, M.L. Fishman, H.K. Chau, and P.J. Wood, "Macromolecular characterisation of three barley beta-glucan standards by size-exclusion chromatography combined with light scattering and viscometry: an inter-laboratory study", *Carbohydr. Polym.* , vol. 45, pp. 11-22, 2001. [http://dx.doi.org/10.1016/S0144-8617(00)00237-X]

[118] J.P. Roubroeks, D.I. Mastromauro, R. Andersson, B.E. Christensen, and P. Aman, "Molecular weight, structure, and shape of oat (1 - > 3),(1 - > 4)-beta-D-glucan fractions obtained by enzymatic degradation with lichenase", In: *Biomacromolecules.* Win 2000, pp. 584-591.

[119] J.P. Roubroeks, R. Andersson, D.I. Mastromauro, B.E. Christensen, and P. Aman, "Molecular weight, structure and shape of oat (1 - > 3),(1 - > 4)-beta-D-glucan fractions obtained by enzymatic degradation with (1 - > 4)-beta-D-glucan 4-glucanohydrolase from Trichoderma reesei", *Carbohydr. Polym.* , vol. 46, pp. 275-285, 2001. [http://dx.doi.org/10.1016/S0144-8617(00)00329-5]

[120] N. Mäkelä, T. Sontag-Strohm, and N.H. Maina, "The oxidative degradation of barley β-glucan in the presence of ascorbic acid or hydrogen peroxide", *Carbohydr. Polym.* , vol. 123, pp. 390-395, 2015. [http://dx.doi.org/10.1016/j.carbpol.2015.01.037] [PMID: 25843872]

[121] F. Qin, M. Kes, and B.E. Christensen, "A study of bioactive, branched (1→3)-β-d-glucans in dimethylacetamide/LiCl and dimethyl sulphoxide/LiCl using size-exclusion chromatography with multi-angle light scattering detection", *J. Chromatogr. A.* , vol. 1305, pp. 109-113, 2013. [http://dx.doi.org/10.1016/j.chroma.2013.07.002] [PMID: 23891208]

[122] F. Qin, F.L. Aachmann, and B.E. Christensen, "Chain length distribution and aggregation of branched (1→3)-β-D-glucans from Saccharomyces cerevisae", *Carbohydr. Polym.* , vol. 90, no. 2, pp. 1092-1099, 2012. [http://dx.doi.org/10.1016/j.carbpol.2012.06.048] [PMID: 22840044]

[123] M. Sletmoen, B.E. Christensen, and B.T. Stokke, "Probing macromolecular architectures of nanosized cyclic structures of (1 > 3)-beta-D-glucans by AFM and SEC-MALLS", *Carbohydr. Res.* , vol. 340, no. 5, pp. 971-979, 2005. [http://dx.doi.org/10.1016/j.carres.2005.01.026] [PMID: 15780261]

[124] C. Augsten, W. Knolle, and K. Maeder, "Characterizing the influence of electron irradiation on scleroglucan", *Carbohydr. Polym.* , vol. 72, pp. 707-718, 2008. [http://dx.doi.org/10.1016/j.carbpol.2007.10.013]

[125] L.N. Zhang, X.F. Zhang, Q. Zhou, P.Y. Zhang, M. Zhang, and X.L. Li, "Triple helix of beta-D-glucan from Lentinus Edodes in 0.5 M NaCl aqueous solution characterized by light scattering", *Polym. J.* , vol. 33, pp. 317-321, 2001. [http://dx.doi.org/10.1295/polymj.33.317]

[126] Y.F. Wang, and L.N. Zhang, "Chain conformation of carboxymethylated derivatives of (1 - > 3)-bet--D-glucan from Poria cocos sclerotium", *Carbohydr. Polym.* , vol. 65, pp. 504-509, 2006.

[http://dx.doi.org/10.1016/j.carbpol.2006.02.014]

[127] M.L. Fishman, P. Cescutti, W.F. Fett, S.F. Osman, P.D. Hoagland, and H.K. Chau, "Screening the physical properties of novel Pseudomonas exopolysaccharides by HPSEC with multi-angle light scattering and viscosity detection", *Carbohydr. Polym.* , vol. 32, pp. 213-221, 1997. [http://dx.doi.org/10.1016/S0144-8617(97)00012-X]

[128] M.U. Beer, P.J. Wood, and J. Weisz, "A simple and rapid method for evaluation of Mark-Houwink-Sakurada constants of linear random coil polysaccharides using molecular weight and intrinsic viscosity determined by high performance size exclusion chromatography: application to guar galactomannan", *Carbohydr. Polym.* , vol. 39, pp. 377-380, 1999. [http://dx.doi.org/10.1016/S0144-8617(99)00017-X]

[129] L. Pitkanen, P. Tuomainen, K.S. Mikkonen, and M. Tenkanen, "The effect of galactose side units and mannan chain length on the macromolecular characteristics of galactomannans", *Carbohydr. Polym.* , vol. 86, pp. 1230-1235, 2011. [http://dx.doi.org/10.1016/j.carbpol.2011.06.018]

[130] C.D. Sommers, H. Ye, R.E. Kolinski, M. Nasr, L.F. Buhse, A. Al-Hakim, and D.A. Keire, "Characterization of currently marketed heparin products: analysis of molecular weight and heparinase-I digest patterns", *Anal. Bioanal. Chem.* , vol. 401, no. 8, pp. 2445-2454, 2011. [http://dx.doi.org/10.1007/s00216-011-5362-z] [PMID: 21901459]

[131] R. Mendichi, and A.G. Schieroni, "Fractionation and characterization of ultra-high molar mass hyaluronan: 2. On-line size exclusion chromatography methods", *Polymer (Guildf.).* , vol. 43, pp. 6115-6121, 2002. [http://dx.doi.org/10.1016/S0032-3861(02)00586-4]

[132] T. Luan, Y. Fang, S. Al-assaf, G. Phillips, and H. Zhang, "Compared molecular characterization of hyaluronan using multiple-detection techniques", *Polymer (Guildf.).* , vol. 52, p. 24, 2011. [http://dx.doi.org/10.1016/j.polymer.2011.10.024]

[133] G.A. Morris, J.G. de al Torre, A. Ortega, J. Castile, A. Smith, and S.E. Harding, "Molecular flexibility of citrus pectins by combined sedimentation and viscosity analysis", *Food Hydrocoll.* , vol. 22, pp. 1435-1442, 2008. [http://dx.doi.org/10.1016/j.foodhyd.2007.09.005]

[134] G.A. Morris, M.C. Ralet, E. Bonnin, J.F. Thibault, and S.E. Harding, "Physical characterisation of the rhamnogalacturonan and homogalacturonan fractions of sugar beet (Beta vulgaris) pectin", *Carbohydr. Polym.* , vol. 82, pp. 1161-1167, 2010. [http://dx.doi.org/10.1016/j.carbpol.2010.06.049]

[135] M.L. Fishman, H.K. Chau, P. Hoagland, and K. Ayyad, "Characterization of pectin, flash-extracted from orange albedo by microwave heating, under pressure", *Carbohydr. Res.* , vol. 323, no. 1-4, pp. 126-138, 2000. [http://dx.doi.org/10.1016/S0008-6215(99)00244-X] [PMID: 10782294]

[136] M.L. Fishman, H.K. Chau, P.H. Cooke, A.T. Hotchkiss Jr, H.K. Chau, P.H. Cooke, and A.T. Hotchkiss Jr, "Global structure of microwave-assisted flash-extracted sugar beet pectin", *J. Agric. Food Chem.* , vol. 56, no. 4, pp. 1471-1478, 2008. [http://dx.doi.org/10.1021/jf072600o] [PMID: 18183951]

[137] M.L. Fishman, H.K. Chau, P.X. Qi, A.T. Hotchkiss Jr, and M.P. Yadav, "Physico-chemical characterization of protein-associated polysaccharides extracted from sugar beet pulp", *Carbohydr. Polym.* , vol. 92, no. 2, pp. 2257-2266, 2013. [http://dx.doi.org/10.1016/j.carbpol.2012.12.001] [PMID: 23399286]

[138] O.H. Idris, P.A. Williams, and G.O. Phillips, "Characterisation of gum from Acacia senegal trees of different age and location using multidetection gel permeation chromatography", *Food Hydrocoll.* , vol. 12, pp. 379-388, 1998. [http://dx.doi.org/10.1016/S0268-005X(98)00058-7]

[139] S. Al-Assaf, T. Katayama, G.O. Philipps, Y. Sasaki, and P.A. Williams, "Quality control of Gum Arabic", *Foods and Food Ingredients Journal of Japan.* , vol. 208, pp. 771-780, 2003.

[140] S. Al-Assaf, J. Cirre, M. Andres-Brull, M. Sakata, C. McKenna, J. Grimsey, and G.O. Philipps, "Correlation of Gum Arabic Molecular Weight Parameters with Emulsification Performance", *Foods and Food Ingredients Journal of Japan.* , vol. 213, pp. 377-383, 2008.

[141] S. Al-Assaf, M. Sakata, C. McKenna, H. Aoki, and G.O. Phillips, "Molecular associations in acacia gums", *Struct. Chem.* , vol. 20, pp. 325-336, 2009. [http://dx.doi.org/10.1007/s11224-009-9430-3]

[142] D. Renard, C. Garnier, A. Lapp, C. Schmitt, and C. Sanchez, "Structure of arabinogalactan-protein from Acacia gum: from porous ellipsoids to supramolecular architectures", *Carbohydr. Polym.* , vol. 90, no. 1, pp. 322-332, 2012. [http://dx.doi.org/10.1016/j.carbpol.2012.05.046] [PMID: 24751048]

[143] D. Renard, E. Lepvrier, C. Garnier, P. Roblin, M. Nigen, and C. Sanchez, "Structure of glycoproteins from Acacia gum: an assembly of ring-like glycoproteins modules", *Carbohydr. Polym.* , vol. 99, pp. 736-747, 2014. [http://dx.doi.org/10.1016/j.carbpol.2013.08.090] [PMID: 24274565]

[144] L. Lopez-Torrez, M. Nigen, P. Williams, T. Doco, and C. Sanchez, "Acacia senegal vs. Acacia seyal gums - Part 1: Composition and structure of hyperbranched plant exudates", *Food Hydrocoll.* , vol. 51, pp. 41-53, 2015. [http://dx.doi.org/10.1016/j.foodhyd.2015.04.019]

[145] I.B. Gashua, P.A. Williams, M.P. Yadav, and T.C. Baldwin, "Characterisation and molecular association of Nigerian and Sudanese Acacia gum exudates", *Food Hydrocoll.* , vol. 51, pp. 405-413, 2015. [http://dx.doi.org/10.1016/j.foodhyd.2015.05.037]

[146] D. Szopinski, W-M. Kulicke, and G.A. Luinstra, "Structure-property relationships of carboxymethyl hydroxypropyl guar gum in water and a hyperentanglement parameter", *Carbohydrate polymers.* , vol. 119, pp. 159-66, Mar 30, 2015.

[147] L. Picton, I. Bataille, and G. Muller, "Analysis of a complex polysaccharide (gum arabic) by multi-angle laser light scattering coupled on-line to size exclusion chromatography and flow field flow fractionation", *Carbohydr. Polym.* , vol. 42, pp. 23-31, 2000. [http://dx.doi.org/10.1016/S0144-8617(99)00139-3]

[148] M. Andres-Brull, S. Al-Assaf, G.O. Phillips, and K. Jackson, "Optimisation of asymmetrical flow-field fractionation for the characterisation of gum arabic (Acacia sengal var senegal) and comparison with gel permeation chromatography", *Anal. Methods.* , vol. 5, pp. 4047-4052, 2013. [http://dx.doi.org/10.1039/c3ay40835c]

[149] I.M. Vold, K.A. Kristiansen, and B.E. Christensen, "A study of the chain stiffness and extension of alginates, in vitro epimerized alginates, and periodate-oxidized alginates using size-exclusion chromatography combined with light scattering and viscosity detectors", *Biomacromolecules.* , vol. 7, no. 7, pp. 2136-2146, 2006. [http://dx.doi.org/10.1021/bm060099n] [PMID: 16827580]

[150] A. Villay, F.L. de Filippis, L. Picton, D. Le Cerf, C. Vial, and P. Michaud, "Comparison of polysaccharide degradations by dynamic high-pressure homogenization", *Food Hydrocoll.* , vol. 27, pp. 278-286, 2012. [http://dx.doi.org/10.1016/j.foodhyd.2011.10.003]

[151] H. Storz, K.J. Müller, F. Ehrhart, I. Gómez, S.G. Shirley, P. Gessner, G. Zimmermann, E. Weyand, V.L. Sukhorukov, T. Forst, M.M. Weber, H. Zimmermann, W-M. Kulicke, and U. Zimmermann, "Physicochemical features of ultra-high viscosity alginates", *Carbohydr. Res.* , vol. 344, no. 8, pp. 985-995, 2009. [http://dx.doi.org/10.1016/j.carres.2009.02.016] [PMID: 19394590]

[152] G. Rousseaux, A. De Wit, and M. Martin, "Viscous fingering in packed chromatographic columns: linear stability analysis", *J. Chromatogr. A.*, vol. 1149, no. 2, pp. 254-273, 2007. [http://dx.doi.org/10.1016/j.chroma.2007.03.056] [PMID: 17420024]

[153] C. Rochas, M. Rinaudo, and S. Landry, "Role of the Molecular-Weight on the Mechanical Properties of Kappa-Carrageenan gels", *Carbohydr. Polym.*, vol. 12, pp. 255-266, 1990. [http://dx.doi.org/10.1016/0144-8617(90)90067-3]

[154] S. K. Singh, and S. P. Jacobsson, "Kinetics of Acid-Hydrolysis of Kappa-Carrageenan as Determined by Molecular-Weight (SEC-MALLS-RI), Gel Breaking Strength and Viscosity Measurements", *Carbohydr. Polym.*, vol. 23, pp. 89-103, 1994.

[155] G. Marcelo, E. Saiz, and M.P. Tarazona, "Unperturbed dimensions of Carrageenans in different salt solutions", *Biophys. Chem.*, vol. 113, no. 3, pp. 201-208, 2005. [http://dx.doi.org/10.1016/j.bpc.2004.09.005] [PMID: 15620505]

[156] G. Berth, J. Vukovic, and M.D. Lechner, "Physicochemical Characterization of Carrageenans-A Critical Reinvestigation", *J. Appl. Polym. Sci.*, vol. 110, pp. 3508-3524, 2008. [http://dx.doi.org/10.1002/app.28937]

[157] F.M. Almutairi, G.G. Adams, M.S. Kök, C.J. Lawson, R. Gahler, S. Wood, T.J. Foster, A.J. Rowe, and S.E. Harding, "An analytical ultracentrifugation based study on the conformation of lambda carrageenan in aqueous solution", *Carbohydr. Polym.*, vol. 97, no. 1, pp. 203-209, 2013. [http://dx.doi.org/10.1016/j.carbpol.2013.04.027] [PMID: 23769538]

[158] K. Bongaerts, H. Reynaers, F. Zanetti, and S. Paoletti, "On the molar mass of kappa-carrageenan in the course of conformational transition from the disordered to the fundamental ordered form", *Macromolecules.*, vol. 32, pp. 675-682, 1999. [http://dx.doi.org/10.1021/ma981203z]

[159] K. Bongaerts, H. Reynaers, F. Zanetti, and S. Paoletti, "Equilibrium and nonequilibrium association processes of kappa-carrageenan in aqueous salt solutions", *Macromolecules.*, vol. 32, pp. 683-689, 1999. [http://dx.doi.org/10.1021/ma981204r]

[160] H.H. Ammar, S. Lajili, R. Ben Said, D. Le Cerf, A. Bouraoui, and H. Majdoub, "Physico-chemical characterization and pharmacological evaluation of sulfated polysaccharides from three species of Mediterranean brown algae of the genus Cystoseira", *DARU-Journal of Pharmaceutical Sciences.*, vol. 23, Jan 13, 2015.

[161] G.A. Morris, J. Castile, A. Smith, G.G. Adams, and S.E. Harding, "Macromolecular conformation of chitosan in dilute solution: a new global hydrodynamic approach", *Carbohydr. Polym.*, vol. 76, pp. 616-621, 2009. [http://dx.doi.org/10.1016/j.carbpol.2008.11.025]

[162] K. Tommeraas, S.P. Strand, B.E. Christensen, O. Smidsrod, and K.M. Varum, "Preparation and characterization of branched chitosans", *Carbohydr. Polym.*, vol. 83, pp. 1558-1564, 2011. [http://dx.doi.org/10.1016/j.carbpol.2010.10.008]

[163] C. Schatz, C. Viton, T. Delair, C. Pichot, and A. Domard, "Typical physicochemical behaviors of chitosan in aqueous solution", *Biomacromolecules.*, vol. 4, no. 3, pp. 641-648, 2003. [http://dx.doi.org/10.1021/bm025724c] [PMID: 12741780]

[164] F.M. Almutairi, T. Erten, G.G. Adams, M. Hayes, P. McLoughlin, M.S. Kök, A.R. Mackie, A.J. Rowe, and S.E. Harding, "Hydrodynamic characterisation of chitosan and its interaction with two polyanions: DNA and xanthan", *Carbohydr. Polym.*, vol. 122, pp. 359-366, 2015. [http://dx.doi.org/10.1016/j.carbpol.2014.09.090] [PMID: 25817680]

[165] O. Ringena, S. Lebioda, R. Lehnen, and B. Saake, "Size-exclusion chromatography of technical lignins in dimethyl sulfoxide/water and dimethylacetamide", *J. Chromatogr. A.*, vol. 1102, no. 1-2, pp. 154-163, 2006.

[http://dx.doi.org/10.1016/j.chroma.2005.10.037] [PMID: 16288767]

[166] S. Baumberger, A. Abaecherli, M. Fasching, G. Gellerstedt, R. Gosselink, B. Hortling, J. Li, B. Saake, and E. de Jong, "Molar mass determination of lignins by size-exclusion chromatography: towards standardisation of the method", *Holzforschung.* , vol. 61, pp. 459-468, 2007. [http://dx.doi.org/10.1515/HF.2007.074]

[167] G.E. Fredheim, S.M. Braaten, and B.E. Christensen, "Molecular weight determination of lignosulfonates by size-exclusion chromatography and multi-angle laser light scattering", *J. Chromatogr. A.* , vol. 942, no. 1-2, pp. 191-199, 2002. [http://dx.doi.org/10.1016/S0021-9673(01)01377-2] [PMID: 11822385]

[168] B. Cathala, B. Saake, O. Faix, and B. Monties, "Association behaviour of lignins and lignin model compounds studied by multidetector size-exclusion chromatography", *J. Chromatogr. A.* , vol. 1020, no. 2, pp. 229-239, 2003. [http://dx.doi.org/10.1016/j.chroma.2003.08.046] [PMID: 14661746]

[169] G. E. Fredheim, S. M. Braaten, and B. E. Christensen, "Comparison of molecular weight and molecular weight distributions of softwood and hardwood lignosulfonates", *J. Wood Chem. Technol.* , vol. 23, no. 2003, pp. 197-215, 2003.

[170] A.V. Gidh, S.R. Decker, C.H. See, M.E. Himmel, and C.W. Williford, "Characterization of lignin using multi-angle laser light scattering and atomic force microscopy", *Anal. Chim. Acta.* , vol. 555, pp. 250-258, 2006. [http://dx.doi.org/10.1016/j.aca.2005.09.023]

[171] A.V. Gidh, S.R. Decker, T.B. Vinzant, M.E. Himmel, and C. Williford, "Determination of lignin by size exclusion chromatography using multi angle laser light scattering", *J. Chromatogr. A.* , vol. 1114, no. 1, pp. 102-110, 2006. [http://dx.doi.org/10.1016/j.chroma.2006.02.044] [PMID: 16566937]

[172] W. Kopaciewicz, and F.E. Regnier, "Nonideal size-exclusion chromatography of proteins: effects of pH at low ionic strength", *Anal. Biochem.* , vol. 126, no. 1, pp. 8-16, 1982. [http://dx.doi.org/10.1016/0003-2697(82)90102-6] [PMID: 7181119]

[173] N.P. Golovchenko, I.A. Kataeva, and V.K. Akimenko, "Analysis of pH-dependent protein interactions with gel filtration medium", *J. Chromatogr. A.* , vol. 591, no. 1-2, pp. 121-128, 1992. [http://dx.doi.org/10.1016/0021-9673(92)80229-N] [PMID: 1613047]

[174] P.L. Dubin, S.L. Edwards, M.S. Mehta, and D. Tomalia, "Quantitation of Nonideal Behavior in Protein Size-Exclusion Chromatography", *J. Chromatogr. A.* , vol. 635, pp. 51-60, 1993. [http://dx.doi.org/10.1016/0021-9673(93)83113-7]

[175] R.D. Ricker, and L.A. Sandoval, "Fast, reproducible size-exclusion chromatography of biological macromolecules", *J. Chromatogr. A.* , vol. 743, no. 1, pp. 43-50, 1996. [http://dx.doi.org/10.1016/0021-9673(96)00283-X] [PMID: 8817873]

[176] G.B. Irvine, "High-performance size-exclusion chromatography of peptides", *J. Biochem. Biophys. Methods.* , vol. 56, no. 1-3, pp. 233-242, 2003. [http://dx.doi.org/10.1016/S0165-022X(03)00061-7] [PMID: 12834979]

[177] S. P. John, "Is Any Measurement Method Optimal for all Aggregate Sizes and Types?", *AAPS J.* , vol. 8, pp. E564-71, 2006.

[178] T. Arakawa, D. Ejima, T. Li, and J.S. Philo, "The critical role of mobile phase composition in size exclusion chromatography of protein pharmaceuticals", *J. Pharm. Sci.* , vol. 99, no. 4, pp. 1674-1692, 2010.

[179] D.M. O'callaghan, W.J. Donnelly, H.M. Slattery, and D.M. Mulvihill, "Non-Size Exclusion Effects During Gel Permeation Chromatography of Milk Protein Hydrolysates on an Fplc Superose 12 Column", *J. Liq. Chromatogr.* , vol. 18, pp. 1543-1562, 1995. [http://dx.doi.org/10.1080/10826079508009294]

[180] R. Tantipolphan, S. Romeijn, Jd. Engelsman, R. Torosantucci, T. Rasmussen, and W. Jiskoot, "Elution behavior of insulin on high-performance size exclusion chromatography at neutral pH", *J. Pharm. Biomed. Anal.* , vol. 52, no. 2, pp. 195-202, 2010.
[http://dx.doi.org/10.1016/j.jpba.2010.01.014] [PMID: 20129751]

[181] C.M. Yu, S. Mun, and N.H. Wang, "Phenomena of insulin peak fronting in size exclusion chromatography and strategies to reduce fronting", *J. Chromatogr. A.* , vol. 1192, no. 1, pp. 121-129, 2008.
[http://dx.doi.org/10.1016/j.chroma.2008.03.055] [PMID: 18405908]

[182] D. Ejima, R. Yumioka, T. Arakawa, and K. Tsumoto, "Arginine as an effective additive in gel permeation chromatography", *J. Chromatogr. A.* , vol. 1094, no. 1-2, pp. 49-55, 2005.
[http://dx.doi.org/10.1016/j.chroma.2005.07.086] [PMID: 16257288]

[183] R. Yumioka, H. Sato, H. Tomizawa, Y. Yamasaki, and D. Ejima, "Mobile phase containing arginine provides more reliable SEC condition for aggregation analysis", *J. Pharm. Sci.* , vol. 99, no. 2, pp. 618-620, 2010.
[http://dx.doi.org/10.1002/jps.21857] [PMID: 19569199]

[184] D.F. Moriarty, C. Fiorillo, C. Miller, and W. Colón, "A truncated peptide model of the mutant P61A FIS forms a stable dimer", *Biochim. Biophys. Acta.* , vol. 1774, no. 1, pp. 78-85, 2007.
[http://dx.doi.org/10.1016/j.bbapap.2006.09.012] [PMID: 17118726]

[185] M. Szulc, R. Mhatre, J. Mazzeo, and I.S. Krull, "“Detection in liquid chromatography,” High Resolution Sep. Anal. Biol. Macromol", *Pt A.* , vol. 270, pp. 175-200, 1996.

[186] J. Wen, T. Arakawa, and J.S. Philo, "Size-exclusion chromatography with on-line light-scattering, absorbance, and refractive index detectors for studying proteins and their interactions", *Anal. Biochem.* , vol. 240, no. 2, pp. 155-166, 1996.
[http://dx.doi.org/10.1006/abio.1996.0345] [PMID: 8811899]

[187] E. Folta-Stogniew, and K.R. Williams, "Determination of molecular masses of proteins in solution: Implementation of an HPLC size exclusion chromatography and laser light scattering service in a core laboratory", *J. Biomol. Tech.* , vol. 10, no. 2, pp. 51-63, 1999.
[PMID: 19499008]

[188] J.L. Carceller, and T. Aussenac, "Size characterisation of glutenin polymers by HPSEC-MALLS", *J. Cereal Sci.* , vol. 33, pp. 131-142, 2001.
[http://dx.doi.org/10.1006/jcrs.2000.0356]

[189] B.S. Kendrick, B.A. Kerwin, B.S. Chang, and J.S. Philo, "Online size-exclusion high-performance liquid chromatography light scattering and differential refractometry methods to determine degree of polymer conjugation to proteins and protein-protein or protein-ligand association states", *Anal. Biochem.* , vol. 299, no. 2, pp. 136-146, 2001.
[http://dx.doi.org/10.1006/abio.2001.5411] [PMID: 11730335]

[190] A. Oliva, M. Llabrés, and J.B. Fariña, "Comparative study of protein molecular weights by size-exclusion chromatography and laser-light scattering", *J. Pharm. Biomed. Anal.* , vol. 25, no. 5-6, pp. 833-841, 2001.
[http://dx.doi.org/10.1016/S0731-7085(01)00359-4] [PMID: 11377066]

[191] K. Ahrer, A. Buchacher, G. Iberer, D. Josic, and A. Jungbauer, "Analysis of aggregates of human immunoglobulin G using size-exclusion chromatography, static and dynamic light scattering", *J. Chromatogr. A.* , vol. 1009, no. 1-2, pp. 89-96, 2003.
[http://dx.doi.org/10.1016/S0021-9673(03)00433-3] [PMID: 13677648]

[192] T. Wang, and J.A. Lucey, "Use of multi-angle laser light scattering and size-exclusion chromatography to characterize the molecular weight and types of aggregates present in commercial whey protein products", *J. Dairy Sci.* , vol. 86, no. 10, pp. 3090-3101, 2003.
[http://dx.doi.org/10.3168/jds.S0022-0302(03)73909-5] [PMID: 14594226]

[193] A. Oliva, J.B. Farina, and M. Llabres, "Measurement of uncertainty in peptide molecular weight determination using size-exclusion chromatography with multi-angle laser light-scattering detection and matrix-assisted laser desorption/ionization time-of-flight mass spectrometry", *Anal. Chim. Acta.* , vol. 512, pp. 103-110, 2004.
[http://dx.doi.org/10.1016/j.aca.2004.02.022]

[194] M. Beaulieu, M. Corredig, S.L. Turgeon, L. Wicker, and J.L. Doublier, "The formation of heat-induced protein aggregates in whey protein/pectin mixtures studied by size exclusion chromatography coupled with multi-angle laser light scattering detection", *Food Hydrocoll.* , vol. 19, pp. 803-812, 2005.
[http://dx.doi.org/10.1016/j.foodhyd.2004.10.025]

[195] T. Arakawa, J.S. Philo, D. Ejima, D.K. Tsumoto, and F. Arisaka, "Aggregation Analysis of Therapeutic Proteins, Part 1", *Bioprocess Int.* , vol. 4, pp. 32-42, 2006.

[196] H. Ye, "Simultaneous determination of protein aggregation, degradation, and absolute molecular weight by size exclusion chromatography-multiangle laser light scattering", *Anal. Biochem.* , vol. 356, no. 1, pp. 76-85, 2006.
[http://dx.doi.org/10.1016/j.ab.2006.05.025] [PMID: 16839514]

[197] W. Qi, R.X. Su, Z.M. He, Y.B. Zhang, and F.M. Jin, "Pepsin-induced changes in the size and molecular weight distribution of bovine casein during enzymatic hydrolysis", *J. Dairy Sci.* , vol. 90, no. 11, pp. 5004-5011, 2007.
[http://dx.doi.org/10.3168/jds.2007-0230] [PMID: 17954739]

[198] S. Rongxin, H. Zhimin, and Q. Wei, "Pancreatic hydrolysis of bovine casein: changes in the aggregate size and molecular weight distribution", *Food Chem.* , vol. 107, pp. 151-157, 2008.
[http://dx.doi.org/10.1016/j.foodchem.2007.07.078]

[199] A. Oliva, M. Llabrés, and J.B. Fariña, "Estimation of uncertainty in size-exclusion chromatography with a double detection system (light-scattering and refractive index)", *Talanta.* , vol. 78, no. 3, pp. 781-789, 2009.
[http://dx.doi.org/10.1016/j.talanta.2008.12.039] [PMID: 19269428]

[200] M. Pothecary, S. Ball, and P. Clarke, "Protein applications for advanced multi-detector size-exclusion chromatography", *Chromatography Today.* , vol. 5, pp. 32-35, 2012.

[201] C.E. Espinosa-de la Garza, M.P. Miranda-Hernández, L. Acosta-Flores, N.O. Pérez, L.F. Flores-Ortiz, and E. Medina-Rivero, "Analysis of therapeutic proteins and peptides using multiangle light scattering coupled to ultra high performance liquid chromatography", *J. Sep. Sci.* , vol. 38, no. 9, pp. 1537-1543, 2015.
[http://dx.doi.org/10.1002/jssc.201400863] [PMID: 25727056]

[202] J. Lauber, R. Handrick, S. Leptihn, P. Dürre, and S. Gaisser, "Expression of the functional recombinant human glycosyltransferase GalNAcT2 in Escherichia coli", *Microb. Cell Fact.* , vol. 14, p. 3, 2015.
[http://dx.doi.org/10.1186/s12934-014-0186-0] [PMID: 25582753]

[203] J.F. Pollock, R.S. Ashton, N.A. Rode, D.V. Schaffer, and K.E. Healy, "Molecular characterization of multivalent bioconjugates by size-exclusion chromatography with multiangle laser light scattering", *Bioconjug. Chem.* , vol. 23, no. 9, pp. 1794-1801, 2012.
[http://dx.doi.org/10.1021/bc3000595] [PMID: 22794081]

[204] D. Berek, "Size exclusion chromatographya blessing and a curse of science and technology of synthetic polymers", *J. Sep. Sci.* , vol. 33, no. 3, pp. 315-335, 2010.
[http://dx.doi.org/10.1002/jssc.200900709] [PMID: 20127919]

CHAPTER 5

Field-Flow Fractionation (FFF)

Agnès Rolland-Sabaté[*, 1], **Serge Battu**[2], **Frédéric Bonfils**[3], **Karim Chelbi**[3] and **Michel Martin**[4]

[1] *UMR408 Sécurité et Qualité des Produits d'Origine Végétale, INRA, Université Avignon, F-84000 Avignon, France; UR1268 Biopolymères Interactions Assemblages, INRA, F-44316 Nantes, France*

[2] *EA3842 Homéostasie et Pathologies Cellulaires, Laboratoire de Chimie Analytique, Faculté de Pharmacie, Univeristé de Limoges, France*

[3] *CIRAD, UMR Ingénierie des Agropolymères et Technologies Emergentes (IATE), F-34398 Montpellier, France*

[4] *Ecole Supérieuure de Physique et de Chimie Industrielles, Laboratoire de Physique et Mécanique des Milieux Hétérogènes (PMMH, UMR 7636 CNRS, ESPCI-ParisTech, Université Pierre et Marie Curie, Université Paris-Diderot), 10 rue Vauquelin, 75231 Paris Cedex 05, France*

Abstract: Field-Flow Fractionation techniques (FFF) are size-based separation methods first described in 1966 by Giddings. They belong to the family of liquid chromatographic techniques, but they are operated without any stationary phase.Yet, they have the unique ability to separate an extremely broad range of molecules, macromolecules and particles, and in particular very large particles, with a high resolution. FFF are versatile: by varying the experimental conditions, the range, speed, and power of the separation could be optimized. FFF techniques can succeed when Size-Exclusion Chromatography (SEC) methods fails, and they have a broad range of applications. In this chapter the theory of FFF is approached together with calibration and determination of some structural parameters such as size and molar mass, the instrumentation and detection of various classic FFF types is described and we show the added value of FFF techniques for the characterization of various biopolymers including polysaccharides, proteins and natural rubber.

Keywords: Aggregation, Field-Flow Fractionation, Multiangle Laser Light Scattering, Natural Rubber, Polysaccharides, Proteins, Starch, Static and Dynamic Light Scattering

[*] **Corresponding author Agnès Rolland-Sabaté:** UMR408 Sécurité et Qualité des Produits d'Origine Végétale, INRA, Université Avignon, F-84000 Avignon, France; Tel: +33 (0)4 32 72 25 22; Fax: +33 (0)4 32 72 24 92; E-mail: agnes.rolland-sabate@inra.fr

Martin Masuelli & Denis Renard (Eds.)

INTRODUCTION

Characterization of biopolymers, from polysaccharides to proteins, is one of the main challenges in biotechnology. Analytical methods should not only give accurate information on the molecule structure, but also have to preserve and investigate its native conformation and supra-molecular, non-covalent interactions. Size-Exclusion Chromatography (SEC) is a routine and a well-established analytical separation method but it has limitations and pitfalls for the analysis of large, polydisperse, complex and branched polymers or biological extracts. For example its large internal surface area sometimes produces interactions, shear degradation, poor analysis recoveries and anomalous elution of branched polymers and it requires preinjection filter (see Chapter 4). An Alternative method is Field-Flow Fractionation (FFF) which was introduced in 1966 [1] and operates in an open channel thus allowing a gentle separation mechanism, without extensional shear and possible separation in native sample environnement.The rate of scientific publications on effective application of FFF to biopolymers and their assemblies is constantly increasing. FFF coupled with multi-detection is indeed a powerful tool for obtaining information on macromolecular structure, such as size and molar mass distributions and conformation, which are all important features for complex biopolymers. This chapter presents first a brief overview of the theory of fractionation by FFF, followed by calibration and determination of some structural parameters, instrumentation and detection of various classic FFF types. Following this, the applications of Sedimentation FFF, Thermal FFF and Flow FFF for the characterization of biopolymers and bioassemblies based on polysaccharides, proteins and natural rubber are developed.

THEORY OF FIELD-FLOW FRACTIONATION (FFF)

The expression FFF encompasses a variety of separation methods which all share the same basic separation principle. They are indeed based on the coupling of a non-uniform flow of a carrier liquid with a force field applied orthogonally to the main flow direction in a conduit filled with the carrier fluid [1]. They differ by the nature of the applied field and are named according to the nature (Fig. **1**). In addition, they may also differ by the geometrical configuration of the separation conduit. Most frequently, the conduits have parallel plate geometry. In some cases (particularly in hollow fiber flow FFF, HF5), a tubular (cylindrical) geometry is used. The fluid flow, in isothermal systems, has a parabolic velocity profile. The field, acting perpendicularly to the flow distributes the molecules (or particles) of a given component non-uniformly across the channel. The average velocity at which this component is displaced along the channel depends on the particular distribution of its molecules (or particles) within the various flow streamlines.

When the distribution profiles of the molecules of two components are different, these components are transported at different velocities along the channel by the non-uniform flow and can eventually become separated when they reach the channel outlet. This is the basis of the separation in FFF.

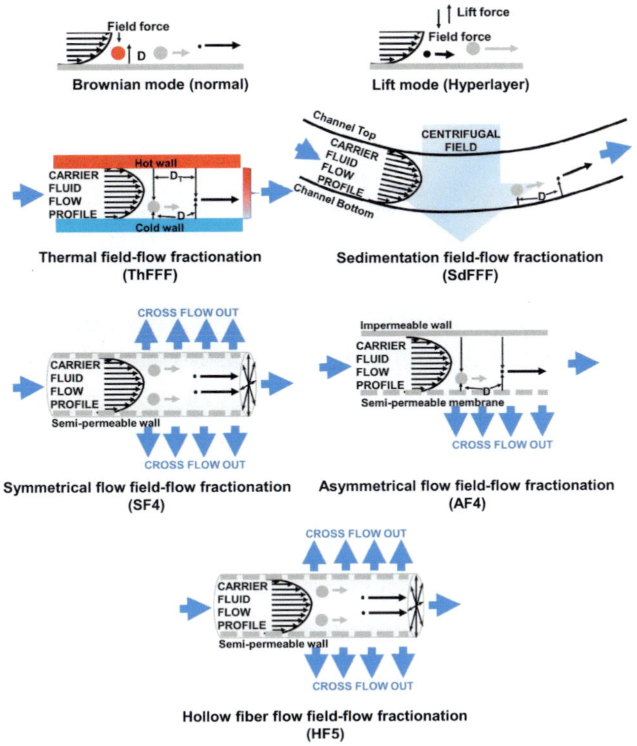

Fig. (1). Schemes of the FFF fractionation modes, ThFFF, SdFFF and Flow FFF systems.

There are several mechanisms that can lead to a non-uniform concentration distribution of the component in the field direction. For species that are relatively small (<1 μm), diffusion (*i.e.* Brownian motion) is the main mechanism that opposes the field force. This retention mode is named as "Brownian mode" (or normal mode). In this mode, a component is characterized from its mean sojourn time in the channel, the retention time t_R, by relating it to a dimensionless parameter, λ, which reflects the strength of its interaction with the applied field. λ, the basic FFF parameter, is expressed as:

$$\lambda = \frac{k_B T}{F w} = \frac{D}{U w} \qquad (1)$$

Where k_B is the Boltzmann constant, T is the absolute temperature, w is the channel thickness, F is the force exerted by the field, D is the component diffusion coefficient, and U is the field-induced velocity. These two expressions of λ are equivalent since $F = f\,U$ (Stokes equation) and $D = k_B\,T\,/\,f$ (Einstein equation) f being the friction factor. The Stokes-Einstein equation relates the diffusion coefficient to the hydrodynamic radius R_h of the component molecules (or particles) as:

$$D = \frac{k_B T}{6\pi\eta R_h} \quad \textbf{(2)}$$

Where η is the viscosity of the carrier liquid. In sedimentation FFF (SdFFF, Fig. **1**), which uses a gravitational or centrifugal force, G being the acceleration of the field (equal to the Earth gravitational acceleration in the first case, and to $\omega^2 R_c$ in the second one, where ω is the angular velocity, R_c the curvature radius of the channel), λ is given by:

$$\lambda = \frac{6k_B T}{\pi\left(\rho_p - \rho_l\right)G w}\frac{1}{d_v^3} \quad \text{(SdFFF)} \quad \textbf{(3)}$$

Where ρ_p and ρ_l are the densities of the component particles and carrier liquid, respectively, and d_v the diameter of a spherical particle having the same volume as that of the component particles. In flow FFF (FlFFF or F4, Fig. **1**), the field is a cross-flow of carrier liquid, U being the velocity of this uniform cross-flow. Therefore λ is expressed as:

$$\lambda = \frac{DV_0}{Q_c w^2} = \frac{k_B T V_0}{6\pi\eta Q_c w^2}\frac{1}{R_h} \quad \text{(F4)} \quad \textbf{(4)}$$

Where Q_c is the cross-flow rate and V_0 the volume of the FFF channel. In thermal FFF (ThFFF, Fig. **1**), a temperature gradient, dT/dx, is imposed across the channel. The component migration in this gradient is due to the Soret effect, characterized by the Soret coefficient, S_T, and occurs generally toward the cold wall. λ is given by:

$$\lambda = \frac{1}{\left(S_T + \gamma\right)\frac{\mathrm{d}T}{\mathrm{d}x} w} \approx \frac{1}{S_T\,\Delta T} = \frac{D}{D_T\,\Delta T} \quad \text{(ThFFF)} \quad \textbf{(5)}$$

ΔT is the temperature difference between the two walls, D_T is the thermodiffusion coefficient (sometimes called thermophoretic mobility) and γ is the thermal expansion coefficient of the carrier liquid. For some component types, *e.g.* homopolymers, D_T is independent of the molar mass, so that their λ values are proportional to D and, thus, according to eq. (2), are inversely proportional to their hydrodynamic radius.

Basic Retention Equation in the Brownian Mode

The relationship between the t_R and λ is complex and depends on the field strength, the flow characteristics as well as on the particular operating conditions. In order to approach it, one refers it to a model corresponding to a simplified set of conditions which grasps the main processes occurring during the migration in the FFF channel. This basic FFF retention model assumes, first, that, during its migration all along the channel, the component concentration in the field direction decreases exponentially from the accumulation wall to the depletion one, λw being the space constant of this exponential distribution, and, secondly, that the axial flow velocity profile is parabolic (Poiseuille flow). In these conditions, the ratio, R, of the mean component axial velocity to the average flow velocity is [2]:

$$R = 6\lambda\left[\coth\left(\frac{1}{2\lambda}\right)-2\lambda\right] = 6\lambda\left[\frac{1+e^{-1/\lambda}}{1-e^{-1/\lambda}}-2\lambda\right] \qquad \textbf{(6)}$$

This is the basic FFF retention equation in the Brownian mode. When the field strength and the average carrier velocity are constant all along the channel, R represents the ratio, t_0/t_R, of the void time t_0 (*i.e.* the mean time spent in the channel by the carrier liquid) to the retention time t_R. From the measurement of t_R and the determination of t_0, R, is determined, followed by λ using eq. (6), from which the relevant component parameter is obtained using eqs. (3), (4) or (5) depending on the nature of the applied force field. To invert eq. (6), the high retention limit (*i.e.* vanishing λ limit) of this equation is frequently used. This limit is expressed as:

$$R = 6\lambda \qquad \text{(high retention limit)} \qquad \textbf{(7)}$$

Then, using this equation, together with eq. (1), it appears that:

$$t_R = \frac{wt_0}{6k_BT}F \qquad \text{(high retention limit)} \qquad \textbf{(8)}$$

i.e. t_R is linearly proportional to the force exerted by the field on the component molecules (or particles). Although frequently used due their simplicity, eqs. (7) and (8), and derived expressions, are approximations from the exact equation of the basic model. They lead to errors larger than 10%, 3% or 1% if t_R/t_0 is smaller than 4, 12 or 34, respectively. A much more correct estimation, λ_{est}, of λ from the retention time is obtained if the following expression is used instead of eq. (7):

$$\lambda_{est} = \frac{1}{4}\left(1-\sqrt{1-\frac{4t_0}{3t_R}}\right) \qquad \textbf{(9)}$$

which corresponds to $R = 6\,\lambda\,(1 - 2\,\lambda)$. Indeed, the error on λ associated to eq. (9) is lower than 1% if t_R/t_0 is larger than 1.53 and lower than 0.027% if t_R/t_0 is larger than 2.

It is important to be aware of the assumptions behind the derivation of eq. (6) and subsequent expressions (eqs. (7)-(9)). Indeed, any failure to satisfy these conditions leads to a more or less significant error in the determination of the component parameter of interest from its retention time. Assuming that the sample concentration decreases exponentially from the accumulation wall implies that: a) the sample starts its migration in near steady-state conditions (hence that the sample relaxation following its injection in the channel is fulfilled); b) that the force acting on the sample molecules or particles is constant all across and along the channel (hence electrostatic and van der Waals interactions between the component and the channel walls are absent; furthermore eqs. (6)-(9) do not apply for field programming operation); c) that the molecules or particles are small enough to be considered as point-like objects, *i.e.* they do not exhibit steric exclusion from the channel walls, they do not lead to hydrodynamic lift forces acting on them and they move at the same velocity as the unperturbed carrier at the position of their center of mass; d) that the component concentration is small enough for particle-particle interactions to be negligible (no sample overloading). In addition, the assumption of a parabolic flow profile implies that: e) the system is operated in isothermal conditions (*i.e.* there is no viscosity gradient in the channel) and f) edge effects are negligible (*i.e.* the channel cross-section aspect ratio is large enough). Equations are generally available when one (or more) of the conditions a) to f) is relaxed, but the derivation of the component parameter of interest then becomes more complex [3].

Peculiarities of the Basic Retention Model in Asymmetrical Flow FFF

Asymmetrical flow FFF (AsFlFFF or AF4) is presently the most widely used FFF method. While so-called symmetrical flow FFF (SF4) systems are parallel plate channels in which the two main walls are permeable to allow the uniform cross-flow of carrier liquid by means of a cross-flow pumping system independent of the axial flow pumping system; AF4 channels contain a single permeable wall, covered by a semi-permeable membrane, the depletion wall being plain (Fig. **1**). Therefore, in the elution process of AF4 systems, a single inlet flow pump is used, providing both the axial flow rate and the cross-flow rate. Since part of the inlet flow rate is crossing the membrane, and the axial flow rate decreases from the inlet to the outlet of the channel, which implies that the average axial flow velocity is not constant along the channel (except in a very particular set of operating conditions rarely implemented). Therefore, t_0 is not equal to V_0/Q_{in}, where Q_{in} is the channel inlet flow rate, as it is in symmetrical systems, but instead given by eq. (10) [3, 4]:

$$t_0 \;=\; \frac{V_0}{Q_c}\ln\left(1+\frac{Q_c}{Q_{out}}\frac{V_0'}{V_0}\right) \tag{10}$$

where Q_c is the cross-flow rate and Q_{out}, the flow rate at the outlet of the channel (note that $Q_{in} = Q_c + Q_{out}$). In AF4 systems, a focusing step is generally used in order to limit the initial width of the sample as well as to start the subsequent elution step in near-steady-state conditions. Subsequently, a correction ratio has to be used as shown in eq. (10) where is the channel volume past the focusing position (t_0 is then the time elapsed since the start of the elution step). This equation is valid with whatever the shape of the channel. Furthermore, the cross-flow velocity is not constant across the channel thickness, but varies from zero at the depletion wall to U_0 at the membrane (which acts as the accumulation wall retaining species which are too large to pass through the porous structure of the membrane). This invalidates the above hypothesis b) underlying the basic retention eq. (6). The error made by neglecting this effect is negligible in high retention conditions, but can reach several %, up to about 10% or some in practical situations. A correction in eq. (6) for estimating the λ value from the experimental R value has been recently computed [5] under the assumption that the cross-flow velocity at the membrane is constant along the channel [6]. When using AF4 to characterize a specie using, eq. (4), special attention should be paid, first, to the estimation of the channel thickness which is usually quite different from the nominal value because the membrane tends to protrude inside the channel [7], and second, electrostatic and van der Waals interactions of the component with membrane do not significantly affect the retention [8].

HF5 is another form of F4 systems, since a single inlet carrier flow pumping device is used to provide both axial and cross-flow rates in the elution step. These systems use hollow fibers of a cylindrical geometry (Fig. **1**). Assuming that the cross-flow velocity at the membrane, U_0, is constant all along the fiber, t_0 is also given by eq. (10), the ratio being then equal to (1 - z_0/L) where z_0 is the sample position at the start of the elution step. t_0/t_R, is then [9]:

$$R = \frac{4\left(e^{-1/4\lambda} - e^{-1/\lambda}\right)}{\sqrt{\pi/\lambda}\left[\operatorname{erf}\left(1/\sqrt{\lambda}\right) - \operatorname{erf}\left(1/\sqrt{4\lambda}\right)\right]} - 2 \qquad \textbf{(11)}$$

where (y) being the error function of the variable y and λ being defined as:

$$\lambda = \frac{D}{|U_0| R_{hf}} = \frac{2\pi L D}{Q_c} \qquad \textbf{(12)}$$

where L and R_{hf} are the fiber length and radius, respectively. The high retention limit (low λ limit) of eq. (11) is then [10]:

$$R = 4\lambda(1-4\lambda) \qquad \textbf{(13)}$$

It can be noted from eq. (12) that, for a given cross-flow rate and fiber length, λ and, thus, R do not depend on the fiber radius.

Retention of Large Species in FFF - Lift Retention Mode

For FFF operating in the Brownian mode, t_R increases when increasing the size of the component particles (or macromolecules) as a result of the stronger compression of their cloud near the accumulation wall. However, when the size of the particles becomes non negligible compared to the space constant of the cloud, equal to λw, the particles can no longer be considered as point-like objects and become sterically excluded from the wall. A retention model taking into account this steric effect has been developed [11]. It shows that, with increasing particle size, the retention ratio reaches a minimum for a particular size, the steric inversion size. The faster elution of the larger particles in this steric mode can be interpreted as due to the fact that the flow velocity at the position of the center of particles rolling along the wall is larger for larger particles, because of the shape of the flow velocity profile. However, even if it correctly predicts the order of elution of particles of various sizes, this interpretation is too simplistic for two reasons. First, due to particle-wall hydrodynamic effects, the migration velocity of a particle is not exactly equal to the unperturbed flow velocity at its center [12]. Second, particles of finite size transported by a flow in a tube or in a parallel plate conduit undergo a transverse hydrodynamic lift force of inertial origin which moves them away from the wall as well as away from the channel center line or plane. Thus, these particles tend to focus at some intermediate position between the wall and channel center [13] where they tend to form a hyperlayer, hence the name hyperlayer mode given to this operating mode. Because the size dependence of this lift force is stronger than the applied field force, the distance of the transverse focusing position from the wall depends on the particle size and increases with the size, which lead to the same elution order *vs.* size as that predicted by the steric model. The three effects (steric exclusion, particle-wall hydrodynamic interactions and hydrodynamic lift) are simultaneously involved in the behavior of the particles and several names have been used to describe the retention mechanism (steric, hyperlayer, lift hyperlayer). Because it is difficult to unravel the individual contributions of these effects and because the migration of the particles along the channel cannot occur in steady-state conditions, the theoretical prediction of the retention of large particles is difficult and we prefer to simply describe the elution of the particles above the inversion point as occurring in the lift retention mode. The particle diameter corresponding to the steric inversion depends on the field strength, the carrier flow rate and can be adjusted by manipulating these parameters. In practice it is around 1 μm. The size based selectivity (defined as the limit of the ratio of the relative difference in t_R of two components to the relative difference of their diameters, when the latter is vanishing) in the lift mode, depends on the experimental conditions and is predicted to be larger in F4 than in SdFFF [14], in agreement with experimental observations. Highly efficient separations of particles can be performed in this mode by combining field and flow programming in F4 [15].

Brownian mode for sub-micron particles and lift mode for micron-sized particles are the main operating modes used in FFF (Fig. **1**). Other mechanisms exist or may be involved for performing separations [3]. Especially, colloidal (electrostatic and van der Waals) interactions between particles and wall (or membrane) are generally present in nearly all FFF systems. Because they induce forces on the particles that vary quickly with the particle-wall distance, they affect the distribution of the particle cloud near the wall, which may have a significant influence on the retention ratio. They should be taken into account for the proper characterization of particles. A semi-empirical model has been developed for this purpose [16].

INSTRUMENTATION AND DETECTION

Sedimentation FFF (SdFFF)

Sedimentation FFF techniques can be divided in two groups, Gravitational FFF (GrFFF/GFFF) using the simple earth gravity; and SdFFF, using centrifugal or multigravitational field created by channel spinning [14, 17, 18]. Where GrFFF device is simple to build and set up, SdFFF is one of the most complex, requiring insertion of separation channel into rotating basket, rotation speed control to select and program field strength [19, 20], and high performance rotating seals allowing mobile phase and sample flowing without leakage [21, 22]. Optimization of elution conditions are based on field strength (value and decay), flow rate and the nature of mobile phase (pH, ionic strength, surfactant, viscosity…). In SdFFF, R is expressed as [14, 17, 23]:

$$\frac{1}{R} = \frac{t_R}{t_0} = \frac{|F|w}{6k_BT} = \frac{\pi {d_v}^3(\rho_p - \rho_l)Gw}{36k_BT} \quad \textbf{(14)}$$

The retention in SdFFF is connected to particles size and density, but also to shape and rigidity. For submicron particles, under Brownian elution mode, SdFFF presents a high size selectivity (≈ 3) [14]. Therefore after proper calibration and using retention parameter, and/or using effective detectors, SdFFF is as a powerful method to measure particle diameter, particle size distribution (PSD), molar mass, aggregate formation, macromolecule structure and organization, molecule adsorption or linking on nanoparticles [14, 23 - 26]. For micron-sized species such as starch granules eluted under the lift elution mode, the size selectivity usually ranges from 0.6 to 0.9 [14]. From retention equation, it could be possible to use SdFFF to determine particle size, PSD or sphericity index [17, 22, 25]. Diameter and PSD are determined after calibration using latex standards of known size and density. The calibration curve presents the following form [22, 27, 28]:

$$log\ t_R = -S_d\ log\ d + S_{\Delta\rho} log\ G\Delta\rho + log\ t_c \quad \textbf{(15)}$$

where S_d and $S\Delta\rho$ are the size and density selectivity respectively, and t_c a constant depending on the elution conditions.

If sample particles, have a density different from that of standards, then a density compensation methodology [28] can be used through eq. (16):

$$rpm_{sample} = rpm_{standard} \frac{\sqrt{\Delta\rho_{standard}}}{\sqrt{\Delta\rho_{sample}}} \quad \textbf{(16)}$$

Eq. (16) allows calculating the rotation per minute (*rpm*) for which t_R of the sample is similar to that of standard, and then determining sample size with the same calibration curve.

Flow FFF (F4)

The separation process in F4 is based on differences in Brownian diffusion coefficients and separation depends then on the size and shape of particles or molecules [29, 30]. All the F4 use similar devices: HPLC injector, pumps, and detector, their instrumental design and set up are yet different (see theoretical section and Fig. **1**). In AF4 working with constant cross-flow in Brownian mode, R_h can be determined *via* the following equation, obtained by combining eq. (4) and (7):

$$t_R = \frac{w^2 \pi \eta \, t_0 \, Q_C}{V_0 \, k_B \, T} R_h \qquad \textbf{(17)}$$

Decaying cross-flow gradients are often used for samples with high size disparities instead of a constant cross-flow. In this case, size distributions cannot be obtained by using eq. (17), but they can be calculated with its modifications for cross-flow gradients [31, 32]. Alternatively R_h can be determined by coupling FFF with online quasi-elastic light scattering (QELS) thus providing experimental size distributions. AF4-QELS coupling allowed confirming the efficiency of the AF4 retention equations theory in Brownian mode with constant cross-flow for the calculation of the size of glycopolymers and to highlight some of their limits [33]. The retention of the analyte on the membrane produces wrong size distributions when using retention equations, whereas size higher than 50 nm produce wrong R_h values with online QELS because of too long correlation times [34].

Detectors that are commonly used online with AF4 are multi-angle laser light scattering (MALLS) detectors which give the molar mass and the radius of gyration (R_g or r_{ms}) of each fraction (see Chapter 4). The weight-average molar mass (M_w) can be obtain provided that the refractive index increment (*dn/dc*) is known for the substance, and a concentration detector it used (UV-Vis or differential refractometric index (DRI) detectors generally). It is possible to distinguish between different conformations within the size distribution obtained from the AF4 with the *log-log* plot of R_g *vs.* M_w. The Flory exponent (v), extracted from the slope, has theoretically a value of 0.33 for a sphere, 1 for a rod and 0.5-0.6 for a random coil. The ratio between R_g and R_h gives valuable information about the conformation and shape of the macromolecules over the size distribution. The R_g/R_h ratio is 0.778 for a hard sphere, whereas typical ratios for macromolecules are between 1 and 2, and ratios >2 generally indicate elongated conformations [35]. R_g/R_h ratios between 0.3 and 0.6 account for "micro-gels" [36]. Furthermore, R_g/R_h together with apparent densities has been proposed as a means to distinguish aggregated *vs.* non-aggregated structures in a population [37,

38]. The apparent density ($\hat{\rho}$) over a size distribution can be obtained by calculation from M_w and R_g or R_h as:

$$\hat{\rho}_{rms,i} = \frac{M_{wi}}{V(R_g)_i N_A}\alpha \quad \text{or} \qquad \hat{\rho}_{h,i} = \frac{M_{wi}}{V(R_h)_i N_A} \tag{18}$$

where M_{wi} and V_i are the molar mass and the volume of fraction i, N_A is the Avogadro number and α=0.465.The apparent density describes the apparent volume of the molecules which can be considered to influence fundamental parameters such as the overlap concentration of a macromolecule. Such relationships have been observed for high amylopectin starch [39]. The apparent density is also linked to branching density as highly branched macromolecules are expected to display a higher density than less branched. The apparent density also describes the distribution of mass in volume for each fraction and, thus, provides information about scaling in a macromolecular population.

Optimization of elution conditions are based on cross-flow (value and decay) and carrier flow rates, membrane (type and cut off), channel thickness, carrier liquid composition (ionic strength, pH, viscosity…), amount of sample injected, sample preparation [30, 40]. A recent paper reviews the principal draw backs of F4 [30], when some of them could be avoided by a thourough optimization. For example, the membrane selection is particularly important in order to minimize unwanted membrane/sample interactions or sample loss. Loss of sample could be due to attractive interactions or too large cut off (which limits the accessible analyte molar mass range), whereas repulsive interactions can interfere with the fractionation. A drawback in comparison to column liquid chromatography is the high sample dilution occurring during migration in the channelup to the detector which limits the mass sensitivity. In addition the Brownian mode could not operate with high concentration as interactions between neighbour particles can modify their spatial distribution. When a sample is injected at a too high concentration, peaks are shifted and distorded. This overloading effect has been reported for different biopolymers (glutenins, amylopectins) [3, 41 - 44]. A variant of AF4, the frit-inlet AF4 (FIAF4) has been designed to reduce the concentration near the membrane and the occurrence of unwanted sample wall interactions caused by the stoppage of migration flow during the focusing period. This system has a specific channel design which allows by-passing the typical focusing/relaxation of classic AF4 channels (see theory section). During sample injection in FIAF4, sample components are pushed toward the accumulation wall by the action of a supplementary flow through a small permeable frit element located near the injection port [45]. Thus the analytes are hydrodynamically relaxed while they are continuously carried to the separation process without stopping the migration. However, the band broadening is higher in this configuration.

Thermal FFF (ThFFF)

ThFFF systems include common associated devices: sampler, pump, and detector, and the ThFFF channel is constructed by clamping a spacer with a geometric cut-out between two metal blocks made of copper and have mirror finished plating applied to the surface. The temperature gradient is achieved by keeping the hot wall at an elevated temperature through the use of software controlled heating rods, and the cold wall at lower temperature using a recirculating chiller. D and D_T act in opposition to each other (eq.(5), Fig. **1**) and allow separation according to molecular size and chemical composition, as the magnitude of the driving force depend on the chemical composition of polymers and the carrier liquid (polymer-solvent combination), and is independent of the molar mass of many homopolymers [29, 46 - 51]. Field programming can be employed to improve the separation of highly dispersed polymers. R can be measured experimentally and related to λ *via* eq. (19), a modification of eq. (6) including a velocity distortion factor term (ν_t) to account for the deviation from the parabolic velocity flow profile due to changes in eluent viscosity and thermal conductivity across the channel thickness due to the applied ΔT [52].

$$R = 6\,\lambda\left\{\nu_t + (1 - 6\lambda\nu_t)\left[coth\left(\frac{1}{2\lambda}\right) - 2\lambda\right]\right\} \tag{19}$$

When ν_t is equal to zero, the parabolic profile is regained. ThFFF is commonly used to measure D_T values for different polymer-solvent combinations *via* the retention time:

$$t_R = \frac{D_T \Delta T\, t_0}{6D} = \frac{D_T \Delta T\, \pi\eta t_0}{k_B T} R_h \tag{20}$$

The eluent used is indeed a crucial parameter and since D_T is unique for each polymer-solvent system, ThFFF can be used, *via* the measurement of the D_T value, to evaluate the chemical composition and the microstructure of a copolymer, provided D is measured independently [53 - 56]. ThFFF also allows determining the absolute molar mass (M) only if $S_T(D_T/D)$ is known, otherwise calibration is required [57 - 59] using narrow polymer standards. By plotting $logD/D_T$ *versus* $logM$, or $logM$ *versus* $logt_R$, the ratio D/D_T could be calculated [29, 57 - 59]. By coupling ThFFF to online MALLS-QELS-DRI detections one can obtain rapidly and simultaneously M, D and D_T (as t_R is proportional to D_T/D, eq. 20) distributions, and as a result, composition distributions [54, 55]. ThFFF is commonly coupled with MALLS, DRI, UV-Vis or viscometric detectors for absolute molar mass or size determination.

If ThFFF could be used to fractionate particles, its main applications are polymers fractionation [60 - 62] and it particularly suits for high molar mass homopolymers (>100 g.mol^{-1}), copolymers, and microgel-containing polymers [56]. Yet when F4 has been used predominantly in applications with aqueous carrier liquids, ThFFF is mainly used with organic solvents because hydrophilic polymers have a weak retention in water (low D_T). That is the reason why there are not, to our knowledge, publications describing the fractionation of water soluble biopolymers in aqueous solvents by ThFFF.

APPLICATIONS TO HYDROSOLUBLE BIOPOLYMERS

Polysaccharides

Starch

Starch is the most abundant storage reserve carbohydrate in plants. It is made up of two α-D glucans built upon mainly α-(1,4) linkages: amylose (~25%) which is essentially linear, and amylopectin, the major component, branched with 5-6% α-(1,6) linkages. Starch polysaccharides are organized within semi-crystalline granular structures of about 2-100 μm diameter, insoluble in cold water. Starch is used in a native or modified form for many industrial applications in food, paper, cosmetics, medical, materials, *etc*. Starch and starch derivatives structure, such as size distribution of the granules and molecular characteristics of amylose and amylopectin, greatly influences their functional properties [63 - 67]. Their characterization is then crucial for a better understanding of structure-properties relations.

A major application of SdFFF, concerns the characterization of starch granules, *i.e.* particles size and mass distribution [64, 65, 68], in particular for those presenting a bimodal PSD such as barley or wheat starches. These starches contain two types of granules, large A-type granules (10 to 40 μm) representing 90% in weight of the mass distribution, and small B-type granules (1 to 10 μm) representing 90% in number of the PSD [63 - 65]. Small granules affect the rate and extent of modification during malting. They have a high proportion of amylopectin and are less sensitive to amylase action leading to a slow surface erosion [63, 69, 70]. Many studies have been conducted to determine the particle size and mass distribution to determine the ratio of A-type/B-type granules, based on the use of calibrated GrFFF or SdFFF device. Chmelík *et al.* [69, 71 - 74] made an important work on barley starch used for malting. They described the optimization of GrFFF methods to sort granules, hyphenated with different characterization methods (Scanning electron microscopy (SEM), Low Angle Laser Light Scattering (LALLS), MALLS, Matrix-Assisted Laser Desorption/Ionisation (MALDI) mass spectrometry (MS)) to determine the PSD, the ratio of A-type/B-type granules, or the molar mass distribution of starch components. Reschiglian *et al.* [75] and Contado & Dondi [76] optimized a protocol associating a pre-fractionation with continuous split flow thin cell (SPLITT cell) with FFF to establish the PSD of the different sub-type of starch granules based on density compensation method [76, 77]. By using calibrated SdFFF to fractionate bimodal starches (barley [28] and wheat [78]), Farmakis *et al.* also studied the size distribution and mass ratio between A-type and B-type granules. They defined optimal elution conditions by taking account of the pH, the ionic strength, the sample concentration and the nature of the surfactants [28, 78, 79]. They also used different mathematical methodologies [78] to determine from the UV signal, the mass ratio of starch granules and compared PSD [78, 80]. They thus measured the impact of these conditions on the value of size and mass distribution compared to the results obtained by SEM or light scattering techniques [28, 78, 79]. Chmelík *et al.* [72] also demonstrated that not only the ratio of large and small starch granules is important for malting properties, but also other factors such as activity of hydrolytic enzymes (α- and β-amylases), or the effects of varieties and harvest year [74]. Amylases (α-amylase in particular) are able to attack the native starch granule by the way of many different processes (exo and endo-corrosion) [70]. This process, called amylolysis, is associated with both granule's size and density changes. Gr-and SdFFF, can be used as easy and fast methods to monitor amylolysis, by the simple comparison of fractograms between native and treated starches. In association with specific methods to characterize and quantify the product of amylolysis (LALLS, MALDI-MS, NMR, microscopy, *etc.*), FFF can be further used to select enzymes, plants, conditions leading to better kinetics of cooking, malting or other starch modification processes [81 - 83]. More recently, Kim *et al.* [84] demonstrated that calibrated GrFFF can be an efficient device to monitor the PSD evolution of potato and sweet potato starches due to changes in swelling properties. Concerning the starch granules composition, Hanselman *et al.* and Aberle *et al.* [84 - 86] described the use of SdFFF coupled with online MALLS and DRI detectors to study the molar mass and radius distributions of starch components. The fractionation of amylopectin clusters has been performed under high field strength (15000g, Brownian mode) with programmed field decay, allowing the study of large range molar mass distribution which is not fully achieved by SEC. Yet, these analyses are not quantitative because particles with an R_h smaller than 50 nm are not fractionated which produces poor elution recoveries [86].

Several studies report on the use of SF4-MALLS-DRI [87 - 89] and AF4-MALLS-DRI [34, 41, 90 - 101] for the macromolecular characterization of native starch polysaccharides and some of their derivatives [39, 44, 102 - 109]. The authors use generally decayed cross-flow as efficient separation of amylose and amylopectin cannot be achieved by applying constant cross flows [88]. A systematic study comparing hydrodynamic chromatography coupled with SEC (HDC-SEC) and AF4 for starch polymers fractionation has shown that AF4 gives a better fractionation of amylose and amylopectin and of the native amylopectin fraction than HDC-SEC (Fig. **2**), without any shear degradation [34]. Yet, as starch macromolecules have an extremely broad size distribution (5 nm to 400 nm), a tendency to aggregation and amylopectin has an extremely high M_w (10^8 $g.mol^{-1}$), a great attention must be payed to the method developement. In fact the recoveries of the AF4 analysis of starch macromolecules are not always good: 50% [88], 84-100% [87], 60-89% [102], 92-95% [106]. These loss of sample are the result of retention on the membrane (large amylopectin fractions are loosen) or too gentle flow methods and high membrane cut off, which aid the fractionation of large molecules but may induce and a loss of sample (mainly amylose) through the membrane. The concentration must be carefully adjusted as amylopectin has a great tendency to overloading [39, 41, 90, 101], and this effect is increased for cationic amylopectins [44]. Moreover an inversion of the fractionation mode from Brownian to lift has been observed when decreasing the channel thickness due to the large R_h of the amylopectin macromolecules [90]. One of the main issues in starch macromolecular characterization is to achieve the amylose and amylopectin solubilisation [112]. Solubilisation in hot water using autoclaving [101, 112 - 114], in dimethylsulfoxide (DMSO) at 100°C [101], or in DMSO and then in water using microwave heating under pressure [34, 41] is generally used but solutions generally still contain aggregates. Perez-Rea *et al.* [101] shown recently, by comparing different solubilisation methods, that solubilizations in DMSO at 100°C or in water by autoclaving at 140°C give the best results for starch polymers structural characterization by AF4-MALLS. However they found that even after a filtration procedure, some aggregates are remaining in the larger fractions of amylopectin. An extensive study of amylopectin from various botanical sources using AF4-MALLS has shown that it is possible to evaluate the branching features of amylopectins *via* a comprehensive treatment of light scattering data [34, 40].

Branching parameters (see Chapter 4) were determined for the amylopectins by comparing the R_g for the branched (amylopectin) and the corresponding linear macromolecule (amylose) enabling the acquisition of branching data *via* the hyperbranched theory model over the size distributions. The influence of various processes on native starch have also been studied by AF4-MALLS such as high-pressure homogenization [39, 105, 107], jet cooking [106], extrusion and γ-ray irradiation [100], pregelatinisation [110], alkaline treatment [111]. A significant degradation was found in studies of mechanical degradations [39, 100, 103, 105]. A decrease of molar massfunction of the homogenization pressure was observed together with drastic changes in conformation, shape (R_g/R_h) and apparent densities. Some chemically modified starches were also studied by AF4-MALLS such as hydroxypropyl starch [102] and octenyl succinic anhydride starch [104, 106] carboxymethylated starch [108] and cationic starches [44, 104, 106]. It is noteworthy that tailing was seen in the elution profile due to attractive interaction between cationic starch molecules and the channel membrane, and repulsion among cationic starch molecules. In order to avoid such effects, the ionic strength of the eluent must be carefully adjusted when working with such polyelectrolytes.

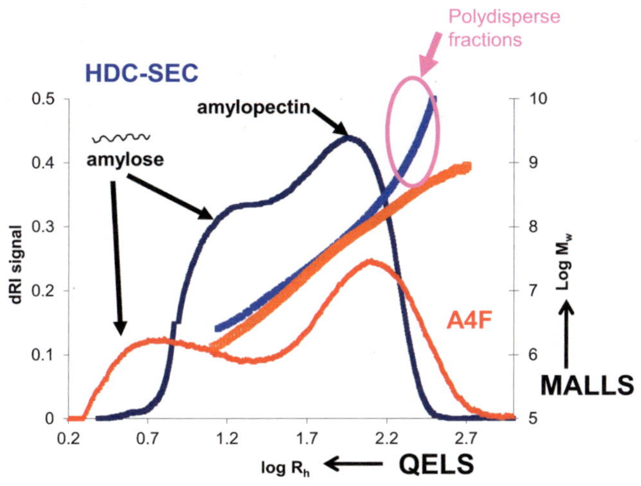

Fig. (2). Comparison of HDC-SEC-MALLS-QELS-DRI and AF4-MALLS-QELS-DRI analyses for the fractionation of wild type rice starch (adapted from [34]).

Non Starchy α-Glucans

Glycogen is the major energy reserve in animals and bacteria. It is mainly made up of α-(1,4) linked α-D-glucosyl units, yet it has a highly branched structure with 7-10% of α-(1,6) branching points evenly distributed within the glycogen particles. AF4-MALLS has proven to be very powerful for the analysis of natural [41, 115] and synthetic glycogens [33, 115, 116]. Size and molar mass distributions of glycogen-type polymers, with various size and branching degree and synthesized enzymatically (with elongation and branching enzymes) *in vitro* from sucrose, were obtained by AF4-MALLS-QELS [33, 116] with a better resolution than with SEC (Fig. **3**). Size distributions obtained by AF4-MALL--QELS were in line with TEM data [33] and molar mass and size distributions have shown to greatly depend on the substrate concentration, the branched/elongation enzymatic activity ratio [33] (Fig. **3**) and time of synthesis

[116]. Structural data were extracted from scaling relationships, R_g/R_h values and apparent densities. The AF4-MALLS-QELS-DRI analysis thus permits to approach the molecular mechanism of the enzymatic polymerization [33, 116] and to point out that natural glycogens have various conformations function not only of their origin but also within their size distribution [33, 34, 41, 115].

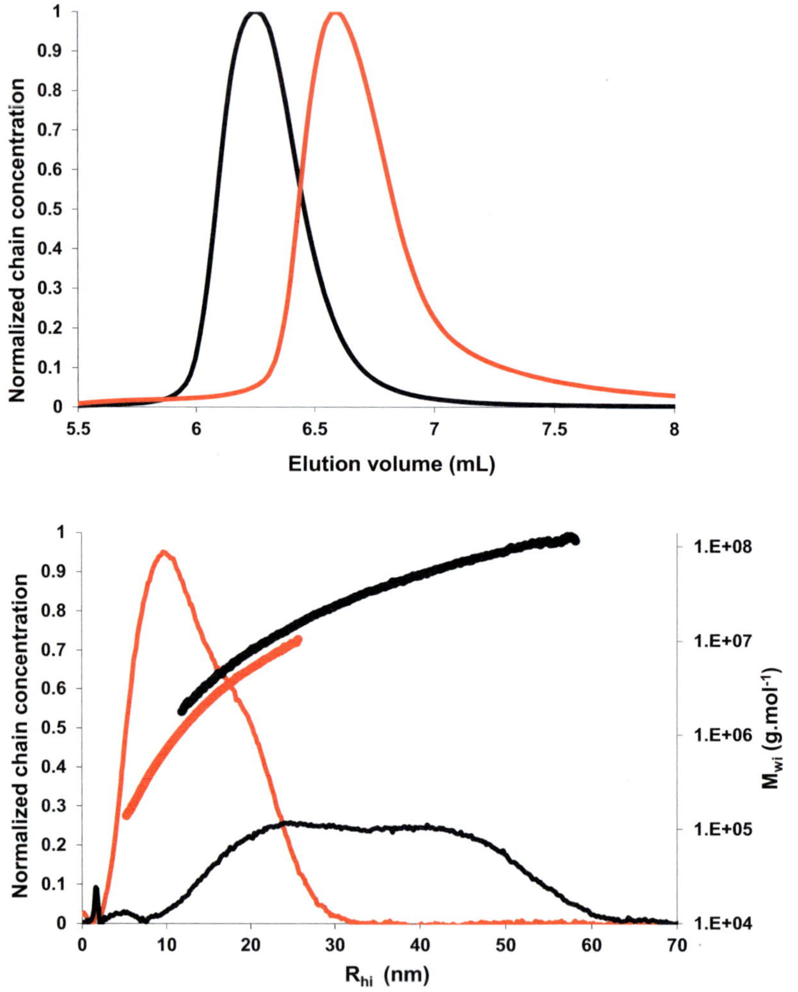

Fig. (3). Comparison of SEC (top) and AF4-MALLS-QELS-DRI (bottom) analyses for the fractionation of synthetic glycogens (adapted from [33, 117]). Glycogens enzymatically synthetized with 17 (red) and 205 (black) g.L^{-1} sucrose, molar mass (thick lines).

Dextrans are produced by lactic acid bacteria or enzymatically by dextransucrases and they are used for example as blood plasma volume expanders or to improve the texture and viscosity of fermented food products. They are based upon a backbone of α-(1,6) linked D-glucosyl units with randomly distributed side chains connected to the backbone by α-(1,3), α-(1,2) or α-(1,4) linkages. Their branching topology is complex and not clearly understood, the determination of the structure and macromolecular properties of high molar mass dextrans is thus challenging. They are only few AF4-MALLS studies on dextrans dealing with commercial dextrans from Leuconostoc spp., sometimes used as standards for calibration [34, 118, 119] and dextrans isolated from other bacteria [119] or synthetized by enzymes [120]. AF4-MALLS has shown its high efficiency for structural determination (M_w, R_g and scaling properties) of extremely high M_w dextrans (0.76 to 6.02 × 10^8 g.mol^{-1}) displaying from 3 to 20% α-(1,3) linkages, and synthesized *in vitro* from sucrose by engineered mutants of dextransucrase [120]. The ν values reveal that the dextrans, which contain 15 and 20% α-(1,3) linkages, are highly dense and contain long ramifications, whereas the dextran with 3% α-(1,3) linkages behaves as a random coil.

Pullulans are linear polymaltotrioses form microbial origin composed of α-(1,6) linked units. As standard substances they have been used to approach AF4 efficiency [121], and determine the effect of various parameters on the separation: carrier liquid composition [122, 123], flow conditions [118, 124], temperature [125], or injected amount [118, 123, 126] and also for calibration [34, 121]. Adolphi *et al.* [127] confirmed the flexible coil behavior of pullulan in good solvent by determining their molar masses, radii, persistence length (stiffness), R_g/R_h and apparent densities from average properties thanks to SF4-MALLS-DRI. SF4-MALLS have also been used to study pullulan derivatives (ampholytic and amphiphilic). Duval *et al.* [128] reported better results and more complete characterization of hydrophobically modified pullulans in aqueous solutions than with SEC-MALLS due to the absence of interactions between the hydrophobic segments and the stationary phase in SF4. Moreover, AF4-MALLS-QELS made possible the detection of degraded products and high-aggregation compounds produced during pullulan modification [129].

β-Glycans

Cellulose is an insoluble linear chain of β-(1,4) linked glucose units. It is widely represented in nature and used for many applications in industry in a derivatized form. AF4-MALLS has been utilized for comprehensive works on ethylhydroxyethyl cellulose [130 - 132], hydroxypropylmethyl cellulose [133, 134], hydroxypropyl cellulose [124, 135]) and carboxymethyl cellulose [136, 137]. Optimizing the experimental conditions, especially flow rates and injected amounts, is of great importance for adequate analysis, as under estimations of polydispersities and molar masses have been reported for modified celluloses in case of high injected quantity and low recovery rates were observed with high flow rates or high injected quantity [133]. Arabinoxylans are plant cell wall hemicelluloses constituted by β-(1,4)- linked D-xylopyranosyl backbone with α-L- arabinofuranosyl units attached to the positions of O-2, O-3, or both. Their structure and substitution pattern vary with the origin, and they have a tendency to aggregation. Pitkanen *et al.* [138] showed that even if the aggregate concentrations were low, AF4 possess a higher fractionating power for the differentiation of aggregates from individual chains than SEC which led to overestimation of molar mass and size.

Physicochemical properties of cereal β-glucans (alternating blocks of β-(1,3) and β-(1,4) linked glucose units), associated with beneficial health effects, are related to their solution behavior and possibly to their propensity to form aggregates. That is why several recent studies have used AF4-MALLS [38, 139, 140] to characterize their behavior in solution. Cereal β-glucans solutions contain aggregates even in alkaline solvents [38] or after heating in dilute salt solutions [38, 140]. By utilizing AF4-MALLS, distinction between single macromolecules and aggregates was achieved [38]. Moreover, by combining data from the different detectors, it was possible to probe different conformations of β-glucan through the calculation of apparent density and the R_g/R_h ratio *versus* the degree of aggregation ($M/M_{w,0}$ with $M_{w,0}$ the M_w of the single macromolecule). Three species were thus identified: rod-like (chains shorter than the persistence length) and random coil (above the persistence length), aggregated structures as hyperbranched, and swollen micro-gel secondary aggregates (*i.e.* aggregates of the primary aggregates). Cereal β-glucan could also be semi-quantitatively identified in complex mixtures by using post-AF4 channel in-line labeling with calcofluor [141] and fluorescence detection [139]. The effects of 0.45 μm filtering of β-glucan solutions before injection and of freeze-thaw cycles have also been investigated by AF4-MALLS [140]. Freeze-thaw cycling resulted in considerable aggregation and a drastic increase in the density of the aggregates when the filtration of solutions influenced the conformation of large aggregates. Moreover, AF4-MALLS allowed following the structural modifications of pure barley and oat β-glucans that occurs through gastrointestinal passage [139]. Makela *et al.* [142] compared the efficiency of AF4-MALLS in aqueous solvent and SEC-MALLS in LiBr-DMSO to follow the effect of oxidative degradation of barley β-glucan. Both SEC and AF4 showed comparable results as long as aggregates (only observed in AF4) were not included in the integration. Oxidation has shown to both cause degradation of β-glucan and to increase aggregation tendency.

Scleroglucan is a branched polysaccharide consisting of β-(1,3) connected D-glucopyranose units with β-(1,6) connections of one glucopyranose monomer approximately every third chain monomer. It is widely used in various industrial applications thanks to its exceptionally high viscosity in aqueous solution and resistance to hydrolysis, temperature and electrolyte changes. Its molecular conformation in solution, a triple helix molecular structure which shows a rigid to flexible behavior function of molar mass, explains these properties. Establishing its molecular size (distribution) is then particularly important in product control. Augsten *et al.* [143] have shown the better suitability of AF4-MALLS than SEC for the broad size and molar mass distributions of scleroglucan samples. In addition, the characterization *via* AF4-MALLS proved the presence of low and high M_w fractions (10^5-10^6 g.mol^{-1} and 10^7-10^8 g.mol^{-1} respectively) in electron irradiated scleroglucan samples, showing that this sterilization process caused scission of the long chains and increased the lower M_w fraction.

The stability of levan, a polyfructan comprised of D-fructofuranosyl monomers, was approached by monitoring changes in its physicochemical properties determined by AF4-MALLS (M_w, R_g,R_h, R_g/R_h and aggregation state) with respect to solution pH and hydrolysis time [144].

Gums

Gum arabic and mesquite gum are based on the association of proteinaceous moieties to a complex polysaccharide structure consisting mainly of an arabinogalactan chain. Their emulsifying properties are due to this association. The comparative fractionation of gum arabic by SF4 and SEC [145, 146] showed reliable data [146] even if SF4 produces a better resolution between the protein rich fraction and the protein poor fraction. The distribution of proteinaceous moieties over the size distribution was obtained by using a fluorescence detector in combination with AF4 on labeled gums [147]. By using AF4-MALLS to determine the conformational properties of the gums, the authors were also able to connect emulsification properties and molecular conformation.

Xanthan gum is an anionic heteropolysaccharide produced by bacteria which consists of a backbone of β-(1,4)-linked D-glucose residues with branches of mainly D-mannopyranosyl and D-glucopyranosyl uronicacid. In solution, it forms double helices with an extended rod-like conformation which is believed to produce its highly pseudo-plastic flow properties. In a study comparing analytical ultra-centrifugation and AF4 for the fractionation of globular proteins and xanthan, Pauck & Colfen [148] concluded to the obtention of a "wrong"diffusion coefficient for elongated and rod-like macromolecules. Yet the flow conditions and overloading effects were not investigated in this paper while the sensitivity to overloading for xanthan gum in AF4 has been reported [149]. This effect is probably due to its extreme molecular conformation and strong viscosity increasing properties.

Marine Biopolymers

Chitosan is a natural cationic polysaccharide based on β-(1,4)-linked poly (N-glucosamine) that finds increasing use in pharmaceutical applications. AF4-MALLS-DRI was successfully applied to characterize various commercial chitosans and batches of pharmaceutical interest with M_w from 10^4 to 10^6 g.mol^{-1} [150, 151]. The conformation was molar mass dependent, ranging from open structures molecules to θ coil conditions or an even more compact conformation. The main advantages of AF4 for these studies were the broad mass range covered and the possibility to analyze samples without purification. Online AF4-UV-Vis-MALLS-QELS investigations of DNA/chitosan complexes yield simu-ltaneously the size distribution of the complexes together with the content of unbound polycation (chitosan). To facilitate the UV-Vis detection, the chitosans were labeled with rhodamine B.

Carrageenan is made up of galactose and 3,6-anhydrogalactose, which can be both sulfated and non-sulfated, and alternatively linked by β-(1,4) and α-(1,3). Comprehensive studies on the coil-helix transition of α and κ-carrageenan occurring in the presence of salts [152, 153] or with temperature [149] were performed using AF4-MALLS. In 0.1 M NaCl, α-carrageenan appeared more elongated (R_g/R_h= 2.7) compared to what was reported for κ-carrageenan previously (R_g/R_h= 1.7) [153] which behaved as a random coil, while formed helical structures in 0.1 M NaI. Wittgren *et al.* [153] and Viebke & Williams [149] also found that the R_g/R_h is overestimated when R_h is calculated *via* the retention time, probably because of insufficient data on the behavior of rod like macromolecules in the separation channel. They suggested therefore that F4 theory has to be modified to be able to account for rod-like macromolecules. Alginate (alginic acid) is a linear copolymer of β-(1,4) linked D-mannuronate and α-(1,4) linked L-guluronate. It is difficult to analyze because of its polyelectrolyte character [154] and strong viscosity. Yet the correct operating conditions were established for the characterization of alginate by AF4-MALLS [154]. Moreover, in a comparative study between SEC and AF4, Storz *et al.* [155] shown that getting molar mass information for ultra-high viscous alginates with $M_w > 10^5$ g.mol^{-1} is only possible with AF4-MALLS, especially when alginates were modified by covalent fixation of poly(ε-caprolactone) [156]. SF4 has also been used for studying alginates behavior under filtration conditions [157]. Conformational changes in the alginates and adsorption at the separation channel membrane function of salt concentration were especially investigated. Alginate-heavy metal interactions have been studied using AF4-MALLS by Alasonati *et al.* [158] who found interesting conformational changes in the presence of metal ions (Pb^{2+} and Cd^{2+}). By coupling AF4 to inductively coupled plasma MS to detect Pb and Cd, their concentration was determined on the whole size distribution of alginates.

Hyaluronan

Hyaluronic acid (HA or hyaluronan) is a linear polysaccharide constituted by a repeating unit composed of β-D-glucuronic acid and N-acetyl-β-D-glucosamine. HA is a polyelectrolyte (carboxylic groups) and occurs in the body as a sodium salt (NaHA). HA and NaHA have important biological function and applications in pharmaceutics and cosmetics, which are linked to their macromolecular structure and conformation in solution. Understanding their structure is then crucial, yet their polyelectrolyte characteristics and their high M_w (up to 10^8 $g.mol^{-1}$) made it difficult to achieve because of the lack of relevant SEC packing materials with sufficient pore sizes and the possible shear degradation of HA chains [159]. AF4-MALLS has been used for the size characterization of HA and NaHA [160], especially with FIAF4, a particularly gentle technique [161 - 163]. The evolution of molar mass distribution and conformation of HA samples in aqueous solution after biodegradation processes such as thermal treatment [164 - 166], irradiation by γ-rays [161, 166], ultrasonication, and enzymatic or alkaline hydrolysis [167] has also been followed by FIAF4-MALLS. Kim *et al.* [163] have used FIAF4-MALS to describe the effect of the ionic strength on the conformation (R_g *vs* M_w), the M_w, and the intermolecular entanglement and disentanglement of NaHA. This has been achieved by varying the ionic strength of the carrier solution by using different NaCl concentrations. NaHA molecules become more extended and form a random coil structure due to the decreased charge screening effect when the NaCl concentration decreases. Nevertheless, in order to obtain a reliable fractionation, a previous optimization of experimental conditions (ionic strength of carrier liquid, sample concentration and injection amount) and field programming (cross-flow rate) should be done. In particular, the salt concentration of the carrier solution appeared to have a large effect on the retention of the (charged) hyaluronate molecules.

Proteins

Proteins are one of the largest applications of FFF technique even if it is not yet widely used for their study. Although SdFFF [168] has been recently applied to protein characterization, F4 has been mostly employed for protein analysis. Probably because *D*, which is related to size and shape, is a fundamental parameter to determine in order to probe changes in the native structure of the proteins which determines their functional efficacy (*e.g.* for antibodies or enzymes).

AF4 can be used for the prefractionation of proteins before their characterization by MS. Proteins from cytoplasmic membrane of prostatic cancer cell lysates have been prefractionated by Kang *et al.* utilizing FIAF4 [169] whereas Li *et al.* [170] used AF4 to clean up protein complexes adsorbed on polystyrene particles and Kim *et al.* [171] showed that HF5 is a promising method for a pre-MS separation technique in proteomics as it allows improving the MS detection.

Characterization of the extent of protein aggregation with respect to their size and amount is crucial because these aggregates play a role in many properties of biological systems. The characterization of protein aggregates in the size range 0.1-1.0 μm is of particular interest but it corresponds to a measurement gap in currently used techniques. F4 techniques are better than SEC for protein aggregation characterization since it covers a broader size range, utilizes low shear rates, and can be performed under a broad range of buffer conditions. It can therefore be used to gain understanding of the susceptibility of protein aggregation to different ionic strengths, pH, buffer, *etc.* while preserving fragile aggregate structures [172, 173].

AF4 has been used for the fractionation of glutenins (*i.e.* ultra-high molar mass wheat proteins) which have a highly aggregated structure [43, 174, 175]. More recently, AF4-MALLS allowed (i) to obtain the molar mass distribution [176, 177] and efficient separation and size characterization of monomeric and polymeric wheat proteins in a single run [178], (ii) to evidence their molecular degradation while sonication is used for the sample preparation [176 - 178]. Casein micelles (*i.e.* natural colloidal protein aggregates in milk) have a major importance on the properties and rheology of many dairy products which are controlled by their size distribution and structure. The size and conformational properties of casein micelles have been investigated utilizing AF4-MALLS [37]. This made possible to point out a large majority of casein micelles of spherical shape and few larger aggregates with drastically different properties corresponding to non-isotropic conformations with lower apparent densities. By comparison with modeling results, it could be concluded that the aggregates should consist of individual casein micelles aggregated in linearly (threadlike) or in a sheet-like structure, rather than forming random spherical clusters. Guyomarc'h *et al.* [179] with AF4-MALLS, and Zhu *et al.* [180] with HF5-MALLS, were able to follow the heat-induced aggregation in the skim milk [179] and of β-lactoglobulin [180] and to separate species in a wide size range: from individual proteins to large aggregates in one run, which has been impossible with SEC [179]. Gelatin can also been investigated with AF4 which is more suitable than SEC for its fractionation as AF4 do not abrade the high molar mass species [181]. By using AF4-MALLS-UV, Rbii *et al.* [182, 183] showed that the gelatin aging resulted in the formation of dense aggregated particles. SF4 and SdFFF have also been utilized to follow the hen egg white aggregation induced by $ZnCl_2$ and/or pH [173], SdFFF has been used to study the formation of aggregates and SF4 to determine the amount of non-aggregated proteins.

The most often used detection for proteins is the UV detection, which is usefull to obtain a chemical composition distribution. However, it is important to be aware that when UV detection is used as a concentration source it complicates the interpretation of the results for large aggregates as the scattering contribution to light extinction increase with the particule size.

AF4-MALLS not only gives information on the native aggregation of proteins, but it may also be a valuable support in monitoring biotechological processes for recombinant protein production and refolding. For example, aggregation and conformational changes of human immunoglobulin (IgG) induced either by thermal stress or freeze-thaw cycling has been evaluated by AF4 and SEC coupled with MALLS, UV, and fluorescence detection [184]. The elution recoveries were twice as more in AF4 and in particular, the amount of larger aggregates (trimers or larger) measured was greater by AF4, indicating a loose of the aggregates in the SEC due to adsorption or shearing. Meanwhile, SEC was shown to give better resolution between the IgG monomers and dimers compared to AF4. AF4-MALLS has also been employed to study (i) prion protein aggregation and thereby find correlation between size and infectivity of the prion particles [185], (ii) the dynamic aggregation process of Aβ proteins and peptides which are crucial in the pathogenesis of Alzheimer's disease [186].

The gentle nature of the separation by AF4 preserves the structure and conformation of analytes, which makes it a very valuable technique for the precise characterization of Virus-like particles (*i.e.* self-assemblies of viral proteins, VLP) [187]. First, they can be fractionated into fragments, monomers, dimers, oligomers, and aggregates using carefully selected cross-flow programs, focusing times, and membrane materials, second, in combination with MALLS a more accurate and faster size distribution information is obtained for heterogeneous samples of VLP than with TEM or QELS [187, 188]. AF4-based analytical characterization therefore becomes indispensable for complete characterization of VLP [189], as M_w, R_h and R_g, the composition and the purity of the final drug product can be obtained from a single analysis of DNA-VLP gene delivery vehicles by AF4-MALLS-DRI-fluorescence-diode array detection. Moreover AF4-MALLS made possible the evaluation of the quaternary (aggregate) size distribution of glutathione S-transferase purified viral protein [190] and influenza virions during vaccine preparation processes [191].

AF4 also allows fractionating bioparticules such as lipoproteins, which are assemblies of proteins and lipids that have a lipids and cholesterols carrier function in blood. As they are involved in coronary artery diseases, their size and shape characterization in the serum of patients, prior to cardiovascular treatment, may improve diagnosis and remedial methodology. Setala *et al.* [192] followed the transfer mechanisms of phospholipids and phospholipid transfer proteins between small unilamellar lipid vesicles and high-density lipoproteins (HDL) using AF4. Miniaturized systems (chip-type AF4) that lower the analysis time, the eluent consumption and the amount of sample load, have proven their efficacy to probe aggregation and fusion after enzyme treatment for low-density lipoprotein (LDL) [193] and for the lipidomic analysis of LDL and HDL from human serum [194]. The size and shape of lipoproteins from a whole serum sample have been determined by AF4-MALLS and HF5-MALLS [195], and recently cholesterol and triglyceride assemblies associated with lipoprotein subclasses was characterized by AF4 with online enzymatic detection [196, 197], using UV-Vis detectors to monitor the reaction products. This opens up perspectives for fast lipoprotein profiling.

APPLICATIONS TO BIOPOLYMERS SOLUBLE IN ORGANIC SOLVENTS

The majority of literature on FFF in organic solvents concerns synthetic polymers [198]. Only very few biopolymers are soluble in organic solvents. The most common are natural polyisoprene, or natural rubber (NR) [199, 200] and polyesters (polyhydroxyalkanoates or PHAs) produced by microorganisms [201, 202]. PHAs are intracellular deposited by bacteria as energy or reserve. Today, microbial polyesters are only analyzed by SEC with chloroform as mobile phase. Yet, one paper reported on the analysis of amylose-polystyrene block copolymers in organic solvent by AF4 [203], and there is a unique publication reporting the separation of water soluble biopolymers by ThFFF which is about polysaccharide (pullulans, dextrans and starch) fractionation in DMSO [204]. However such separations have not been replicated, one of the reasons may be the poor solubility of these materials and the incomplete fractionation of the amylose and amylopectin starch components. Only one study reported on the structural properties of native and oxidized condensed tannins (a major class of plant polyphenols) obtained by AF4-MALS-UV in ethanol [205], the main advantage of AF4 being less interactions with tannins. Finally, to our knowledge, the only biopolymer analyzed by FFF techniques in organic solvent is NR.

Among the many plants producing polyisoprene, some are able to biosynthetize very long polyisoprene chains [206]. The most known plants are *H. brasiliensis*, the only commercial source of NR, *P. argentatum* (guayule) and *T. koksaghyz*

(Russian dandelion) [207]. NR is an irreplaceable elastomer used to make many items and especially tires (~60% of the world production). The NR very specific properties (crystallization under strain, low heat build-up, *etc.*) cannot be mimicked by synthetic elastomers. NR is composed of about 94% of poly(*cis*-1, 4-isoprene) and about 6% of biochemical compounds, called non-isoprene, or non-rubber, compounds (proteins, lipids, carbohydrates and minerals). The presence of certain of these non-isoprene compounds are responsible of the very specific associative structure of NR, compared to its synthetic counterparts [208 - 210]. NR exhibit a complex associative structure, a mixture of polyisoprene chains (55-70%), microaggregates (5-15%) and macroaggregates (5-25%) [211, 212]. It is very important to fully characterize the NR structure to better understand the mechanisms of its complex structuration but also the relationships between its structure and its mechanical properties.

SEC is beyond doubt the most popular and used technique to analyze the macromolecular structure of NR. However, a recent study has highlighted some difficulties for analyzing the macromolecular structure of NR with SEC-MALLS [213]. Indeed, a so-called abnormal elution phenomenon was observed [213]. This abnormal elution is due to co-elution of microaggregates with short chains. These microaggregates have a very compact structure more comparable to a sphere than a random coil [213]. The solution to this SEC limitation for NR sample analysis might be the use of techniques based on FFF. Today, the most popular FFF techniques used for polymer fractionation in organic solvents are ThFFF and AF4 [198, 214]. NR had been analyzed by both techniques, the first papers published concern ThFFF [215 - 217] and, more recently, AF4-MALLS has been used by Dubascoux *et al.* [218].

Characterization by Thermal Field-Flow Fractionation (ThFFF)

Very few papers concern the analysis of NR by ThFFF, all published by the same team. Lee *et al.* [215, 216] showed the capability to analyze different NR samples by ThFFF with tetrahydrofuran as mobile phase (flow rate 0.2-0.3 mL min^{-1}). The ThFFF channel used had a thickness of 0.0127 cm, a breadth of 2 cm, and a length of 45.6 cm. A power programming was employed for all ThFFF analysis of NR samples, where the temperature gradient (ΔT) was gradually decreased from 90°C to 3°C during a run. However, they did not use a MALLS detector but only concentration detectors (DRI or evaporative light scattering detector). They determined the number-average molar mass (M_n) and the M_w thanks to a polystyrene standard calibration curve. The chemical structure of polyisoprene chains in NR is rather complex [209]. Both chain-ends are supposed to be linked to a protein (or a peptide) and a lipid. Therefore, the separation in ThFFF of the polyisoprene chains of NR should be rather a complex process. Indeed, the separation in ThFFF varies not only because of the molar mass of the chains but also because of different chemical composition. This chemical structure dependence in the separation process in ThFFF of NR samples should be exemplified for microaggregates. Unfortunately, there is no study yet published on ThFFF-MALS characterization of NR samples.Despite the limitations reported above, Kim *et al.* [217] showed very interesting correlation between the rheological properties of the NR samples analyzed and their M_n and M_w determined by ThFFF.

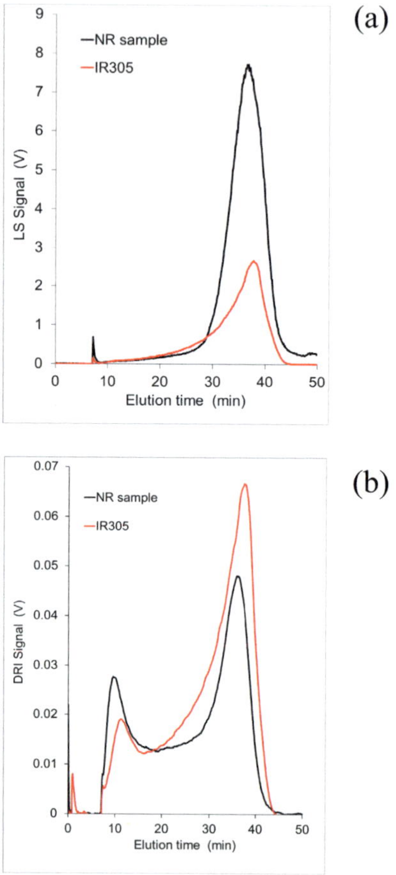

Fig. (4). Fractograms from the light scattering (LS) detector at 90° (**Top**) and the refractometer (DRI) (**Bottom**) for a natural rubber (NR) sample and a synthetic polyisoprene (IR305) (initial concentration before filtration and injection was ~ 2 mg.mL^{-1}).

Characterization by Asymmetrical Flow-Field Flow Fractionation (AF4)

Dubascoux *et al.* [218] characterized NTR samples by AF4-MALLS and compared the results to SEC-MALLS. They used a channel (350 μm spacer) adapted for organic solvents. The membrane was made of cellulose material treated for compatibility with organic solvents with a cut-off of 5 kg mol^{-1}. The crossflow was starting at 1 $mL.min^{-1}$ and decreased linearly in 30 min to 0.08 $mL.min^{-1}$. The solvent used was tetrahydrofuran.

The light scattering (LS) fractograms display a long front tailing (*i.e.* a long and low signal increase) and an abrupt increase in the LS signal at about 29 min (Fig. **4 Top**). The DRI signal shows a first peak at the beginning of the elution (8-15 min) then a rather constant signal from 15 to 29 min and a second peak at the end of the fractogram (Fig. **4 Bottom**).

The fractograms obtained for the NR samples are compared to the ones for a commercial synthetic poly(*cis*-1,4-isoprene) (IR305). The DRI signals are rather close, in terms of intensity, for the NR samples and the synthetic rubber IR305 (Fig. **4 Bottom**). However, the LS signals are dramatically different at the end of the fractogram (30-42 min). The LS signal for the NR sample is about 4 times more intense than the LS signal for the synthetic rubber. It is seen from the R_g *vs* M_w plot (Fig. **5**) of the NR sample, that M_w increased with R_g with a slope v of 0.6 (the theoretical value for random coils). From a M_w of about 3 x 10^6 $g.mol^{-1}$ the slope changed and R_g reached a quasi-plateau, contrary to the synthetic polyisoprene sample (IR305) (Fig. **5**). All these results should indicate that more and more compact microaggregates elute in the terminal zone of the fractogram (30 to 45 min). As proposed by Kim *et al.* [212, 213], the soluble part of NR samples is made with two entities: i) macromolecules of polyisoprene with different lengths (random coil structure) and ii) compact microaggregates (sphere-like structure).

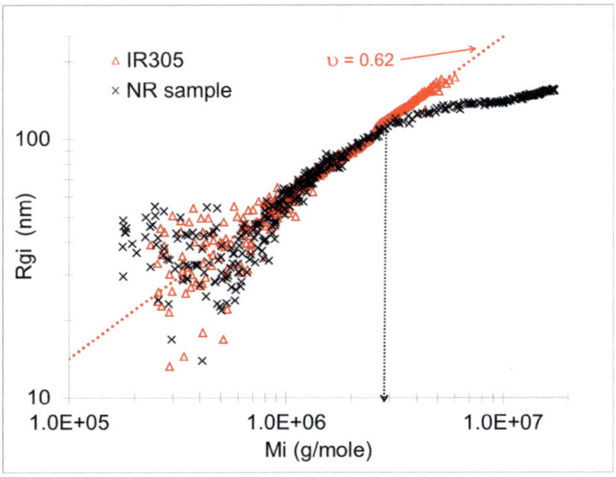

Fig. (5). Variation of R_g *versus* M_w for the natural rubber and the synthetic polyisoprene samples (v: Flory exponent).

CONCLUSION

Though the special associative structure of NR generating aggregates, it is clear that FFF techniques are under-used for the characterization of this biopolymer. Today, ThFFF is the most used but it is necessary to couple this technique with special detector such as MALLS. Promising study has been done on the characterization of NR with AF4-MALLS, but the separation between the polyisoprene random coils and the microaggregates is not optimal and needs to be improved, if possible.

CONCLUDING REMARKS

FFF techniques encounter a substantial instrumental and theoretical development as well as advancement in research and applications. Although most FFF techniques have been applied to biopolymers, today's applications are mainly focused on F4. Used in combination with with MALLS, QELS, dRI, UV-vis, MALDI-TOF-MS, ICP-MS, fluorescence or viscometric detection, FFF is a powerful tool for both separation and structural characterization of biopolymer (and bioassemblies) molar mass, size, structural and chemical composition distributions. FFF can overcome many of the limitations of current separation techniques, especially concerning ultrahigh large and branched biopolymers, aggregates and microgels because of the absence of a stationary phase. Yet FFF currently lacks the maturity of liquid chromatography as far as published methods and easy-to-handle commercially available instruments. And it also has some pitfalls: sample loss and adsorption, erroneous size distributions due to membrane fouling and/or problems with non-spherical sample particles, sample mass overloading, aggregation and dilution, and long method development. However, it can be used as a valuable complementary technique for biopolymer characterization.

CONFLICT OF INTEREST

The author (editor) declares no conflict of interest, financial or otherwise.

ACKNOWLEDGEMENTS

The authors thank Sophie Guilois for her help in formatting the manuscript.

REFERENCES

[1] J.C. Giddings, "A new separation concept based on a coupling of concentration and flow nonuniformities", *Sep. Sci.,* vol. 1, pp. 123-125, 1966.

[2] M.E. Hovingh, G.E. Thompson, and J.C. Giddings, "Column parameters in thermal field-flow fractionation", *Anal. Chem.,* vol. 42, pp. 195-203, 1970. [http://dx.doi.org/10.1021/ac60284a003]

[3] M. Martin, "Theory of field flow fractionation", In: *Advances in Chromatography.,* P.R. Brown, E. Grushka, Eds., vol. 39. Marcel Dekker: New York, 1998, pp. 1-138.

[4] K.G. Wahlund, and J.C. Giddings, "Properties of an asymmetrical flow field-flow fractionation channel having one permeable wall", *Anal. Chem.,* vol. 59, no. 9, pp. 1332-1339, 1987. [http://dx.doi.org/10.1021/ac00136a016] [PMID: 3605623]

[5] P.S. Williams, "Retention ratio and nonequilibrium bandspreading in asymmetrical flow field-flow fractionation", *Anal. Bioanal. Chem.,* vol. 407, no. 15, pp. 4327-4338, 2015. [http://dx.doi.org/10.1007/s00216-015-8734-y] [PMID: 25953429]

[6] P. Déjardin, "Design of an asymmetrical flow field-flow fractionation cell with both mean channel and membrane velocities constant", *J. Chromatogr. A,* vol. 1187, no. 1-2, pp. 209-215, 2008. [http://dx.doi.org/10.1016/j.chroma.2008.01.085] [PMID: 18302960]

[7] M. Martin, and M. Hoyos, "On the no-field method for void time determination in flow field-flow fractionation", *J. Chromatogr. A,* vol. 1218, no. 27, pp. 4117-4125, 2011. [http://dx.doi.org/10.1016/j.chroma.2011.01.010] [PMID: 21256498]

[8] H. Dou, E.C. Jung, and S. Lee, "Factors affecting measurement of channel thickness in asymmetrical flow field-flow fractionation", *J. Chromatogr. A,* vol. 1393, pp. 115-121, 2015. [http://dx.doi.org/10.1016/j.chroma.2015.03.025] [PMID: 25817708]

[9] H-L. Lee, J.F. Reis, J. Dohner, and E.N. Lightfoot, "Single-phase chromatography: solute retardation by ultrafiltration and electrophoresis", *AIChE J.,* vol. 20, pp. 776-784, 1974. [http://dx.doi.org/10.1002/aic.690200420]

[10] J.Å. Jönsson, and A. Carlshaf, "Flow field flow fractionation in hollow cylindrical fibers", *Anal. Chem.,* vol. 61, pp. 11-18, 1989. [http://dx.doi.org/10.1021/ac00176a004]

[11] J.C. Giddings, G.C. Lin, and M.N. Myers, "Fractionation and Size Distribution of Water-Soluble Polymers by Flow Field-Flow Fractionation", *J. Liq. Chromatogr.,* vol. 1, pp. 1-20, 1978. [Errata in Separation Science and Technology, vol. 14, pp. 869-870, 1979.]. [http://dx.doi.org/10.1080/01483917808068374]

[12] L. Pasol, M. Martin, M.L. Ekiel-Jezewska, E. Wajnryb, J. Blawzdziewicz, and F. Feuillebois, "Motion of a sphere parallel to plane walls in a Poiseuille flow. Application to field-flow fractionation and hydrodynamic chromatography", *Chem. Eng. Sci.,* vol. 66, pp. 4078-4089, 2011. [Corrigendum, *Chem. Eng. Sci.,* vol. 90, pp. 51-52, 2013.]. [http://dx.doi.org/10.1016/j.ces.2011.05.033]

[13] M. Martin, and P.S. Williams, "Theoretical basis of field-flow fractionation", In: *Theoretical Advancement in Chromatography and Related Separation Techniques,* D. Dondi, G. Guiochon, Eds., vol. 383. Kluwer: Dordrecht, 1992, pp. 513-580. NATO ASI Series C: Mathematical and Physical Sciences [http://dx.doi.org/10.1007/978-94-011-2686-1_18]

[14] M. Martin, and R. Beckett, "Size selectivity in field-flow fractionation: lift mode of retention with near-wall lift force", *J. Phys. Chem. A,* vol. 116, no. 25, pp. 6540-6551, 2012. [http://dx.doi.org/10.1021/jp212414e] [PMID: 22506593]

[15] S.K. Ratanathanawongs, and J.C. Giddings, "Dual-field and flow-programmed lift hyperlayer field-flow fractionation", *Anal. Chem.,* vol. 64, no. 1, pp. 6-15, 1992. [http://dx.doi.org/10.1021/ac00025a003] [PMID: 1736679]

[16] P.S. Williams, Y. Xu, P. Reschiglian, and J.C. Giddings, "Colloid characterization by sedimentation field-flow fractionation: correction for particle-wall interaction", *Anal. Chem.,* vol. 69, no. 3, pp. 349-360, 1997. [http://dx.doi.org/10.1021/ac9606012] [PMID: 21639187]

[17] J.C. Giddings, "Field-flow fractionation: analysis of macromolecular, colloidal, and particulate

materials", *Science,* vol. 260, no. 5113, pp. 1456-1465, 1993.
[http://dx.doi.org/10.1126/science.8502990] [PMID: 8502990]

[18] J.C. Giddings, "The field-flow fractionation family: Underlying principles", In: *Field-Flow Fractionation Handbook,* M.E. Schimpf, K. Caldwell, J.C. Giddings, Eds., John Wiley & Sons, Inc.: New York, 2000, pp. 3-30.

[19] J.F. Ranville, D.J. Chittleborough, F. Shanks, R.J. Morrison, T. Harris, F. Doss, and R. Beckett, "Development of sedimentation field-flow fractionation-inductively coupled plasma mass-spectrometry for the characterization of environmental colloids", *Anal. Chim. Acta,* vol. 381, pp. 315-329, 1999.
[http://dx.doi.org/10.1016/S0003-2670(98)00702-8]

[20] P.S. Williams, and J.C. Giddings, "Power programmed field-flow fractionation: a new program form for improved uniformity of fractionating power", *Anal. Chem.,* vol. 59, no. 17, pp. 2038-2044, 1987.
[http://dx.doi.org/10.1021/ac00144a007] [PMID: 3674422]

[21] P. Cardot, S. Battu, and R. Sarrazin, Field flow fractionation separation device with a separation channel and a counter channel, 2006. Ep Patent 1679124

[22] M.H. Moon, "Sedimentation Field-Flow Fractionation", In: *Field-Flow Fractionation Handbook.,* M.E. Schimpf, K. Caldwell, J.C. Giddings, Eds., John Wiley & Sons Inc.: New York, 2000, pp. 225-237.

[23] S.K. Williams, J.R. Runyon, and A.A. Ashames, "Field-flow fractionation: addressing the nano challenge", *Anal. Chem.,* vol. 83, no. 3, pp. 634-642, 2011.
[http://dx.doi.org/10.1021/ac101759z] [PMID: 21090634]

[24] F. Dondi, and M. Martin, "Physicochemical measurements and distributions from Field Flow Fractionation", In: *Field-Flow Fractionation Handbook.,* M. Schimpf, K.D. Caldwell, C. Giddings, Eds., John Wiley & Sons, Inc.: New York, 2000, pp. 103-132.

[25] J.C. Giddings, "Measuring colloidal and macromolecular properties by FFF", *Anal. Chem.,* vol. 67, no. 19, pp. 592A-598A, 1995.
[http://dx.doi.org/10.1021/ac00115a001] [PMID: 8686907]

[26] J.C. Giddings, and P.S. Williams, "Multifaceted analysis of 0.01- to 100 µm particles by sedimentation field-flow fractionation", *Am. Lab.,* vol. 25, pp. 88-95, 1993.

[27] L. Farmakis, J. Sakellaraki, A. Koliadima, D. Gavril, and G. Karaiskakis, "Size analysis of barley starch granules by sedimentation/steric field flow fractionation", *Starke,* vol. 52, pp. 275-282, 2000.
[http://dx.doi.org/10.1002/1521-379X(20009)52:8/9<275::AID-STAR275>3.0.CO;2-M]

[28] M.H. Moon, and J.C. Giddings, "Rapid separation and measurement of particle size distribution of starche granules by sedimentation/steric field-flow fractionation", *J. Food Sci.,* vol. 58, pp. 1166-1171, 1993.
[http://dx.doi.org/10.1111/j.1365-2621.1993.tb06139.x]

[29] M. Schimpf, C. Caldwell, and J.C. Giddings, *Field flow fractionation handbook.*New York, USA, 2000.

[30] K-G. Wahlund, "Flow field-flow fractionation: critical overview", *J. Chromatogr. A,* vol. 1287, pp. 97-112, 2013.
[http://dx.doi.org/10.1016/j.chroma.2013.02.028] [PMID: 23510956]

[31] A. Håkansson, E. Magnusson, B. Bergenståhl, and L. Nilsson, "Hydrodynamic radius determination with asymmetrical flow field-flow fractionation using decaying cross-flows. Part I. A theoretical approach", *J. Chromatogr. A,* vol. 1253, pp. 120-126, 2012.
[http://dx.doi.org/10.1016/j.chroma.2012.07.029] [PMID: 22835686]

[32] E. Magnusson, A. Håkansson, J. Janiak, B. Bergenståhl, and L. Nilsson, "Hydrodynamic radius determination with asymmetrical flow field-flow fractionation using decaying cross-flows. Part II. Experimental evaluation", *J. Chromatogr. A,* vol. 1253, pp. 127-133, 2012.

[http://dx.doi.org/10.1016/j.chroma.2012.07.005] [PMID: 22818771]

[33] A. Rolland-Sabaté, S. Guilois, F. Grimaud, C. Lancelon-Pin, X. Roussel, S. Laguerre, A. Vikso-Nielsen, J.L. Putaux, C. D'Hulst, G. Potocki-Véronèse, and A. Buléon, "Characterization of Hyperbranched Glycopolymers produced in vitro using enzymes", *Anal. Bioanal. Chem,* vol. 406, pp. 1607-1618, 2014.

[34] A. Rolland-Sabaté, S. Guilois, B. Jaillais, and P. Colonna, "Molecular size and mass distributions of native starches using complementary separation methods: asymmetrical flow field flow fractionation (A4F) and hydrodynamic and size exclusion chromatography (HDC-SEC)", *Anal. Bioanal. Chem.,* vol. 399, no. 4, pp. 1493-1505, 2011.
[http://dx.doi.org/10.1007/s00216-010-4208-4] [PMID: 20878394]

[35] S. Hansen, "Translational friction coefficients for cylinders of arbitrary axial ratios estimated by Monte Carlo simulation", *J. Chem. Phys.,* vol. 121, no. 18, pp. 9111-9115, 2004.
[http://dx.doi.org/10.1063/1.1803533] [PMID: 15527378]

[36] W. Burchard, "Solution properties of branched macromolecules", In: *Adv. Polym. Sci.*vol. 143. SPRINGER-VERLAG BERLIN: Berlin, 1999, pp. 113-194.

[37] M. Glantz, A. Håkansson, H. Lindmark Månsson, M. Paulsson, and L. Nilsson, "Revealing the size, conformation, and shape of casein micelles and aggregates with asymmetrical flow field-flow fractionation and multiangle light scattering", *Langmuir,* vol. 26, no. 15, pp. 12585-12591, 2010.
[http://dx.doi.org/10.1021/la101892x] [PMID: 20666417]

[38] A. Hakansson, M. Ulmius, and L. Nilsson, "Asymmetrical flow field-flow fractionation enables the characterization of molecular and supramolecular properties of cereal beta-glucan dispersions", *Carbohydr. Polym.,* vol. 87, pp. 518-523, 2012.
[http://dx.doi.org/10.1016/j.carbpol.2011.08.014]

[39] C.C. Rojas, K.G. Wahlund, B. Bergenståhl, and L. Nilsson, "Macromolecular geometries determined with field-flow fractionation and their impact on the overlap concentration", *Biomacromolecules,* vol. 9, no. 6, pp. 1684-1690, 2008.
[http://dx.doi.org/10.1021/bm800127n] [PMID: 18537296]

[40] J. Gigault, J.M. Pettibone, C. Schmitt, and V.A. Hackley, "Rational strategy for characterization of nanoscale particles by asymmetric-flow field flow fractionation: a tutorial", *Anal. Chim. Acta,* vol. 809, pp. 9-24, 2014.
[http://dx.doi.org/10.1016/j.aca.2013.11.021] [PMID: 24418128]

[41] A. Rolland-Sabaté, P. Colonna, M.G. Mendez-Montealvo, and V. Planchot, "Branching features of amylopectins and glycogen determined by asymmetrical flow field flow fractionation coupled with multiangle laser light scattering", *Biomacromolecules,* vol. 8, no. 8, pp. 2520-2532, 2007.
[http://dx.doi.org/10.1021/bm070024z] [PMID: 17645307]

[42] H. Colfen, and M. Antonietti, "Field-flow fractionation techniques for polymer and colloid analysis", In: *New Developments in Polymer Analytics I*vol. 150. 2000, pp. 67-187.

[43] C. Arfvidsson, and K.G. Wahlund, "Mass overloading in the flow field-flow fractionation channel studied by the behaviour of the ultra-large wheat protein glutenin", *J. Chromatogr. A,* vol. 1011, no. 1-2, pp. 99-109, 2003.
[http://dx.doi.org/10.1016/S0021-9673(03)01145-2] [PMID: 14518767]

[44] S. Lee, P.O. Nilsson, G.S. Nilsson, and K.G. Wahlund, "Development of asymmetrical flow field-flow fractionation-multi angle laser light scattering analysis for molecular mass characterization of cationic potato amylopectin", *J. Chromatogr. A,* vol. 1011, no. 1-2, pp. 111-123, 2003.
[http://dx.doi.org/10.1016/S0021-9673(03)01144-0] [PMID: 14518768]

[45] M.H. Moon, H. Kwon, and I. Park, "Stopless flow injection in asymmetrical flow field-flow fractionation using a frit inlet", *Anal. Chem.,* vol. 69, no. 7, pp. 1436-1440, 1997.
[http://dx.doi.org/10.1021/ac960897b] [PMID: 21639349]

[46] J.J. Kirkland, L.S. Boone, and W.W. Yau, "Retention effects in thermal field-flow fractionation", *J. Chromatogr. A,* vol. 517, pp. 377-393, 1990. [http://dx.doi.org/10.1016/S0021-9673(01)95735-8]

[47] J.J. Gunderson, and J.C. Giddings, "Chemical composition and molecular-size factors in polymer analysis by Thermal Field-Flow Fractionation and size exclusion chromatography", *Macromolecules,* vol. 19, pp. 2618-2621, 1986. [http://dx.doi.org/10.1021/ma00164a027]

[48] A.C. van Asten, E. Venema, W.T. Kok, and H. Poppe, "Use of Thermal Field-Flow Fractionation for the fractionation of Polybutadiene in Various Organic-Solvents", *J. Chromatogr. A,* vol. 644, pp. 83-94, 1993. [http://dx.doi.org/10.1016/0021-9673(93)80122-O]

[49] J.R. Runyon, and S.K. Williams, "A theory-based approach to thermal field-flow fractionation of polyacrylates", *J. Chromatogr. A,* vol. 1218, no. 39, pp. 7016-7022, 2011. [PMID: 21872869]

[50] G.E. Kassalainen, and S.K. Williams, "Coupling thermal field-flow fractionation with matrix-assisted laser desorption/ionization time-of-flight mass spectrometry for the analysis of synthetic polymers", *Anal. Chem.,* vol. 75, no. 8, pp. 1887-1894, 2003. [http://dx.doi.org/10.1021/ac020594j] [PMID: 12713047]

[51] W. Hiller, W. van Aswegen, M. Hehn, and H. Pasch, "Online ThFFF-NMR: A Novel Tool for Molar Mass and Chemical Composition Analysis of Complex Macromolecules", *Macromolecules,* vol. 46, pp. 2544-2552, 2013. [http://dx.doi.org/10.1021/ma400350y]

[52] J.E. Belgaied, M. Hoyos, and M. Martin, "Velocity profiles in thermal field flow fractionation", *J. Chromatogr. A,* vol. 678, pp. 85-96, 1994. [http://dx.doi.org/10.1016/0021-9673(94)87077-2]

[53] S.K. Williams, "Addressing new challenges in polymer analysis using field-flow fractionation and MALDI-MS", In: *Macro 2006-41st International Symposium on Macromolecules*, Rio de Janeiro: Brazil, 2006.

[54] J.R. Runyon, and S.K. Williams, "Thermal field-flow fractionation of acrylic copolymers", *13th International Symposium on Field- and Flow-based Separations,* 2007 Salt Lake City: Utah, USA

[55] C.A. Ponyik, D.T. Wu, and S.K. Williams, "Separation and composition distribution determination of triblock copolymers by thermal field-flow fractionation", *Anal. Bioanal. Chem.,* vol. 405, no. 28, pp. 9033-9040, 2013. [http://dx.doi.org/10.1007/s00216-013-7282-6] [PMID: 23975085]

[56] N. Ngaza, M. Brand, and H. Pasch, "Multidetector-ThF3 as a Novel Tool for the Investigation of Solution Properties of Amphiphilic Block Copolymers", *Macromol. Chem. Phys.,* vol. 216, pp. 1355-1364, 2015. [http://dx.doi.org/10.1002/macp.201500070]

[57] S. Lee, "Application of thermal field-flow fractionation for characterization of industrial polymers", *J. Microcolumn Sep.,* vol. 9, pp. 281-286, 1997. [http://dx.doi.org/10.1002/(SICI)1520-667X(1997)9:4<281::AID-MCS5>3.0.CO;2-Y]

[58] L. Pasti, D. Melluci, C. Contado, F. Dondi, and I. Mingozzi, "Calibration in thermal field-flow fractionation with polydisperse standards: application to polyolefin characterization", *J. Sep. Sci.,* vol. 25, pp. 691-701, 2002. [http://dx.doi.org/10.1002/1615-9314(20020701)25:10/11<691::AID-JSSC691>3.0.CO;2-L]

[59] W. Cao, P.S. Williams, M.N. Meyers, and J.C. Giddings, "Thermal field-flow fractionation universal calibration: extension for consideration of cold wall temperature", *Anal. Chem.,* vol. 71, pp. 1597-1609, 1999.

[http://dx.doi.org/10.1021/ac981094m]

[60] Y.S. Gao, K.D. Caldwell, M.N. Myers, and J.C. Giddings, "Extension of Thermal Thermal Field-Flow Fractionation to Ultra-High Molecular Weight Polystyrenes", *Macromolecules,* vol. 18, pp. 1272-1277, 1985.
[http://dx.doi.org/10.1021/ma00148a039]

[61] J. Janca, *Microthermal Field-Flow Frationation.*HNB Publishing: New York, NY, USA, 2008.

[62] F.A. Messaud, R.D. Sanderson, J.R. Runyon, T. Otte, H. Pasch, and S.K. Williams, "An overview on field-flow fractionation techniques and their applications in the separation and characterization of polymers", *Prog. Polym. Sci.,* vol. 34, pp. 351-368, 2009.
[http://dx.doi.org/10.1016/j.progpolymsci.2008.11.001]

[63] N. Lindeboom, P.R. Chang, and R.T. Tyler, "Analytical, biochemical and physicochemical aspects of starch granule size, with emphasis on small granule starches: A review", *Starke,* vol. 56, pp. 89-99, 2004.
[http://dx.doi.org/10.1002/star.200300218]

[64] R.P. Ellis, M.P. Cochrane, M.F. Dale, C.M. Duffus, A. Lynn, I.M. Morrison, R.D. Prentice, J.S. Swanston, and S.A. Tiller, "Starch production and industrial use", *J. Sci. Food Agric.,* vol. 77, pp. 289-311, 1998.
[http://dx.doi.org/10.1002/(SICI)1097-0010(199807)77:3<289::AID-JSFA38>3.0.CO;2-D]

[65] J.P. Davis, N. Supatcharee, R.L. Khandelwal, and R.N. Chibbar, "Synthesis of novel starches in planta: Opportunities and challenges", *Starke,* vol. 55, pp. 107-120, 2003.
[http://dx.doi.org/10.1002/star.200390036]

[66] A. Buléon, P. Colonna, V. Planchot, and S. Ball, "Starch granules: structure and biosynthesis", *Int. J. Biol. Macromol.,* vol. 23, no. 2, pp. 85-112, 1998.
[http://dx.doi.org/10.1016/S0141-8130(98)00040-3] [PMID: 9730163]

[67] J.R. Stark, and A. Lynn, "Starch granules large and small", *Biochem. Soc. Trans.,* vol. 20, no. 1, pp. 7-12, 1992.
[http://dx.doi.org/10.1042/bst0200007] [PMID: 1633990]

[68] D. Clédat, S. Battu, R. Mokrini, and P.J. Cardot, "Rice starch granule characterization by flow cytometry scattering techniques hyphenated with sedimentation field-flow fractionation", *J. Chromatogr. A,* vol. 1049, no. 1-2, pp. 131-138, 2004.
[http://dx.doi.org/10.1016/S0021-9673(04)01271-3] [PMID: 15499925]

[69] J. Chmelík, A. Krumlova, M. Budinska, T. Kruml, V. Psota, I. Bohacenko, P. Mazal, and H. Vydrova, "Comparison of size characterization of barley starch granules determined by electron and optical microscopy, low angle laser light scattering and gravitational field-flow fractionation", *J. Inst. Brew.,* vol. 107, pp. 11-17, 2001.
[http://dx.doi.org/10.1002/j.2050-0416.2001.tb00074.x]

[70] A.C. Dona, G. Pages, R.G. Gilbert, and P.W. Kuchel, "Digestion of starch: In vivo and in vitro kinetic models used to characterise oligosaccharide or glucose release", *Carbohydr. Polym.,* vol. 80, pp. 599-617, 2010.
[http://dx.doi.org/10.1016/j.carbpol.2010.01.002]

[71] J. Chmelík, A. Krumlova, and J. Caslavsky, "Characterization of starch materials from barley by gravitational field-flow fractionation and matrix-assisted laser desorption/ionization mass spectrometry", *Chemical Papers-Chemicke Zvesti,* vol. 52, pp. 360-361, 1998.

[72] J. Chmelík, K. Mazanec, I. Bohacenko, and V. Psota, "Relationship between the ratio of large and small starch granules determined by gravitational field-flow fractionation and malting quality of barley varieties", *J. Liq. Chromatogr.,* vol. 30, pp. 1289-1301, 2007.
[http://dx.doi.org/10.1080/10826070701274767]

[73] J. Chmelík, and V. Psota, "Characterization of starch materials from barley", *Proc. Congr. - Eur.*

Brew. Conv., 1999pp. 421-428

[74] V. Psota, I. Bohacenko, J. Hartmann, M. Budinska, and J. Chmelik, "Comparison of the GFFF and LALLS methods for the measurement of starch granule size distribution in spring barley caryopses", *J. Inst. Brew.,* vol. 108, pp. 200-203, 2002.
[http://dx.doi.org/10.1002/j.2050-0416.2002.tb00541.x]

[75] P. Reschiglian, A. Zattoni, S. Casolari, A. Krumlova, M. Budinska, and J. Chmelík, "Size characterization of barley starch granules by gravitational field-flow fractionation: a rapid, low-cost method to assess the brewing capability of different strains", *Ann. Chim.,* vol. 92, no. 4, pp. 457-467, 2002.
[PMID: 12073892]

[76] C. Contado, and F. Dondi, "Barley starch granules subject to SPLITT cell fractionation and Sd/StFFF size characterization", *Starke,* vol. 53, pp. 414-423, 2001.
[http://dx.doi.org/10.1002/1521-379X(200109)53:9<414::AID-STAR414>3.0.CO;2-N]

[77] C. Contado, P. Reschiglian, S. Faccini, A. Zattoni, and F. Dondi, "Continuous split-flow thin cell and gravitational field-flow fractionation of wheat starch particles", *J. Chromatogr. A,* vol. 871, no. 1-2, pp. 449-460, 2000.
[http://dx.doi.org/10.1016/S0021-9673(99)01191-7] [PMID: 10735325]

[78] L. Farmakis, A. Koliadima, G. Karaiskakis, A. Zattoni, and P. Reschiglian, "Study of the influence of surfactants on the size distribution and mass ratio of wheat starch granules by sedimentation/steric field-flow fractionation", *Food Hydrocoll.,* vol. 22, pp. 961-972, 2008.
[http://dx.doi.org/10.1016/j.foodhyd.2007.05.001]

[79] L. Farmakis, G. Karaiskakis, and A. Koliadima, "Investigation of the variation of the mass ratios for large and small starch granules with the pH and ionic strength of the dispersing medium by sedimentation/steric field-flow fractionation", *J. Liquid Chromatogr. Relat. Technol.,* vol. 25, pp. 2135-2152, 2002.
[http://dx.doi.org/10.1081/JLC-120013998]

[80] A. Zattoni, D. Melucci, G. Torsi, and P. Reschiglian, "Quantitative analysis in field-flow fractionation using ultraviolet-visible detectors: an experimental design for absolute measurements", *J. Chromatogr. Sci.,* vol. 38, no. 3, pp. 122-128, 2000.
[http://dx.doi.org/10.1093/chromsci/38.3.122] [PMID: 10702920]

[81] K. Mazanec, F. Dycka, and J. Bobalova, "Monitoring of barley starch amylolysis by gravitational field flow fractionation and MALDI-TOF MS", *J. Sci. Food Agric.,* vol. 91, no. 15, pp. 2756-2761, 2011.
[http://dx.doi.org/10.1002/jsfa.4518] [PMID: 21717465]

[82] X. Morelon, S. Battu, C. Salesse, G. Begaud-Grimaud, D. Cledat, and P.J. Cardot, "Sedimentation field flow fractionation monitoring of rice starch amylolysis", *J. Chromatogr. A,* vol. 1093, no. 1-2, pp. 147-155, 2005.
[http://dx.doi.org/10.1016/j.chroma.2005.07.084] [PMID: 16233880]

[83] C. Salesse, S. Battu, G. Begaud-Grimaud, D. Cledat, J. Cook-Moreau, and P.J. Cardot, "Sedimentation field flow fractionation monitoring of bimodal wheat starch amylolysis", *J. Chromatogr. A,* vol. 1129, no. 2, pp. 247-254, 2006.
[http://dx.doi.org/10.1016/j.chroma.2006.06.104] [PMID: 16870195]

[84] S.T. Kim, S.Y. Seo, and S. Lee, "Study on swelling of starch granules using gravitational field-flow fractionation (GrFFF)", *Analytical Science and Technology,* vol. 24, pp. 249-255, 2011.
[http://dx.doi.org/10.5806/AST.2011.24.4.249]

[85] T. Aberle, W. Burchard, G. Galinsky, R. Hanselmann, R.W. Klingler, and E. Michel, "Particularities in the structure of amylopectin, amylose and some of their derivatives in solution", *Macromol. Symp.,* vol. 120, pp. 47-63, 1997.
[http://dx.doi.org/10.1002/masy.19971200108]

[86] R. Hanselmann, M. Ehrat, and H.M. Widmer, "Sedimentation field flow fractionation combined with

multi angle laser light scattering applied for characterization of starch polymers", *Starke,* vol. 46, pp. 345-349, 1995.
[http://dx.doi.org/10.1002/star.19950470905]

[87] P. Roger, B. Baud, and P. Colonna, "Characterization of starch polysaccharides by flow field-flow fractionation-multi-angle laser light scattering-differential refractometer index", *J. Chromatogr. A,* vol. 917, no. 1-2, pp. 179-185, 2001.
[http://dx.doi.org/10.1016/S0021-9673(01)00689-6] [PMID: 11403469]

[88] S. You, S.G. Stevenson, M.S. Izydorczyk, and K.R. Preston, "Separation and characterization of barley starch polymers by a flow field-flow fractionation technique in combination with multiangle light scattering and differential refractive index detection", *Cereal Chem.,* vol. 79, pp. 624-630, 2002.
[http://dx.doi.org/10.1094/CCHEM.2002.79.5.624]

[89] W.J. Kim, C.H. Eum, S.T. Lim, J.H. Han, S.G. You, and S. Lee, "Separation of amylose and amylopectin in corn starch using dual-programmed flow field-flow Fractionation", *B. Kor. Chem. Soc.,* vol. 28, pp. 2489-2492, 2007.
[http://dx.doi.org/10.5012/bkcs.2007.28.12.2489]

[90] M. van Bruijnsvoort, K.G. Wahlund, G. Nilsson, and W.T. Kok, "Retention behaviour of amylopectins in asymmetrical flow field-flow fractionation studied by multi-angle light scattering detection", *J. Chromatogr. A,* vol. 925, no. 1-2, pp. 171-182, 2001.
[http://dx.doi.org/10.1016/S0021-9673(01)01020-2] [PMID: 11519803]

[91] S. Juna, P.A. Williams, and S. Davies, "Determination of molecular mass distribution of amylopectin using asymmetrical flow field-flow fractionation", *Carbohydr. Polym.,* vol. 83, pp. 1384-1396, 2011.
[http://dx.doi.org/10.1016/j.carbpol.2010.09.045]

[92] E. Pérez, O. Gibert, A. Rolland-Sabaté, Y. Jiménez, T. Sánchez, A. Giraldo, B. Pontoire, S. Guilois, M-C. Lahon, M. Reynes, and D. Dufour, "Physicochemical, functional, and macromolecular properties of waxy yam starches discovered from Mapuey (Dioscorea trifida) genotypes in the Venezuelan Amazon", *J. Agric. Food Chem.,* vol. 59, no. 1, pp. 263-273, 2011.
[http://dx.doi.org/10.1021/jf100418r] [PMID: 21158430]

[93] K-G. Wahlund, M. Leeman, and S. Santacruz, "Size separations of starch of different botanical origin studied by asymmetrical-flow field-flow fractionation and multiangle light scattering", *Anal. Bioanal. Chem.,* vol. 399, no. 4, pp. 1455-1465, 2011.
[http://dx.doi.org/10.1007/s00216-010-4438-5] [PMID: 21181137]

[94] S. Juna, and A. Huber, "Molecular characteristics of native sago starch and isolated fractions determined using asymmetrical flow field-flow fractionation", *Starke,* vol. 64, pp. 171-180, 2012.
[http://dx.doi.org/10.1002/star.201100100]

[95] S. Juna, and A. Huber, "Effect of varying flow regimes upon elution behaviour, apparent molecular characteristics and hydrodynamic properties of amylopectin isolated from normal corn starch using asymmetrical flow field-flow fractionation", *J. Chromatogr. A,* vol. 1219, pp. 161-172, 2012.
[http://dx.doi.org/10.1016/j.chroma.2011.11.024] [PMID: 22142969]

[96] S. Juna, and A. Huber, "Determination of molar mass distribution of tapioca starch using asymmetrical flow field flow fractionation", *Starke,* vol. 64, pp. 87-96, 2012.
[http://dx.doi.org/10.1002/star.201100066]

[97] S. Juna, and A. Huber, "Characterisation of normal corn starch using asymmetrical flow field-flow fractionation", *Starke,* vol. 64, pp. 18-26, 2012.
[http://dx.doi.org/10.1002/star.201100068]

[98] A. Rolland-Sabaté, T. Sanchez, A. Buléon, P. Colonna, B. Jaillais, H. Ceballos, and D. Dufour, "Structural characterization of novel cassava starches with low and high-amylose contents in comparison with other commercial sources", *Food Hydrocoll.,* vol. 27, pp. 161-174, 2012.
[http://dx.doi.org/10.1016/j.foodhyd.2011.07.008]

[99] A. Rolland-Sabaté, T. Sanchez, A. Buléon, P. Colonna, H. Ceballos, S.S. Zhao, P. Zhang, and D.

Dufour, "Molecular and supra-molecular structure of waxy starches developed from cassava (Manihot esculenta Crantz)", *Carbohydr. Polym.,* vol. 92, no. 2, pp. 1451-1462, 2013. [http://dx.doi.org/10.1016/j.carbpol.2012.10.048] [PMID: 23399176]

[100] A. Beilvert, F. Chaubet, L. Chaunier, S. Guilois, G. Pavon-Djavid, D. Letourneur, A. Meddahi-Pellé, and D. Lourdin, "Shape-memory starch for resorbable biomedical devices", *Carbohydr. Polym.,* vol. 99, pp. 242-248, 2014. [http://dx.doi.org/10.1016/j.carbpol.2013.08.015] [PMID: 24274502]

[101] D. Perez-Rea, B. Bergenståhl, and L. Nilsson, "Development and evaluation of methods for starch dissolution using asymmetrical flow field-flow fractionation. Part I: Dissolution of amylopectin", *Anal. Bioanal. Chem.,* vol. 407, no. 15, pp. 4315-4326, 2015. [http://dx.doi.org/10.1007/s00216-015-8611-8] [PMID: 25925852]

[102] B. Wittgren, K.G. Wahlund, M. Andersson, and C. Arfvidsson, "Polysaccharide characterization by flow field-flow fractionation-multiangle light scattering: Initial studies of modified starches", *Int. J. Polym. Anal. Ch.,* vol. 7, pp. 19-40, 2002. [http://dx.doi.org/10.1080/10236660214599]

[103] S.E. Bowen, D.A. Gray, C. Giraud, M. Majzoobi, C.E. Testa, L.A. Perez, and S.E. Hill, "Lipid oxidation and amylopectin molecular weight changes occurring during storage of extruded starch samples", *J. Cereal Sci.,* vol. 43, pp. 275-283, 2006. [http://dx.doi.org/10.1016/j.jcs.2005.08.010]

[104] D.O. Krentz, C. Lohmann, S. Schwarz, S. Bratskaya, T. Liebert, J. Laube, T. Heinze, and W.M. Kulicke, "Properties and flocculation efficiency of highly cationized starch derivatives", *Starke,* vol. 58, pp. 161-169, 2006. [http://dx.doi.org/10.1002/star.200500431]

[105] G. Modig, L. Nilsson, B. Bergenstahl, and K.G. Wahlund, "Homogenization-induced degradation of hydrophobically modified starch determined by asymmetrical flow field-flow fractionation and multi-angle light scattering", *Food Hydrocoll.,* vol. 20, pp. 1087-1095, 2006. [http://dx.doi.org/10.1016/j.foodhyd.2005.12.011]

[106] G. Modig, PO. Nilsson, and KG. Wahlund, "Influence of Jet-Cooking Temperature and Ionic Strength on Size and Structure of Cationic Potato Amylopectin Starch as Measured by Asymmetrical Flow Field-Flow Fractionation Multi-Angle Ligth Scattering", *Starch-Stärke,* vol. 58, pp. 55-65, 2006.

[107] L. Nilsson, M. Leeman, K.G. Wahlund, and B. Bergenståhl, "Mechanical degradation and changes in conformation of hydrophobically modified starch", *Biomacromolecules,* vol. 7, no. 9, pp. 2671-2679, 2006. [http://dx.doi.org/10.1021/bm060367h] [PMID: 16961331]

[108] S. Lee, S.T. Kim, B.R. Pant, H.D. Kwen, H.H. Song, S.K. Lee, and S.V. Nehete, "Carboxymethylation of corn starch and characterization using asymmetrical flow field-flow fractionation coupled with multiangle light scattering", *J. Chromatogr. A,* vol. 1217, no. 27, pp. 4623-4628, 2010. [http://dx.doi.org/10.1016/j.chroma.2010.04.082] [PMID: 20537343]

[109] S. Juna, and A. Huber, "Formation of nano- and micro-structures of various botanical sources of native starches investigated employing asymmetrical flow field-flow fractionation", *Starke,* vol. 65, pp. 1029-1037, 2013. [http://dx.doi.org/10.1002/star.201300059]

[110] S. Juna, S. Hayden, M. Damm, C.O. Kappe, and A. Huber, "Influence of temperature on the apparent molar masses and sizes of pregelatinized wx corn in aqueous media determined using asymmetrical flow field-flow fractionation", *Starke,* vol. 65, pp. 954-961, 2013. [http://dx.doi.org/10.1002/star.201300021]

[111] S. Juna, S. Hayden, M. Damm, C.O. Kappe, and A. Huber, "Nanoprecipitation of native pea starches treated in alkaline media at various temperatures employing a dedicated microwave reactor", *Starke,* vol. 66, pp. 124-131, 2014.

[http://dx.doi.org/10.1002/star.201300066]

[112] M.J. Gidley, I. Hanashiro, N.M. Hani, S.E. Hill, A. Huber, J-L. Jane, Q. Liu, G.A. Morris, A. Rolland-Sabaté, A.M. Striegel, and R.G. Gilbert, "Reliable measurements of the size distributions of starch molecules in solution: Current dilemmas and recommendations", *Carbohydr. Polym.,* vol. 79, pp. 255-261, 2010.

[113] T. Aberle, W. Burchard, W. Vorwerg, and S. Radosta, "Conformational Contributions of Amylose and Amylopectin to the Structural-Properties of Starches from Various Sources", *Starke,* vol. 46, pp. 329-335, 1994.
[http://dx.doi.org/10.1002/star.19940460903]

[114] R. Hanselmann, W. Burchard, M. Ehrat, and H.M. Widmer, "Structural properties of fractionated starch polymers and their dependence on the dissolution process", *Macromolecules,* vol. 29, pp. 3277-3282, 1996.
[http://dx.doi.org/10.1021/ma951452c]

[115] C. Fernandez, C.C. Rojas, and L. Nilsson, "Size, structure and scaling relationships in glycogen from various sources investigated with asymmetrical flow field-flow fractionation and 1H NMR", *Int. J. Biol. Macromol.,* vol. 49, no. 4, pp. 458-465, 2011.
[http://dx.doi.org/10.1016/j.ijbiomac.2011.05.016] [PMID: 21640751]

[116] J. Ciric, A. Rolland-Sabaté, S. Guilois, and K. Loos, "Characterization of enzymatically synthesized amylopectin analogs via asymmetrical flow field flow fractionation", *Polymer (Guildf.),* vol. 55, pp. 6271-6277, 2014.
[http://dx.doi.org/10.1016/j.polymer.2014.09.059]

[117] F. Grimaud, C. Lancelon-Pin, A. Rolland-Sabaté, X. Roussel, S. Laguerre, A. Viksø-Nielsen, J.L. Putaux, S. Guilois, A. Buléon, C. DHulst, and G. Potocki-Véronèse, "In vitro synthesis of hyperbranched α-glucans using a biomimetic enzymatic toolbox", *Biomacromolecules,* vol. 14, no. 2, pp. 438-447, 2013.
[http://dx.doi.org/10.1021/bm301676c] [PMID: 23311582]

[118] B. Wittgren, and K.G. Wahlund, "Fast molecular mass and size characterization of polysaccharides using asymmetrical flow field-flow fractionation-multiangle light scattering", *J. Chromatogr. A,* vol. 760, pp. 205-218, 1997.
[http://dx.doi.org/10.1016/S0021-9673(96)00777-7]

[119] N.H. Maina, L. Pitkänen, S. Heikkinen, P. Tuomainen, L. Virkki, and M. Tenkanen, "Challenges in analysis of high-molar mass dextrans: comparison of HPSEC, AsFlFFF and DOSY NMR spectroscopy", *Carbohydr. Polym.,* vol. 99, pp. 199-207, 2014.
[http://dx.doi.org/10.1016/j.carbpol.2013.08.021] [PMID: 24274497]

[120] R. Irague, A. Rolland-Sabaté, L. Tarquis, J.L. Doublier, C. Moulis, and P. Monsan, "M. ERemaud-Siméon, G. Potocki-Véronèse, and A. Buléon, “Structure and property engineeringof alpha--Glucanssynthesized bydextransucrase mutants", *Biomacromolcules,* vol. 13, pp. 187-195, 2012.
[http://dx.doi.org/10.1021/bm201453r]

[121] E. Ji, S.H. Choi, K.R. Yoon, J.H. Chun, and S. Lee, "Elution behavior of protein and pullulan in asymmetrical flow field-flow fractionation (AsFlFFF)", *B. Kor. Chem. Soc.,* vol. 27, pp. 1433-1438, 2006.
[http://dx.doi.org/10.5012/bkcs.2006.27.9.1433]

[122] M.A. Benincasa, G. Cartoni, and C. Delle Fratte, "Flow field-flow fractionation and characterization of ionic and neutral polysaccharides of vegetable and microbial origin", *J. Chromatogr. A,* vol. 967, no. 2, pp. 219-234, 2002.
[http://dx.doi.org/10.1016/S0021-9673(02)00756-2] [PMID: 12685569]

[123] M.A. Benincasa, and C. Delle Fratte, "Influence of ionic strength, sample size, and flow conditions on the retention behavior of pullulan in flow field-flow fractionation", *J. Chromatogr. A,* vol. 1046, no. 1-2, pp. 175-184, 2004.

[http://dx.doi.org/10.1016/S0021-9673(04)01019-2] [PMID: 15387187]

[124] M. Leeman, K.G. Wahlund, and B. Wittgren, "Programmed cross flow asymmetrical flow field-flow fractionation for the size separation of pullulans and hydroxypropyl cellulose", *J. Chromatogr. A,* vol. 1134, no. 1-2, pp. 236-245, 2006. [http://dx.doi.org/10.1016/j.chroma.2006.08.065] [PMID: 16963058]

[125] C. Viebke, and P.A. Williams, "The influence of temperature on the characterization of water-soluble polymers using asymmetric flow field-flow-fractionation coupled to multiangle laser light scattering", *Anal. Chem.,* vol. 72, no. 16, pp. 3896-3901, 2000. [http://dx.doi.org/10.1021/ac991205x] [PMID: 10959979]

[126] A. Litzén, and K.G. Wahlund, "Improved separation speed and efficiency for proteins, nucleic acids and viruses in asymmetrical flow field flow fractionation", *J. Chromatogr. A,* vol. 476, pp. 413-421, 1989. [http://dx.doi.org/10.1016/S0021-9673(01)93885-3] [PMID: 2777988]

[127] U. Adolphi, and W.M. Kulicke, "Coil dimensions and conformation of macromolecules in aqueous media from flow field-flow fractionation/multi-angle laser light scattering illustrated by studies on pullulan", *Polymer (Guildf.),* vol. 38, pp. 1513-1519, 1997. [http://dx.doi.org/10.1016/S0032-3861(96)00675-1]

[128] C. Duval, D. Le Cerf, L. Picton, and G. Muller, "Aggregation of amphiphilic pullulan derivatives evidenced by on-line flow field flow fractionation/multi-angle laser light scattering", *J. Chromatogr. B Biomed. Sci. Appl.,* vol. 753, no. 1, pp. 115-122, 2001. [http://dx.doi.org/10.1016/S0378-4347(00)00517-X] [PMID: 11302436]

[129] Z. Souguir, S. Roudesli, E.L. Picton, D. Le Cerf, and E. About-Jaudet, "Novel cationic and amphiphilic pullulan derivatives I: Synthesis and characterization", *Eur. Polym. J.,* vol. 43, pp. 4940-4950, 2007. [http://dx.doi.org/10.1016/j.eurpolymj.2007.09.017]

[130] M. Andersson, B. Wittgren, and K.G. Wahlund, "Ultrahigh molar mass component detected in ethylhydroxyethyl cellulose by asymmetrical flow field-flow fractionation coupled to multiangle light scattering", *Anal. Chem.,* vol. 73, no. 20, pp. 4852-4861, 2001. [http://dx.doi.org/10.1021/ac0104734] [PMID: 11681461]

[131] M. Andersson, B. Wittgren, H. Schagerlöf, D. Momcilovic, and K.G. Wahlund, "Size and structure characterization of ethylhydroxyethyl cellulose by the combination of field-flow fractionation with other techniques. Investigation of ultralarge components", *Biomacromolecules,* vol. 5, no. 1, pp. 97-105, 2004. [http://dx.doi.org/10.1021/bm030051z] [PMID: 14715014]

[132] B. Porsch, M. Andersson, B. Wittgren, and K.G. Wahlund, "Molecular mass distribution analysis of ethyl(hydroxyethyl)cellulose by size-exclusion chromatography with dual light-scattering and refractometric detection", *J. Chromatogr. A,* vol. 946, no. 1-2, pp. 69-81, 2002. [http://dx.doi.org/10.1016/S0021-9673(01)01572-2] [PMID: 11873984]

[133] B. Wittgren, and K.G. Wahlund, "Effects of flow-rates and sample concentration on the molar mass characterisation of modified celluloses using asymmetrical flow field-flow fractionation-multi-angle light scattering", *J. Chromatogr. A,* vol. 791, pp. 135-149, 1997. [http://dx.doi.org/10.1016/S0021-9673(97)00831-5]

[134] B. Wittgren, and K.G. Wahlund, "Size characterisation of modified celluloses in various solvents using flow FFF-MALS and MB-MALS", *Carbohydr. Polym.,* vol. 43, pp. 63-73, 2000. [http://dx.doi.org/10.1016/S0144-8617(99)00193-9]

[135] W.M. Kulicke, C. Clasen, and C. Lohman, "Characterization of water-soluble cellulose derivatives in terms of the molar mass and particle size as well as their distribution", *Macromol. Symp.,* vol. 223, pp. 151-174, 2005. [http://dx.doi.org/10.1002/masy.200550511]

[136] M.A. Benincasa, G. Cartoni, and C. Delle Fratte, "Flow field-flow fractionation and characterization of ionic and neutral polysaccharides of vegetable and microbial origin", *J. Chromatogr. A,* vol. 967, no. 2, pp. 219-234, 2002.
[http://dx.doi.org/10.1016/S0021-9673(02)00756-2] [PMID: 12685569]

[137] C. Clasen, and W.M. Kulicke, "Determination of viscoelastic and rheo-optical material functions of water-soluble cellulose derivatives", *Prog. Polym. Sci.,* vol. 26, pp. 1839-1919, 2001.
[http://dx.doi.org/10.1016/S0079-6700(01)00024-7]

[138] L. Pitkänen, M. Tenkanen, and P. Tuomainen, "Behavior of polysaccharide assemblies in field-flow fractionation and size-exclusion chromatography", *Anal. Bioanal. Chem.,* vol. 399, no. 4, pp. 1467-1472, 2011.
[http://dx.doi.org/10.1007/s00216-010-4160-3] [PMID: 20820993]

[139] M. Ulmius, S. Adapa, G. Ãnning, and L. Nilsson, "Gastrointestinal conditions influence the solution behaviour of cereal Î²-glucans *in vitro*", *Food Chem.,* vol. 130, pp. 536-540, 2012.
[http://dx.doi.org/10.1016/j.foodchem.2011.07.066]

[140] M. Ulmius, G. Önning, and L. Nilsson, "Solution behavior of barley β-glucan as studied with asymmetrical flow field-flow fractionation", *Food Hydrocoll.,* vol. 26, pp. 175-180, 2012.
[http://dx.doi.org/10.1016/j.foodhyd.2011.05.004]

[141] K.G. Jorgensen, "Quantification of high molecular weight (1-3)(1-4)-beta-D-glucan using calcofluor complex formation and flow-injection analysis", *Carlsberg Res. Commun.,* vol. 53, pp. 287-296, 1988.
[http://dx.doi.org/10.1007/BF02904433]

[142] N. Mäkelä, T. Sontag-Strohm, and N.H. Maina, "The oxidative degradation of barley β-glucan in the presence of ascorbic acid or hydrogen peroxide", *Carbohydr. Polym.,* vol. 123, pp. 390-395, 2015.
[http://dx.doi.org/10.1016/j.carbpol.2015.01.037] [PMID: 25843872]

[143] C. Augsten, W. Knolle, and K. Maeder, "Characterizing the influence of electron irradiation on scleroglucan", *Carbohydr. Polym.,* vol. 72, pp. 707-718, 2008.
[http://dx.doi.org/10.1016/j.carbpol.2007.10.013]

[144] J.R. Runyon, L. Nilsson, M. Ulmius, A. Castro, R. Ionescu, C. Andersson, and C. Schmidt, "Characterizing changes in levan physicochemical properties in different pH environments using asymmetric flow field-flow fractionation", *Anal. Bioanal. Chem.,* vol. 406, no. 6, pp. 1597-1605, 2014.
[http://dx.doi.org/10.1007/s00216-013-7388-x] [PMID: 24121433]

[145] L. Picton, I. Bataille, and G. Muller, "Analysis of a complex polysaccharide (gum arabic) by multi-angle laser light scattering coupled on-line to size exclusion chromatography and flow field flow fractionation", *Carbohydr. Polym.,* vol. 42, pp. 23-31, 2000.
[http://dx.doi.org/10.1016/S0144-8617(99)00139-3]

[146] M. Andres-Brull, S. Al-Assaf, G.O. Phillips, and K. Jackson, "Optimisation of asymmetrical flow-field fractionation for the characterisation of gum arabic (Acacia sengal var senegal) and comparison with gel permeation chromatography", *Anal. Methods,* vol. 5, pp. 4047-4052, 2013.
[http://dx.doi.org/10.1039/c3ay40835c]

[147] J. Alftrén, J.M. Peñarrieta, B. Bergenståhl, and L. Nilsson, "Comparison of molecular and emulsifying properties of gum arabic and mesquite gum using asymmetrical flow field-flow fractionation", *Food Hydrocoll.,* vol. 26, pp. 54-62, 2011.
[http://dx.doi.org/10.1016/j.foodhyd.2011.04.008]

[148] T. Pauck, and H. Colfen, "Hydrodynamic analysis of macromolecular conformation. A comparative study of flow field flow fractionation and analytical ultracentrifugation", *Anal. Chem.,* vol. 70, pp. 3886-3891, 1998.
[http://dx.doi.org/10.1021/ac980228t]

[149] C. Viebke, and P.A. Williams, "Determination of molecular mass distribution of kappa- carrageenan and xanthan using asymmetrical flow field-flow fractionation", *Food Hydrocoll.,* vol. 14, pp. 265-270,

2000.
[http://dx.doi.org/10.1016/S0268-005X(99)00066-1]

[150] C. Augsten, and K. Mäder, "Characterizing molar mass distributions and molecule structures of different chitosans using asymmetrical flow field-flow fractionation combined with multi-angle light scattering", *Int. J. Pharm.*, vol. 351, no. 1-2, pp. 23-30, 2008.
[http://dx.doi.org/10.1016/j.ijpharm.2007.09.013] [PMID: 17964095]

[151] S. Mao, C. Augsten, K. Mäder, and T. Kissel, "Characterization of chitosan and its derivatives using asymmetrical flow field-flow-fractionation: a comparison with traditional methods", *J. Pharm. Biomed. Anal.*, vol. 45, no. 5, pp. 736-741, 2007.
[http://dx.doi.org/10.1016/j.jpba.2007.08.012] [PMID: 17890038]

[152] A. Bourgoin, E. Zablackis, and J.B. Poli, "Characterization of alpha-carrageenan solution behavior by field-flow fractionation and multiangle light scattering", *Food Hydrocoll.*, vol. 22, pp. 1607-1611, 2008.
[http://dx.doi.org/10.1016/j.foodhyd.2007.11.001]

[153] B. Wittgren, J. Borgstrom, L. Piculell, and K.G. Wahlund, "Conformational change and aggregation of kappa-carrageenan studied by flow field-flow fractionation and multiangle light scattering", *Biopolymers,* vol. 45, pp. 85-96, 1998.
[http://dx.doi.org/10.1002/(SICI)1097-0282(199801)45:1<85::AID-BIP7>3.0.CO;2-V]

[154] E. Alasonati, M.A. Benincasa, V.I. Slaveykova, M-A. Benincasa, and V.I. Slaveykova, "Asymmetrical flow field-flow fractionation coupled to multiangle laser light scattering detector: optimization of crossflow rate, carrier characteristics, and injected mass in alginate separation", *J. Sep. Sci.,* vol. 30, no. 14, pp. 2332-2340, 2007.
[http://dx.doi.org/10.1002/jssc.200700211] [PMID: 17683047]

[155] H. Storz, K.J. Müller, F. Ehrhart, I. Gómez, S.G. Shirley, P. Gessner, G. Zimmermann, E. Weyand, V.L. Sukhorukov, T. Forst, M.M. Weber, H. Zimmermann, W-M. Kulicke, and U. Zimmermann, "Physicochemical features of ultra-high viscosity alginates", *Carbohydr. Res.,* vol. 344, no. 8, pp. 985-995, 2009.
[http://dx.doi.org/10.1016/j.carres.2009.02.016] [PMID: 19394590]

[156] I. Colinet, V. Dulong, T. Hamaide, D. Le Cerf, and L. Picton, "New amphiphilic modified polysaccharides with original solution behaviour in salt media", *Carbohydr. Polym.,* vol. 75, pp. 454-462, 2009.
[http://dx.doi.org/10.1016/j.carbpol.2008.08.002]

[157] W. van de Ven, I. Punt, A. Kemperman, and M. Wessling, "Unraveling ultrafiltration of polysaccharides with flow field flow fractionation", *J. Membr. Sci.,* vol. 338, pp. 67-74, 2009.
[http://dx.doi.org/10.1016/j.memsci.2009.04.008]

[158] E. Alasonati, B. Stolpe, M.A. Benincasa, M. Hassellov, and V.I. Slaveykova, "Asymmetrical flow field flow fractionation - Multidetection system as a tool for studying metal-alginate interactions", *Environ. Chem.,* vol. 3, pp. 192-198, 2006.
[http://dx.doi.org/10.1071/EN06024]

[159] R. Mendichi, and A.G. Schieroni, "Fractionation and characterization of ultra-high molar mass hyaluronan: 2. On-line size exclusion chromatography methods", *Polymer (Guildf.),* vol. 43, pp. 6115-6121, 2002.
[http://dx.doi.org/10.1016/S0032-3861(02)00586-4]

[160] T. Luan, Y. Fang, S. Al-assaf, G. Phillips, and H. Zhang, "Compared molecular characterization of hyaluronan using multiple-detection techniques", *Polymer (Guildf.),* vol. 52, p. 24, 2011.
[http://dx.doi.org/10.1016/j.polymer.2011.10.024]

[161] M. Ali, E. Hwang, I.H. Cho, and M.H. Moon, "Characterization of sodium hyaluronate blends using frit inlet asymmetrical flow field-flow fractionation and multiangle light scattering", *Anal. Bioanal. Chem.,* vol. 402, no. 3, pp. 1269-1276, 2012.

[http://dx.doi.org/10.1007/s00216-011-5531-0] [PMID: 22101460]

[162] H. Lee, H. Kim, and M.H. Moon, "Field programming in frit inlet asymmetrical flow field-flow fractionation/multiangle light scattering: application to sodium hyaluronate", *J. Chromatogr. A,* vol. 1089, no. 1-2, pp. 203-210, 2005. [http://dx.doi.org/10.1016/j.chroma.2005.06.069] [PMID: 16130788]

[163] B. Kim, S. Woo, Y-S. Park, E. Hwang, and M.H. Moon, "Ionic strength effect on molecular structure of hyaluronic acid investigated by flow field-flow fractionation and multiangle light scattering", *Anal. Bioanal. Chem.,* vol. 407, no. 5, pp. 1327-1334, 2015. [http://dx.doi.org/10.1007/s00216-014-8379-2] [PMID: 25542570]

[164] H. Lee, I.H. Cho, and M.H. Moon, "Effect of dissolution temperature on the structures of sodium hyaluronate by flow field-flow fractionation/multiangle light scattering", *J. Chromatogr. A,* vol. 1131, no. 1-2, pp. 185-191, 2006. [http://dx.doi.org/10.1016/j.chroma.2006.07.049] [PMID: 16899247]

[165] J.H. Kwon, E. Hwang, I.H. Cho, and M.H. Moon, "Depolymerization study of sodium hyaluronate by flow field-flow fractionation/multiangle light scattering", *Anal. Bioanal. Chem.,* vol. 395, no. 2, pp. 519-525, 2009. [http://dx.doi.org/10.1007/s00216-009-2952-0] [PMID: 19649622]

[166] D.Y. Shin, E. Hwang, I.H. Cho, and M.H. Moon, "Molecular weight and structure characterization of sodium hyaluronate and its gamma radiation degradation products by flow field-flow fractionation and on-line multiangle light scattering", *J. Chromatogr. A,* vol. 1160, no. 1-2, pp. 270-275, 2007. [http://dx.doi.org/10.1016/j.chroma.2007.05.047] [PMID: 17537445]

[167] M.H. Moon, D.Y. Shin, N. Lee, E. Hwang, and I.H. Cho, "Flow field-flow fractionation/multiangle light scattering of sodium hyaluronate from various degradation processes", *J. Chromatogr. B Analyt. Technol. Biomed. Life Sci.,* vol. 864, no. 1-2, pp. 15-21, 2008. [http://dx.doi.org/10.1016/j.jchromb.2008.01.023] [PMID: 18282750]

[168] S. Saeseaw, J. Shiowatana, and A. Siripinyariond, "Sedimentation field-flow fractionation: Size characterization of food materials", *Food Res. Int.,* vol. 38, pp. 777-786, 2005. [http://dx.doi.org/10.1016/j.foodres.2005.04.001]

[169] D. Kang, J.S. Yoo, M.O. Kim, and M.H. Moon, "A soft preparative method for membrane proteome analysis using frit inlet asymmetrical flow field-flow fractionation: application in a prostatic cancer cell line", *J. Proteome Res.,* vol. 8, no. 2, pp. 982-991, 2009. [http://dx.doi.org/10.1021/pr800689y] [PMID: 19140673]

[170] J. Li, J. Ge, Y. Yin, and W. Zhong, "Multiplexed affinity-based protein complex purification", *Anal. Chem.,* vol. 80, no. 18, pp. 7068-7074, 2008. [http://dx.doi.org/10.1021/ac801251y] [PMID: 18715017]

[171] K. H. Kim, J. Y. Lee, and M. H. Moon, "Effect of sodium dodecyl sulfate on protein separation by hollow fiber flow field-flow fractionation", *Analyst,* vol. 136, pp. 388-392, 2011. [http://dx.doi.org/10.1039/C0AN00172D]

[172] P. Reschiglian, B. Roda, A. Zattoni, M. Tanase, V. Marassi, and S. Serani, "Hollow-fiber flow field-flow fractionation with multi-angle laser scattering detection for aggregation studies of therapeutic proteins", *Anal. Bioanal. Chem.,* vol. 406, no. 6, pp. 1619-1627, 2014. [http://dx.doi.org/10.1007/s00216-013-7462-4] [PMID: 24337135]

[173] A. Samontha, C. Nipattamanon, J. Shiowatana, and A. Siripinyanond, "Toward better understanding of salt-induced hen egg white protein aggregation using field-flow fractionation", *J. Agric. Food Chem.,* vol. 56, no. 19, pp. 8809-8814, 2008. [http://dx.doi.org/10.1021/jf801458d] [PMID: 18788749]

[174] K.G. Wahlund, M. Gustavsson, F. MacRitchie, T. Nylander, and L. Wannerberger, "Size characterisation of wheat proteins, particularly glutenin, by asymmetrical flow field flow fractionation", *J. Cereal Sci.,* vol. 23, pp. 113-119, 1996.

[http://dx.doi.org/10.1006/jcrs.1996.0011]

[175] L. Daqiq, C.M. Fellows, F. Bekes, and E. Lees, "Methodologies for symmetrical-flow field-flow fractionation analysis of polymeric glutenin", *J. Texture Stud.,* vol. 38, pp. 273-296, 2007. [http://dx.doi.org/10.1111/j.1745-4603.2007.00098.x]

[176] C. Arfvidsson, K.G. Wahlund, and A.C. Eliasson, "Direct molecular weight determination in the evaluation of dissolution methods for unreduced glutenin", *J. Cereal Sci.,* vol. 39, pp. 1-8, 2004. [http://dx.doi.org/10.1016/S0733-5210(03)00038-9]

[177] L. Pitkanen, T. Sontag-Strohm, and P. Kanerva, "Enhanced separation and characterization of gluten polymers by asymmetrical flow field-flow fractionation coupled with multiple detectors", *J. Cereal Sci.,* vol. 59, pp. 126-131, 2014. [http://dx.doi.org/10.1016/j.jcs.2013.11.009]

[178] E. Lemelin, T. Aussenac, F. Violleau, L. Salvo, and V. Lein, "Impact of cultivar and environment on size characteristics of wheat proteins using asymmetrical flow field-flow fractionation and multi-angle laser light scattering", *Cereal Chem.,* vol. 82, pp. 28-33, 2005. [http://dx.doi.org/10.1094/CC-82-0028]

[179] F. Guyomarch, F. Violleau, O. Surel, and M-H. Famelart, "Characterization of heat-induced changes in skim milk using asymmetrical flow field-flow fractionation coupled with multiangle laser light scattering", *J. Agric. Food Chem.,* vol. 58, no. 24, pp. 12592-12601, 2010. [http://dx.doi.org/10.1021/jf102808f] [PMID: 21090649]

[180] R. Zhu, W. Frankema, Y. Huo, and W.T. Kok, "Studying protein aggregation by programmed flow field-flow fractionation using ceramic hollow fibers", *Anal. Chem.,* vol. 77, no. 14, pp. 4581-4586, 2005. [http://dx.doi.org/10.1021/ac050398n] [PMID: 16013876]

[181] W. Fraunhofer, G. Winter, and C. Coester, "Asymmetrical flow field-flow fractionation and multiangle light scattering for analysis of gelatin nanoparticle drug carrier systems", *Anal. Chem.,* vol. 76, no. 7, pp. 1909-1920, 2004. [http://dx.doi.org/10.1021/ac0353031] [PMID: 15053651]

[182] K. Rbii, F. Violleau, S. Guedj, and O. Surel, "Analysis of aged gelatin by AFlFFF-MALS: Identification of high molar mass components and their influence on solubility", *Food Hydrocoll.,* vol. 23, p. 1024, 2009. [http://dx.doi.org/10.1016/j.foodhyd.2008.08.004]

[183] K. Rbii, O. Surel, N. Brambati, A-M. Buchert, and F. Violleau, "Study of gelatin renaturation in aqueous solution by AFlFFF-MALS: Influence of a thermal pre-treatment applied on gelatin", *Food Hydrocoll.,* vol. 25, pp. 511-514, 2011. [http://dx.doi.org/10.1016/j.foodhyd.2010.08.001]

[184] A. Hawe, W. Friess, M. Sutter, and W. Jiskoot, "Online fluorescent dye detection method for the characterization of immunoglobulin G aggregation by size exclusion chromatography and asymmetrical flow field flow fractionation", *Anal. Biochem.,* vol. 378, no. 2, pp. 115-122, 2008. [http://dx.doi.org/10.1016/j.ab.2008.03.050] [PMID: 18455994]

[185] J.R. Silveira, A.G. Hughson, and B. Caughey, "Fractionation of prion protein aggregates by asymmetrical flow field-flow fractionation", In: *Amyloid, Prions, and Other Protein Aggregates, Pt B.,* I. Kheterpal, R. Wetzel, Eds., vol. 412. 2006, pp. 21-33. [http://dx.doi.org/10.1016/S0076-6879(06)12002-9]

[186] D.C. Rambaldi, A. Zattoni, P. Reschiglian, R. Colombo, and E. De Lorenzi, "In vitro amyloid Abeta(142) peptide aggregation monitoring by asymmetrical flow field-flow fractionation with multi-angle light scattering detection", *Anal. Bioanal. Chem.,* vol. 394, no. 8, pp. 2145-2149, 2009. [http://dx.doi.org/10.1007/s00216-009-2899-1] [PMID: 19562325]

[187] Y.P. Chuan, Y.Y. Fan, L. Lua, and A.P. Middelberg, "Quantitative analysis of virus-like particle size and distribution by field-flow fractionation", *Biotechnol. Bioeng.,* vol. 99, no. 6, pp. 1425-1433, 2008.

[http://dx.doi.org/10.1002/bit.21710] [PMID: 18023039]

[188] L.F. Pease III, D.I. Lipin, D-H. Tsai, M.R. Zachariah, L.H. Lua, M.J. Tarlov, and A.P. Middelberg, "Quantitative characterization of virus-like particles by asymmetrical flow field flow fractionation, electrospray differential mobility analysis, and transmission electron microscopy", *Biotechnol. Bioeng.,* vol. 102, no. 3, pp. 845-855, 2009.
[http://dx.doi.org/10.1002/bit.22085] [PMID: 18958863]

[189] A. Citkowicz, H. Petry, R.N. Harkins, O. Ast, L. Cashion, C. Goldmann, P. Bringmann, K. Plummer, and B.R. Larsen, "Characterization of virus-like particle assembly for DNA delivery using asymmetrical flow field-flow fractionation and light scattering", *Anal. Biochem.,* vol. 376, no. 2, pp. 163-172, 2008.
[http://dx.doi.org/10.1016/j.ab.2008.02.011] [PMID: 18342613]

[190] D. I. Lipin, L. H. L. Lua, and A. P. J. Middelberg, "Quaternary size distribution of soluble aggregates of glutathione-S-transferase-purified viral protein as determined by asymmetrical flow field flow fractionation and dynamic light scattering", *J. Chromatogr. A,* vol. 1190, pp. 204-214, 2008.
[http://dx.doi.org/10.1016/j.chroma.2008.03.032]

[191] Z. Wei, M. McEvoy, V. Razinkov, A. Polozova, E. Li, J. Casas-Finet, G.I. Tous, P. Balu, A.A. Pan, H. Mehta, and M.A. Schenerman, "Biophysical characterization of influenza virus subpopulations using field flow fractionation and multiangle light scattering: correlation of particle counts, size distribution and infectivity", *J. Virol. Methods,* vol. 144, no. 1-2, pp. 122-132, 2007.
[http://dx.doi.org/10.1016/j.jviromet.2007.04.008] [PMID: 17586059]

[192] N.L. Setälä, J.M. Holopainen, J. Metso, S.K. Wiedmer, G. Yohannes, P.K. Kinnunen, C. Ehnholm, and M. Jauhiainen, "Interfacial and lipid transfer properties of human phospholipid transfer protein: implications for the transfer mechanism of phospholipids", *Biochemistry,* vol. 46, no. 5, pp. 1312-1319, 2007.
[http://dx.doi.org/10.1021/bi0621866] [PMID: 17260960]

[193] G. Yohannes, M. Sneck, S.J. Varjo, M. Jussila, S.K. Wiedmer, P.T. Kovanen, K. Oörni, and M-L. Riekkola, "Miniaturization of asymmetrical flow field-flow fractionation and application to studies on lipoprotein aggregation and fusion", *Anal. Biochem.,* vol. 354, no. 2, pp. 255-265, 2006.
[http://dx.doi.org/10.1016/j.ab.2006.04.018] [PMID: 16750506]

[194] K.H. Kim, J.Y. Lee, S. Lim, and M.H. Moon, "Top-down lipidomic analysis of human lipoproteins by chip-type asymmetrical flow field-flow fractionation-electrospray ionization-tandem mass spectrometry", *J. Chromatogr. A,* vol. 1280, pp. 92-97, 2013.
[http://dx.doi.org/10.1016/j.chroma.2013.01.025] [PMID: 23375771]

[195] D.C. Rambaldi, A. Zattoni, S. Casolari, P. Reschiglian, D. Roessner, and C. Johann, "An analytical method for size and shape characterization of blood lipoproteins", *Clin. Chem.,* vol. 53, no. 11, pp. 2026-2029, 2007.
[http://dx.doi.org/10.1373/clinchem.2007.091942] [PMID: 18035594]

[196] R.N. Qureshi, W.T. Kok, and P.J. Schoenmakers, "Fractionation of human serum lipoproteins and simultaneous enzymatic determination of cholesterol and triglycerides", *Anal. Chim. Acta,* vol. 654, no. 1, pp. 85-91, 2009.
[http://dx.doi.org/10.1016/j.aca.2009.06.060] [PMID: 19850173]

[197] D.C. Rambaldi, P. Reschiglian, A. Zattoni, and C. Johann, "Enzymatic determination of cholesterol and triglycerides in serum lipoprotein profiles by asymmetrical flow field-flow fractionation with on-line, dual detection", *Anal. Chim. Acta,* vol. 654, no. 1, pp. 64-70, 2009.
[http://dx.doi.org/10.1016/j.aca.2009.06.016] [PMID: 19850170]

[198] S. Podzimek, *Light Scattering, Size Exclusion Chromatography and Asymmetric Flow Field Flow Fractionation: Powerful Tools for the Characterization of Polymers, Proteins and Nanoparticles.* John Wiley & Sons, Inc: Hoboken, New Jersey, 2011.
[http://dx.doi.org/10.1002/9780470877975]

[199] L. Vaysse, F. Bonfils, J. Sainte-Beuve, and M. Cartault, "Natural Rubber", In: *Polymer Science, A Comprehensive Reference.,* K. Matyjaszewski, M. Möller, Eds., vol. 10. Elsevier BV: Amsterdam, 2012, pp. 281-293.
[http://dx.doi.org/10.1016/B978-0-444-53349-4.00267-3]

[200] K.N. Fuller, "Rheology of raw rubber", In: *Natural Rubber Science and Technology.,* A. Roberts, Ed., Oxford University Press: Oxford, 1988, pp. 141-176.

[201] S.J. Park, T.W. Kim, M.K. Kim, S.Y. Lee, and S.C. Lim, "Advanced bacterial polyhydroxyalkanoates: towards a versatile and sustainable platform for unnatural tailor-made polyesters", *Biotechnol. Adv.,* vol. 30, no. 6, pp. 1196-1206, 2012.
[http://dx.doi.org/10.1016/j.biotechadv.2011.11.007] [PMID: 22137963]

[202] S. Tomizawa, M. Hyakutake, Y. Saito, J. Agus, K. Mizuno, H. Abe, and T. Tsuge, "Molecular weight change of polyhydroxyalkanoate (PHA) caused by the PhaC subunit of PHA synthase from Bacillus cereus YB-4 in recombinant Escherichia coli", *Biomacromolecules,* vol. 12, no. 7, pp. 2660-2666, 2011.
[http://dx.doi.org/10.1021/bm2004687] [PMID: 21618968]

[203] K. Loos, A. Boker, H. Zettl, A.F. Zhang, G. Krausch, and A.H. Muller, "Micellar aggregates of amylose-block-polystyrene rod-coil block copolymers in water and THF", *Macromolecules,* vol. 38, pp. 873-879, 2005.
[http://dx.doi.org/10.1021/ma0345549]

[204] J. Lou, M.N. Myers, and J.C. Giddings, "Separation of polysaccharides by thermal Field-Flow Fractionation", *J. Liq. Chromatogr.,* vol. 17, pp. 3239-3260, 1994.
[http://dx.doi.org/10.1080/10826079408013201]

[205] A. Vernhet, S. Dubascoux, B. Cabane, H. Fulcrand, E. Dubreucq, and C. Poncet-Legrand, "Characterization of oxidized tannins: comparison of depolymerization methods, asymmetric flow field-flow fractionation and small-angle X-ray scattering", *Anal. Bioanal. Chem.,* vol. 401, no. 5, pp. 1559-1569, 2011.
[http://dx.doi.org/10.1007/s00216-011-5076-2] [PMID: 21573842]

[206] C.L. Swanson, R.A. Buchanan, and F.H. Othey, "Molecular weights of natural rubbers from selected temperate zone plants", *J. Appl. Polym. Sci.,* vol. 23, pp. 743-748, 1979.
[http://dx.doi.org/10.1002/app.1979.070230309]

[207] J.B. van Beilen, and Y. Poirier, "Establishment of new crops for the production of natural rubber", *Trends Biotechnol.,* vol. 25, no. 11, pp. 522-529, 2007.
[http://dx.doi.org/10.1016/j.tibtech.2007.08.009] [PMID: 17936926]

[208] S. Amnuaypornsri, J. Sakdapipanich, S. Toki, B. Hsiao, N. Ichikawa, and Y. Tanaka, "Strain-induced crystallization of natural rubber: Effect of proteins and phospholipids", *Rubber Chem. Technol.,* vol. 81, pp. 753-766, 2008.
[http://dx.doi.org/10.5254/1.3548230]

[209] Y. Tanaka, and L. Tarachiwin, "Recent advances in structural characterization of natural rubber", *Rubber Chem. Technol.,* vol. 82, pp. 283-314, 2009.
[http://dx.doi.org/10.5254/1.3548250]

[210] S. Wisunthorn, L. Vaysse, S. Liengprayoon, J. Sainte Beuve, and F. Bonfils, "SEC-MALS Study of Dynamic Structuring of Natural Rubber: Comparative Study of Two Hevea brasiliensis Genotypes", *J. Appl. Polym. Sci.,* vol. 124, pp. 1570-1577, 2012.
[http://dx.doi.org/10.1002/app.35099]

[211] P.W. Allen, and G.M. Bristow, "The gel phase in natural rubber", *Rubber Chem. Technol.,* vol. 36, pp. 1024-1034, 1963.
[http://dx.doi.org/10.5254/1.3539624]

[212] C. Kim, J. Sainte Beuve, S. Guilbert, and F. Bonfils, "Study of chain branching in natural rubber using

size-exclusion chromatography coupled with a multi-angle light scattering detector (SEC-MALS)", *Eur. Polym. J.,* vol. 45, pp. 2249-2259, 2009.
[http://dx.doi.org/10.1016/j.eurpolymj.2009.05.015]

[213] C. Kim, M.H. Morel, J.S. Beuve, S. Guilbert, A. Collet, and F. Bonfils, "Characterization of natural rubber using size-exclusion chromatography with online multi-angle light scattering. Study of the phenomenon behind the abnormal elution profile", *J. Chromatogr. A,* vol. 1213, no. 2, pp. 181-188, 2008.
[http://dx.doi.org/10.1016/j.chroma.2008.10.052] [PMID: 18992896]

[214] T. Otte, H. Pasch, T. Macko, R. Brüll, F.J. Stadler, J. Kaschta, F. Becker, and M. Buback, "Characterization of branched ultrahigh molar mass polymers by asymmetrical flow field-flow fractionation and size exclusion chromatography", *J. Chromatogr. A,* vol. 1218, no. 27, pp. 4257-4267, 2011.
[http://dx.doi.org/10.1016/j.chroma.2010.12.072] [PMID: 21238968]

[215] A.M. Seungho Lee, "Determination of Molecular Weight and Gel Content of Natural Rubber Using Thermal Field-Flow Fractionation", *Macromolecules,* vol. 28, pp. 6354-6356, 1995.
[http://dx.doi.org/10.1021/ma00122a050]

[216] C.H.E. Seungho Lee, and Anthony R. Plepys, "Capability of thermal field-flow fractionation for analysis of processed natural rubber", *Bull. Korean Chem. Soc.,* vol. 21, pp. 69-74, 2000.

[217] W-S. Kim, C.H. Eum, A. Molnár, J-S. Yu, and S. Lee, "Repeatability and reproducibility of thermal field-flow fractionation in molecular weight determination of processed natural rubber", *Analyst (Lond.),* vol. 131, no. 3, pp. 429-433, 2006.
[http://dx.doi.org/10.1039/B513773J] [PMID: 16496053]

[218] S. Dubascoux, C. Thepchalerm, E. Dubreucq, S. Wisunthorn, L. Vaysse, S. Kiatkamjornwong, C. Nakason, and F. Bonfils, "Comparative study of mesostructure of natural and synthetic polyisoprene by SEC-MALS and AF4-MALS", *J. Chromatogr. A,* vol. 1224, pp. 27-34, 2012.
[http://dx.doi.org/10.1016/j.chroma.2011.12.010] [PMID: 22245173]

CHAPTER 6

Rheology of Recent Vegetal-Based Biopolymers

Laura Patricia Martínez-Padilla*

Laboratorio de Propiedades Reológicas y Funcionales en Alimentos, Facultad de Estudios Superiores Cuautitlán, Universidad Nacional Autónoma de México, México

Abstract: In this chapter, the rheology of polysaccharides extracted from cactus and agave plants is summarized. These vegetal sources of polysaccharides (*Opuntia* mucilage, *Opuntia* pectin, and agave fructans) were selected for their functional properties (thickening, gelling, or emulsifying) and their bioactive role in the the prevention and/or treatment of disease. The source, production, extraction processes, structure, and functional properties (related to conformation) of these polysaccharides are briefly described before the rheology of these biopolymers is discussed, and recommended uses are suggested.

Keywords: Agave plant, Carreau model, Characteristic time, Cross model, Fructans, Intrinsic viscosity, Loss modulus, Mucilage, Neolevans, Newtonian, Nopal, *Opuntia*, Pectin, Rheology, Shear rate, Shear-thinning, Storage modulus, Viscoelasticity, Viscosity, Zero shear viscosity.

INTRODUCTION

Biopolymers are polysaccharides and proteins with diverse applications in the food and pharmaceutical industries. These biomacromolecules are often referred to as hydrocolloids [1] and are added to perform functions including thickening and gelling of processed foods. Generally, polysaccharides are able to form viscous solutions or three-dimensional gel structures at low concentration, considerably modifying the rheology of the aqueous systems to which they are added.

To date, the vegetal-based biopolymers that predominate in industrial applications are obtained from various sources and include cellulose and its derivatives (major structural polysaccharides in the cell walls of higher plants): microcrystalline cellulose, carboxymethylcellulose, methylcellulose; starch (corn, potato, sweet

* **Corresponding author L. P. Martínez-Padilla**: Laboratorio de Propiedades Reológicas y Funcionales en Alimentos, Facultad de Estudios Superiores Cuautitlán, Universidad Nacional Autónoma de México; Tel/Fax: (52)5556232038; E-mail: lpmp@unam.mx

Martin Masuelli & Denis Renard (Eds.)

potato, wheat, sago palm, tapioca, rice, banana, soybean, arrowroot, bracken, lotus, cassava, and waxy maize); hemicelluloses, including, mannans and galactomannans (guar and locust or carob bean and taro), xyloglucans, glucomannans, arabinoxylans, *β*-D-glucans, and arabinogalactan; pectins (citrus, apple, and sugar beet); exudate gums (arabic, tragacanth, karaya, and ghatti) and seaweed gums (alginate, carrageenans, agar) [1 - 3]. Some of these biopolymers have been modified physically or chemically to enhance their functional properties, and their rheology has been extensively studied and reported elsewhere [4 - 7].

Vegetal sources provide an abundant supply of biopolymers. In this chapter, we discussed the rheology of more recent sources of vegetal polysaccharides in detail that are consumed for their health benefits, such as *Opuntia* mucilage and pectins and agave fructans, which have received less attention.

Opuntia Polysaccharides

Sources

Members of the genus *Opuntia* from the Cactaceae family characteristically produce polysaccharide mucilages and pectins. The Cactaceae family is reported to consist of nearly 1,500 species from approximately 130 genera [8]. Cactus pear, commonly known as prickly pear and produced by the perennial *Opuntia* cactus, is well adapted to arid and semiarid zones, where water for cultivation can be limited. This genus is endemic to the Americas and is found in many countries (Argentina, Bolivia, Brazil, Chile, Egypt, Eritrea, Ethiopia, Greece, India, Israel, Italy, Mexico, Peru, Morocco, Spain, Tunisia, South Africa, and the state of Texas (USA)) [8, 9].

The cactus (*Opuntia ficus-indica, Ofi*) that produces the prickly pear is native to Mexico and can be consumed as a fruit or a green vegetable, and young leaves can be used in salad dishes. Cacti produce edible stems known as pads, phylloclades, or cladodes; synonyms are "nopales" or "pencas." Nopal is consumed as a vegetable at different stages but primarily as fresh, tender nopal stems called "nopalitos," which correspond to the first stage of development (less than 32 days of growth) [9, 10].

Many species of *Opuntia* cacti also produce a desirable fruit known as cactus pear, a many-seeded berry with a thick peel that encloses a flavorful pulp [11]. For example, *Opuntia ficus-indica Mill* is cultivated in Chile for fruit production [12].

Chemical characterizations of nopal showed that the presence of carbohydrates, dietary fiber, proteins, glycoproteins, vitamins, and minerals [8 - 10, 13]. Recently, non-nutritive components of cactus pear have been shown to be potentially active antioxidant phytochemicals. The composition of phenolic acids and flavonoids from *Opuntia* spp. have been reported in cladodes [10] and in cactus pears, including conjugated flavonoids such as quercetin, kaempferol, isorhamnetin, ascorbic acid, carotenoids, phenolic acids, and flavonoids [11]. Antioxidants are recognized protective against chronic disease by preventing the formation of free radicals. Sugars and acids are primarily responsible for the flavor of fruits, and the sugar:acid ratio is often used in harvest and quality indices of different fruit commodities [8].

Mucilage Production

Cladodes secrete a slimy fluid known as mucilage, which can also be found inside cladodes, fruit or fruit peel. In general, to obtain the mucilage, precipitation by alcohol (ethanol or 2-propanol) is performed after water extraction (homogenization of the cactus pads with water). However, very low yields are achieved (1-1.5% fresh weight, 19-22% dry weight) from fruit peel or pads [13].

A simple laboratory-scale extraction method was reported, where diced cladodes were blended in water, filtered, and centrifuged repeatedly; the extract was freeze-dried and resuspended in water, and trichloroacetic acid (TCA) was added before the mixture was left to stand prior to further centrifugation. The precipitate was resuspended and recentrifuged. The TCA-treated supernatant was dialyzed against distilled water for 24 h, precipitated with absolute ethanol and centrifuged to yield 0.7 g/kg of cladodes. Before extraction, the cladodes were boiled in 80% ethanol to inactivate enzymes [14].

Another method of extracting mucilage involves macerating the stems into a pulp and then centrifuging, decanting, and precipitating with acetone (1:2). The precipitate is then washed with isopropyl alcohol (1:1) and dried [15]. Alternatively, petroleum ether can be used to degrease the cladodes, and the pads are then macerated in deionized water and filtered under vacuum. The filtered product is directly freeze-dried or purified by ultrafiltration to eliminate salts and compounds of low molecular weight (Mw) [16].

The mucilage from peeled fruits is blended with water and filtered. The filtrate is centrifuged, and the supernatant is concentrated (vacuum) and dialyzed against distilled water (48 h). The resulting solution is concentrated under reduced pressure and poured into ethanol (1:5 v/v), and the resulting precipitate is separated by centrifugation, dissolved in distilled water, and freeze-dried [12].

To date, there is no available industrial-scale method of extracting mucilage, but a market for dietary supplements based on dried nopal tablets and soft drinks is expanding worldwide [9, 13, 17].

Pectin Production

In this section, we will review the complex extraction methods currently being used to extract pectins from *Opuntia* cladodes and fruits, a calcium-sensitive gelling biopolymer [17].

Fully de-esterified pectin was extracted from fresh nopal pads using an alkaline extraction medium in the presence of a sequestering agent [18]. In this process, fresh pads were cleaned, cut into small pieces, heated in water (85°C for 20 min to inactivate the enzymes), and left to cool (~23°C). The mixture was neutralized to pH 7.5 from an initial pH 4 to induce de-esterification of methoxy groups and filtered through cloth. The solid residue was suspended in a NaOH (50 mM) sodium hexametaphosphate (7.5 g/L) solution (pH 12) and stirred continuously for 1 h. The resulting cactus pectin alkaline extract was filtered through cloth, and the supernatant was recovered. Pectin was precipitated (pH 2) and left to stand overnight (5°C). The precipitate was recovered by centrifugation and resuspended in water, and the pH was adjusted to 8 to redissolve the pectin. To purify the pectin, the polymer solution was filtered (3, 1.2 and 0.8 μm membranes), precipitated with absolute ethanol (50% v/v), washed successively with ethanol/water mixtures and absolute ethanol, and left to dry at ambient temperature (0.6% w/w yield of the fresh weight).

More recently, low methoxyl pectin was extracted from prickly pear peel [19] by removing and milling the peel for a few minutes. Ground peels were steam-blanched for 20 min to inactivate enzymes and subjected to sequential extraction in a glass tank with constant stirring.

During pectin extraction, a solid:solvent ratio of 300 g L^{-1} was used. Peels were leached with deionized water for 2 h (~23°C), and the supernatant was discarded. The wet peels were leached again with deionized water for 2 h (60°C), resuspended in an aqueous solution (1% ethylenediamine tetraacetic acid) at pH 4.0 for 2 h at 70°C, and the liquid phase was filtered through a cloth. After cooling, the polymer extracted was precipitated using an equal volume of 96% ethanol, acidified to remove free salts and sugars and left to stand overnight (5°C). The polysaccharide was separated by filtration using a cloth, dispersed in water and dialyzed (12 kDa) against deionized water (48 h). The polysaccharide in the continuous phase was precipitated by adding cold ethanol, washed successively with 96% ethanol and acidified using a 60-mm porous glass filter connected to a vacuum. Finally, the biopolymer was dried in a convection oven (45°C) and ground and sieved to produce a homogeneous powder.

Structure

The structures of the mucilages that have been proposed for the two distinctive water-soluble fractions share similarities in the composition of their neutral constituent sugar residues; one mucilage lacks gelling properties, and the other is a pectin with gelling properties (Ca^{2+}). The latter has significantly more linear polygalacturonic acid [17, 18]. The sugar compositions of mucilage and pectins from cladodes, peeled nopales, prickly pears, peel and seeds were all shown to be very similar [16 - 22]. A summary of the recently reported polysaccharide compositions is presented in Table **1**.

Earlier studies reported that the mucilage of *Ofi* is a neutral polysaccharide composed of L-arabinose, L-rhamnose, D-galactose, D-xylose, and D-galacturonic acid. The first suggested structure described the molecule as a linear repeating core chain of (1-4) linked α-D-galacturonic acid and (1-2) linked β-L-rhamnose with lateral chains of (1-6)- β-D-galactose attached to O-4 rhamnose residues.

The galactose residues are present at the O-3 position or as a double substitution at the O-3 and O-4 positions. The composition of the outlying chains is complex, including at least 20 identified oligosaccharides (mostly disaccharides and trisaccharides). These invariably contained (1 $\rightarrow$ 5)-linked L-arabinose residues and D-xylose as terminal groups in a near 1:2 ratio of xylose:arabinose. Rhamnose and galacturonic acid were reported to be confined to the acid-resistant backbone of the mucilage because they were not apparent after hydrolysis [9].

The Mw of mucilage varies from 23 and 4300 kDa. Differences in isolation techniques and contamination with other cell compounds may explain the wide spectrum of results [9, 14, 15]. An extraction of two fractions yielded average high and low molar masses of 13 x 10^6 and 3.9 x 10^3 g/mol, respectively and proteins (mainly albumins) constituted the bulk of the latter (potentially emulsifying) fraction [16].

Infrared spectra of fully de-esterified pectin obtained from fresh pads (alkaline extraction) was nearly identical to those of commercial citrus pectin and homogalacturonan-rich pectin isolated from prickly pear, lime peel, and sugar beet pulp [18].

Table 1. Composition of mucilage and pectin sugar from *Opuntia* plants and fruits.

Source	Type of polysaccharide	Composition
Peeled prickly pear nopal (*Ofi*)	Pectins high Mw	Galacturonic acid (9.6%), Glucuronic acid (1.0%), rhamnose (6.9%), arabinose (33.1%), galactose (20.3%), glucose (1%), xylose (18.7%) [16].
Fresh nopal pads (*Ofi*)	Pectins	Uronic acids (56.3%), rhamnose (0.5%), arabinose (5.6%), galactose (6.5%), xylose (0.9%) [17].
Fresh nopal pads (*Ofi*)	Mucilage	Uronic acids (11%), rhamnose (1.75%), arabinose (17.93%), galactose (20.99%), xylose (3.06%) [17].
Fresh nopal pads (*Ofi*)	Literature comparison (pectins obtained by various alkaline processes)	Uronic acids (6.4-64%), rhamnose (0-7%), arabinose (6-44%), galactose (20.3-22%), glucose (1-2.6%), xylose (2.1-22.1%) [17].
Fresh nopal pads (*Ofi*)	Fully de-esterified pectin	Uronic acid (85.4%), galactose (7%), arabinose (6%), and minor quantities of rhamnose and xylose [18].
Prickly pear fruit peel (*Opuntia albircapa Scheinvar*)	Pectins	Galacturonic acid, rhamnose, arabinose, galactose, and glucose [19].
Prickly pear seeds (*Ofi*)	Pectic polysaccharides	Uronic acid (9%), rhamnose (0.6%), arabinose (3.1%), galactose (1%), glucose (40.6%), xylose (44.8%), and mannose (1%) [20].
Prickly pear skin (*Ofi*)	Three fractions obtained by sequential extraction	Uronic acid (35.6-64.5%), rhamnose (1.4-3.7%), arabinose (12.3-17.2%), galactose (4-10.8%), glucose (3.3-29.3%), xylose (0.9-4.5%), and mannose (0.5-1.6%) [21].
Prickly pear spines (*Ofi*)	Untreated spines	Galacturonic acid (6.3%), rhamnose (1.1%), arabinose (44.2%), galactose (2.2%), glucose (47.3%), and xylose (0.9%) [22].

Functional Properties

In general, the thickening, gelling, and emulsifying properties of extracted mucilage and pectin will dictate their technological applications. Because mucilage from *Opuntia* exists as a calcium salt in cells, Ca^{2+} has a significant effect on its water-holding capacity and other biophysical properties [17]. However, the water-soluble polysaccharide fraction with thickening properties in *Ofi* represents less than 10% of the water-soluble material [16].

Pectins from mucilage show excellent gelling properties in the presence of $CaCl_2$. As in citrus pectins, depending on the degree of methylation (DM), mucilage pectins can form aqueous gels in acidic media, under high sugar concentrations (high DM pectins, DM >50%) or by interaction with divalent cations, particularly Ca^{2+} (low DM pectins, DM <50%). In the latter case, the sol-gel transition is accompanied by a coil-to-helix conformational change, similar to other food polysaccharides (agarose, kappa carrageenan, and gellan). In these systems, gelling occurs as a highly cooperative process governed by the establishment of ordered, helical structures stabilized by a profuse hydrogen-bonded network. Occasionally, the onset temperature of gelation for nopal pectin gels in these systems is controlled by the concentration of Ca^{2+} [17].

Ofi mucilage exhibited emulsifying capabilities in dilute oil-water emulsions, and reduced surface and interfacial tensions caused the formation of small oil droplets and a film at the oil-water interface [9, 23, 24]. Although this could be attributed to protein residues in the biopolymer [16], the mucilage does not contribute to the viscosity of the system but instead stabilizes the emulsion [23]. Similarly, reduced surface tension facilitates the formation of foam, as reported for egg foams that were stabilized with less syneresis and greater volume than the control up to 48 h (0.5% and 0.8% nopal mucilage) [9].

Mucilage Rheology

Numerous groups have defined the rheological properties of various *Ofi* mucilage extracts using rotational and oscillatory shear techniques. Here, only the rheology of mucilages that increase solution viscosity, not those that form gels, will be described.

A mucilage biopolymer (13×10^6 g/mol) solution is characterized by the flexible random-coil conformation of a highly solvated polymer, with a measured $[\eta]$ of ~ 850 mL/g [16]. A limit of 1.55 g/L was set to separate the dilute and semi-dilute regimes. In the dilute regime, viscosity depends only on the dimensions of the polymer chains. However, the conformation of the chain is dependent on salinity due to the negative charge of the carboxylic groups on galacturonic and glucuronic residues.

Intrinsic viscosity diminished when 0.5-mM NaCl (~820 mL/g) or $CaCl_2$ (~740 mL/g) was added [16]. These values are similar to those reported (613 mL/g) for apple pectin (~1.55×10^5 average Mw) with a 72.9% degree of esterification (0.155-M NaCl) [5]. Higher $[\eta]$ was obtained in xanthan gum aqueous solutions (0.01-M NaCl) (10300 mL/g). Xanthan gum is a very popular biopolymer thickening agent [25], reported here for comparison.

At intermediate concentrations (semi-dilute regime) and under steady-shear tests, all mucilage solutions showed non-Newtonian shear-thinning behavior described by the Carreau [16] or Cross [14, 15, 19, 26, 27] models, including four parameters: (1) Newtonian viscosity at low shear rates, η_0; (2) Newtonian viscosity at high shear rates, η_∞; (3) a characteristic time; and (4) a parameter describing the shear-thinning zone. Moreover, non-Newtonian shear-thinning behavior of aqueous, reconstituted (spray or freeze-dried) mucilage samples (3-6% p/v) was also observed [26, 27]. The Cross model was used to describe the flow curves of spray-dried samples, and the Ostwald-de Waele model was used to describe those of freeze-dried samples [26].

Table **2** summarizes some literature η_0 values. As in entangled polymers, zero-shear viscosity values increase as polymer concentration increases; the Newtonian plateau simultaneously shifts to the low-shear-rate region [16]. Values of η_0 are highly dependent on the extraction method, purity, Mw, and dehydration method. For comparison, mucilage η_0 values were lower than those observed in purified xanthan gum dissolutions, which varied from 1 x 10^2 to 4 x 10^4 Pa.s [25], and commercial xanthan samples (0.033-2.89 Pa.s) at very low concentrations (0.05-0.15%) [28, 29].

Table 2. Zero-shear viscosity (η_0) of mucilage solutions.

Source		η_0 ***(Pa.s)***
Ofi mucilage solution [14]	0.5%	~ 0.1
Ofi mucilage solution [14]	1.6%	~ 9
Ofi mucilage solution [14]	5.8%	~ 200
Biopolymer in water (*Ofi*) [16]	30 g/L	11.49
Biopolymer in water (*Ofi*) [16]	30 g/L, 1M, $LiNO_3$	8.28
Biopolymer in water (*Ofi*) [16]	30 g/L, 1M, $CaCl_2$	6.62
Pectin from prickly pear skin(*Opuntia albicarpa Scheinvar*) [19]	20 g/L	2.44
Pectin from prickly pear skin(*Opuntia albicarpa Scheinvar*) [19]	10 g/L	0.128
Reconstituted mucilage solution(*Ofi*, spray dried) [26]	3%	0.0103-0.0175
Reconstituted mucilage solution(*Ofi*, spray dried) [26]	6%	0.0537-0.0761
Reconstituted mucilage solution(*Ofi*, spray dried and reconstituted spray dried) [27]	6%	0.048-0.049

Compared to the η_0 of solutions in pure water, a loss in viscosity was observed in the presence of a monovalent cation (Li^+), but this loss is less marked than that corresponding to the same solution (*i.e.*, identical polymer and electrolyte concentration) in the presence of a divalent cation (Ca^{2+}). The authors interpret this to result from complex formation between calcium ions and carboxylic groups located along the backbone of the polymer [16].

In several cases, the Newtonian viscosity values at η_∞ were small compared to η_0 and varied from 0.001-0.005 Pa.s in reconstituted mucilage dissolutions [26]. These values are similar to the viscosity of the solvent (water) and were generally neglected in the Cross or Carreau models.

The characteristic time for spray-dried mucilage to reconstitute, as it relates to the critical shear rate at which shear-thinning flow behavior starts or at which the entangled density begins to decrease, was small and varied from 0.003 to 0.026 s (3 and 6%) (Cross model) [26].

Again, these values were lower than those obtained for xanthan gum (0.56 and 1.14 s for 0.05 and 0.1%, respectively) (Carreau model) [28, 29]. This time constant was often of the same order of magnitude as that of the Rouse time, derived from the dissolution viscosity (bead-spring theory) [5].

The flow behavior indexes of the Ostwald-de Waele model in reconstituted mucilage solutions of spray-dried samples (1%) were close 0.90, while that of freeze-dried mucilage (1-6%) was lower, indicating a high dependence on shear rate (shear-thinning behavior) (0.50-0.74).

As in other macromolecules, shear-thinning was attributed to an orientation effect when the shear rate increased because long-chain polymer molecules (randomly positioned) become aligned in the direction of the flow, resulting in less interaction between adjacent polymer chains. Such behavior is characteristic of biopolymers [26].

In a similar study, the rheological behavior of reconstituted spray-dried mucilage (3 and 6%) isolated from the cladodes of pitahaya (*Hylocereus undatus*) was obtained.

The same non-Newtonian shear-thinning behavior was evaluated, characterized by low η_0 (0.0048-0.132 Pa.s) and low characteristic time (0.001-0.06 s) [30].

In this mucilage concentration effect, viscosity curves deviated from the typical "master curve" for linear random-coil behavior that is applied for many polysaccharides used as industrial thickeners [4]. This behavior was attributed to the formation of colloidal aggregates [14].

The viscoelastic properties of concentrated mucilage solutions from a polysaccharide (3000 kDa) isolated from cladodes were also measured in the linear viscoelastic region [14].

Mechanical spectra at 1, 1.6, and 2% (0.1-M NaCl at 20 °C) confirmed liquid-like behavior due to the high frequency (ω) dependence of the mechanical moduli and the predominance of the loss modulus (G") over the storage modulus (G') at low ω. At higher ω, G'>G". However, a characteristic crossover (G'=G") was detected, whereby both moduli increased when the concentration was increased and the crossover diminished as concentration increased (1-2%) from ~ 30 1/s to ~10 1/s.

This corresponded to characteristic times of ~0.03 s and ~ 0.1 s, respectively [14]. Moreover, the mechanical response in a time-temperature superposition master curve of another nopal mucilage extract (1.3%, 20 °C) showed a crossover close to 0.1 rad/s, corroborating the idea that mechanical spectra correspond with that of an entangled network of disordered polymer coils [6, 17]. These values were smaller when compared to the crossover of high-methoxy pectin solution of ~60 rad/s (0.1-M NaCl, pH 7, T 25°C) [5].

The dehydration method can affect the viscoelastic properties of the product. Spray-drying produced a stable powdered product with rheological properties (viscous and elastic) lower than those obtained by freeze-drying. In spray-dried samples, predominantly viscous behavior was reported (G'>G"), with liquid like-behavior similar to mucilage solutions.

In freeze-dried samples (3%), the storage modulus dominated over the loss modulus (G'>G") [26]. Freeze-dried samples showed patterns similar to a concentrated biopolymer solution or a weak gel, as was described for a 0.5% aqueous dissolution of xanthan gum [25].

The crossover (~8.5 1/s) was very different for spray-dried samples (7.2%), which showed liquid-like behavior (G">G') compared to that of freeze-dried samples (6%), similar to the behavior of an entangled network of disordered polymer coils (G'>G"; ~100 1/s).

In reconstituted mucilage solutions (3-7%), G' and G" were between ~0.1 and 6 Pa (3-100 rad/s). However, in reconstituted spray-dried mucilage dissolutions (3 and 6%) isolated from the cladodes of pitahaya (*Hylocereus undatus*), the moduli G' and G" were lower (0.002-1.05 Pa; 1-60 rad/s) [30].

Pectin Rheology

A lower value of [η] than was obtained for mucilage was reported in low methoxyl pectin extracted from fresh nopal by the alkaline process (234 mL/g in 0.1-M NaCl at 20°C) [18].

The rheology of low-methoxy pectin gels with added $CaCl_2$, obtained in dynamic oscillatory tests, showed characteristic gel-like mechanical spectra. A slight frequency (ω) dependence of the mechanical moduli and a predominance of the storage modulus (G') over the loss modulus (G") without crossover (G'=G") were observed [18, 19]. In gels (4 g/kg and 0.75-mM $CaCl_2$), G' varied from ~1000 to 2000 Pa and G" varied from ~100 to 300 Pa [19].

When the stoichiometric equivalence level R ($=2[Ca^{2+}]/[COO^-]$) was increased in low-methoxy gels (16 g/L), both moduli improved: the storage modulus changed from ~50-70 Pa to ~3000-4000 Pa, and the loss modulus changed from ~2-20 Pa to ~50-400 Pa. Addition of calcium at high temperatures followed by cooling led to the formation of elastic gels under a highly cooperative process, where the critical gelling temperature upon cooling increased directly with R [18].

The gelling process is partially thermoreversible because gels, which have high thermal hysteresis (~50°C) with R>0.35, melt when heated. The gelling behavior of this system is interpreted in terms of the formation of two distinct junctions mediated by R. Specifically, short 'egg-box' type junctions form directly at high temperatures upon the addition of calcium (limited zones are related to high mobility of the chains), and highly cooperative 2_1 helix junctions are followed by aggregation at lower temperatures under a thermal conformational transition driven by charge neutralization and lower chain mobility [18].

Few studies have explored mucilage gum mixtures of *Opuntia ficus-indica* and other polysaccharides. Mixed gels of *ι*-carrageenan and mucilage gum (2% polymer concentration of a binary system) showed no enhancement in mechanical properties compared to pure *ι*-carrageenan, while a k-carrageenan/mucilage gum system in a 4:1 ratio with 12 mM KCl showed higher G' and gel rigidity (resistance to compression test, 15% deformation) than pure k-carrageenan gels. These results suggest that a mucilage/k-carrageenan mixture exhibits a synergistic effect [31].

Uses

The physicochemical and functional properties previously described indicate potential applications of *Opuntia* polysaccharides in many fields. As a rheological modifier, mucilage and pectins can be used as thickening or gelling agents (food additives) in the food, pharmaceutical, and cosmetic industries. The simplest application of pectin extracted from *Opuntia* is in fruit preserves (jams, jellies, and marmalades), where the concentration used depends on the degree of methylation, fruit type, and total solids content [17].

As also discussed, the market for dietary supplements (*e.g.*, nopal powder, dietary fiber, and as antioxidants) based on dried nopal in tablets or soft drinks is expanding [9, 10, 17]. In traditional medicine, mucilage was used to treat diseases. It is considered a source of dietary fiber and has been associated with physiological effects such as decreased blood cholesterol levels (hypocholesterolemic effects) and control of glucose levels (hypoglycemic effect). Mucilage may also reduce the risk of colon cancer, protect gastric mucosa and reduce the symptoms of chronic constipation, diverticular disease and hemorrhoids. Anti-inflammatory activity for the treatment of enteritis, antiviral properties, water-holding capacity, and relief of symptoms caused by benign prostatic hypertrophy have also been reported [9, 17].

In foods, mucilage can be included as a fat replacer, flavor binder, surfactant (emulsifying or foaming agent), suspension stabilizer, syneresis inhibitor, crystallization controller, or edible film. In addition, mucilage has been studied as a water purifier in combination with lime (organic adhesive) and as a method of improving the infiltration of water into soil [9].

Agave Fructans

Source

Fructans are fructose-based oligosaccharides and polysaccharides representing the major reserve carbohydrate in flowering plant species. They are present in vegetables, fruits, grains (leek, onion, garlic, asparagus, oat, wheat, barley) and tubers (chicory, Jerusalem artichoke, dahlia, yacon, jicama) [32 - 35]. Agave plants are a more recently discovered source of vegetal fructans [36 - 39] that can also be produced by certain species of bacteria [2].

Inulin is a linear fructan that is obtained mainly from chicory roots [32, 33]. Levans are the most commonly ramified fructans and are produced in grasses and grains as well as by fermentation (bacteria and yeast) [2, 40]. A complex mixture of fructans with branched and linear linkages has been reported in agave plants [36 - 39].

Fructans can be extracted from agave plants, which are cultivated in several regions of Mexico mainly for tequila production. In addition, some species of cacti such as the Agavaceae family are distributed and well adapted to arid and semiarid regions of Mexico, which is considered the origin and site of biodiversification of the *Agave* genus [32, 36 - 39].

Production

Ninety percent of the world's fructan (inulin-type) industrial production is processed in Belgium, the Netherlands, and France [32]. In Mexico, several industries produce agave fructans and organic agave syrups (mainly fructose syrups). Most commercial powders from agave are referred to as inulin, but not all fructans extracted from agave have linear linkages similar to chicory inulin [36], and thus, the term should be avoided; the more specific nomenclature "agavins" has been proposed [37].

The process for obtaining fructans from agave is simple. In general, the leaves of the agave plant are removed and the resulting hearts are fragmented into small pieces and pulverized to produce a pulp. The pulp is extracted with hot water and filtered. The fructan solution can be clarified, refiltrated, and demineralized, and various concentrations and drying steps can be applied to obtain a white powder [32].

The solubility of fructans in aqueous media is governed by their molecular structure and variations in solvent temperature, where higher temperature solvents produce fructans with better solubility. Symmetrical fructans are less soluble than branched fructans, and highly symmetrical fructans such as inulin tend to form crystalline structures such that elevated temperatures between 50 and 80°C are needed to obtain a solution. However, once dissolved, such solutions are stable, even at room temperature [32].

A summary of selected laboratory extractions and purifications of agave fructans follows. In the first, 30 g of extracted agave pine is combined with 100 mL of 80% w/v ethanol and shaken for 1 h at 75 °C. The sample is filtered and re-extracted twice with 50 mL of water at 70 °C for 30 min. The supernatants are combined, reduced to 10 mL, desalted by passage through a 200-400 mesh ion exchange column, and the pH is adjusted to neutral. The eluent is passed through an activated charcoal column and concentrated by rotary evaporation under reduced pressure. A high degree of polymerization (DP) of precipitated fructans was achieved by adding 30 mL of absolute ethanol. Samples were freeze-dried and stored [36].

In another method, the leaves were removed from the agave plant and the resulting hearts were fragmented into small pieces of 5 to 10 cm and pulverized. The pulp was extracted with water (90°C, 40 min) and filtered. The fructan solution was then clarified with diatomaceous earth, passed through a bed of activated charcoal (120-cm thick), demineralized by passage through an ion exchange column, and spray-dried (140 ºC) [41].

The agave fructans from 5 kg of shredded pine were extracted with hot water for 2 h at 70°C with eventual agitation using an agave:water ratio of 1:1 (w/v). The aqueous extract was centrifuged to eliminate fibers and spray-dried [42].

Alternatively the fructo-oligosaccharide (FOS) extraction method has been reported using different plant parts [39]. FOSs are important because they appear to be biologically active compounds that can be used as prebiotics [32 - 35]. First, 2.5 g of ground, freeze-dried, white inner-core leaf-material was dispersed in 30 mL of distilled water with NaN_3 (0.005%) at room temperature to avoid microbial hydrolysis of glycosidic linkages. After stirring for 5 h, the dispersion was centrifuged (3800 rpm, 30 min), and after washing the solid residue twice with 20 mL of water, the pooled supernatants containing extracted FOS were pooled and freeze-dried. Next, 10 g of ground white freeze-dried pieces of stem material was suspended in 80 mL of methanol at room temperature and stirred for 14 h. The FOS-containing methanol was aspirated, and the residue was washed twice with 20 mL of methanol. If the methanol is evaporated from stable FOSs/methanol-solutions at elevated temperatures of 50°C in a vacuum (150 mbar), the extracted FOSs that remain can then be dissolved in 20 mL water and freeze-dried.

Structure

Fructans are biopolymers consisting of two or more fructose (Fru) units. A “starting" or internal glucose (Glc) may or may not be present [32 - 35, 43]. Fructans can be linear or branched poly-fructose molecules (fructosyl-fructose linkages) that are either β-(2,1) linked, as in inulin, or β-(2,6) linked, as in levans.

Fructan DP ranges from two to a few hundred, depending on the species, developmental stage, and environmental conditions [43]. For example, inulin, the most common fructan, has a DP that varies from 2 to ~60 units [32 - 35]. Low-DP polymers (3-10) are FOSs and can be obtained by extraction from plants [39] by hydrolysis or enzymatic methods [42, 44]. DP distribution profiles vary with agave species [36 - 39].

In general, mixed-type fructans found in agave plants are similar to those found in Liliaceae (*e.g.*, fructans in garlic) [32]. Fructans present in *Agave tequilana* have a DP ranging from 3 to 29 [36 - 39]. Meanwhile, bacterial levans are much larger than those produced by plants, with multiple branches and high Mw ($\sim 2 \times 10^6$ Da) [40].

Fructans are classified according to differences in glycosidic linkages [β(2,1), β(2,6), or both]. The best studied fructans are linear inulin-type fructans consisting of β(2,1)-linked fructose units attached to a sucrose (GLc α1, 2β Fru) starter unit, where trisaccharide 1-kestotriose (Glc α1,2 βFru1,2 βFru) is the essential building block. The smallest representatives inulo-n-oses are inulobiose (Fru1,2 βFru) and inulotriose (Fru1,2 βFru1,2 βFru) [43].

The building block for linear levan-type fructans is 6-kestotriose (Glc α1,2 βFru6,2 βFru), which is further elongated to polymers with β(2,6)-linkages. Graminan-type fructans contain both β(2,1) and β(2,6) linkages. Even more complex fructans based on the 6G-kestotriose backbone (Fru2,6Glc α1,2 βFru), have additional Fru elongations on both sides (*e.g.*, oat and asparagus). These are termed neo-inulin-type [pre-dominantly β(2,1) linkages] and neo-levan-type [predominantly β(2,6) linkages] fructans, respectively [43].

Fructans from *Agave tequilana* were characterized as a complex mix of branched neo-levans with β(2,1) and β(2,6) linkages [36 - 39]. *Agave veracruz* and *Agave deserti* fructans were found to be similar to those from *Agave tequilana* because they have extra internal glucose units [36]. However, an extra internal glucose unit and only one $\beta(2\rightarrow1)$ linkage were reported in FOSs from leaves of *Agave tequilana* (six years old); a strong dependence of the structure on biosynthesis at different plant locations, transport activities, and general environmental impacts can explain this result [39].

Water-soluble fructans from *Agave americana* grown in South Africa were reported to contain a single glucose per fructan chain (average DP of 14) that is predominantly internal rather than the terminal, as observed in inulin [45].

The content of total sugars and fructans in *Agave tequilana* depends on the age of the plant (*e.g.*, in a 6.5-year-old plant, 97.3% of the sugars present in agave were found to be fructans) [32, 38]. Due to their content of monosaccharides and disaccharides, low Mw fructans FOSs are typically sweet in taste [32].

Functional Properties

Inulin-type fructans have been applied widely in the food and pharmaceutical industries (functional foods, nutritional composites, and drugs) [35]. They have been widely used as food ingredients because of their gelling and thickening capacity. Although the emulsifying properties of inulin have not been reported in the literature, there are some studies of hydrophobically modified inulin (INUTEC®SP1) and a modified inulin from alkenyl succinic anhydride (octenyl and dodecenyl succinic esters of inulin) [46, 47] reported as alternative biodegradable surfactants.

Inulin has been proposed as a substitute for sugar or fat because of its low caloric value and ability to retain product texture (*e.g.*, in ice creams, yogurts, fermented milks, and baked products) [32]. As a fat replacement, inulin has been reported to improve the taste and texture of dairy products [48]. In contrast, few reports have investigated the effect of agave fructans on the physicochemical and sensory characteristics of foods [49].

Other applications include use as a prebiotic and a supplement to increase dietary fiber in commercial products, where it can contribute to improved gastrointestinal conditions [32 - 35]. The functional properties of agave fructans can be related to their structural conformation and high solubility [32, 41] because they allow the formation of a very low-viscosity solution instead of a gel. This property could be useful for new technological applications that are different from those reported for commercial inulin.

Fructans from agave can easily be degraded by thermal or enzymatic treatments to produce agave syrup, functional prebiotic oligosaccharides and sweetening powder, leading to free sugars (mainly fructose) [32 - 34]. Patents for raw materials have been approved for many purposes [50]. Moreover, high fructose syrup (70°Brix, ~88% fructose), obtained from agave fructans has lower viscosity (~ 212 mPa.s) and surface tension ~ (45 dyn/cm) than corn syrup because of its highest fructose:glucose ratio [51].

Agave plants are a valuable source of raw material due to the fibrous and complex sugar content of their leaves and core. It can function as an immunomodulator and regulator of intestinal microbiota, promote the development of satietogenic/incretin peptides, reduce cholesterol and glucose levels, and promote lipid homoeostasis [34, 52]. In addition, their chaff waste can be used for several biotechnological purposes [32, 35].

Rheological Behavior

Shear viscosity and [η] using rotational rheometers and capillary viscometers, respectively, have been used to characterize the flow properties of some fructan solutions. Few studies related to viscosity of agave fructans are available in the literature.

A low [η] of linear fructans (*i.e.*, inulin) in aqueous solvent was reported (~7.1 mL/g, ~7000 Da) [53]. Similar fructans in dimethyl sulfoxide solution showed a higher [η] of ~9.1 mL/g (samples extracted from Jerusalem artichoke, ~3400 Da) and ~10.7 in mL/g (samples extracted from chicory root, ~6200 Da). Variations were attributed to differences in Mw. An extended spheroidal conformation was suggested, assuming that the exponent of the Mark-Houwink-Kuhn-Sakurada (MHKS) power law relationship ([η] = f(Mw)) was close to 0.3 [54]. These [η] values were small when compared to the high-Mw polysaccharides, even for *ofi* mucilage.

In aqueous solutions of 2,1-*β*-D-fructan treated with ultrasound, a MHKS exponent of 0.71 for Mw < 4 x 10^4 was reported, indicating that the fructan chain behaves as a random coil expanded by an excluded volume. For those with a Mw > 10^5, [η] exhibits a weak dependence on Mw and finally becomes almost independent of Mw. This behavior was interpreted in terms of a globular conformation for the high-Mw fructan molecules [55].

Higher [η] were reported for ramified fructans (levan, Mw ~2 × 10^6 Da) produced by microorganisms (*Bacillus* sp.), ranging from 14 to 17 mL/g [40]. The concentration (c) effect was studied as a function of volume occupancy ([η]c), a dimensionless parameter. Four distinct regions (ranges of concentrations) in the classical double logarithmic plot of specific viscosity *versus* [η]c were reported, and the slopes of each region were 1.2, 2.0, 3.6 and 9.7 [40]. Two distinct regions were found in the most disordered polysaccharide solutions [4], where the slopes before and after a volume occupancy of 4 were 1.4 and 3.3, respectively. Although higher [η] was established in levan polysaccharides from *Microbacterium laevaniformans* (38 mL/g) [56], all were lower than that of high-Mw polysaccharides (*i.e.*, xanthan gum, ~10300 mL/g [25]).

There are few reports concerning the shear viscosity of aqueous solutions of agave fructans. No gel formation was reported in high mechanical-thermal-treated samples (11000 rpm; 25 °C, 5 min; 80 °C, 2 min), but Newtonian behavior was observed at 30 °C for almost all concentrations studied (5-80%) [41].

The low-Mw agave fructans analyzed (Mw ~3.2 kDa, DP ~18) develop low viscosity at 30 °C: below 5 mPa.s for concentrations of 30% or less, 10 mPa.s for concentrations of 40%, and below 100 mPa.s for concentrations of 40-60% [41]. Similarly, levan fructans from *Bacillus* sp. showed Newtonian behavior at 20 °C of 4 and 800 mPa.s for concentrations of 9.2 to 46%, respectively. These differences can be attributed to the Mw of bacterial levans, which are much larger than those produced by plants [40].

A power dependence of viscosity on concentration was calculated for agave fructans. For concentrations lower than 35%, the 0.4 exponent obtained was attributed to isolated fructan chains present as ideally isolated coils, characteristic of a diluted regime. For more concentrated systems (35-60%), molecular interactions increased and the power exponent was 6.2, suggesting entangled dissolution (overlapped coils) [41].

Fig. (**1**) shows an analysis of different viscosities reported for fructans. The viscosity of agave fructans was compared to the respective levan fructans from *Bacillus* sp.; data were interpolated from original figures [40, 41].

The viscosities of D-glucose ($C_6H_{12}O_6$), D-fructose ($C_6H_{12}O_6$), and sucrose ($C_{12}H_{22}O_{11}$) solutions at different concentrations were included for comparison [57].

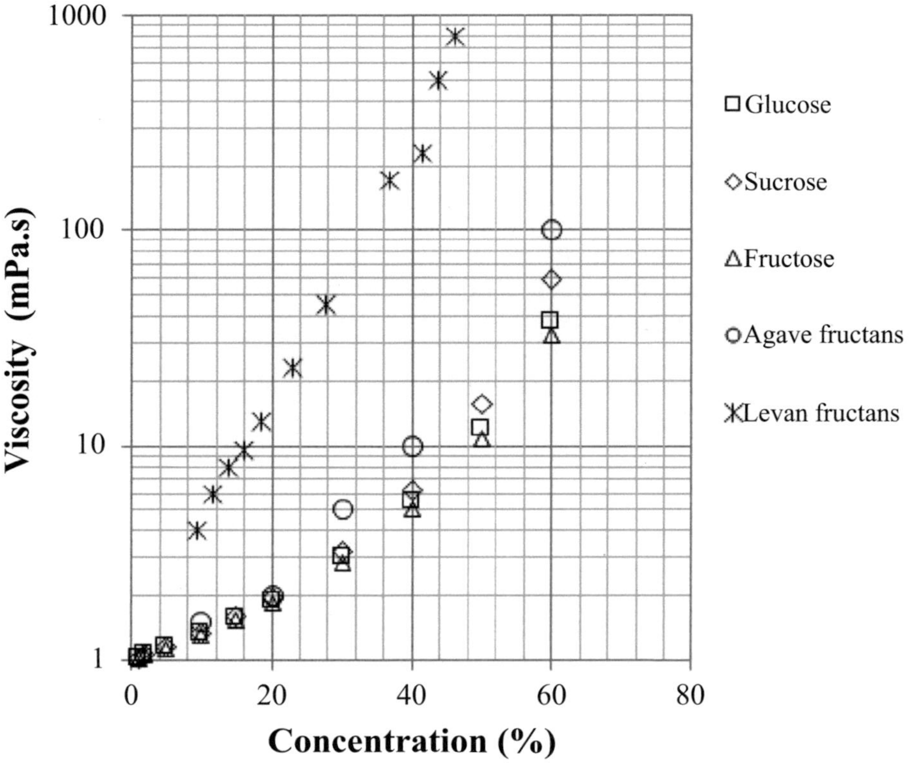

Fig. (1). Viscosity as a function of concentration of agave and levan fructans (*Bacillus* sp.) and selected sugar solutions.

Agave fructan viscosity was more similar to that of fructose or sucrose solutions than to that of levan fructans, implying that the rheological behavior of agave fructans is closer to that of low-Mw (monosaccharides or disaccharides) than high-Mw (polysaccharides) carbohydrates. In addition, the agave fructan conformation can be considered more similar to an extended spheroidal conformation than an extended random coil conformation.

Shear-thinning behavior and highly viscous samples were reported in concentrated samples of agave fructans (90%) at 30°C [41] and levan fructans from *Microbacterium laevaniformans* [56]. However, there is uncertainty whether agave fructans (90%) show true liquid-like behavior without elastic properties at these very high concentrations.

Moreover, levan fructans with high Mw showed a flow behavior index with near-Newtonian behavior (0.88-0.94) for concentrations between 0.1 and 10% [56].

Shear viscosity of samples of agave fructans decreases with temperature according to the Arrhenius model. The activation energy reported for agave fructans ranges from 15 to 18.5 kJ/mol (5-30%; 30-60°C) [41] and for levan fructans from 24 to 13.5 kJ/mol (0.1-10%).

Uses

Agave has a negligible content of monosaccharides (glucose and fructose less than 0.2%) and sucrose (0.5 to 1.5%), making it an ideal raw material for obtaining highly pure fructan [32].

As mentioned earlier, there is an expanding global market for dietary supplements such as *ofi* biopolymers (*e.g.*, in prebiotics and as antioxidants). Fructans are also being considered as a source of dietary fiber because their linkages are associated with physiological effects including stimulation of beneficial colon flora, improved mineral absorption, rebalanced metabolic activities (lipid homeostasis), and protection against colon cancer [33, 34, 52].

Hydrolyzed agave fructan is a main component of agave syrup, contributing up to 70% of its fructose, minerals (Ca and Mg), and vitamins. It also makes a large contribution to sweetening power (1.8-fold sweeter than sucrose), which can be useful in formulated functional foods (*e.g.*, healthy, diabetic, or low-calorie) or for use directly as a final product [32].

CONCLUDING REMARKS

Knowledge of the rheology of biopolymers extracted from cactus or agave can be useful in the development of new functional products. Mucilage-type biopolymers may act as thickening or gelling agents, beneficially modifying the rheology of the final products. It is well known that rheology greatly impacts the texture of products, has implications for processing operations, and contributes to quality control.

When the viscosity of a product is slightly modified by a biopolymer (*e.g.*, agave fructans), larger amounts can be added to a new product, such as those required to improve health (*e.g.*, dietary fiber or prebiotics). This strategy should be approached with caution by professionals, because as biopolymers interact with other food constituents, mainly macromolecules, they can modify or destabilize the final products.

Modern extraction methods must be developed for more recently discovered vegetal biopolymers to motivate their industrial production. Furthermore, more investigation is necessary to reveal other functional properties of these biopolymers, such as surface properties useful to emulsifying, or foaming properties. Their rheological behavior, alone and in combination with other biopolymers, and interactions with various internal or external conditions must be studied to expand their potential for exploitation in the food industry (*e.g.*, fructan agave biopolymers).

CONFLICT OF INTEREST

L.P. Martinez-Padilla certifies that she has no conflict of interest (financial or otherwise) in the subject matter or materials discussed in this manuscript.

ACKNOWLEDGEMENTS

This research was financed in part by a grant obtained from DGAPA-UNAM (project IN118214). The author is grateful to Carolyn Unck for her careful correction of English language and style.

REFERENCES

[1] P.A. Williams, "Natural polymers: Introduction and overview", In: *Renewable Resources for Functional Polymers and Biomaterials. Polysaccharides, Proteins and Polyesters,* P.A. Williams, Ed., RSC Publishing: London, 2011, pp. 1-14.

[2] J. Milani, and G. Maleki, "Hydrocolloids in food industry", In: *Food Industrial Processes-Methods and Equipment,* B. Valdez, Ed., InTech: Croatia, 2012. [E book] [http://dx.doi.org/10.5772/32358]

[3] M. Niaounakis, *Biopolymers: Reuse, Recycling, and Disposal.* Elsevier Inc.: Oxford, 2013.

[4] A.N. Morris, S.B. Cutler, D.A. Ross-Murphy, Rees, and J. Price, "Concentration and shear rate dependence of viscosity in random coil polysaccharide solutions", *Carbohydr. Polym,* vol. 1, no. 1, pp. 5-21, 1981. [http://dx.doi.org/10.1016/0144-8617(81)90011-4]

[5] J.A. da Silva, and M.A. Rao, "Viscoelastic properties of food hydrocolloid dispersions", In: *Viscoelastic Properties of Foods,* M.A. Rao, J.F. Steffe, Eds., Elsevier Science Publisher LTD: London, 1992, pp. 285-315.

[6] A.H. Clark, and S.B. Ross-Murphy, "Structural and mechanical properties of biopolymer gels", In: *Biopolymers (Advances in Polymer Science, vol. 83),* A.H. Clark, K. Kamide, S.B. Ross-Murphy, M. Sato, Eds., Heidelberg: Springer-Verlag, 1987, pp. 57-192. [http://dx.doi.org/10.1007/BFb0023332]

[7] R. Lapasin, S. Pricl, Ed., *Rheology of Industrial Polysaccharides: Theory and Applications.* Chapman & Hall: London, 1999.

[8] G. Singh, "General Review of Opuntias in India", *J. Prof. Assoc. Cactus,* vol. 5, pp. 30-46, 2003.

[9] C. Sáenz, E. Sepúlveda, and B. Matsuhiro, "Opuntia spp mucilage's: A functional component with industrial perspectives", *J. Arid Environ,* vol. 57, no. 3, pp. 275-290, 2004. [http://dx.doi.org/10.1016/S0140-1963(03)00106-X]

[10] T. Guevara-Figueroa, H. Jiménez-Islas, M.L. Reyes-Escogido, A.G. Mortensen, B.B. Laursen, L-W. Lin, A. de León-Rodríguez, I.S. Fomsgaard, and A.P. Barba de la Rosa, "Proximate composition, phenolic acids, and flavonoids characterization of commercial and wild nopal (Opuntia spp.)", *J. Food Compos. Anal,* vol. 23, no. 6, pp. 525-532, 2010. [http://dx.doi.org/10.1016/j.jfca.2009.12.003]

[11] J.O. Kuti, "Antioxidant compounds from four *Opuntia* cactus pear fruit varieties", *Food Chem,* vol. 85, no. 4, pp. 527-533, 2004. [http://dx.doi.org/10.1016/S0308-8146(03)00184-5]

[12] B. Matsuhiro, L.E. Lillo, C. Sáenz, C.C. Urzúa, and O. Zárate, "Chemical characterization of the mucilage from fruits of *Opuntia ficus indica*", *Carbohydr. Polym,* vol. 63, no. 2, pp. 263-267, 2006. [http://dx.doi.org/10.1016/j.carbpol.2005.08.062]

[13] E. Sepúlveda, C. Sáenz, E. Aliaga, and C. Aceituno, "Extraction and characterization of mucilage in Opuntia spp", *J. Arid Environ,* vol. 68, no. 4, pp. 534-545, 2007. [http://dx.doi.org/10.1016/j.jaridenv.2006.08.001]

[14] A. Cárdenas, I. Higuera-Ciapara, and F.M. Goycoolea, "Rheology and aggregation of cactus (*Opuntia ficus-indica*) mucilage in solution", *J. Prof. Assoc. Cactus,* vol. 2, pp. 152-157, 1997.

[15] L. Medina-Torres, E. Brito-De la Fuente, B. Torrestiana-Sanchez, and R. Katthain, "Rheological properties of the mucilage gum (*Opuntia ficus indica*)", *Food Hydrocoll,* vol. 14, pp. 417-424, 2000. [http://dx.doi.org/10.1016/S0268-005X(00)00015-1]

[16] H. Majdoub, S. Roudesli, L. Picton, D. Le Cerf, G. Muller, and M. Grisel, "Prickly pear nopals pectin from *Opuntia ficus indica* . Physicochemical study in dilute and semidilute solutions", *Carbohydr. Polym,* vol. 46, no. 1, pp. 69-79, 2001. [http://dx.doi.org/10.1016/S0144-8617(00)00284-8]

[17] F.M. Goycoolea, and A. Cárdenas, "Pectins from Opuntia spp.: A Short Review", *J. Prof. Assoc. Cactus,* vol. 5, pp. 17-29, 2003.

[18] A. Cárdenas, F.M. Goycoolea, and M. Rinaudo, "On the gelling behaviour of 'nopal' (*Opuntia ficus indica*) low methoxyl pectin", *Carbohydr. Polym,* vol. 73, no. 2, pp. 212-222, 2008. [http://dx.doi.org/10.1016/j.carbpol.2007.11.017]

[19] A.L. Lira-Ortiz, F. Reséndiz-Vega, E. Ríos-Leal, J.C. Contreras-Esquivel, N. Chavarría-Hernández, A. Vargas-Torres, and A. Rodríguez-Hernández, "Pectins from waste of prickly pear fruits (*Opuntia albicarpa Scheinvar* 'Reyna'): Chemical and rheological properties", *Food Hydrocoll,* vol. 37, pp. 93-99, 2014. [http://dx.doi.org/10.1016/j.foodhyd.2013.10.018]

[20] Y. Habibi, M. Mahrouz, and M.R. Vignon, "Isolation and structure of D-xylans from pericarp seeds of *Opuntia ficus-indica* prickly pear fruits", *Carbohydr. Res,* vol. 337, no. 17, p. 15938, 2002. [http://dx.doi.org/10.1016/S0008-6215(02)00186-6] [PMID: 12350330]

[21] Y. Habibi, A. Heyraud, M. Mahrouz, and M.R. Vignon, "Structural features of pectic polysaccharides from the skin of *Opuntia ficus-indica* prickly pear fruits", *Carbohydr. Res,* vol. 339, no. 6, pp. 1119-1127, 2004. [http://dx.doi.org/10.1016/j.carres.2004.02.005] [PMID: 15063200]

[22] M.R. Vignon, L. Heux, M.-E. Malainine, and M. Mahrouz, "Arabinan-cellulose composite in *Opuntia ficus-indica* prickly pear spines", *Carboh. Res,* vol. 339, no. 1, 2, pp. 123-131, 2004. [http://dx.doi.org/10.1016/j.carres.2003.09.023]

[23] N. Garti, and M.E. Leser, "Emulsification properties of hydrocolloids", *Polym. Adv. Technol,* vol. 12, no. 1, pp. 123-135, 2001. [http://dx.doi.org/10.1002/1099-1581(200101/02)12:1/2<123::AID-PAT105>3.0.CO;2-0]

[24] D. Abdessemed, M. Nezari, A.R. Mohamed Hadj, Dj. Abdessemed, F. Bahadi, and A. Halitim, "Emulsifying effect of pectin from *Opuntia ficus-indica* cladode", *J. Chem. Pharmaceu.,* vol. 6, pp.

198-201, 2014.

[25] M. Milas, M. Rinaudo, M. Knipper, and J.L. Schuppiser, "Flow and viscoelastic properties of xanthan gum solutions", *Macromolecules,* vol. 23, no. 9, pp. 2506-2511, 1990. [http://dx.doi.org/10.1021/ma00211a018]

[26] F.M. León-Martínez, J. Rodríguez-Ramírez, L.L. Medina-Torres, L.L. Mendez-Lagunas, and M.J. Bernad-Bernad, "Effects of drying conditions on the rheological properties of reconstituted mucilage solutions (*Opuntia ficus-indica*)", *Carbohydr. Polym,* vol. 84, no. 1, pp. 439-445, 2011. [http://dx.doi.org/10.1016/j.carbpol.2010.12.004]

[27] L. Medina-Torres, E.E. García-Cruz, F. Calderas, R.F. González Laredo, G. Sánchez-Olivares, J.A. Gallegos-Infante, N.E. Rocha-Guzmán, and J. Rodríguez-Ramírez, "Microencapsulation by spray drying of gallic acid with nopal mucilage (*Opuntia ficus indica*)", *LWT Food Sci. Technol.,* vol. 50, no. 2, pp. 642-650, 2013. [http://dx.doi.org/10.1016/j.lwt.2012.07.038]

[28] L.P. Martínez-Padilla, F. López-Araiza, and A. Tecante, "Steady and oscillatory shear behavior of fluid gels formed by binary mixtures of xanthan and gellan", *Food Hydrocoll,* vol. 18, no. 3, pp. 471-481, 2004. [http://dx.doi.org/10.1016/j.foodhyd.2003.07.002]

[29] L.P. Martínez-Padilla, J.L. García-Rivera, V. Romero-Arreola, and N.B. Casas-Alencáster, "Effects of xanthan gum rheology on the foaming properties of whey protein concentrate", *J. Food Eng,* vol. 156, pp. 22-30, 2015. [http://dx.doi.org/10.1016/j.jfoodeng.2015.01.018]

[30] E.E. García-Cruz, J. Rodríguez-Ramírez, L.L. Méndez Lagunas, and L. Medina-Torres, "Rheological and physical properties of spray-dried mucilage obtained from *Hylocereus undatus* cladodes", *Carboh. Polym,* vol. 91, no. 1, 2, pp. 349-402, 2013. [http://dx.doi.org/10.1016/j.carbpol.2012.08.048]

[31] L. Medina-Torres, E. Brito-De La Fuente, B. Torrestiana-Sanchez, and S. Alonso, "Mechanical properties of gels formed by mixtures of mucilage gum (*Opuntia ficus indica*) and carrageenans", *Carbohydr. Polym,* vol. 52, no. 1, pp. 143-150, 2003. [http://dx.doi.org/10.1016/S0144-8617(02)00269-2]

[32] W. Praznik, E. Cieslik, and A. Huber, "Fructans: occurrence and application in food", In: *Chemical and Functional Properties of Food Saccharides,* P. Tomasik, Ed., CRC Press: Boca Raton, 2004, pp. 201-219.

[33] M.B. Roberfroid, "Introducing inulin-type fructans", *Br. J. Nutr,* vol. 93, no. S1, suppl. 1, pp. S13-S25, 2005. [http://dx.doi.org/10.1079/BJN20041350] [PMID: 15877886]

[34] M.B. Roberfroid, "Inulin-type fructans: functional food ingredients", *J. Nutr,* vol. 137, no. 11, suppl. , pp. 2493S-2502S, 2007. [PMID: 17951492]

[35] A.C. Apolinário, B.P. de Lima Damasceno, N.E. de Macêdo Beltrão, A. Pessoa, A. Converti, and J.A. da Silva, "Inulin-type fructans: a review on different aspects of biochemical and pharmaceutical technology", *Carbohydr. Polym,* vol. 101, pp. 368-378, 2014. [http://dx.doi.org/10.1016/j.carbpol.2013.09.081] [PMID: 24299785]

[36] M.G. Lopez, N.A. Mancilla-Margalli, and G. Mendoza-Diaz, "Molecular structures of fructans from *Agave tequilana* Weber var. azul", *J. Agric. Food Chem,* vol. 51, no. 27, pp. 7835-7840, 2003. [http://dx.doi.org/10.1021/jf030383v] [PMID: 14690361]

[37] N.A. Mancilla-Margalli, and M.G. López, "Water-soluble carbohydrates and fructan structure patterns from Agave and Dasylirion species", *J. Agric. Food Chem,* vol. 54, no. 20, pp. 7832-7839, 2006. [http://dx.doi.org/10.1021/jf060354v] [PMID: 17002459]

[38] J. Arrizon, S. Morel, A. Gschaedler, and P. Monsan, "Comparison of the water-soluble carbohydrate composition and fructan structures of *Agave tequilana* plants of different ages", *Food Chem,* vol. 122, no. 1, pp. 123-130, 2010. [http://dx.doi.org/10.1016/j.foodchem.2010.02.028]

[39] W. Praznik, R. Löppert, J.M. Cruz Rubio, K. Zangger, and A. Huber, "Structure of fructo-oligosaccharides from leaves and stem of *Agave tequilana* Weber, var. azul", *Carbohydr. Res,* vol. 381, pp. 64-73, 2013. [http://dx.doi.org/10.1016/j.carres.2013.08.025] [PMID: 24071527]

[40] S. Arvidson, T.B. Rinehart, and F. Gadala-María, "Concentration regimes of solutions of levan polysaccharide from Bacillus sp", *Carbohydr. Polym,* vol. 65, no. 2, pp. 144-149, 2006. [http://dx.doi.org/10.1016/j.carbpol.2005.12.039]

[41] J.A. Ponce-S., E.R. Macías-B., J.F. Soltero-M., V.V. Fernández-E., V. Zúñiga-P., and H.B. Escalona-B., "Physical-Chemical and non-linear rheological properties of aqueous solutions of agave fructans", *e-Gnosis,* vol. 6, Nov. 2008pp. 1-23.Accessed: 7th June, 2015 Available at: http://www.redalyc.org [Accessed: 7th June, 2015]

[42] A. Ávila-Fernández, N. Galicia-Lagunas, M.E. Rodríguez-Alegría, C. Olvera, and A. López-Munguía, "Production of functional oligosaccharides through limited acid hydrolysis of agave fructans", *Food Chem,* vol. 129, no. 2, pp. 380-386, 2011. [http://dx.doi.org/10.1016/j.foodchem.2011.04.088] [PMID: 25212145]

[43] W.V. Den Ende, "Multifunctional fructans and raffinose family oligosaccharides", *Front. Plant Sci,* vol. 4, no. 247, pp. 1-11, 2013. [Online]. [Available from: www.frontiersin.org]. [PMID: 23346092]

[44] A. Avila-Fernández, X. Rendón-Poujol, C. Olvera, F. González, S. Capella, A. Peña-Alvarez, and A. López-Munguía, "Enzymatic hydrolysis of fructans in the tequila production process", *J. Agric. Food Chem,* vol. 57, no. 12, pp. 5578-5585, 2009. [http://dx.doi.org/10.1021/jf900691r] [PMID: 19473003]

[45] N. Ravenscroft, P. Cescutti, M.A. Hearshaw, R. Ramsout, R. Rizzo, and E.M. Timme, "Structural analysis of fructans from *Agave americana* grown in South Africa for spirit production", *J. Agric. Food Chem,* vol. 57, no. 10, pp. 3995-4003, 2009. [http://dx.doi.org/10.1021/jf8039389] [PMID: 19348427]

[46] M. Nedyalkov, L. Alexandrova, D. Platikanov, B. Levecke, and T.F. Tadros, "Wetting properties of aqueous solutions of hydrophobically modified inulin polymeric surfactant", *Colloid Polym. Sci,* vol. 286, no. 6, pp. 713-719, 2008. [http://dx.doi.org/10.1007/s00396-007-1823-5]

[47] J. Morros, B. Levecke, and M.R. Infante, "Hydrophobically modified inulin from alkenyl succinic anhydride in aqueous media", *Carbohydr. Polym,* vol. 84, no. 3, pp. 1110-1116, 2011. [http://dx.doi.org/10.1016/j.carbpol.2010.12.077]

[48] K.J. Aryana, S. Plauche, R.M. Rao, P. McGrew, and N.P. Shah, "Fat-free plain yogurt manufactured with inulins of various chain lengths and *Lactobacillus acidophilus* ", *J. Food Sci,* vol. 72, no. 3, pp. M79-M84, 2007. [http://dx.doi.org/10.1111/j.1750-3841.2007.00302.x] [PMID: 17995804]

[49] G. Crispín-Isidro, C. Lobato-Calleros, H. Espinosa-Andrews, J. Alvarez-Ramirez, and E.J. Vernon-Carter, "Effect of inulin and agave fructans addition on the rheological, microstructural and sensory properties of reduced-fat stirred yogurt.", *LWT Food Sci. Technol.,* vol. 62, no. 1, pp. 438-444, 2015. [http://dx.doi.org/10.1016/j.lwt.2012.07.038]

[50] J.A. Narváez-Zapata, and L.F. Sánchez-Teyer, "Agaves as a raw material: recent technologies and applications", *Recent Pat. Biotechnol,* vol. 3, no. 3, pp. 185-191, 2009. [http://dx.doi.org/10.2174/187220809789389144] [PMID: 19747148]

[51] J.L. Montañez-Soto, J. Venegas-González, A.B. Nicanor, and E.G. Ramos-Ramírez, "Enzymatic production of high fructose syrup from *Agave tequilana* fructans and its physicochemical characterization", *Afr. J. Biotechnol,* vol. 10, no. 82, pp. 19137-19143, 2011. Available online at: http://www.academicjournals.org/AJB
[http://dx.doi.org/10.5897/AJB11.2704]

[52] I.G. Carabin, and W.G. Flamm, "Evaluation of safety of inulin and oligofructose as dietary fiber", *Regul. Toxicol. Pharmacol,* vol. 30, no. 3, pp. 268-282, 1999.
[http://dx.doi.org/10.1006/rtph.1999.1349] [PMID: 10620476]

[53] C.F. Phelps, "The physical properties of inulin solutions", *Biochem. J,* vol. 95, no. 1, pp. 41-47, 1965.
[http://dx.doi.org/10.1042/bj0950041] [PMID: 14333566]

[54] B.H. Azis, B. Chin, M.P. Deacon, S.E. Harding, and G.M. Pavlov, "Size and shape of inulin in dimethyl sulphoxide solution", *Carbohydr. Polym,* vol. 38, no. 3, pp. 231-234, 1999.
[http://dx.doi.org/10.1016/S0144-8617(98)00096-4]

[55] S. Kitamura, T. Hirano, K. Takeo, M. Mimura, K. Kajiwara, B.T. Stokke, and T. Harada, "Conformation of (2→1)-β-D-fructan in aqueous solution", *Int. J. Biol. Macromol,* vol. 16, no. 6, pp. 313-317, 1994.
[http://dx.doi.org/10.1016/0141-8130(94)90062-0] [PMID: 7727345]

[56] I.Y. Bae, I-K. Oh, S. Lee, S-H. Yoo, and H.G. Lee, "Rheological characterization of levan polysaccharides from *Microbacterium laevaniformans* ", *Int. J. Biol. Macromol,* vol. 42, no. 1, pp. 10-13, 2008.
[http://dx.doi.org/10.1016/j.ijbiomac.2007.08.006] [PMID: 17931701]

[57] W.F. Haynes, Ed., *CRC Handbook of Chemistry and Physics*94th ed CRC Press: Boca Raton, 2013.

CHAPTER 7

Gels of Ferulated Arabinoxylans: Rheology, Structural Parameters and Microstructure

Mayra A. Méndez-Encinas[a], **Elizabeth Carvajal-Millan**[a,*], **Madhav Yadav**[b], **Elisa M. Valenzuela-Soto**[c], **Ciria G. Figueroa-Soto**[c], **Orlando Tortoledo-Ortiz**[d] and **Guillermina García-Sánchez**[e]

[a] *Biopolymers, CTAOA, Research Center for Food and Development, Hermosillo, Sonora, México*

[b] *Sustainable Biofuels and Coproducts Research Unit. Eastern Regional Research Center, Agricultural Research Service, US Department of Agriculture, Wyndmoor, PA, USA*

[c] *Biochemistry of proteins, CCA, Research Center for Food and Development, Hermosillo, Sonora, México*

[d] *Instrumental Analysis, CN, Research Center for Food and Development, Hermosillo, Sonora, México*

[e] *Quality of marine products, CTAOA, Research Center for Food and Development, Hermosillo, Sonora, México*

Abstract: One of the major by-products of bioethanol production is distillers dried grains with solubles (DDGS). Maize is one of the main sources for the production of this biofuel. In this way, dietary fiber represents the principal fraction of DDGS, which could be a potential source of added-value biomolecules such as ferulated arabinoxylans (AX). In this chapter, ferulated arabinoxylans extracted from DDGS (DDGAX) were gelled and the gels were studied in terms of rheology, structural parameters and microstructural characteristics. The DDGAX formed gels at 2% (w/v) induced by laccase. The mechanical spectrum and strain sweep of DDGAX gel presented the typical behavior of a solid-like material. The gels swelling ratio (43 g water/g DDGAX) suggested the formation of a compact polymeric network which decreased the water uptake of the gel. DDGAX gels presented an average mesh size value of 96 nm. The surface of the gels was analyzed by scanning electron microscopy revealing a heterogeneous microstructure resembling an imperfect honeycomb. These results indicate that ferulated arabinoxylans from DDGS form elastic and macroporous gels, presenting a microstructure with irregular pore sizes.

Keywords: Arabinoxylan gels, Ferulic acid, Gelling capability, Microstructure, Structural properties, Swelling.

* **Corresponding author Elizabeth Carvajal-Millan:** Biopolymers, CTAOA, Research Center for Food and Development, Hermosillo, Sonora, México; Tel/Fax: +52 662 2892400; E-mail: ecarvajal@ciad.mx

Martin Masuelli & Denis Renard (Eds.)

INTRODUCTION

One of the major by-products of the bioethanol industry is the distillers dried grains with soluble (DDGS). Because of the high amounts of DDGS generated during the process, new alternatives for the use of this by-product are needed.

Since dietary fiber represents a large fraction of the DDGS composition, DDGS is becoming a potential source of added-value biomolecules as arabinoxylans (AX). AX, also called pentosans, are non-starch polysaccharides present in various tissues of cereal grains. The structure of AX consists of linear (1-4)-β -D - xylopyranosyl chains to which α-L-arabinofuranosyl units are attached at the O-2 and/or O-3 positions [1]. Some arabinose residues are ester-linked on O-5 position to ferulic acid (3-methoxy-4-hydroxycinnamic acid) [2].

Hydrogels are hydrophilic polymer networks that are capable of imbibing large amounts of water without dissolution [3, 4]. In this way, ferulated AX can form gels by covalent cross-linking through dimerization of ferulic acid (FA) under oxidative conditions (*e.g.* use of free radical generating agents as laccase and peroxidase/H_2O_2) [5, 6]. Covalently cross-linked AX gels present important hydration properties, have absence of pH or electrolyte susceptibility and exhibit no syneresis after long periods of storage [7]. These gels have several potential applications, ranging from the food to pharmaceutical industries. Potential applications of AX gels as matrices for controlled release of biomolecules [8 - 10, 16] and cells [11] have been previously reported.

The gelling capability and gel properties of AX depend on their structural characteristics (FA content, A/X ratio, molecular weight) [12]. Gels with different rheological and structural characteristics can be obtained by modifying the source of the polysaccharide [13]. The structural parameters of nejayote, wheat and maize bran AX gels have been recently reported [5, 6, 13, 14]. In this chapter, DDGAX gels were characterized in terms of rheological characteristics, equilibrium swelling degree and microstructure. Knowledge of the equilibrium swelling degree allowed the calculation of gel structural parameters like molecular weight between cross-links (M_c), mesh size (ξ), and cross-linking density (ρ_c).

EXPERIMENTAL

Materials

AX from distillers dried grains with solubles (DDGAX) were obtained as previously reported and presented 69% dry basis (d.b.) of pure AX, a ferulic acid content of 6.05 μg/mg polysaccharide, a protein content of 8.2%, an A/X ratio of

1.08 and a molecular weight of 209 kDa [15]. Laccase (benzenediol:oxygen oxidoreductase, E.C.1.10.3.2) from *Trametes versicolor* and other chemical products were purchased from Sigma Co. (St. Louis, MO, US).

METHODS

Amino Acids Profile

The amino acid profile was obtained by HPLC according to the method reported by Vásquez-Ortiz *et al.*, 1995 [24]. The samples were hydrolyzed using 6 M HCl at 150 °C for 6 h. After cooling, the hydrolysates were buffered to pH 2.2 prior to analysis by HPLC. The instrument used was an HPLC (Hewlett Packard Model 1100) equipped with Microsorb RP-C18 (10 cm x 4.6 mm ID) column and fluorescent detector.

Hydroxyproline and proline were determined by analyzing the mixture of 125 μL of sample with 500 μL of borate buffer and 250 μL of NBD-Cl. The mixture was heated at 60 °C for 5 min and stopped with 50 μL of 1 M HCl. The amino acid standard mixture used was L-Amino Acids Standard (Sigma- Aldrich).

Gel Preparation

DDGAX solutions (2% w/v) were prepared in 0.05 M citrate-phosphate buffer, pH 5.5. Laccase (1.675 nkat per mg polysaccharide) was added to solutions as cross-linking agent for the formation of gels. Gels were allowed to develop for 1 h at 25 °C.

Rheology

Small amplitude oscillatory shear was used to follow the gelation process of DDGAX solutions. DDGAX solutions were mixed with laccase and immediately placed on the parallel-plate geometry (40 mm in diameter) of a strain controlled rheometer (Discovery HR-2 rheometer; TA Instruments, New Castle, DE, US). Exposed edges were recovered with silicone to prevent evaporation. The dynamic rheological parameters used to evaluate the gel network were the storage modulus (G'), loss modulus (G") and crossover point (G'>G"). DDGAX gelation was followed at a frequency of 0.25 Hz and 5% strain. At the end of the network formation, a frequency sweep (0.01-10 Hz) and a strain sweep (0.02-20%) were carried out. The rheological measurements were performed in duplicate.

Swelling Experiment

A total of 2 mL of DDGAX solution at 2% (w/v) containing laccase was quickly transferred to a 5 mL tip-cutoff syringe (diameter 1.5 cm) and allowed to gel overnight at 25 °C. After gelation, the gels were removed from the syringes, placed in glass vials, and weighed. The gels were allowed to swell in 20 mL of 0.02% (w/v) sodium azide solution to prevent microbial contamination. During 36 h, the samples were taken out, blotted, and weighed. The equilibrium swelling was reached when the weight of the samples changed by no more than 3% (0.03 g). After weighing, a new aliquot of sodium azide solution was added to the gels. Gels were maintained at 25 °C during the test. The swelling ratio (q) was calculated as [eq. (1)]:

$$q = (W_s - W_{DDGAX})/W_{DDGAX} \tag{1}$$

where W_s is the weight of swollen gels and W_{DDGAX} is the weight of AX in the gel. The AX weight in the gel was calculated by taking into account the AX concentration in the DDGAX solution (2% w/v) and the fresh weight of the gel after removing from the syringe [17].

Structural Parameters of the Gel

From swelling measurements, the molecular weight between two cross-links (M_c), the cross-linking density (ρ_c) and the mesh size (ξ) values of DDGAX gels were obtained as reported elsewhere [17]. M_c was calculated using the model of Flory & Rehner (1943) [18] modified by Peppas & Merrill (1976) [19] for gels where the cross-links are introduced in solution [eq. (2)]:

$$\frac{1}{M_c} - \frac{2}{M_n} = \frac{(\upsilon/V_1)\left[ln(1-\upsilon_{2,s})+\upsilon_{2,s}+X_1(\upsilon_{2,s})^2\right]}{\upsilon_{2,r}\left[(\upsilon_{2,s}/\upsilon_{2,r})^{1/3}-\frac{1}{2}(\upsilon_{2,s}/\upsilon_{2,r})\right]} \tag{2}$$

Where M_n is the average molecular weight of DDGAX (considering only the xylose backbone). The M_n value was calculated from the DDGAX M_w value knowing the A/X ratio. In eq. (2), V_1 is the molar volume of water (18 cm^3/g), $\upsilon_{2,r}$ and $\upsilon_{2,s}$ are the polymer volume fraction of the gel in a relaxed state (directly after gel formation) and at equilibrium swelling, respectively. $\upsilon_{2,r}$ and $\upsilon_{2,s}$ were calculated from the weight of the gels before exposure to the sodium azide solution and at equilibrium swelling, respectively, assuming volume additivity of water and DDGAX. χ_1 is the Flory polymer-solvent interaction parameter. The arabinoxylans-water system was calculated to correspond to theta conditions [20]. χ_1 was therefore taken as equal to 0.5 for the present system.

After M_c calculation by eq. (2), the average mesh size (ξ) of the DDGAX gels was obtained as reported elsewhere [21] [eq. (3)]:

$$\xi = \upsilon_{2,s}{}^{-1/3} \left(\frac{2C_n M_c}{M_r}\right)^{1/2} l \qquad (3)$$

with M_r representing the molecular weight of the repeating unit (xylose, 132 g/mol), C_n the characteristic ratio for arabinoxylans (11.5) [22] and l the bond length between two xyloses (0.286 nm). The cross-linking density in DDGAX gels (ρ_c) has been calculated from M_c as previously reported by Peppas, Moynihan, and Lucht (1985) [23] [eq. (4)]:

$$\rho_c = \frac{1}{\upsilon M_c} \qquad (4)$$

Scanning Electron Microscopy (SEM)

DDGAX gels were frozen using carbon dioxide and lyophilized at -40 °C/5 atm overnight in a freeze dryer Temp Lyph-Lock 6 freeze drier (Labconco, Kansas, MO, USA). The microstructure of the freeze-dried gels was analyzed by SEM using a microscope FEI Quata 200F (FEI Corporation, Hillsboro, OR) at 10 kV accelerating voltage. The images were obtained in SE (secondary electron, aka Everhart-Thornley Detector).

Statistical Analysis

The measurements were made in triplicate and the coefficients of variation were lower than 8%. The rheology, swelling and amino acids profile analysis were made in duplicate. All the results were expressed as mean values.

RESULTS AND DISCUSSION

AX gelling capability depends on their composition and structure (ferulic acid content, arabinose to xylose ratio, molecular weight). DDGAX used in the present study registered 69% dry basis (d.b.) of pure AX, a ferulic acid content of 6.05 µg/mg polysaccharide, a protein content of 8.2%, an A/X ratio of 1.08 and a molecular weight of 209 kDa [15], which are in the range reported for other AX [1, 5, 7, 9].

AX from different sources, especially from maize, are associated with small but significant amount of protein. In order to get additional information about the protein present in DDGAX, the amino acid composition of this residue was determined (Table **1**). The most abundant amino acids in DDGAX were histidine, threonine and asparagine. In contrast, Cirre *et al.* (2014) [31] reported glutamine and proline as the most abundant amino acids in maize fiber AX. However, these amino acids are not desirables in an AX extract as it has been reported that in some celiac disease patients, maize prolamins (proteins having a high proline content) could induce a celiac-like immune response [32, 33]. This difference in amino acids composition could be related to the source and the method of extraction used in the present study.

Table 1. Amino acid content (nmol/mg) of DDGAX.

Amino acid	DDGAX
Asp	49.06
Glu	33.48
Ser	22.88
His	50.49
Gly	25.20
Thr	56.36
Arg	12.50
Ala	21.44
Tyr	15.44
Met	17.99
Val	15.99
Phe	14.68
Ile	13.06
Leu	21.55
Lys	15.96
Pro	n.d.
Hyp	n.d.
Protein (%)	8.24

All values are reported as means of duplicates.
n.d. = not detected

Viscoelastic Characteristics

DDGAX presented a kinetic of gelation similar to those reported by several authors for maize AX induced by laccase [6, 7] (Fig. **1a**). However, the G' value at the end of gelation was higher (G'= 767 Pa) than those reported for other AX gels at the same polysaccharide concentration (G' ranging from 3 up to 44 Pa) [7, 12]. This difference in gels G' value could be related with the high ferulic acid content in DDGAX (6.05 μg/mg polysaccharide) in comparison to those reported for other AX (from 0.25 up to 2 μg/mg polysaccharide) [7, 12]. The mechanical spectrum and strain sweep of the cured DDGAX gel presented the typical behavior of a solid-like material, observing a linear G' independent of frequency, and a G" dependent of frequency and strain, and much smaller than G' [25] (**Fig. 1b, c**). The same behavior has been previously reported in wheat and maize AX gels induced by laccase [26, 27]. Mechanical spectrum and strain sweep show how the gel formed is within the linear behavior at 0.25 Hz and 5% strain, respectively, which are the experimental conditions used in the kinetic of gelation experiment.

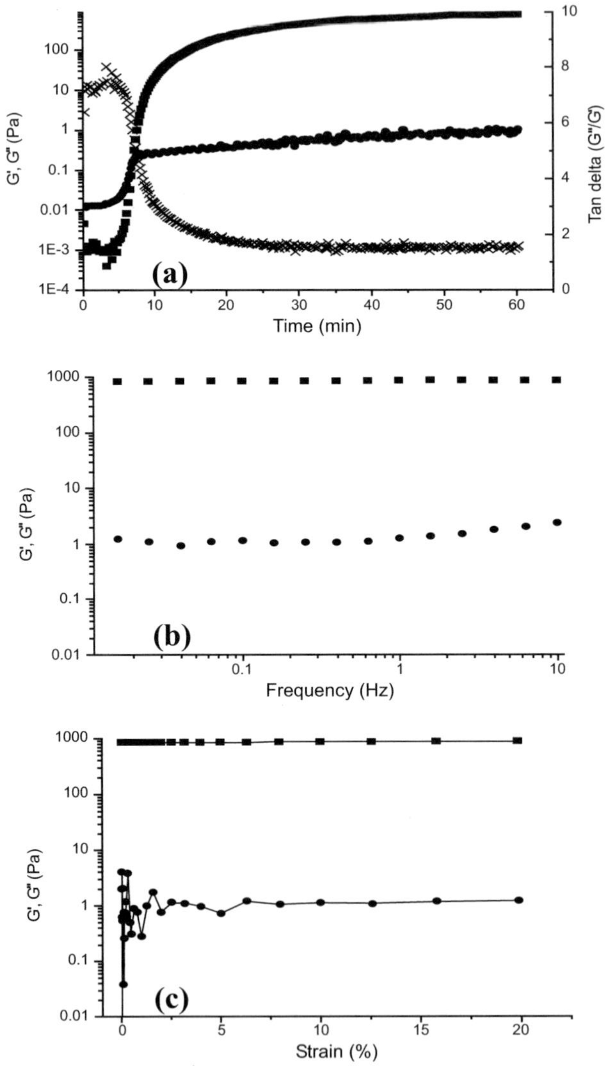

Fig. (1). Kinetic of gelation of DDGAX solution at 2% (w/v) treated with laccase, measurement at 0.25 Hz and 5% strain (a). Mechanical spectrum of the DDGAX gel formed, registered at 5% strain (b). Strain sweep of the DDGAX gel formed, registered at 0.25 Hz (c). G' (■), G" (●), Tan delta (X). All rheological tests made at 25°C.

Structural Characteristics of DDGAX Gel

The swelling behavior of the DDGAX gels was followed for 12 h at 25 °C, with the equilibrium swelling being reached between 8-12 h (Figs. **2** and **3**). The swelling ratio value (q) in the DDGAX gels at 2% (w/v) was 43 g water/g DDGAX (Table **2**). The same q value (43 g water/g AX) was obtained for maize AX gels at 4% (w/v) [5]. In contrast, Berlanga-Reyes *et al.* (2009) [6] reported a higher q value (82 g water/ g AX) for maize AX gels at 2.5% (w/v). The lower water uptake of DDGAX gels could be explained in terms of the higher FA content in DDGAX (6.05 µg/mg polysaccharide) in relation to the FA content reported by Berlanga-Reyes *et al.* (2009) [6] (0.34 µg/mg polysaccharide).

Table 2. Structural characteristics of DDGAX gel at 2% (w/v).

Characteristic	Value
q (g water/g DDGAX) [a]	43 ± 1.22
M_c x 10^3 (g/mol) [b]	34 ± 0.15
ρ_c x 10^{-6} (mol/cm^3) [c]	67 ± 3
ξ (nm) [d]	96 ± 0.84

[a]Swelling equilibrium ratio; [b]Molecular weight between two cross-links; [c]Cross-linking density; [d]Mesh size. All values are reported as means of duplicates.

A higher FA content in AX increase the formation of cross-linking structures such as dimers (di-FA), trimers (tri-FA) and possibly other more complex structures of FA, resulting in a more compact polymeric network that decreases the water uptake of the gel.

The molecular weight between two cross-links (M_c), the cross-linking density (ρ_c) and the mesh size (ξ) of the DDGAX gels are presented in Table **2**. These values were in the range reported for maize AX gels induced by laccase at 2.5% (w/v) (31 x 10^3 g/mol, 49 x 10^{-6} mol/cm^3 and 80 nm for M_c, ρ_c, and ξ, respectively [6]. Higher ξ (201 nm) and M_c (119 x 10^3 g/mol) and lower ρ_c (14.05 x 10^{-6} mol/cm^3) values have been reported for wheat AX gels at the same concentration [17]. This behavior could be related to the high molecular weight reported in AX from wheat flour (438 kDa) [17] in comparison to the alkali-extracted AX from maize bran (209 kDa) used in the present study. The formation of complex ferulate cross-links and/or the involvement of physical interactions between polysaccharide chains could also be responsible of these differences [17, 28].

Because the DDGAX molecular weight (209 kDa) and the A/X ratio (1.08) are known, theoretical calculation of the number of xylose residues per FA molecule can be made. As 48% of the molecule (100 kDa) is composed of xylose residues, the average number of xylose residues per molecule is approximately 758 (xylose molecular weight 132 g/mol). The FA content in DDGAX was 6.05 μg/mg polysaccharide, thus 6.5 FA will be present per DDGAX molecule (FA molecular weight 194 g/mol). Then, in our DDGAX molecule 8.6 FA will be present per 1000 xylose residues. Using the value of 0.5 nm for the length of a xylose residue [12], the length of a fully extended xylan backbone is calculated to be 379 nm and the distance between two adjacent FA will be 45 nm assuming an equidistant position of the FA molecules through the xylan backbone. Previous studies have reported 2-6 FA per 1000 xylosc residues in AX from wheat, triticale, rye, and barley [12].

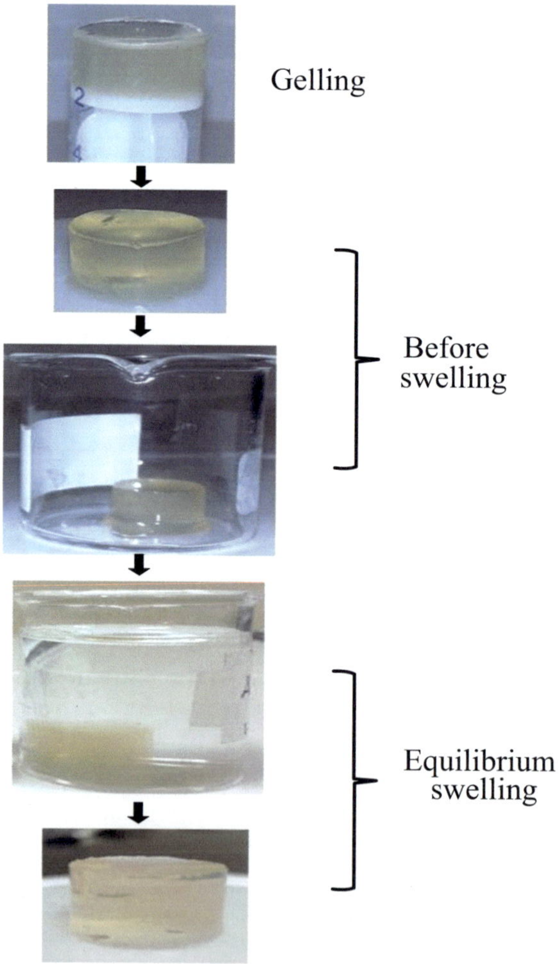

Fig. (2). Swelling of DDGAX gel (2% w/v) at 25°C.

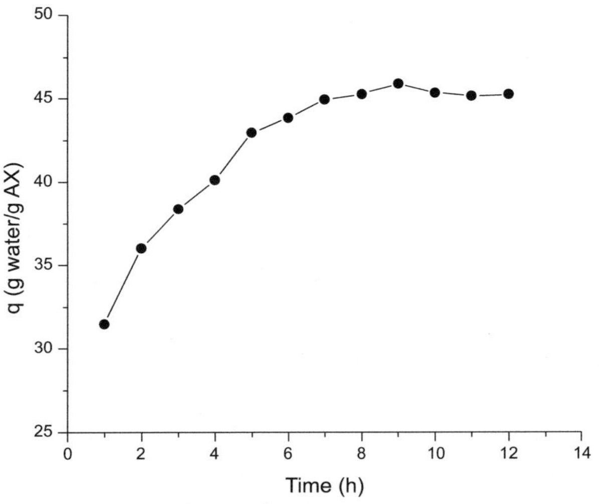

Fig. (3). Swelling kinetic curve of DDGAX gel (2% w/v) in sodium azide solution at 0.02% (w/v) and 25°C.

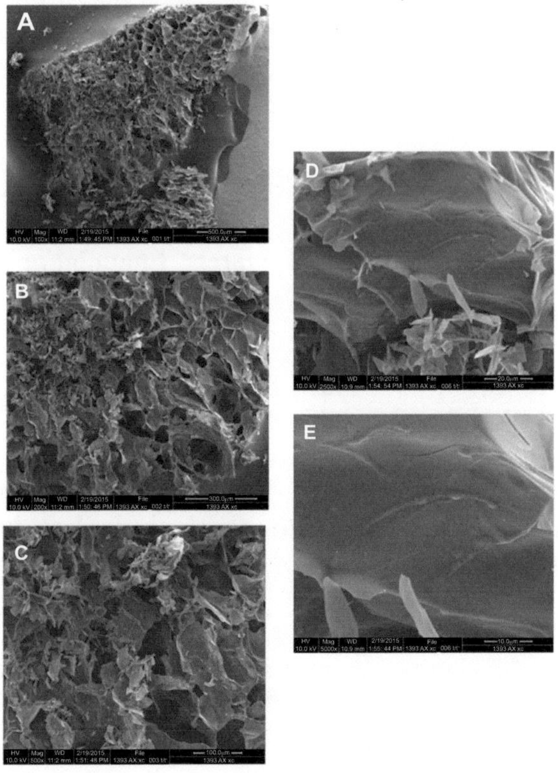

Fig. (4). SEM micrographs of the surface of lyophilized DDGAX gel (2% w/v) at 100x (A), 250x (B), 500x (C), 2500x (D) and 5000x (E) magnification.

DDGAX Gel Microstructure

The surface microstructure of lyophilized DDGAX gels was studied by SEM (Fig. **4**) at 100x (A), 250x (B), 500x (C), 2500x (D) and 5000x (E) magnification. SEM images displayed a porous structure with irregular pore sizes (~ 50-150 µm). Similar results have been found for xylan composites based hydrogels [30]. In general, the gel shows a heterogeneous microstructure with many connections, resembling an imperfect honey comb-like structure, similar to that reported for other lyophilized wheat and maize AX gels [11, 26, 29].

CONCLUDING REMARKS

Distillers dried grains with solubles (DDGS) can be an alternative source of ferulated arabinoxylans (DDGAX) presenting a high ferulic acid content. DDGAX form elastic gels presenting a macroporous network structure and an imperfect honeycomb like microstructure. These gels may find useful applications as encapsulation agents in the development of innovative products, for example in food, biomedical or pharmaceutical industry. Future work should include encapsulation and controlled release experiments of model biomolecules or cells using DDGAX gels.

CONFLICT OF INTEREST

The author (editor) declares no conflict of interest, financial or otherwise.

ACKNOWLEDGEMENTS

This research was supported by Fondo Sectorial de Investigación en Salud y Seguridad Social SSA/IMSS/ISSSTE-CONACYT, Mexico (Grant 179746 to E. Carvajal-Millan). The authors are pleased to acknowledge Alma C. Campa-Mada, Karla G. Martínez-Robinson, Alma R. Toledo Guillén (CIAD) and Stefanie Simon (USDA, Wyndmoor PA, USA) for the technical assistance, and Joseph Uknalis (USDA, Wyndmoor PA, USA) for the microscopy analysis.

REFERENCES

[1] M.S. Izydorczyk, and J.E. Dexter, "Barley β-glucans and arabinoxylans: Molecular structure, physicochemical properties, and uses in food products-a Review", *Food Res. Int.*, vol. 41, pp. 850-868, 2008.
[http://dx.doi.org/10.1016/j.foodres.2008.04.001]

[2] M.M. Smith, and R.D. Hartley, "Occurrence and nature of ferulic acid substitution of cell-wall polysaccharides in graminaceous plants", *Carbohydr. Res.*, vol. 118, pp. 65-80, 1983.
[http://dx.doi.org/10.1016/0008-6215(83)88036-7]

[3] N.A. Peppas, and A.R. Khare, "Preparation and diffusional behavior of hydrogels in controlled release", *Adv. Drug Deliv. Rev.*, vol. 11, pp. 1-35, 1993.
[http://dx.doi.org/10.1016/0169-409X(93)90025-Y]

[4] J. Kopecek, "Smart and genetically engineered biomaterials and drug delivery systems", *Eur. J. Pharm. Sci.,* vol. 20, no. 1, pp. 1-16, 2003.
[http://dx.doi.org/10.1016/S0928-0987(03)00164-7] [PMID: 13678788]

[5] A.L. Martínez-López, E. Carvajal-Millan, J. Lizardi-Mendoza, Y.L. López-Franco, A. Rascón-Chu, E. Salas-Muñoz, C. Barron, and V. Micard, "The peroxidase/H2O2 system as a free radical-generating agent for gelling maize bran arabinoxylans: rheological and structural properties", *Molecules,* vol. 16, no. 10, pp. 8410-8418, 2011.
[http://dx.doi.org/10.3390/molecules16108410] [PMID: 21986519]

[6] C.M. Berlanga-Reyes, E. Carvajal-Millán, G. Caire-Juvera, A. Rascón-Chu, J.A. Marquez-Escalante, and A.L. Martínez-López, "Laccase induced maize bran arabinoxylan gels: Structural and rheological properties", *Food Sci. Biotechnol.,* vol. 18, pp. 1027-1029, 2009.

[7] M.S. Izydorczyk, and C.B. Biliaderis, "Cereal arabinoxylans: Advances in structure and physicochemical properties", *Carbohydr. Polym.,* vol. 28, pp. 33-48, 1995.
[http://dx.doi.org/10.1016/0144-8617(95)00077-1]

[8] C.M. Berlanga-Reyes, E. Carvajal-Millán, J. Lizardi-Mendoza, A. Rascón-Chu, J.A. Marquez-Escalante, and A.L. Martínez-López, "Maize arabinoxylan gels as protein delivery matrices", *Molecules,* vol. 14, no. 4, pp. 1475-1482, 2009.
[http://dx.doi.org/10.3390/molecules14041475] [PMID: 19384279]

[9] E. Carvajal-Millan, S. Guilbert, J-L. Doublier, and V. Micard, "Arabinoxylan/protein gels: Structural, rheological and controlled release properties", *Food Hydrocoll.,* vol. 20, pp. 53-61, 2006.
[http://dx.doi.org/10.1016/j.foodhyd.2005.02.011]

[10] A.B. Hernández-Espinoza, M.I. Piñón-Muñiz, A. Rascón-Chu, V.M. Santana-Rodríguez, and E. Carvajal-Millan, "Lycopene/arabinoxylan gels: rheological and controlled release characteristics", *Molecules,* vol. 17, no. 3, pp. 2428-2436, 2012.
[http://dx.doi.org/10.3390/molecules17032428] [PMID: 22374314]

[11] A. Morales-Ortega, E. Carvajal-Millan, F. Brown-Bojorquez, A. Rascón-Chu, P. Torres-Chavez, Y.L. López-Franco, J. Lizardi-Mendoza, A.L. Martínez-López, and A.C. Campa-Mada, "Entrapment of probiotics in water extractable arabinoxylan gels: rheological and microstructural characterization", *Molecules,* vol. 19, no. 3, pp. 3628-3637, 2014.
[http://dx.doi.org/10.3390/molecules19033628] [PMID: 24662078]

[12] E. Carvajal-Millan, V. Landillon, M.H. Morel, X. Rouau, J.L. Doublier, and V. Micard, "Arabinoxylan gels: impact of the feruloylation degree on their structure and properties", *Biomacromolecules,* vol. 6, no. 1, pp. 309-317, 2005.
[http://dx.doi.org/10.1021/bm049629a] [PMID: 15638534]

[13] C.M. Berlanga-Reyes, E. Carvajal-Millan, G. Niño-Medina, A. Rascón-Chu, B. Ramírez-Wong, and E. Magaña-Barajas, "Low-value maize and wheat by-products as a source of ferulated arabinoxylans", In: *Waste Water - Treatment and Reutilization.,* F.S. García-Einschlag, Ed., INTECH: Rijeka, Croatia, 2011, pp. 341-352.
[http://dx.doi.org/10.5772/15840]

[14] C.M. Berlanga-Reyes, E. Carvajal-Millan, K.B. Hicks, M.P. Yadav, A. Rascón-Chu, J. Lizardi-Mendoza, A.R. Toledo-Guillén, and A.R. Islas-Rubio, "Protein/arabinoxylans gels: effect of mass ratio on the rheological, microstructural and diffusional characteristics", *Int. J. Mol. Sci.,* vol. 15, no. 10, pp. 19106-19118, 2014.
[http://dx.doi.org/10.3390/ijms151019106] [PMID: 25338049]

[15] M.A. Mendez-Encinas, *"Asociación arabinoxilano-proteína y efecto de la fracción proteica en la capacidad gelificante del polisacárido"*, MSc thesis, Centro de Investigación en Alimentación y Desarrollo AC, Mexico, 2015.

[16] E. Vansteenkiste, C. Babot, X. Rouau, and V. Micard, "Oxidative gelation of feruloylated arabinoxylan as affected by protein. Influence on protein enzymatic hydrolysis", *Food Hydrocoll.,* vol.

18, pp. 557-564, 2004.
[http://dx.doi.org/10.1016/j.foodhyd.2003.09.004]

[17] E. Carvajal-Millan, B. Guilbert, M.H. Morel, and V. Micard, "Impact of the structure of arabinoxylan gels on their rheological and protein transport properties", *Carbohydr. Polym.,* vol. 60, pp. 431-438, 2005.
[http://dx.doi.org/10.1016/j.carbpol.2005.02.014]

[18] P.J. Flory, and J. Rehner, "Statistical mechanics of cross-linked polymer networks. II. Swelling", *J. Chem. Phys.,* vol. 11, pp. 521-526, 1943.
[http://dx.doi.org/10.1063/1.1723792]

[19] N.A. Peppas, and E.W. Merrill, "Poly (vinyl alcohol) hydrogels: Reinforcement of radiation-crosslinked networks by crystallization", *J. Polym. Sci., Polym. Phys. Ed.,* vol. 14, pp. 441-457, 1976.
[http://dx.doi.org/10.1002/pol.1976.170140215]

[20] J.W. Qian, M. Wang, D.L. Han, and R.S. Cheng, "A novel method for estimating unperturbed dimension [η]θ of polymer from the measurement of its [η] in a non-theta solvent", *Eur. Polym. J.,* vol. 37, pp. 1403-1407, 2001.
[http://dx.doi.org/10.1016/S0014-3057(00)00254-8]

[21] N.A. Peppas, P. Bures, W. Leobandung, and H. Ichikawa, "Hydrogels in pharmaceutical formulations", *Eur. J. Pharm. Biopharm.,* vol. 50, no. 1, pp. 27-46, 2000.
[http://dx.doi.org/10.1016/S0939-6411(00)00090-4] [PMID: 10840191]

[22] D.R. Picout, and S.B. Ross-Murphy, "On the chain flexibility of arabinoxylans and other β-(1>4) polysaccharides", *Carbohydr. Res.,* vol. 337, no. 19, pp. 1781-1784, 2002.
[http://dx.doi.org/10.1016/S0008-6215(02)00281-1] [PMID: 12423958]

[23] N.A. Peppas, H.J. Moynihan, and L.M. Lucht, "The structure of highly crosslinked poly(2-hydroxyethyl methacrylate) hydrogels", *J. Biomed. Mater. Res.,* vol. 19, no. 4, pp. 397-411, 1985.
[http://dx.doi.org/10.1002/jbm.820190405] [PMID: 4055823]

[24] F.A. Vásquez-Ortiz, G. Caire, I. Higuera-Ciapara, and G. Hernández, "High performance liquid chromatographic determination of free amino acids in Shrimp", *J. Liq. Chrom. Rel. Technol,* vol. 18, pp. 2059-2068, 1995.
[http://dx.doi.org/10.1080/10826079508013960]

[25] J.L. Doublier, and G. Cuvelier, "Gums and hydrocolloids: Functional aspects", In: *Carbohydrates in Food.,* A.C. Eliasson, Ed., Marcel Dekker: New York, 1996, pp. 283-318.

[26] A. Morales-Ortega, E. Carvajal-Millan, Y. López-Franco, A. Rascón-Chu, J. Lizardi-Mendoza, P. Torres-Chavez, and A. Campa-Mada, "Characterization of water extractable arabinoxylans from a spring wheat flour: rheological properties and microstructure", *Molecules,* vol. 18, no. 7, pp. 8417-8428, 2013.
[http://dx.doi.org/10.3390/molecules18078417] [PMID: 23863779]

[27] A.L. Martinez-López, E. Carvajal-Millan, and A. Rascón-Chu, "J. Márquez-Escalante K. Martínez-Robinson. "Gels of ferulated arabinoxylans extracted from nixtamalized and non-nixtamalized maize bran: rheological and structural characteristics". CYTA-J", *Food,* vol. 11, pp. 22-28, 2013.

[28] E. Carvajal-Millan, A. Rascón-Chu, J.A. Márquez-Escalante, V. Micard, N.P. de León, and A. Gardea, "Maize bran gum: Extraction, characterization and functional properties", *Carbohydr. Polym.,* vol. 69, pp. 280-285, 2007.
[http://dx.doi.org/10.1016/j.carbpol.2006.10.006]

[29] S. Iravani, C.S. Fitchett, and D.M. Georget, "Physical characterization of arabinoxylan powder and its hydrogel containing a methyl xanthine", *Carbohydr. Polym.,* vol. 85, pp. 201-207, 2011.
[http://dx.doi.org/10.1016/j.carbpol.2011.02.017]

[30] S. Zhang, Y. Guan, G-Q. Fu, B-Y. Chen, F. Peng, C-L. Yao, and R-C. Sun, "Organic/Inorganic superabsorbent hydrogels based on xylan and montmorillonite", *J. Nanomater.,* 2014. Article ID

675035, 11 pages
[http://dx.doi.org/10.1155/2014/675035]

[31] J. Cirre, "S, Al-Assaf, G.O. Phillips, M. Yadav, K. Hicks. "Improved emulsification performance of corn fibre gum following maturation treatment", *Food Hydrocoll.,* vol. 35, pp. 122-128, 2014.
[http://dx.doi.org/10.1016/j.foodhyd.2013.05.002]

[32] I. Comino, Mde.L. Moreno, A. Real, A. Rodríguez-Herrera, F. Barro, and C. Sousa, "The gluten-free diet: testing alternative cereals tolerated by celiac patients", *Nutrients,* vol. 5, no. 10, pp. 4250-4268, 2013.
[http://dx.doi.org/10.3390/nu5104250] [PMID: 24152755]

[33] J.P. Ortiz-Sánchez, F. Cabrera-Chávez, and A.M. de la Barca, "Maize prolamins could induce a gluten-like cellular immune response in some celiac disease patients", *Nutrients,* vol. 5, no. 10, pp. 4174-4183, 2013.
[http://dx.doi.org/10.3390/nu5104174] [PMID: 24152750]

CHAPTER 8

Sol and Gel Based on Polysaccharide: Characterization and Structure-properties Relationship

Marguerite Rinaudo*

Biomaterials Applications, 6, rue Lesdiguiéres Grenoble, 38000, France

Abstract: The objective of this chapter is to describe the main techniques adapted to characterize ionic or neutral polysaccharides, especially when they are water soluble. The steric exclusion chromatography (SEC), rheology and NMR spectroscopy are, in our opinion, the most important techniques to study the main characteristics of polysaccharides. The same techniques and approaches may also be valid for other water soluble biopolymers such as proteins and nucleic acids. The main characteristics of polyelectrolytes are recalled and it is shown that the ionic behavior of the polymers can be used to establish the nature of the polymer conformation (coil, helix). In this case, it is necessary to combine thermodynamic characteristics with optical rotary power (or differential scanning calorimetry DSC and circular dichroism CD) and molar mass determinationss. The semi-rigid characterization of the polysaccharide chains depending on their conformation is introduced. The persistence length L_p which characterizes the local stiffness of the chain determined by steric exclusion chromatography in application of the worm-like chain model and validated by molecular modelling. Rheology of polysaccharides is of great interest due to fundamental and applied point of views (especially in food, cosmetic or biomedical applications). The solutions even at low polymer concentration are often non Newtonian due to the stiffness of the polysaccharides which also depresses the critical overlap concentration. The main relationships relating viscosity to molar mass and/or polymer concentration are given; the influence of the shear rate for measurement is pointed out; and, the flow and dynamic measurements are described for sol systems. Originality of polysaccharides is that they may associate to form 3-D network, or physical network. This gel is controlled by the thermodynamic conditions and environmental conditions (ionic concentration, nature of ions, temperature, pH…). The main mechanisms of physical gelation are recalled and characterization of the sol-gel transition is introduced. The techniques to follow this transition as well as those for gel rheology are also introduced.

Keywords: Biopolymer, Characterization, Ionic selectivity, Polyelectrolyte, Polysaccharide, Rheology, Sol-gel transition, Steric exclusion chromatography,

* **Corresponding author Marguerite Rinaudo:** Biomaterials Applications, 6, rue Lesdiguiéres Grenoble, 38000, France; Tel: +33611434806; E-mail: marguerite.rinaudo@sfr.fr

Martin Masuelli & Denis Renard (Eds.)

Viscosity, Water-soluble systems.

INTRODUCTION

In this chapter, we intend to expose the general properties of biopolymers and techniques used for their characterization. Especially, polysaccharides that I investigated for a long time will be discussed in this chapter [1, 2]. They are natural polymers or source of many derivatives which may be used in many industrial domains such as food, cosmetic, biomaterial, or pharmaceutics [3 - 6]. Natural polysaccharides produced by biosynthesis are based on sugar repeating units. They are often stereoregular and rich in –OH functional groups which form H-bonds. In stereoregular polysaccharides (carrageenans or bacterial polysaccharides like gellan or succinoglycan), intrachain H-bond networks often stabilize helical conformations. In addition, interchain H-bond networks stabilize physical gels or solid state materials with interesting physical properties (chitin and cellulose are semi-crystalline polysaccharides having good film forming characteristics) [7 - 9]. The stereoregularity is the basis of the cooperative helix-coil transition depending on the thermodynamic conditions (temperature, pH, ionic concentration, nature of the counterions…). As we will show later, helix –coil transition is often also related to the cooperative gel-sol transitions in aqueous polysaccharide systems [10].

The origins of polysaccharides are mainly plants, animals, fungi or bacteria. The polymers are neutral (as cellulose, amylose, curdlan, scleroglucan, agarose…) or ionic (xanthan, gellan, alginate, pectin…) becoming polyelectrolytes [11, 12]. The choice of the solvent composition used to prepare solutions and characterize polysaccharides is one of the most important step. For ionic systems, water is the best solvent, the solubility being promoted by electrostatic repulsions, and for neutral polymers, water often gives interacting systems (aggregates formation and low solubility) but dimethylsulfoxide or H-bond breakers are better. The problem of aggregation in such solutions, probably often induced after drying of the purified samples, is perhaps the most difficult problem to solve before going further in the polymer characterization.

After analysis of polyelectrolyte behavior, and indications on few techniques of polymer characterization (nuclear magnetic resonance NMR, steric exclusion chromatography SEC), we will focus on the rheological properties of these systems in sol and gel states, which are often the properties involved in applications such as thickening or gelling performances.

A-Polyelectrolyte Main Properties

When a macromolecule has ionic groups regularly fixed along its polymeric chain, it is named as "polyelectrolyte". The parameter which controls its thermodynamic properties in solution is the charge parameter λ which is proportional to the linear charge density and introduced in the polyelectrolyte theories proposed by Katchalsky *et al.* [13] and later by Manning [14, 15]. It is expressed by the relationship:

$$\lambda = (\nu/h)\,(e^2/\,DkT) \qquad \textbf{(1)}$$

where ν is the number of ionic charges along a chain with a contour length h, e is the electronic charge, D is the dielectric constant taken as that of water (D = 78) and kT the Boltzmann term. The charge parameter λ is also written as $\lambda = (\nu/h)Q$ with Q the Bjerrum length, *i.e.* 7.2 Å at 25 °C in aqueous solution. λ is directly imposed by the distance b corresponding to the projection of two successive ionic sites on the axis of the chain (b is the length of a monomeric unit if each monomer has a dissociated ionic charge in the extended conformation). The distance b is reduced in case of a single chain helical conformation. The ionic sites may be $-COO^-$ (in pectins, alginates or hyaluronan), $-NH_3^+$ (for chitosan in acidic medium), and $-SO_3^-$ (in carrageenan). In the case of polyelectrolytes with $-COO^-$ and $-NH_3^+$ functions, the net charge (ν) depends strongly on the pH (in connection with the dissociation equilibrium) [16]. In addition, the electrostatic potential of the polyelectrolyte (or charged oligomers) increases progressively when the degree of polymerization increases and reaches a limit as soon as the number of charges (or degree of polymerization DP) of a charged block of ionized monomeric units becomes larger than around 15-20 when no change in the conformation occurs [17, 18]. At that time, thermodynamic properties become independent of the molecular weight (MW) but depend on the regularity of the charge distribution along the chains (example was discussed for blockwise distribution of galacturonic acid units along the chains in pectins) [19]. In addition, b is directly related to the conformation of the polymer; single or double chain helix formation implies a change of the length b and eventually of the molecular weight. This helical conformation often exists in stereoregular polysaccharides (or polypeptides and DNA). These single or double helices are stabilized by intra-chain and/or inter-chain H-bonded network formation [17, 20 - 22]. Following this, the study of electrostatic properties will help to characterize the conformation of these biopolymers in various experimental conditions (pH, ionic concentration, temperature) [10, 20].

A-1. Activity Coefficient of Counterions

Most useful information comes from the determinations of the free fraction of counterions to characterize a polyelectrolyte in the absence of external salt. Its theoretical value is directly related to the charge parameter and to the valency of the counterions. The relations used to determine the activity coefficient γ depend on the charge parameter value compared to 1 (or 0.5 with divalent counterions) and on the valency of the counterions. One of the most usefull experiments for the determinations of activity coefficient γ of counterions is obtained by potentiometry or the transport coefficient f by conductimetry [14] to get information on the charge density, conformation and blockwise distribution of charged blocks as found in low methoxyl pectins. The main relations proposed by Manning at infinite dilution in the absence of external salt and for $\lambda>1$ (or 0.5 with divalent) are:

$$\gamma_1 \sim (2\lambda)^{-1} \text{ for monovalent counterions} \quad \textbf{(2)}$$

$$\gamma_2 \sim (4\lambda)^{-1} \text{ for divalent counterions} \quad \textbf{(3)}$$

In these relations, γ is the activity coefficient related to the free fraction of counterions at equilibrium and inversely proportional to b, the distance between two charges [14]. It allows to establish if the polymer adopts a single helical chain, a double helix on itself or a double helix (or a helical dimer) made of two chains. For this discussion, thermodynamic experiments are combined with additional experimental determinationss such as optical rotation which indicates that an ordered conformation is formed in given conditions and molar mass determinations (to demonstrate if one or two chains are associated). This γ parameter was used to obtain information on the conformation of the chain in carrageenan, xanthan or gellan [20, 21, 23]. In Fig. (**1**), the conformational transition demonstrated by viscometry is clearly located around 32°C (Tm) for gellan in the presence of 0.025M electrolytes.

This figure shows that a conformational transition, associated with a large stiffening of the molecule occurs at Tm=32°C in the tetramethylammonium (TMA) salt form. Following this, under Na and more commonly under potassium form, the viscosity increases when temperature is lower than 32-34°C due to interchain interactions. The Tm values may be related with the ionic selectivity and depend also on the ionic concentration as discussed later.

An example of structural characteristics is given in Table (**1**) for deacylated gellan [20]. Activity coefficients of monovalent counterions (Na^+) and molar mass are given on both sides of the conformational transition allowing to confirm the formation of a double helix at low temperature [10]. The coiled conformation is characterized by $\lambda_c = 0.37$ (calculated from the chemical structure) and the helical conformation ($\lambda_h = 0.75$) is calculated for the double helix established by Chandrasekaran *et al.* [22]. A double helix was also demonstrated previously with κ-carrageenan [21, 24].

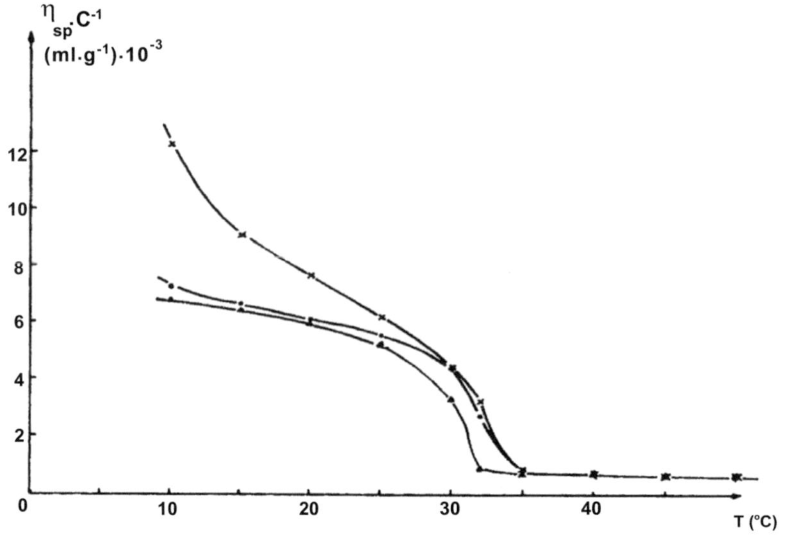

Fig. (1). Reduced viscosity as a function of temperature for deacylated gellan at 0.3 g/L in presence of 0.025 M external salts. Ionic salt forms: K^+ (x) in KCl; Na^+ (•) in NaCl; TMA^+ (▲) in TMACl [20]. Reproduced from *Biopolymers*, vol.30, pp.451-464, 1990 with the permission of John Wiley and Sons.

Table 1. Polyelectrolyte characteristics of deacylated gellan in relation with its conformation in 0.025 M TMACl external salt [10, 20].

Conformation	Calculated γ+ Relation [2]	Experimental γ+	Mw	Lp nm	[η] mL/g
Helical conformation λ_h= 0.75; T=24°C	0.687	0.675	490000	72	3500
Coiled conformation λ_c= 0.37; T= 37°C	0.831	0.765	250000	6	580

DSC, circular dichroïsm, NMR are convenient methods to demonstrate the existence of a helical conformation (at least the existence of an ordered conformation in given conditions) and to determine the thermodynamic conditions for the helix-coil transition. The more valuable techniques to determine the activity coefficient of counterions (or osmotic coefficient) are potentiometry (with a specific electrode), conductimetry or vapour phase osmometry in absence of external salt [16, 25].

It is necessary to add that the activity coefficient of counterions for the polyelectrolyte (*i.e.* the net charge of the polyion) is nearly independent on the external salt concentration in NaCl (Cs). Assuming dilute solution with $\gamma_0 \sim 1$, an additivity rule may be proposed to determine the activity of all counterions such as

$$\gamma_0 \, Ct = \gamma_0 \, (\gamma \, Cp + Cs) \tag{4}$$

Then, in large salt excess, for Cp<< Cs, it comes Ct~Cs.

In this relationship, Ct is the total counterion concentration and Cp is the counterion contribution from the polyelectrolyte in exchange with the external salt in the polyelectrolyte solution. The polyelectrolyte has nearly the same ionic charge in the presence of external salt. This behavior was also studied by Katchalsky and Alexandrowicz [26, 27] and by Manning [15].

A-2. Intrinsic pK, pK_0

It is not so easy to determine the pK_0 of a polyelectrolyte characterizing the carboxylic functions (or –NH2 functions present in chitosan). It is necessary to determine the pK at zero net charge of the polyion such as to screen the inter-site electrostatic interactions. The acid form of the carboxylic polyelectrolyte is prepared by passing a purified sample isolated under the sodium form through an ion exchanger column (under $-SO_3H$ form). Following this, titration with NaOH can be used to get the pK as a function of the degree of neutralization (related to the net charge of the polyion). For that purpose, the pH is determined as a function of the volume of NaOH added up to exact neutralization (around pH= 7.5), obtained at maximum of the first derivative. For each degree of neutralization, α (fraction of –COO- sites neutralized with NaOH; α=1 at pH= 7.5), the pH allows calculating the apparent pKa of the carboxylic function following the relation:

$$pKa = pH + \log (1-\alpha)/\alpha = pK_{0} + \Delta \, pKa(\alpha) \tag{5}$$

where ΔpKa(α) is the electrostatic contribution directly related with the electrostatic potential Ψ(a) (at the minimum distance of approach to the polyelectrolyte (a) and depending on the charge distribution along the chain). The distance “a” is around 0.6 nm for a single chain. ΔpKa(α) can be calculated from Katchalsky theory [13] and compared with the experimental values allowing to determine pK_0, by extrapolation of pKa(α) to α = 0 [16]. An example is given for hyaluronan in water in Fig. (**2**).

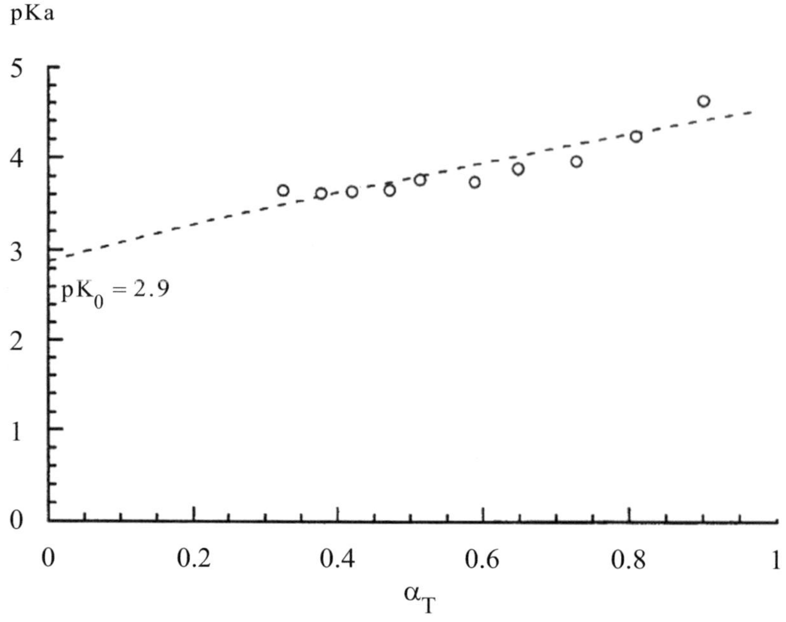

Fig. (2). Evolution of the apparent pK as a function of the total degree of neutralization (α_T) [12, 28]. Reproduced from *Polysaccharides: Structural Diversity and Functional Versatility*, S. Dumitriu Ed., New-York, USA: Marcel Dekker, 1998, pp.313-334 with the permission of Taylor & Francis Group, LLC, a division of Inform plc.

From experimental potentiometric titration, apparent pK were calculated for each α allowing to determine $pK_0 = 2.9 \pm 0.1$ by linear extrapolation (due to the low charge parameter of hyaluronan $\lambda = 0.71$) to a null charge (Fig. **2**). A value between 3 and 3.5 is often found for polycarboxylic acid such as carboxymethylcelluloses [16]. Larger charge parameters impose curvilinear extrapolation as discussed for carboxymethylcelluloses.

For polycationic polysaccharides exemplified by chitosan, the most direct way to get pK_0 is to isolate the polymer under the hydrochloride form in excess of HCl, dissolve directly in water (α=1) and add progressively NaOH up to around pH=7 (α=0). Apparent pK is easily determined as a function of the fraction of dissociate ionic sites to get pK_0 ~6.5 [29].

In Fig. (**3**), the evolution of the apparent pK as a function of the degree of neutralization is given for a series of α-L-glutamic acid oligomers and corresponding polymers. The pK varies linearly for the low DP oligomers and are extrapolated to pK_0= 4.45 independently of the DP [17]. The slopes of pKa(α) increases when the oligomer DP increases due to the electrostatic interaction between ionic sites on the same chain.

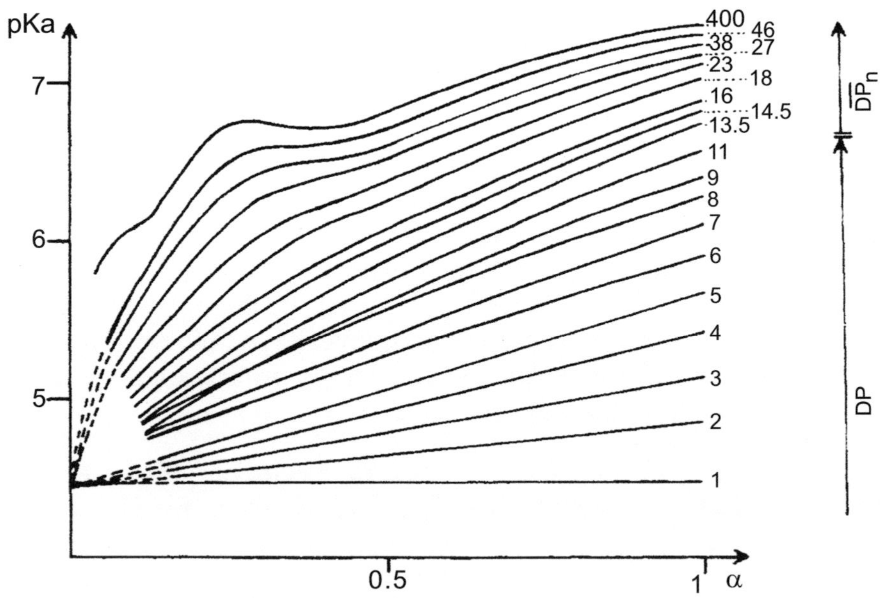

Fig. (3). Influence of the chain length (reflected by the degree of polymerization DP) and degree of neutralization α on the apparent pK for α-L-glutamic acid oligomers and the corresponding polymers PLGA at 25°C [17]. Reproduced from *Biopolymers*, vol.14, pp.2035-2048, 1975 with the permission of John Wiley and Sons.

When DP increases, the β-structure is stabilized for 8<DP<12 and for larger DP, the α-helix formation started at low degree of neutralization. In these conditions, a curvilinear extrapolation is needed due to higher electrostatic potential to obtain the intrinsic pK. In the range of DP = 23 to 46, the helix-coil transition is observed (and confirmed separately by circular dichroism) for a critical degree of neutralization α ~0.4 at 25°C. For the higher DP, aggregates of helices are formed at low degrees of neutralization (α < 0.1) and dissociate to single helices when the net charge increases followed by coil conformation at high degree of neutralization.

A-3. Influence of Ionic Concentration on Polymer Conformation

The dimensions of polyelectrolytes depend on the composition of the solvent, on the net charge of the polyion and especially on the ionic concentration including polyelectrolyte and external salt concentrations as well as the nature of the electrolyte. In addition, the temperature also plays a role when stereoregular polyelectrolytes are able to adopt helical conformation stabilized by H-bonds. First, consider a flexible charged chain. Due to inter-sites electrostatic repulsion, in the absence of external salt, the chain expands when its concentration (Cp, expressed in equivalent of charges/L, the polyelectrolyte concentration) decreases. The Debye length ($\kappa^{-1} = (Cp/\lambda)^{-1/2}$ or $Cs^{-1/2}$ in the presence of excess of external salt Cs) expressed the length over which the electrostatic repulsion is efficient. In excess of salt (over 0.05 M and better 0.1 M 1-1 electrolyte), only short range intrachain electrostatic repulsions still take place. Then, all the determinations of dimensions of polyelectrolyte and molecular weight have to be performed in the presence of such excess of electrolyte; we estimate that 0.1 M 1-1 electrolyte represents the best conditions (with $\kappa^{-1} \approx 1$ nm). In these conditions, for linear coiled polymer, only small variations of the hydrodynamic volume is shown in relation with the inter-site electrostatic repulsions progressively screened when the ionic concentration continues to increase.

For a stereoregular polymer such as κ-carrageenan or gellan, a competition between these electrostatic repulsions and stabilization by H-bonds must be considered. It should be the same for ionic polypeptides or DNA. Then, in presence of external salt, attractions based on H-bond dominate electrostatic repulsions and allow the molecule to adopt a helical conformation (in single or double helix). This conformation is more rigid than that of an irregular coiled polymer, and a worm-like chain model must be considered to characterize the local stiffness and the dimensions of such macromolecule (see part A-4). NMR spectroscopy is a convenient method to demonstrate the existence of a helical conformation in solution (at least the existence of an ordered conformation characterized by low mobility of the different atoms under given conditions). An example is given in Figs. (**4** and **5**) for the bacterial polysaccharide named succinoglycan.

This stereoregular biosynthetised polysaccharide is based on a repeat unit of eight sugars with few substituents: these substituents are puruvyl and succinyl groups in various amount depending on the source and purification techniques used. Proton NMR spectroscopy allows a quantitative determinations of these substituents (protons from $-CH_3$ and $-CH_2-$ groups) when realized at temperature higher than the conformational transition *i.e.* in the disordered conformation. The helix-coil transition for this polysaccharide is very cooperative as shown in Fig. (**5**) [30].

When the proton NMR spectrum is obtained at different temperatures, the integral of the characteristic signals (1 $-CH_3$ from pyruvate and 2 $-CH_2-$ from succinate units respectively) are compared with that of the standard (5 mM sodium acetate). If the polymer and standard amounts are known, the ratio of both integrals is plotted as a function of temperature (Fig. **5A**). The mobility of the protons is very low in the helical conformation but increases sharply when temperature increases at the helix-coil cooperative transition. The coiled conformation is needed to get quantitative determinations or control of the chemical structure of such stereoregular polysaccharides. The same situation is valid for gellan, xanthan, κ-carrageenan, hyaluronan *etc.* [20, 31 - 33].

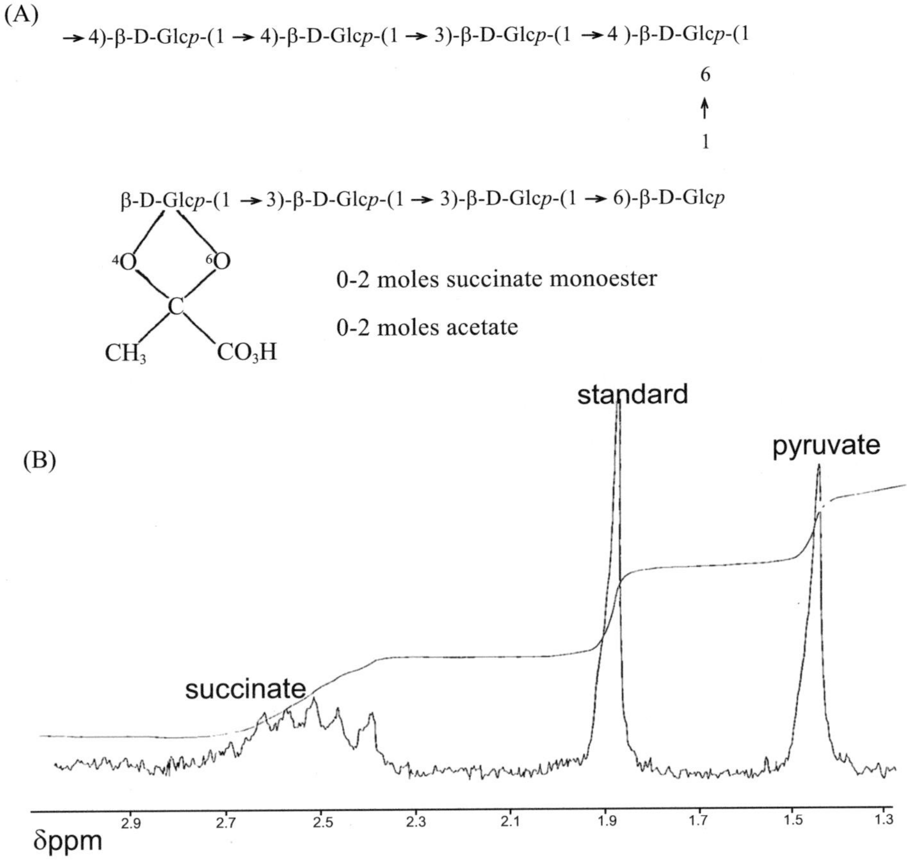

Fig. (4). Characterization of succinoglycan at 5 g/L in D_2O by proton NMR. (A) Molecular structure of the repeat unit; (B) NMR spectra at 85°C in presence of a standard for quantitative analysis of substituents [30]. Reproduced from *Int. J. Biol. Macromol.*, vol. 12, pp.195-200, 1990 with the permission of Elsevier.

This figure shows that a transition from an ordered conformation (no detectable NMR signal) to a disordered conformation takes place when temperature increases. In addition, the specific temperature for conformational transition increases in presence of salt excess (or increase of the polymer concentration), due to the screening of electrostatic interactions. In water at low polymer concentration, the specific temperature for conformational transition Tm= 50°C increases up to 70 °C in presence of excess of external salt (0.1M NaCl). DSC, circular dichroism and optical rotary power are different techniques allowing to evidence the actual conformation and to follow the helix-coil transition in relation with the thermodynamic parameters. Example of the optical rotary power as a function of temperature is given in Fig. (**5B**) for succinoglycan in aqueous solution in presence or absence of external salt.

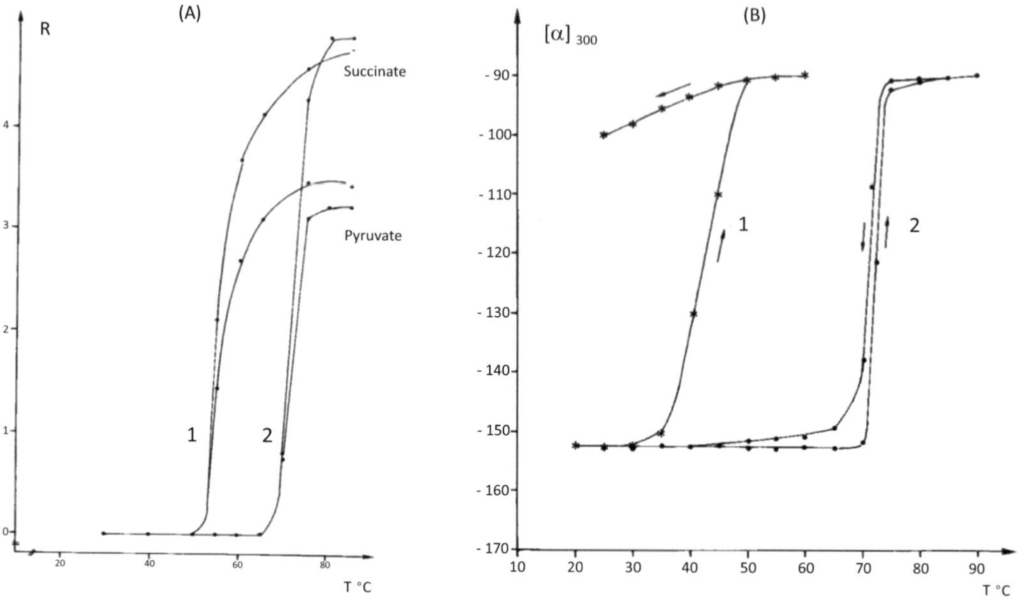

Fig. (5). Helix-coil transition of succinoglycan. (**A**) Relative integrals of succinate and pyruvate signals (R) as a function of temperature (1) in D_2O in the presence of the standard and (2) same as (1) added of 0.1M NaCl. (**B**) Variation of the optical rotary power at 300nm in the same experimental conditions [30]. Reproduced from *Int. J. Biol. Macromol.*, vol.12, pp.195-200, 1990 with the permission of Elsevier.

A-4. Chain Stiffness and Persistence Length

The worm-like chain model is interesting to analyze the behavior of ionic polysaccharides and to characterize their local stiffness. In this model, the rigidity is characterized by a persistence length Lt, corresponding to the value of the persistence length at the ionic concentration considered. Lt = Lp + Le in which Lp is the intrinsic persistence length obtained when Lt is extrapolated to infinite salt concentration such as to screen the electrostatic contribution (Le). The proposed treatment allows us to predict the dimensions of the chain (and particularly the radius of gyration Rg_s obtained by light scattering) and the intrinsic viscosity ([η]).The main relation to analyze the radius of gyration at a given ionic concentration is the following for high molar masses:

$$Rg_s^2 = \alpha_s^2 (L/L_t)/3 \quad \textbf{(6)}$$

in which L is the contour length (proportional to the molecular weight) which is large compared to Lt; αs is the excluded volume which contains a large contribution of the electrostatic interaction.

In salt excess, it is assumed that the Φ-conditions of the corresponding neutral polymer are approached. For intermediate molar masses, it is necessary to use the development proposed by Benoit and Doty for the Φ-state [34]. The model of worm-like chain polyelectrolyte taking into account the electrostatic parameters was developed by Odijk [35, 36] and an example was exposed in detail by Reed for hyaluronan (Fig. **6**) [37].

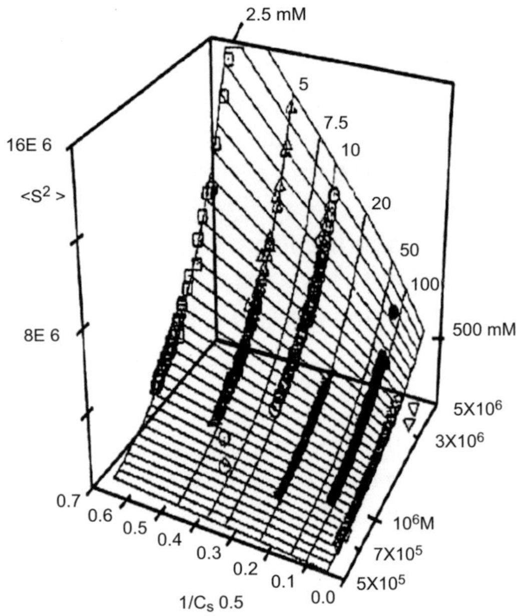

Fig. (6). 3D-plot relating the calculated square of the radius of gyration $<S^2>$, the inverse of the root square of ionic salt concentration ($Cs^{0.5}$) and molar mass (M). The points result from SEC/MALLS experiments on HA for iso-ionic conditions [37]. Reproduced from *Macro-ion characterization from dilute solutions to complex fluids, ACS Symp. Series 548*, K.S. Schmitz Ed., Washington DC, USA: American Chemical Society, 1994, 297-314 with permission of American Chemical Society.

In this figure, it is clearly shown that the treatment adopted allows to predict the square of the radius of gyration ($<S^2>= Rg^2$) as a function of the molar mass and of the ionic salt concentrations (from 2.5 mM up to 500 mM). A very good agreement is obtained for HA using the experimental data from SEC/MALLS with different eluents in a large range of concentrations. From this treatment, it is easy to determine the persistence length Lt variation with the ionic concentration of environment and to obtain L_p.

On another side, the persistence length was recently calculated for few polysaccharides by molecular modeling. This approach is fruitful as soon as the chemical structure is perfectly known; it was proposed on hyaluronan, chitosan with different degrees of acetylation and galactomannan, a neutral polymer, with different mannose/galactose ratios (M/G) [38 - 41]. The values of Lp obtained are assumed to be a good approximation of the unperturbed value allowing to predict not only the intrinsic viscosity of a given polyelectrolyte but also to interpret the high viscosity of solution obtained at given molecular weight and concentration for semi-rigid polysaccharides (such as xanthan) compared to flexible polymers. This approach allows also justifying the low sensitivity of solution viscosity (as for xanthan) to external salt concentration in relationship with the large intrinsic persistence length of stereoregular polymers in the ordered conformation. Especially, in this case, it is usually found $L_p >> Le$. Molecular modeling gives L_p values in good agreement with the value determined experimentally on these three polysaccharides studied [38, 40, 42 - 44]. The parameter L_p (which controls $[\eta]$) is approached experimentally (when the ionic concentration of the eluent is known) from the curve Rg(M) (relation [6] with α_s=1) established from steric exclusion chromatography (SEC) experiments using 0.1 M 1-1 eluent.

Some values of Lp were given in Table (**2**). Lt was also obtained by AFM on succinoglycan (Fig. **7**) [45]. This figure shows a single succinoglycan chain evidenced by a height around 0.5 nm and allows to determine the contour length and the persistence length. L_p=36±13 nm in 0.01 M KCl in agreement with the value obtained by light-scattering (Lt=30-35 nm in 0.1 M KCl) [44, 46].

In addition, the viscosity of a polysaccharide solution can be predicted from L_p: L_p allows to calculate $[\eta]$ for given values of Cs and M and a given polymer. Then, also the zero shear rate specific viscosity is able to be calculated and compared with the experimental data [43]. This point is discussed later with the rheology of polysaccharides.

For such investigations, an equipment of steric exclusion chromatography with 3 detectors on line is necessary with: a differential refractometer, a multiangle light scattering detector and a viscometer as described later.

Table 2. Intrinsic persistence length of some polysaccharides.

	L_p (nm)	Conditions	References
Succinoglycan coil	5	0.1M NaCl	[44]
Succinoglycan single chain helix	30,35 36	0.1M NaCl AFM	[44, 46, 45]
Gellan coil	6.1	36°C; 0.025 M TMACl	[20]
Gellan double helix	72	25°C; 0.025 M TMACl	[20]

(Table 2) contd.....

	L_p (nm)	Conditions	References
Hyaluronan	8- 9 8. 5	NaCl 0.1M 25°C Modeling	[33, 43, 39]
Xanthan renaturated	160	0.1M NaCl; 30°C	[44, 47]
Xanthan single chain helix	46	0.1 M NaCl; 30°C	[44, 47]
Xanthan coil	5	0.01M NaCl; 80°C	[44]
Chitosan DA<0.2	11 9	0.3M acetic acid/ 0.2M sodium acetate; 30°C Modeling	[42, 38]
Chitosan DA=0.6 and Chitin	12.5	Modeling	[38]
Galactomannan M/G~1	9.3 9.6	Water Modeling	[40, 40, 41]

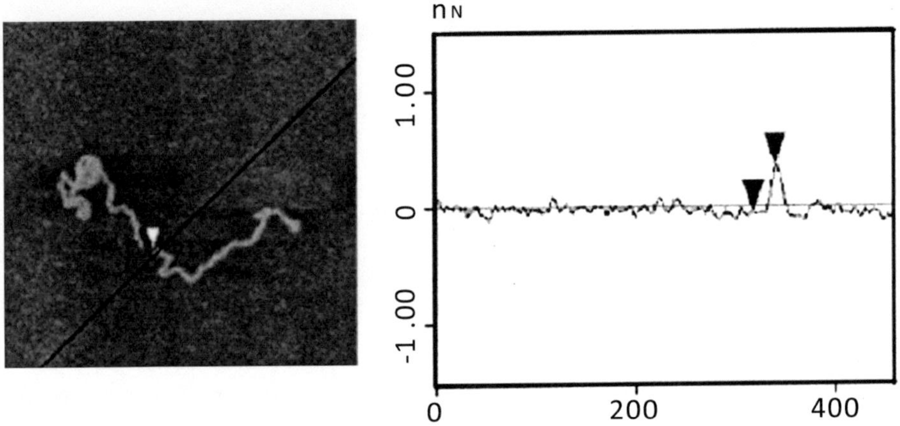

Fig. (7). Section analysis of AFM image of succinoglycan chain in 0.01M KCl [[45]]. Reproduced from *Macromolecules,* vol.33, no 20, pp.7440-7447, 2000 with the permission of American Chemical Society.

A-5. Interchain Electrostatic Interactions and Viscometric Behavior

An original behavior in salt-free solution is observed by viscometry or radiation scattering (light and neutron). A peak is observed when the reduced viscosity is plotted as a function of polymer concentration as shown in Fig. (**8**). These results were obtained using the sodium salt form of a short chain of sodium polygalacturonate. First, each curve passes through a maximum which position depends on the external salt concentration, this maximum being located at Cp = 2Cs (for λ<1) or 2λCs (when λ is larger than 1) [48].

It must be pointed out directly that the viscosity of a polyelectrolyte solution in water and in the absence of external salt or at low concentration of external salt cannot be directly interpreted in terms of dimensions of the polymer. In that respect, a polyelectrolyte is completely different from a neutral polymer. The position of the viscosity peak was shown to be independent of the molar mass of the polymer with hyaluronan in solution [49].

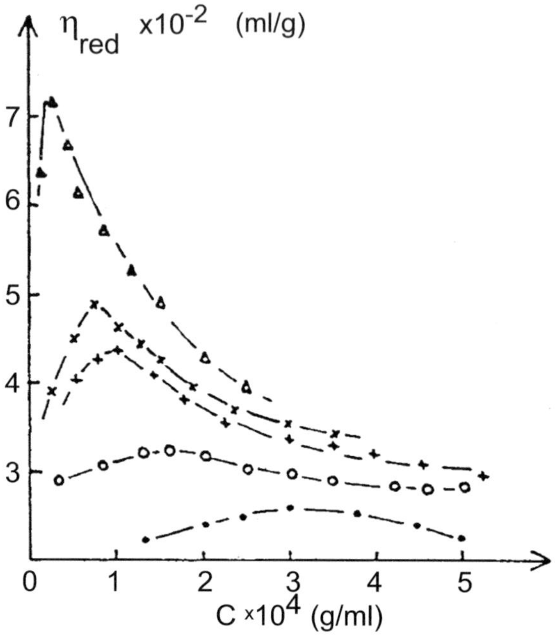

Fig. (8). Variation of the reduced viscosity of sodium polygalacturonate (Mw=25500 g/mol) as a function of polymer concentration at different NaCl concentrations. • 5 x10^{-4} M; o 2x10^{-4} M; +1 x10^{-4} M; x 5x10^{-5} M; Δ 1x10^{-5} M [48]. Reproduced from *Macro-ion characterization from dilute solutions to complex fluids, ACS Symp. Series 548,* K.S. Schmitz Ed., Washington DC, USA: American Chemical Society, 1994, pp.315-321 with permission of the American Chemical Society.

On the right side of the viscometric peak, long range interchain electrostatic interactions are established between the chains in solution, but on the left side, chains are progressively diluted and the environment is controlled by the external salt concentration which remains constant and high in comparison with the ionic polymer concentration at high dilution of the polymer (Fig. **8**).

Corresponding to the right side of the peak, at the same experimental conditions, a peak located at q_{max} (q being the wave vector) is observed in light or neutron scattering as a function of the scattering wave vector q (the q vector range necessary to evidence this peak depends on the range of polymer concentrations covered) (Fig. **9**). It indicates that a preferential distance (d *~ q_{max}^{-1}) exists in solution in the absence (or at low concentration) of external salt and that q_{max} increases when the polymer concentration increases as a function of $C_P^{1/2}$

(Cp being here the polymer concentration expressed in mass) as predicted for a hexagonal packing (see figure 9B) [46, 51].

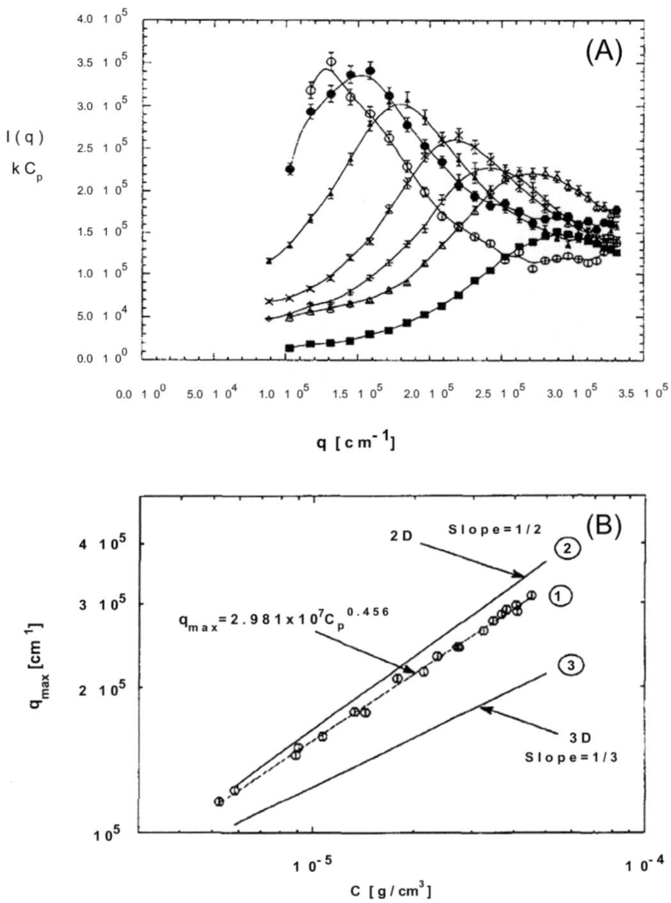

Fig. (9). (A) Variation of the reduced light scattering intensity I(q)/ kCp as a function of the wave vector q for succinoglycan in salt-free solutions at different polyelectrolyte concentrations Cp: (o) 0.65 x 10-5 g/mL; (•) 1.01 x 10-5 g/ mL; (▲) 1.59 x 10-5 g/ mL; (x) 2.38 x 10-5 g/ mL; (+) 3.03 x 10-5 g/ mol; (Δ) 3.84 x 10-5 g/ mL; (■) 4.53 x 10-5 g/ mL; (B) Dependence of the position of peak position q_{max} as a function of polymer concentration [46]. Reproduced from *J. Phys. II,* vol.4, no 6, pp.1001-1019, 1994 with the permission of EDP Sciences.

The peak amplitude is very sensitive to the presence of external salt as shown in Fig. (**10**) for a semi-rigid polysaccharide (xanthan) in relation with the screening effect. Its position remains unchanged at given polymer concentration and it disappears over a critical ratio Cp/Cs~2 or 2λ (for λ >1) (Cp being the polymer concentration expressed in charge equiv/L). These conditions have to be connected with the viscometric behavior discussed previously [48]. At Cs ≥ 0.03

M NaCl, the scattering peak disappears in the experimental conditions of Fig. (**10**).

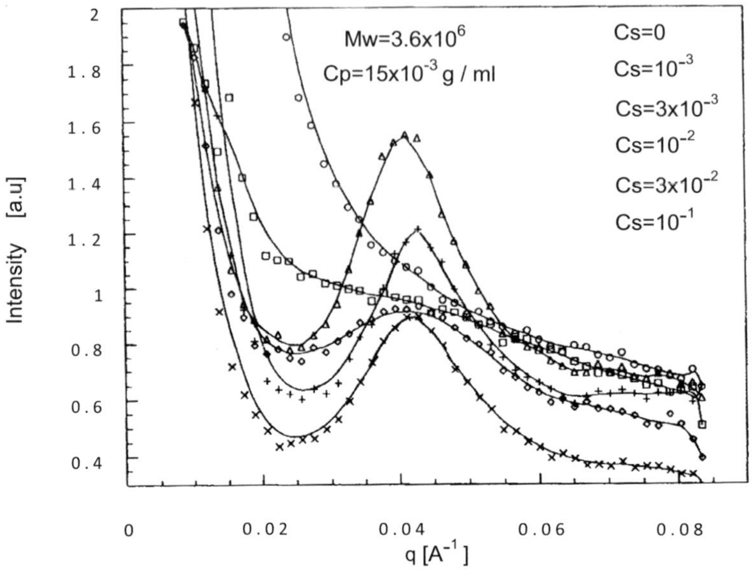

Fig. (10). Variation of I(q) as a function of the neutron wave vector q for xanthan (Mw= $3.6x10^{-6}$, $Cp=15x10^{-3}$g/mL) at different external salt concentration Cs. T=25°C [51]. Reproduced from *Macromolecules*, vol.28, pp.3119-3124, 1995 with the permission of American Chemical Society.

Thus, to characterize a polyelectrolyte in aqueous solution by its molar mass and/or dimensions, it is necessary to isolate the chains by screening the long range electrostatic repulsions; this situation is obtained in the presence of 0.05 M or better 0.1 M monovalent external salt (NaCl, $NaNO_3$...).

A-6. Ionic Selectivity

An original behavior also concerns the ionic selectivity in some polysaccharides; this ionic selectivity is related to ion pair formation and occurs in the range of a charge parameter over 1 corresponding to the limit for counterion condensation in the Manning theory [14, 15]. Nevertheless, some exceptions were obtained; even for $\lambda<1$, selectivity was shown for the coiled conformation for κ-carrageenan under K^+ and Na^+ forms (in the coiled conformation $\lambda c = 0.68$). This conclusion was obtained from ultrasound absorption experiments: K^+ forms more ion pairs than Na^+ favoring the double helix formation [52]. In addition, for monovalent series of counterions (from which tetraalkylammonium, R_4N^+), it was found a selectivity order [21]:

$Rb^+ > K^+, Cs^+ > Na^+ > Li^+, NH_4^+ > R_4N^+$

For κ-carrageenan, at low temperature, a helical conformation is stabilized and it was shown that the critical temperature (Tm) for helix-coil transition increases when the ionic concentration increases ; the Tm values determined at a constant ionic concentration (0.1M) follows the same ionic sequence (the same selectivity is also obtained for the sol-gel transition). A phase diagram was established (Fig. **11**) relating the conformation with the total ionic concentration and the temperature.

The phase diagram obtained for κ-carrageenan and a given counterions (K^+ or Na^+ in this example) allows to locate the conditions for coil conformation on the left side (high temperature), the double helix on the right side under the critical ionic concentration C* on the straight line indicating the helix-coil transition [53]. Over C*, helices associate to give a gel which melts at a temperature higher than the temperature corresponding to the helix-coil transition, characterized by the well-known hysteresis. C* depends on the counterions, corresponding to the ionic selectivity in which K^+ forms more ion pairs than Na^+ playing a role also on the stability of κ-carrageenan gels.

For divalent counterions a small ionic selectivity is observed with the sequence [21]:

$Ba^{2+} > Sr^{2+} \sim Ca^{2+} > Mg^{2+} \sim Zn^{2+} > Co^{2+}$

On carboxymethylcellulose [54, 55], it was found by ultrasound absorption that a sequence of selectivity among monovalent counterions is:

Li> Na> K>Rb>Cs ~TMA

as shown in Fig. (**12**). The tetramethylamonium (TMA) chloride electrolyte having a large hydrophobic cation is used as reference. It is recognized for not forming ion pairs with polyelectrolyte.

On gellan, a polycarboxylic system with low charge parameter under the coil conformation, no ionic selectivity among monovalent counterions is obtained for the coil-helix transition ($\lambda_h = 0.75$ for the double helix). The ionic selectivity occurs under the helical conformation in relation with gel formation following the selectivity:

K> Na> Li> TMA

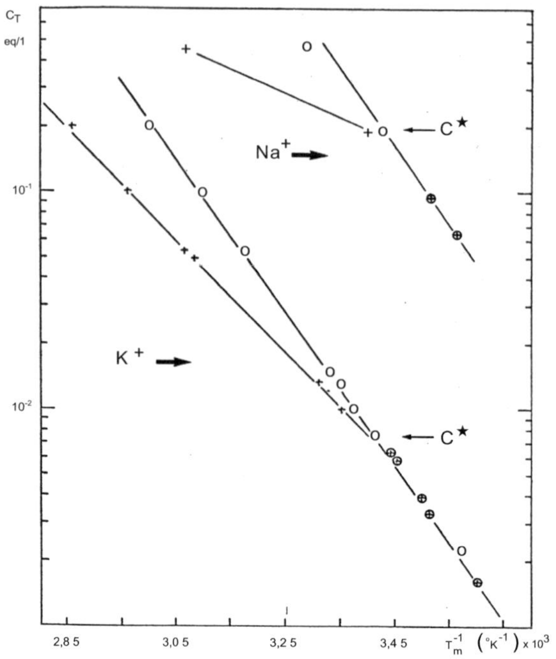

Fig. (11). Influence of the counterion concentration and the inverse of temperature (K) on the conformation of κ-carrageenan. Influence of the nature of counterions on the phase diagram [53]. Reproduced from *Solution properties of polysaccharides,ACS Symp. Series 150*, D. Brant Ed., 1981, pp.367-378 with permission of the American Chemical Society.

looking the sequence obtained with coil-helix transition with sulphated carrageenan with $\lambda_c = 0.68$ [20].

From all these data, it is concluded that to characterize a charged polymer (polyelectrolyte) by its macromolecular parameters (MW, [η]) in aqueous solution, it is necessary to isolate the chains by screening of the long range electrostatic repulsions as mentioned previously. This situation is obtained in presence of monovalent external salt (NaCl, $NaNO_3$...) for coils. For stereoregular polymers, it is necessary to avoid the chain association (and aggregation or gelation) and/or stabilisation of an ordered conformation by control of the temperature, nature of the counterions, and ionic concentration.

In these conditions, viscosity and SEC are used as usually for neutral polymer. The ionic concentration of the eluent (larger than 0.05M with 1-1 electrolyte to prevent electrostatic exclusion from the SEC porous gels)is the only parameter which still plays a role on the hydrodynamic volume (due to the contribution of the electrostatic persistence length contribution Le). Taking into account the stiffness of many polysaccharides as well as DNA, the universal calibration must be used if SEC equipment is not connected with a light scattering detector. Besides, it is important to point out that another very important step, the purification, is crucial when studying a polysaccharide [1, 8]

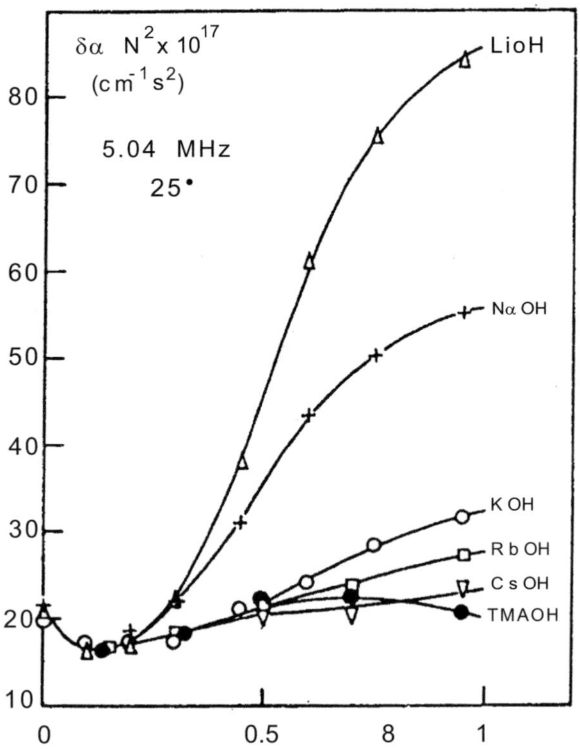

Fig. (12). Variation of the ultrasound absorption (δα) at frequency N=5.04 MHz and 25°C during neutralization (θ) of a carboxymethylcellulose (DS= 2.49; 0.1 $-COO^-/L$) by different monovalent alkaline hydroxide. Reproduced from *J. Chim. Phys.*, vol.68, pp.1258-1260, 1971, with the permission of Société Chimique de France.

B. Characterization of Water Soluble Polymer

The first step before going further is to isolate the biopolymer under purified form. For an anionic polymer, the better seems to isolate the polymer under the sodium salt form, after exchanging the multivalent counterions in excess of NaCl and ethanol precipitation [1]. Cationic polymers with $-NH_2$ functions are soluble in acidic conditions; it may be isolated in slightly alkaline conditions under the $-NH_2$ form (as example for chitosan soluble at pH<6.5 but insoluble in neutral medium) or under the hydrochloric form ($-NH_3^+$, Cl^- obtained in HCl excess) which is directly water soluble. For water soluble neutral polysaccharides, they are purified by precipitation from aqueous solution in presence of ethanol or isopropanol.

B-1. NMR Spectroscopy

Proton NMR is a very convenient and rapid technique to control chemical structure of homogeneous polymer, especially stereoregular bacterial polysaccharides. For unknown polysaccharides, a very long and difficult analysis must be realized to establish the composition, the arrangement of sugar units, position of the linkages *etc.* [56]. For NMR on water soluble polymers, it must be dissolved in D_2O at dilute concentration (lower than 10g/L). The spectrum is performed at high temperature (around 85°C) (Fig. **13**). It is important to adopt a high temperature especially in proton NMR to decrease the viscosity of the solution but also (and it is connected) to break the H-bond network which stabilizes the local conformation such as a helical conformation as demonstrated for hyaluronan, xanthan or gellan [20, 31, 33, 57].

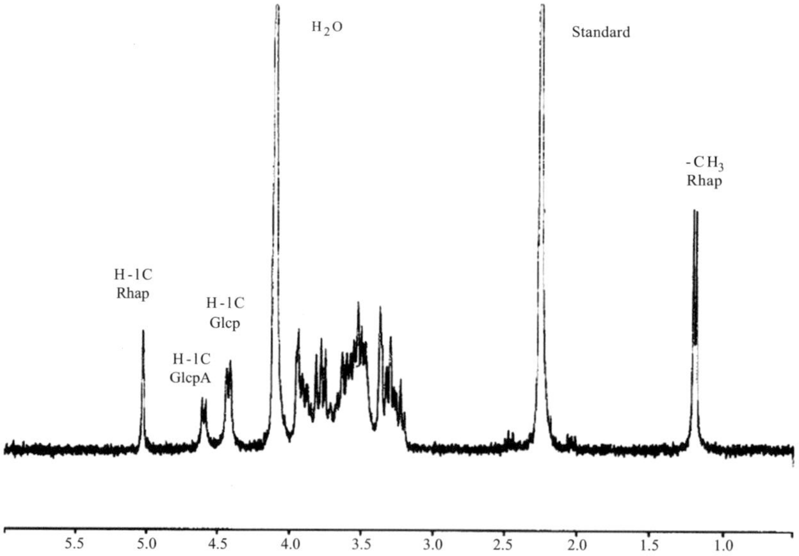

Fig. (13). Proton NMR of deacylated gellan in D_20 at 85°C [57]. Reproduced from *Novel Macromolecules in Food Systems*, G.Doxastakis and V. Kiosseoglou, Eds., Amsterdam, The Netherlands: Elsevier Science B.V. 2000, pp.239-263 with permission from Elsevier.

In Fig. (**13**), the chemical structure of the repeat unit of deacylated gellan is given. The proton NMR spectrum at high temperature allows to confirm the chemical structure and to control the deacylation (no glyceryl, no acetyl substituents along the chain are evidenced in the spectrum) [57]. When the NMR is realized in presence of a standard (known amount of small molecule with $-CH_3$ groups such as DMSO or sodium acetate), the influence of the temperature on the conformation may be tested. An example is given on hyaluronan (Fig. **14**). In this specific case, even at 85°C, as demonstrated in presence of a standard, the protons of hyaluronan are not completely mobile for quantitative approach. This is reflected by the ratio X = amplitude of the signal for one proton for H-1 of hyaluronan divided by the amplitude of the signal of one proton of the standard taking into account the respective concentration. Fig. (**14B**) shows the intrachain H-bond network involved in stabilization of hyaluronan in aqueous solution; the H-bonds progressively dissociate when temperature increases.

When hydrophobic interactions are involved such as in methylcellulose, NMR spectroscopy must be performed at low temperature in D_2O to prevent association or gelation discussed later. Otherwise, a good solvent for neutral polysaccharides may be DMSO-d_6 for proton and carbon NMR [58, 59].

B-2. SEC Chromatography Analysis

Steric exclusion chromatography is one of the most powerful techniques to characterize a water soluble polymer in solution. In our work, we used a SEC equipment from Waters (Alliance GPCV2000) associated with a MALLS from Wyatt Co (Dawn DSP-F). There are three detectors in line: a viscometer giving a signal proportional to $C_i[\eta]_i$, a differential refractometer giving C_i and a multiangle light scattering detector giving CiMi (for each eluted species i corresponding to a given elution volume Vi and extrapolation at zero angle) (Fig. **15**).

At each elution volume, the angular dependence of the scattered light is also obtained to determine the radius of gyration of the species i (Rgi). The influence of electrostatic interaction and external salt concentration were analyzed on SEC chromatography of hyaluronan, an anionic polysaccharide [60]. For polyelectrolytes, the eluent chosen has to be a 0.05M or better 0.1M 1-1 electrolyte to screen long range electrostatic repulsions causing exclusion from the chromatographic support [61]. Analysis of the data gives the weight-average molecular weight Mw, the number-average molecular weight Mn, the intrinsic viscosity [η], and the weight fraction distribution w_i (M_i) or the cumulative distribution function (Fig. **16**). In addition, one gets the Mark-Houwink parameters (K and a) relating the intrinsic viscosity and the molar mass and the parameters relating the radius of gyration and the molar mass in the following relations:

$$[\eta]\ \text{mL/g} = KM^a \quad (7)$$

and

$$Rg = K' M^b \quad (8)$$

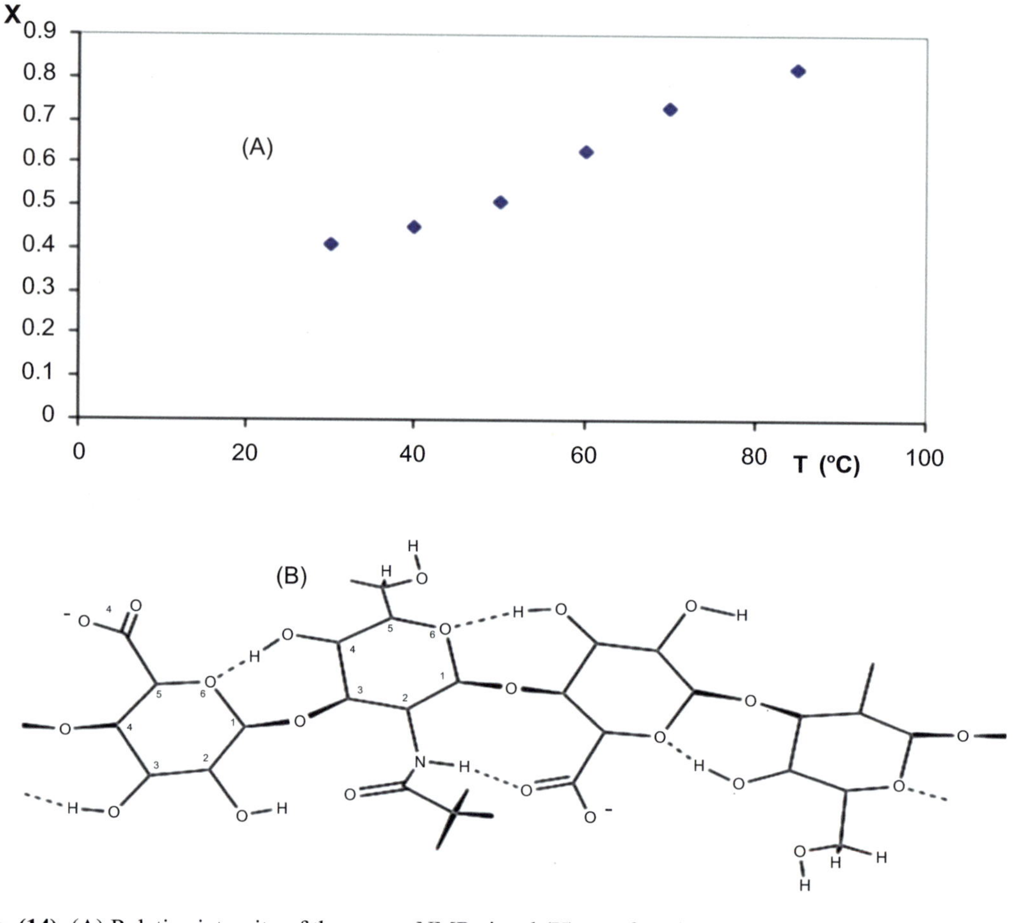

Fig. (14). (**A**) Relative intensity of the proton NMR signal (X) as a function of temperature for hyaluronan in D_2O. (**B**) Example of structure stabilized by intrachain H-bonds in hyaluronan (HA) [33]. Reproduced from *Hyaluronan, vol. 1: Chemical, biochemical and biological aspects,* J.F. Kennedy, G.O. Phillips, P.A. Williams and V.C. Hascall Eds., Cambridge, England: Woodhead Pub but now Elsevier. 2002, pp.37-46 with permission of Elsevier.

The curve Rg (M) allows also determining the intrinsic persistence length (Lp) when the ionic concentration is known to be able to take into account the role of the electrostatic contribution (Le) on Lt (relation [6]). This analysis demonstrates the advantage of SEC technique especially to get the Mark-Houwink parameters. In that respect, an important question concerns the experimental data published in the literature: the values of L_p deduced from Rg or $[\eta]$ data must often be reconsidered due to the polydispersity of the samples used, to the presence of aggregates or interactions between chains in solution in relation with the conditions of polymer purification, to the choice of the solvent, to the chemical structure of the polysaccharides. Polysaccharides being rich in -OH are able to form H-bonds causing aggregation especially after drying. In this case, light scattering signal is perturbed by aggregates and the molar masses are overestimated. In addition, the intrinsic viscosity is modified by chain association. It may increase or decrease in relation with the aggregate structure for a given system.

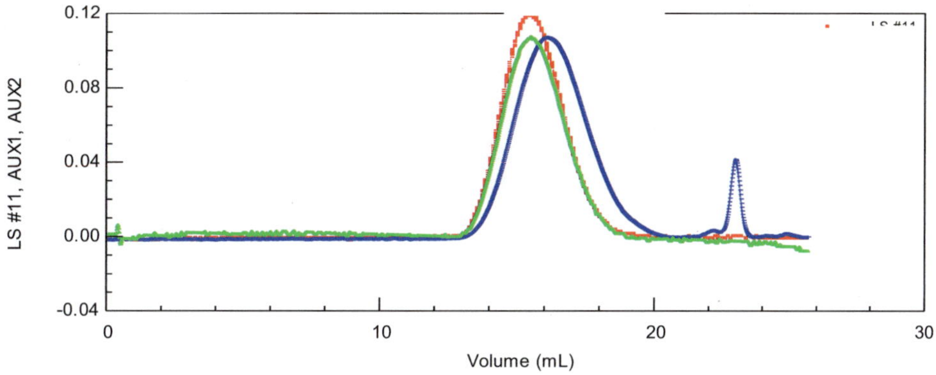

Fig. (15). SEC chromatogram of hyaluronan in 0.1M NaCl at 30°C with 3 traces: LS(red) = light scattering; Aux 1(blue) = differential refractometer ; Aux 2 (green)= viscometer.

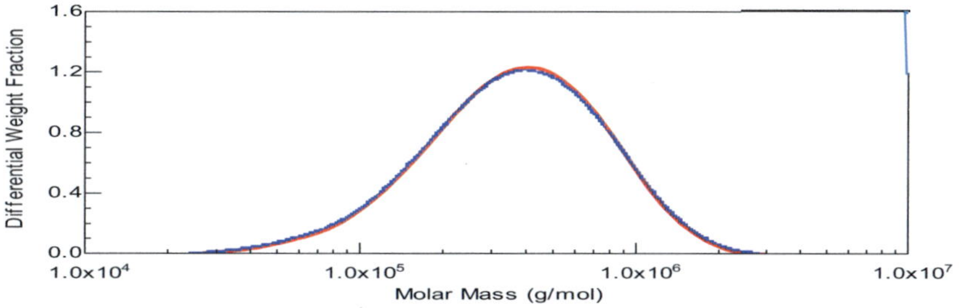

Fig. (16). Differential molar mass distribution obtained for a hyaluronan sample. The two curves (red and blue) show the good reproducibility of experimental data when the polymer is perfectly soluble.

This technique was applied recently for analysis of synovial fluids from different patients [62]. Due to the difference between stiffness of hyaluronan and proteins which compose the synovial fluids, it is possible to quantify hyaluronan (content 1-2 g/L and MW=2-3 x 10^6) and proteins (yield ~30/L and average MW~10^5) in each synovial fluid [63]. An example is given in Fig. (**17**).

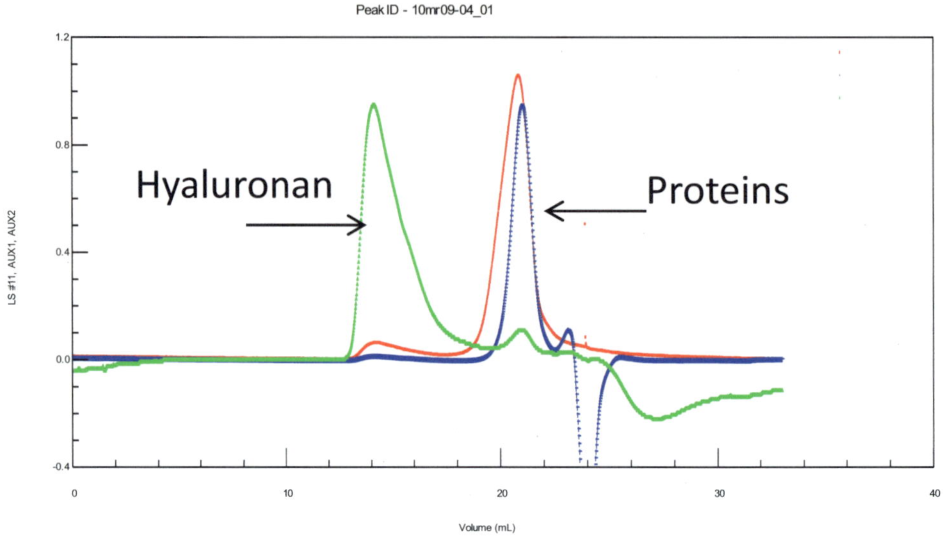

Fig. (17). The SEC of synovial fluid for determinations of soluble proteins and HA molecular weights and concentrations. Temperature = 30 °C; eluent = 0.1 mol/L $NaNO_3$. LS red= light-scattering signal; AUX 1 blue= refractive index signal; AUX 2 green= viscometric signal.

C- Rheology of Biopolymers in Aqueous Solutions

Two types of experiments (steady flow and dynamic) are usually performed in direct relation with the fundamental properties of the systems. Polysaccharides are usually good thickeners, which means that, when dissolve in a solvent, they increase particularly its viscosity. This effect is reflected by the viscosity which has to be examined as a function of the polymer concentration, molar mass, stiffness of the polymer, temperature, shear rate adopted and the composition of the solvent (especially ionic concentration in aqueous medium, nature of the ions and pH for ionic polymers).

C-1. Flow Behavior. Influence of MW and C

Due to their characteristics, the flow curves plotted as a function of the shear rate show a non-Newtonian character (characterized by a critical shear rate $\dot{\gamma}_c$ over which the viscosity decreases when the shear rate increases). An example is given at (Fig. **18**).

The first part of the curve is called the Newtonian plateau (η_0) and it is the domain where the viscosity has to be taken to determine the intrinsic viscosity. The critical γ value (γ_c) separates the Newtonian and non-Newtonian regimes. The critical value γ_c expresses the inverse of longest relaxation time of the macromolecule (τ) in solution. This critical shear rate γ_c decreases when the molar mass increases or when the polymer concentration increases. It is related to disentanglement and orientation of the chains in the flow. γ_c remains constant when polymer concentration is lower than overlap concentration C* (assumed to be approached by the relationship C*~ $[\eta]^{-1}$). Extensive study on rheology of hyaluronan with different concentration and molar masses was developed previously [8, 9, 28, 64, 65].

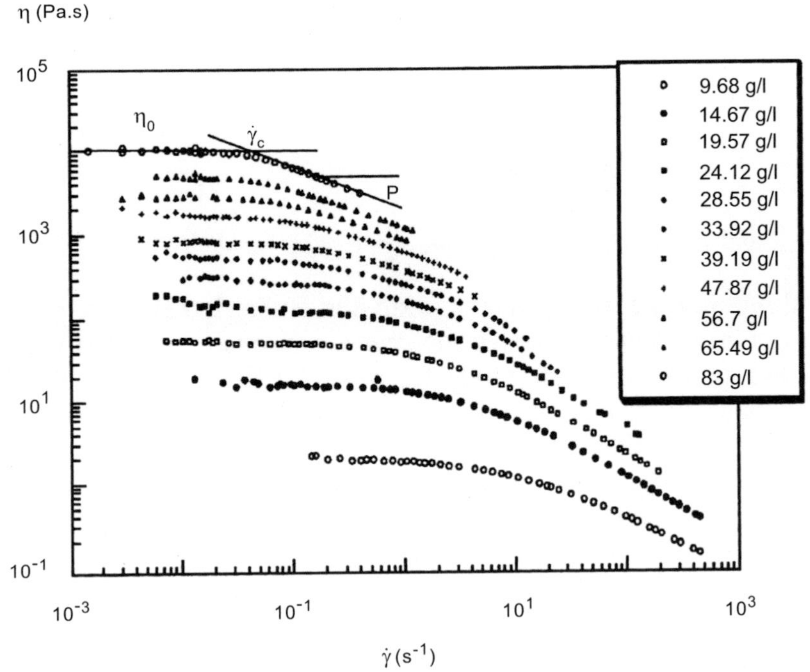

Fig. (18). Variation of the viscosity as a function of the shear rate and polymer concentration for hyaluronan solutions (Mw= $1.3x10^6$) in 0.1M NaCl at 25°C [9, 64]. Reproduced from *Polysaccharides: Structural Diversity and Functional Versatility, 2nd ed.* S. Dimitriu Ed., New-York, USA: 2005, pp.535-549 with the permission of Taylor & Francis Group, LLC, a division of Inform plc.

This type of curve must be performed in presence of monovalent external salts for a charged polymer. Then, following the Huggins relation, the reduced viscosity plotted as a function of polymer concentration C is linear in the dilute domain at low γ (in the Newtonian plateau):

$$\eta_{sp}/C= [\eta] + k'[\eta]^2 C \quad \textbf{(9)}$$

with k', the Huggins constant 0.3< k'<0.7; the specific viscosity $\eta_{sp}= (\eta-\eta_s)/\eta_s$ with η_s the solvent viscosity.

Over C* (with the overlap parameter C[η]>1), the following development as a function of the fundamental parameter c[η] has to be adopted at zero shear rate

[66, 67]:

$$\eta_{sp} = [\eta]C + k' [\eta]C^2 + B ([\eta]C)^n \quad \textbf{(10)}$$

with the Huggins constant around 0.4, B=2x10^{-2} and n~4 for xanthan or hyaluronan

or

$$\eta_{sp} = [\eta]C [1 + k_1[\eta]C + k_2 ([\eta]C)^2 + k_3 ([\eta]C)^3] \quad \textbf{(11)}$$

proposed by Kwei *et al.* [68] with k_1=0.4, $k_2 = k_1^2/2!$, $k_3 = k_1^3/3!$

In the semi-dilute regime, the chains overlap and entanglements are progressively formed giving a larger dependency of the specific viscosity with C[η]. Exponent n is predicted to increase to a value varying from 3.4 to 4. It is also the regime in which the viscosity decreases seriously when the shear rate increases (slope p in log-log plot) due to disentanglements and/or orientation characterizing the viscoelastic character (see Fig. **18**). Relationship [11] allows drawing a master curve which relates the specific viscosity at zero shear rate with the overlap parameter C[η]. This curve is given in Fig. (**19**). It is valid for all the soluble polymers whatever their chemical structure, their molar mass, the polymer concentration and solvent.

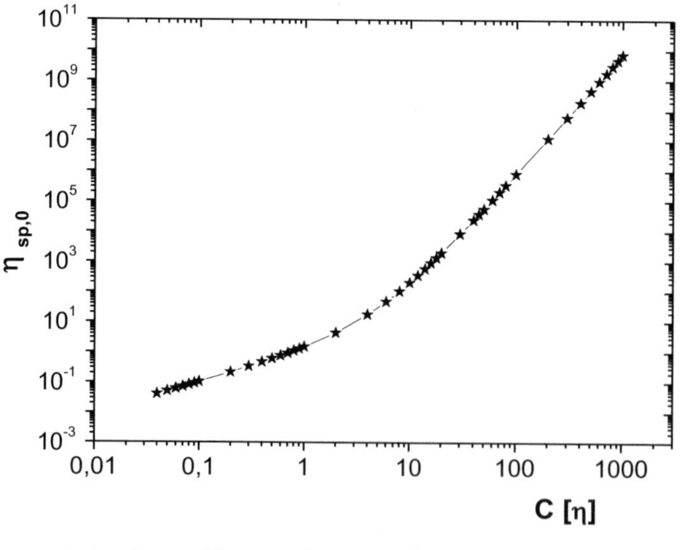

Fig. (19). Master curve relating the specific viscosity at zero shear rate ($\eta_{sp,0}$) to the overlap parameter C[η].

An example of such behavior is given in Fig. (**20**) for HA having different MW and concentrations.

As soon as loose interactions exist in solution, as frequently with polysaccharides (especially with neutral polymers) in aqueous medium, the slope n is larger than 3.4 – 4 at high concentration. It increases progressively with the increase of the interactions. It is clearly demonstrated with galactomannans: solubility of galactomannan increases when the yield in galactose as side chain units increases. In the same time, the interchain interactions decrease as well as the slope n in the semi dilute regime [2, 69 - 71]. So, deviation from the predicted curve (Fig. **19**) relating the specific viscosity at zero shear rate and the overlap parameter (relation [11]) indicates the existence of interchain interactions (*i.e.* a lack of solubility). In the same time, the Newtonian plateau usually disappears in the low γ range values (and often also at low polymer concentration).

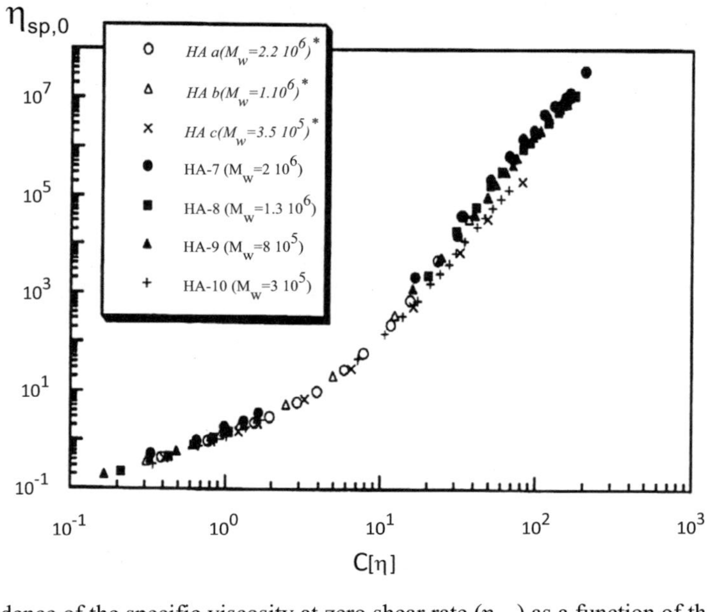

Fig. (20). Dependence of the specific viscosity at zero shear rate ($\eta_{sp,0}$) as a function of the overlap parameter C[η] for hyaluronan solutions with different MW and concentrations at 25°C in 0.1M NaCl [9]. Reproduced from *Polysaccharides: Structural Diversity and Functional Versatility, 2nd edit.* S. Dimitriu Ed., New-York, USA: 2005, pp.535-549 with the permission of Taylor & Francis Group, LLC, a division of Inform plc.

C-2. Dynamic Measurements

The second series of experiments concerns the dynamic measurements. A sinusoidal deformation is imposed to the solution in a large range of frequencies, and the response is a complex modulus decomposed in an in-phase response (G' reflecting the elastic character) and an out off-phase response (G" reflecting the viscous response). This study allows the analysis of the viscoelastic behavior of a polymeric solution. An example is given in Fig. (**21**).

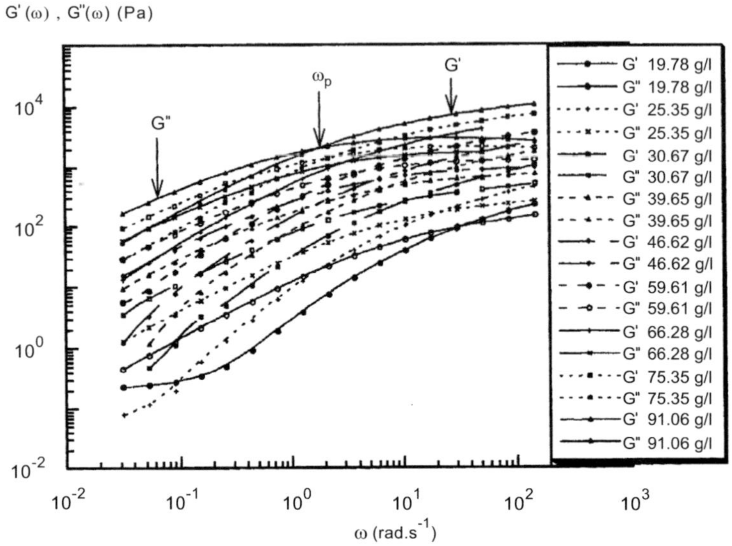

Fig. (21). Rheological experiments on HA with Mw= $8x10^5$ at different concentrations in 0.1M NaCl at 25°C. Storage and loss moduli (G' and G" in Pa) as a function of the pulsation ω (rad.s^{-1}) [9, 64]. Reproduced from *Polysaccharides: Structural Diversity and Functional Versatility, 2nd edit.* S. Dimitriu Ed., New-York, USA: 2005, pp.535-549 with the permission of Taylor & Francis Group, LLC, a division of Inform plc.

For a solution, at low frequency, G" is larger than G' but over a critical frequency ω_p, G' becomes larger than G" corresponding to the presence of entanglements, *i.e.* transitory crosslink points. ω_p is displaced to lower frequency when the polymer concentration increases; in the same time, the value of G (where G'=G") at ω_p increases. For a perfect solution of linear polymer, shifts along the two axes allow to get a master curve (all the points are on the same reference curve). This was clearly obtained for xanthan [72] and for hyaluronan [8, 64]. It is interesting to mention that, for perfectly soluble polymers, the Cox Merz transformation [73] is valid allowing to compare the complex dynamic viscosity as a function of ω and the steady state flow viscosity as a function of the shear rate [8, 9, 72]. With interacting systems, the complex dynamic viscosity values are higher than the flow viscosity.

D- Mechanisms of Gelation

In many cases, polysaccharides are able to form physical and reversible gels. Their formation is based on the thermodynamic characteristics of the systems in relation with their chemical structure. The different mechanisms were discussed previously [74].

D-1. Gel made of a Single Polymer

It can be recognized three main types of gels when a single polysaccharide is involved [74].

D-1.1. Ionic Interactions:

Some gels are stabilized by cooperative interaction with calcium counterions as in pectins with low degree of methylation and in alginates. The role of the nature of the saccharidic unit and its configuration are very specific and involved in the calcium complex based on guluronic acid in alginate as well as galacturonic acid in pectins. The model proposed is called "egg-box model"; the gels can be destroyed in the presence of excess of monovalent counterions (by ion exchange mechanism) and addition of complexing agents (oxalate, citrate) [75]. In the following figure, the activity coefficients of calcium counterions as a function of the DP of different oligosaccharides are given: (1) oligomannuronates as found in alginates; (2) oligogalactuconates as found in pectins; (3) oligoguluronate as found in alginates too (see Fig. **22**) [76].

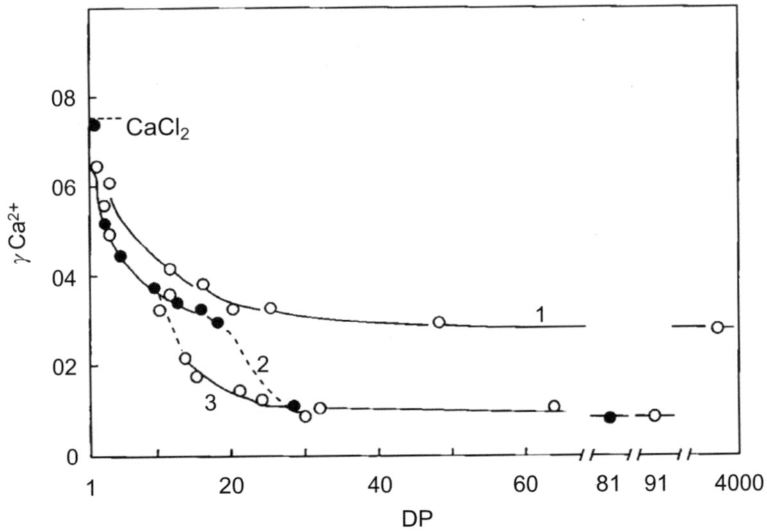

Fig. (22). Activity coefficient of calcium counterions for some oligouronates as a function of the polymerization degree (DP) [75a]. Reproduced from *Pure and Applied Chemistry*, vol.*42*, no 3, pp.371-397, 1975 with permission of IUPAC.

This figure demonstrates that a critical DP between 10 and 20 characterizes oligoguluronate and galacturonate where dimers are cooperatively formed as the first step of gelation in alginate and low methoxyl-pectins respectively. In these cases, the gelation is favoured by the presence of blocks of guluronic and galacturonic units respectively. The model given in Fig. (**23**) shows the complex formed in presence of calcium involving the dimer formation.

The activities of different divalent counterions were investigated on partially hydrolysed pectins [77, 78]. These results confirm the formation of dimers with Ca^{2+}, Ba^{2+} or Sr^{2+} in dilute solution followed by gelation. With Mg^{2+} counterions, no dimer nor gel is formed (Fig. **24**). In addition, using different techniques, the role of the carboxylic sites distribution (from comparison of blockwise distribution obtained by pectin esterase hydrolysis and random distribution obtained by alkaline hydrolysis) was also demonstrated on gelation [77 - 79].

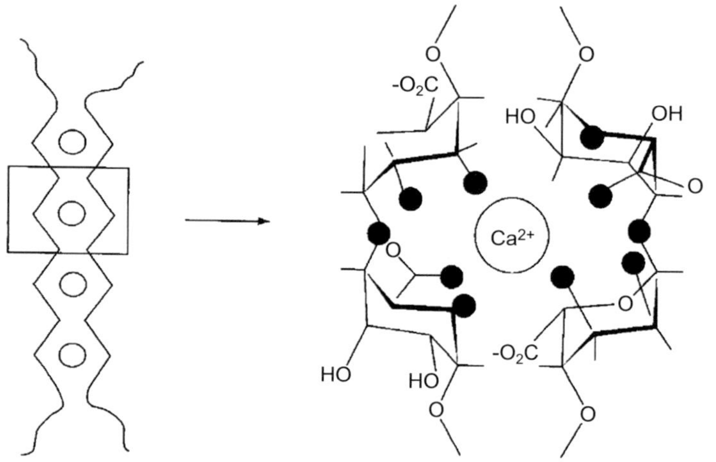

Fig. (23). Specific interaction of calcium with oligouronic acid in the egg-box model and dimer formation [75b]. Reproduced from *Biomacromolecules* vol.2, pp.1089-1096, 2001 wih the permission of American Chemical Society.

Ability to form gel can be tested from viscosity or light scattering experiments in dilute solution. A pseudo percolation point is observed for progressive addition of divalent counterions in the polymeric solution as shown in Fig. (**24**) [79].

In this case, the apparent degree of chain association is expressed by the ratio between the actual increase of light scattered $(I-I_0)$ for each divalent counterion addition and the initial light scattered $(I-I_0)_K$ (I_0 being the intensity scattered by the solvent). Ionic selectivity for gelation is clearly shown with Ba>Sr>Ca. A new copolymer forming gel in presence of calcium was recently prepared. Poly(HEMAm)-graft-(1-4)-α-L-guluronan consists of a synthetic flexible backbone grafted with well-defined oligoguluronic acids (DP=20). Advantage of this biohybrid glycopolymer is to form a softer gel compared to those obtained with alginate [80, 81] (Fig. **25**).

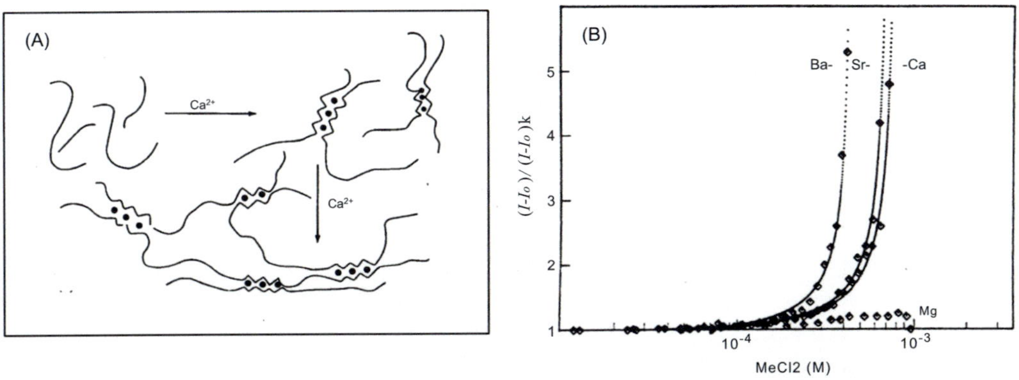

Fig. (24). (**A**) Progressive association of pectin chains in the presence of calcium counterions [74]. Reproduced from *J. Intell. Mater. Syst. Struct.,* vol.4, pp.210-215, 1993 with the permission of SAGE Publications. (**B**) Variation of light intensity scattered by potassium pectate solution $(I-I_0)_K$ during divalent counterion addition (degree of pectin esterification < 2%; concentration Cp=0.15 g/L) [79]. Reproduced from *Br. Polym. J.*, vol.17, no 2, pp.181-184, 1985 with permission of John Wiley and Sons.

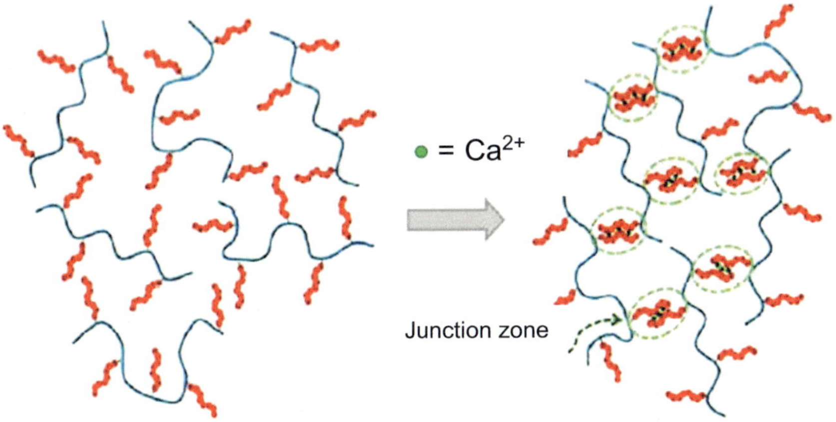

Fig. (25). Mechanism of ionotropic gelation of poly(HEMAm-g-GulA20) [80]. Reproduced from *Biomacromolecules*, vol.13, pp.3108-3119, 2012 with the permission of American Chemical society.

D-1.2. H-bond Interactions:

They are able to stabilize gels as found in highly methylated pectins in acidic medium in presence of saccharose, in carrageenans, or in gellan. These gels form at low temperature and melt on heating. A mechanism was proposed for gellan and carrageenan based on a two steps process: double helices are formed which associate to give connected aggregates. A schematic representation of the phase diagram corresponding to carrageenan and gellan is given in Fig. (**26**) [74].

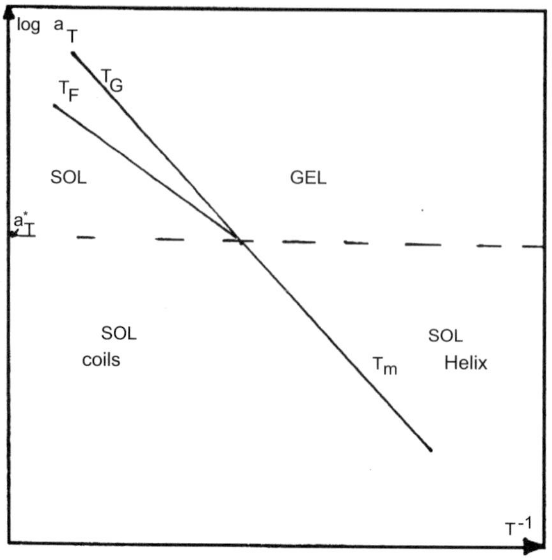

Fig. (26). Schematic phase diagram relating the total ionic concentration (a_T) and the inverse of T (K) in log-log scale for gellan or carrageenan. The straight line gives Tm, the temperature for helix-coil transition, T_G is the temperature for gelation and T_F the gel melting temperature [74]. Reproduced from *J. Intell. Mater. Syst. Struct.*, vol.4, pp.210-215, 1993 with permission of SAGE Productions.

This phase diagram shows the influence of the total ionic concentration (including external salt and polyelectrolyte concentrations) and temperature on the structure: i) the sol domain of coiled chains (at higher temperatures and low ionic concentration); ii) isolated double helices at lower temperature and ionic concentration lower than C*; iii) the gel forms in the domain over a critical ionic concentration C*. Over this concentration C*, a hysteresis exists between melting and formation of the gel (from coil conformation at high temperature, gel appears at lower temperature in the same time as helices are formed).

On gellan and carrageenan, differential scanning calorimetry (DSC) gives one peak on cooling but two on heating; these two peaks correspond successively to the melting of isolated (or loosely associated) helices followed by the melting of the more stable aggregates of helices [2, 82]. The same mechanism was also demonstrates on carrageenan by optical rotary power and conductivity analysis [53]. Induction of gel is related to the nature of the counterions: for gellan and κ-carrageenan, K^+ is known to promote gelation better than Na^+ and Li^+. The phase diagram obtained for κ-carrageenan (λ_h= 1.65), was given in Fig. (**9**): gel with a hysteresis in temperature appears at a lower KCl salt concentration (C*= $7x10^{-3}$ equiv./L) than in presence of NaCl (C*= $2x10^{-1}$ equiv/L). For gellan which has a lower charge parameter (λ_h= 0.75), no ionic selectivity exists for the helix-coil transition with monovalent counterions but it occurs for the sol-gel transition [10,

82, 83]. The critical ionic concentrations for gelation are C*= $2x10^{-2}$ equiv/L under the K^+ form and 4.5 $x10^{-2}$ equiv/L under Na+ form. The ionic selectivity discussed is the same as that obtained for the helix-coil transition especially for κ-carrageenan [84]. The counterions also play a role on the mechanical rigidity of the gels [1, 10].

In relation with the mechanism involved, these gels are often more rigid than that obtained by chemical cross-linkage of flexible synthetic polymers; the swelling and deflating are also very different from chemically crosslinked gels [85, 86]. It was shown that κ-carrageenan gel is nearly not sensitive to non-solvent environment (ethanol or acetone). On the contrary, ι-carrageenan which has a higher charge parameter gives a softer gel which deflats in presence of ethanol in relation with lower degree of chain association [85].

Gels are also formed with highly methylated pectins [87], hyaluronan [9] or alginates in acidic medium [88]. Cooperative H-bond network stabilizes thermoreversible gels. Fig. (**27**) gives an example of influence of the pH (and consequently of the net charge) on the rheological behavior of alginates at a given ratio mannuronic/guluronic units (M/G).

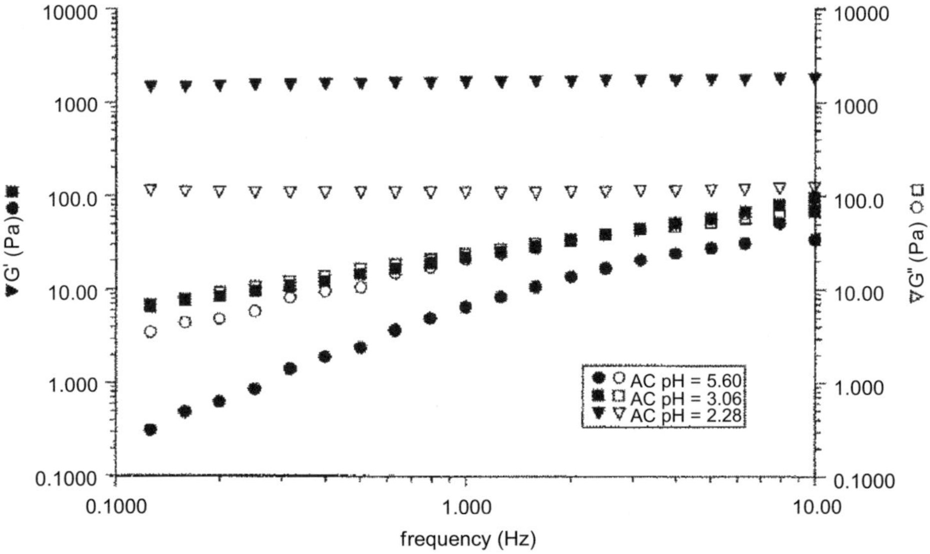

Fig. (27). Influence of pH on the dynamic rheological behavior of a 20 g L^{-1} alginate sample (M/G=1.3) solution in 0.15 mol L^{-1} NaCl at 25 °C [88]. Reproduced from *Polym. Int., vol.* 59, no 11, pp.1531-1541, 2010 with the permission of John Wiley and Sons.

At pH = 5.60, the rheological behavior is that of a solution. When pH decreases down to 3.05, the critical behavior is obtained with G'=G" independently of the frequency (corresponding to the Winter and Chambon criteria discussed later) [89]. At pH=2.28, one gets G'>G" and both are nearly independent on the frequency characterizing a gel behavior. Gelation of stereoregular ionic polysaccharides occurs in many cases (example was also obtained with hyaluronan).

It is necessary to mention that co-ions also play a role on the gelation and physical properties of carrageenan gels [90, 91]. With this polymer, the iodide was recognized to associate specifically preventing gelation [92].

D.1.3. Hydrophobic Interactions:

A third type of gel concerns thermo-inducible gels which form on temperature increase. It is the case of amphiphilic polymers as well known for commercial methylcelluloses. These polymers are characterized by a low critical temperature (LCST) around 30°C [93] as shown in the phase diagram (Fig. **28**) [59].

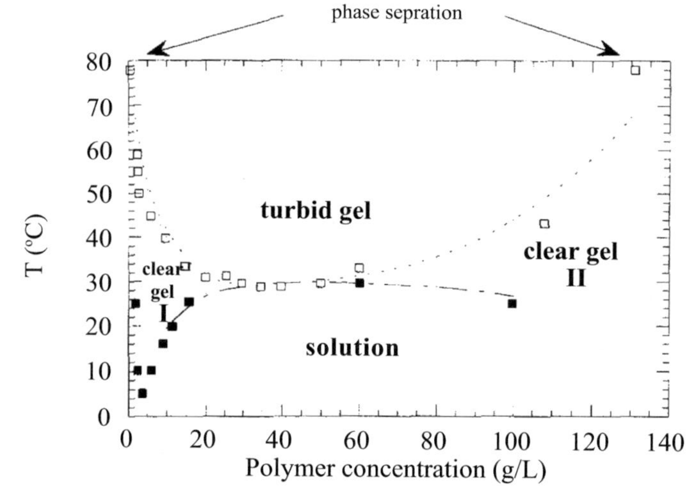

Fig. (28). Phase diapram of an aqueous A4C methylcellulose (DS=1.7) solution including the cloud point (□) and the sol-gel line (■) determined from oscillatory measurements [59]. Reproduced from *Polymer*, vol.39, no 25, pp.6251-6259, 1998 with the permission of Elsevier.

The gelation of methylcelluloses is followed easily by rheology. Strong gel is formed over 60°C. In fact, it was shown that the mechanism consists of two steps: when temperature increases, association of highly methylated zones occurs (it is the clear gel or pre-gel over 35°C for a degree of substitution DS=1.8) followed by a phase separation over around 60°C and formation of a strong turbid gel (Fig. **29**) [6, 94]. This figure shows that over around 30°C, it comes G'>G" in all the range of temperature covered.

Alkylchitosans were obtained by controlled random chemical modification of chitosan in homogeneous phase. They are obtained by reductive amination corresponding to the reaction of chitosan $–NH_2$ with aldehyde having different alkyl chains. For alkylated chains with a number of $–CH_2-$ groups higher than eight, thermogelation is obtained in aqueous medium in presence of external salt to repress the electrostatic repulsion between positively charged chitosan chains in acidic medium [95]. The association between alkyl side chains is looser than association of blocks in methylcellulose. The same type of modification was also performed on hyaluronan allowing to produce new viscosupplements with interesting rheological behavior [96].

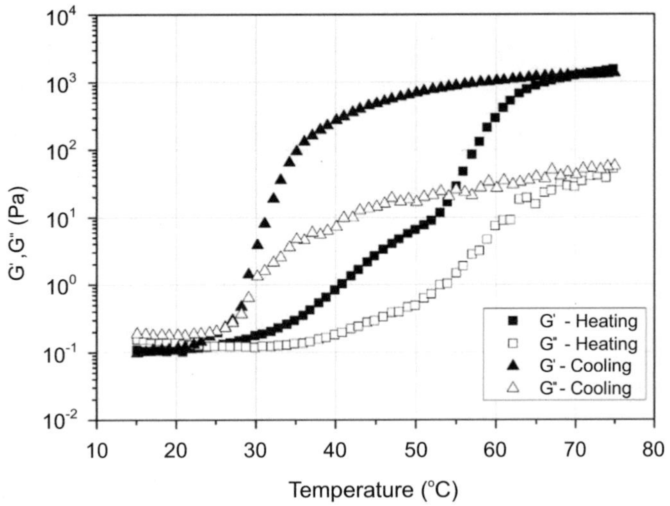

Fig. (29). Temperature dependence of dynamic viscoelasticity for methylcellulose A4M at 10 g/L in aqueous solution. Constant frenquency of 0.5 Hz, strain of 1% and temperature rate of 0.5 °C/min. ■,□ heating; ▲,Δ cooling. G' full symbols and G" open symbols [94]. Reproduced from *Inter. J. Polym. Anal. Ch.*, vol.20, pp.1–9, 2015 [22]with the permission of Taylor & Francis.

To conclude, all the physical gels are stiff when they are based on junction zones involving cooperative interactions. In alginate, it needs blocks of guluronic acids, in low methoxy pectins, blocks of galacturonic acids, in methylcellulose, blocks of trimethylglucoses. Looser gels are obtained when the association is based on short side chains regularly grafted along polymeric chains as in alkylchitosans or poly (HEMAm-g-GulA20).

D-2. Two Polymers- based Gels

The mixture of polysaccharides or polysaccharides- proteins was discussed by Morris [97]. The mechanisms involved are often difficult to establish.

Nevertheless, these systems have extensive food applications.

D-2.1. Two non-gelling Polymers

Networks are formed by cooperative specific interactions between the two water soluble polymers. A well-known case is the mixture galactomannan/xanthan or synergistic gelling system [98 - 101]. It was proved that the interaction requests the helical conformation of xanthan allowing cooperative H-bonds interaction. In addition, deacetylation of xanthan favored gel formation. Such mixed gel was named coupled network by Cairns *et al.* [102]. Another type of binary gel was found when mixing two chitosan derivatives: chitosan grafted with hydrophobic group (adamantane) and chitosan grafted with cyclodextrin. Specific interaction occurs by encapsulation of adamantane in the cyclodextrin cavity [103].

A more popular type of gels are obtained in some conditions when polycation is mixed with polyanion. It is named electrostatic polyelectrolyte complexes (PEC). The mixing often goes to precipitation but, in some experimental conditions (pH, external salt), one gets a gel. Such gels are formed between polysaccharide-polysaccharide and for polysaccharide-proteins. A good example is when gelatin at a pH~3.9 (lower than the isoelectric pH) is mixed with alginate [97, 104, 105]. The main applications of these electrostatic complexes are: antithrombogenic materials, controlled release systems, encapsulation of drugs, immobilization of enzymes and cells, and gene carriers (with DNA and chitosan for example) [106].

D-2.2 Two Gelling Polymers

An original interpenetrated network for a mixture of two independent gelling polymers was studied using the agarose/carrageenan mixture [107]. In such system, it was shown that melting of κ-carrageenan is independent of that of agarose. The melting of agarose is nearly independent on the ionic concentration but that of carrageenan is modified by the total counterion concentration.

Another system is the alginate - pectin mixed gels. These two polymers separately form gels in acidic conditions. But alginate with a high content of poly-L-guluronate forms firm gels in the presence of pectins with a high degree of methyl ester at low pH (lower than 3.6, the pH being decreased in this case with glucono-γ-lactone) [97]. These gels are generally thermoreversible and the temperature of melting increases when the pH decreases if they are produced by the cold set procedure. The nature of the interaction has to be established but it seems that a heterologous association forms between specific sequences of the two polymers.

These two polymers are also able to form mixed gels in presence of calcium. The 2-fold structure of the specific blocks (guluronic and galacturonic blocks) should allow formation of sterically heterologous junctions [108, 109].

D-2.3. One Gelling and One non-gelling Polymer

Phase separated network occurring when phase separation (or polymer exclusion) of the two polymers is followed by gelation. It occurs when the interactions between the two polymers are less favorable than those between like segments of each polymer [97]. Then, as most frequently occurring, the incompatibility causes an exclusion of each polymer from the other with an increase of the effective concentration of both polymers. A biphasic gel is obtained. An example is given with gelatin in presence of the flexible dextran. Presence of dextran accelerates conversion of the disordered sol state to the gel formation involving the formation of the stiff triple helical gelatin [105, 110]. Same types of gel are formed in gelatin-maltodextrins and gelatin-soluble starch [97].

E- Sol-gel Transition and Rheological Characterization of Gels

E-1. Sol-gel Transition

Rheology is a convenient technique for gel characterization. Nevertheless, it is relatively difficult to analyse the results correctly but it gives good informations in terms of applications. Gels are viscoelastic systems having both elastic (solid) and viscous (liquid) properties [111 - 113]. Physical gels as those formed on the basis of interacting zones are stabilized by H-bonds, ionic or hydrophobic interactions. They are usually stiffer than chemically crosslinked gel as schematized in Fig. (**30**). The term "gel" is often discussed as in reference [114] and relatively difficult to characterize conveniently.

The behavior of chemical gels (covalent linkages) is mainly entropic, based on rubber elasticity theory. On the opposite, physical gels are mainly enthalpic with a reversible character depending on the environment and slowly evolving with time.

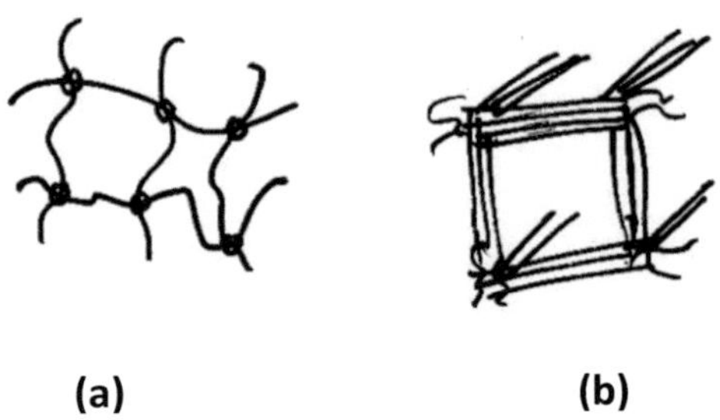

Fig. (30). Schematic representation of (**a**) chemically crosslinked gel and (**b**) physical gel.

The sol-gel transition conditions depend on temperature, time and polymer and/or salt concentrations. In gelling conditions, gels form when the polymer concentration is larger than a critical concentration (usually larger than the overlap concentration depending on the molar mass) needed to form a continuous gel. At lower concentration, microgels are formed.

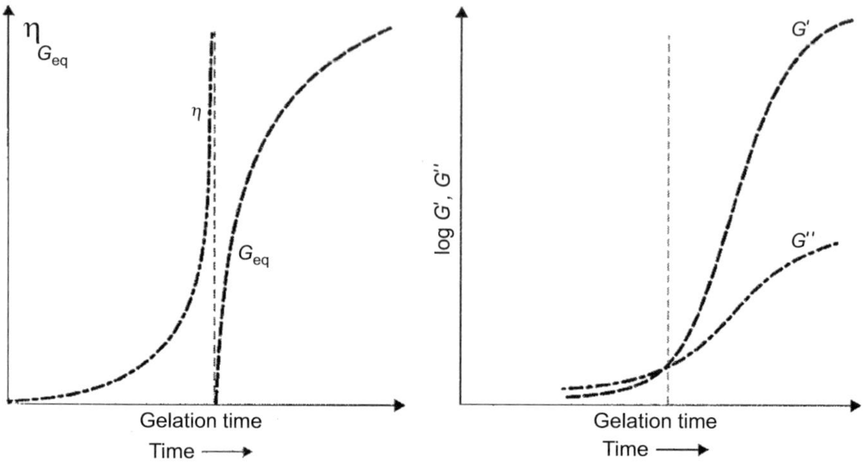

Fig. (31). (**A**) Time growth of equilibrium gel modulus Geq rises rapidly at/or just after the gelation time, before reaching a final asymptotic level. (**B**) Because of the finite frequency effect and the contribution of non-ideal network assembly contributions, both G' and G" will tend to rise before the true gelation point [112]. Reproduced from *Physical gels from Biological and synthetic polymers*, Cambridge, UK: Cambridge University Press, 2013 with the permission of Cambridge University Press, UK.

In Fig. (**31**), the determinations of gel time as a function of time at given temperature and at concentration larger than the critical concentration is shown. Firstly, from a solution, the viscosity increases sharply and after, corresponding to increase number of the crosslink points, the equilibrium gel modulus Geq (usually G' at low frequency) rises rapidly before reaching a final asymptotic level (Fig. **31A**). Secondly, the gel time is often determined at a given frequency, when the evolution of G' and G" are plotted as a function of time (Fig. **31B**). The time where G'=G" is named the gel time (it depends slightly on the frequency adopted).

The same type of behavior is obtained when i) the moduli are plotted as a function of the temperature to determine the temperature of gelation for gellan or carrageenan but it is needed to take into account time to attain equilibrium; ii) the moduli are plotted as a function of the number of divalent cations added in alginates.

The Winter and Chambon criteria is also used to determine the gel point: when G' and G" are plotted as a function of the frequency in log-log plot for increasing crosslinks formed, it occurs a particular case where G'(ω) and G''(ω) varies as $\omega^{0.5}$ and where G'(ω) = G''(ω) [89]. The 3-D networks are also analyzed in some case using the theory of percolation [112].

Physical measurements for sol-gel transition in gelling systems are mainly realized with oscillatory shear rheometry using commercial equipment with controlled stress using cone-plate or plate-plate geometry and temperature control. A sinusoidal oscillation of strain γ and frequency ω is applied to the sample in the linear regime. If the sample is elastic, the resultant stress wave is in phase with the strain wave independent on the frequency. This allows determining the elastic modulus G as the ratio of stress and strain. If the material is liquid, the sinusoidal stress wave is 90° out of phase with the imposed strain. For viscoelastic material, the stress wave has a frequency-dependent phase difference δ or tan δ reflecting the ratio of viscous/elastic responses (tan δ = G'/G'') at a given frequency, ω. The rheological behavior is characterized by the general figure giving G' and G'' as a function of ω in log-log plot (see Fig. **21** and Fig. **27**) determined in the domain where the moduli are independent on the strain applied (*i.e.* linear viscoelastic region LVR). For a gel, in LVR domain, constant moduli values with G'>G'' are usually obtained at all frequencies. For larger strain, the moduli decrease due to fracture.

When strong gel is at equilibrium, G' and G'' become nearly independent on the frequency with G'>> G'' obtained from dynamic rheology. In some cases, a slow process occurs at rest with deflating of the gel and free water expulsion. This mechanism is named syneresis. It especially occurs with strong gels like κ-carrageenans at low temperature in presence of K+ counterions or methylcellulose at high temperature and concentration.

Self-standing gels may be casted in a cylindrical cell and compression at low rate with a tensile test machine allows to determine the elastic Young modulus E of gels [10, 88]. It is important to apply slowly low strains. For the stiff physical gels, water is expelled from the gel in case of large deformation changing the gel structure. For some softer self-standing gels, the deformation remains reversible even for large deformation. For many polysaccharide gels, it was established that E varies with the polymer concentration following the relation $E \sim C^2$ [10, 74].

E-2. Ionic Selectivity in Ionic Polymers

The mechanical properties of ionic gels (carrageenan or gellan) at equilibrium are related to the nature of the counterions. For κ-carrageenan, the ionic selectivity for sol-gel transition is the same as for the helix-coil transition. The role of the counterions was clearly demonstrated also with gellan. In this case, no selectivity among monovalent counterions was observed for the helix-coil transition but the ionic selectivity (Li<Na<K) controls the total counterion concentration needed for gelation (as schematized in Fig. **26**) as well as the elastic modulus E of the gels formed at constant ionic concentration and polymer concentration (Table **3**).

Table 3. Elastic modulus (E x 10^{-4}Pa) of gellan gel in presence of different counterions at 0.1M external salt concentration. Polymer concentration 13g/L [10].

KCl	NaCl	LiCl
31.4	11.9	0.44

E-3. Influence of MW and Ionic Concentration

On alginate and carrageenan, it was shown that the gel elastic modulus increases rapidly when molar mass increases up to around 2x10^5 and goes to a plateau. The critical MW for gel formation depends slightly on the polymer concentration. The data are given in Figs. (**32** and **33**).

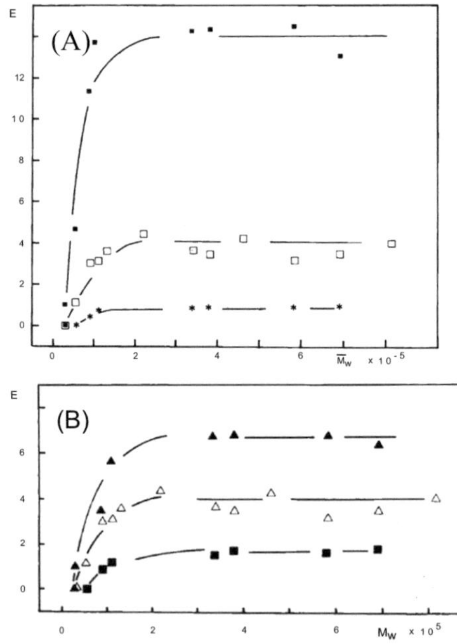

Fig. (32). Elastic modulus (x10^{-4} Pa) as a function of Mw: (**A**) for κ-carrageenan in 0.1M KCl at (★) 5 g/L; (□) 10 g/L; (■) 20 g/L and (**B**) for κ-carrageenan at 10 g/L in different KCl concentrations: 0.05M (■); 0.1M (Δ) and 0.5M (▲) [115]. Reproduced from *Carbohydr. Polym.*, vol.12, pp. 255-266, 1990 with the permission of Elsevier.

The elastic modulus of the κ-carrageenan gel also depends strongly on polymer concentration and on ionic concentration of the medium in which the gel is formed [115]. When E is plotted as a function of the polymer concentration in

KCl, in a log-log plot, the slope remains around 2 but the modulus increases when the external salt concentration KCl increases at a given polymer concentration.

The alginate sample used has a high content in guluronic acid (75% or F_G=0.75) and guluronic unit blocks with an average block length obtained by NMR indicated by $N_{G>1}$= 17.5 (Fig. **33**). In these conditions, stronger gels are obtained in presence of calcium when polymer concentration increases. In addition, it is clear that the critical concentration for alginates gel formation in presence of calcium depends also on the molar mass of the polymer (which controls the overlap concentration C*) [116]. Extrapolation to critical conditions for gel formation corresponds to C[η]~1.5.

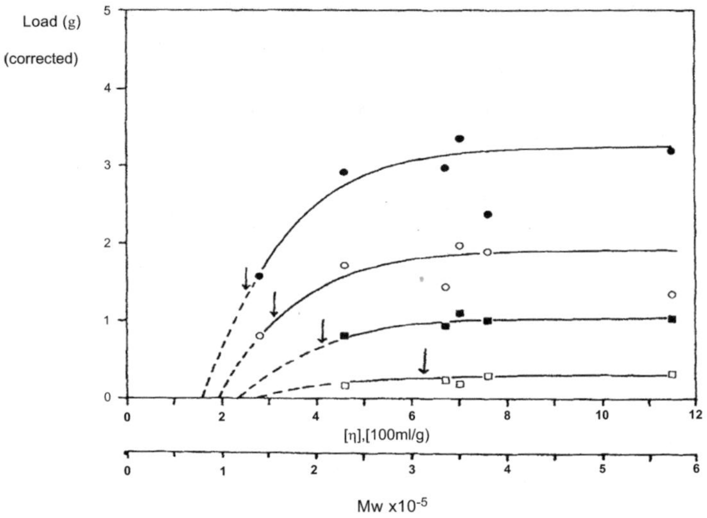

Fig. (33). Influence of the molar mass (Mw), intrinsic viscosity [η] and polymer concentration on the gel strength for alginates from L. hyperborea (outer cortex). F_G=0.75, $N_{G>1}$= 17.5. Concentrations: (•) 1%; (o) 0.8%; (■) 0.6%; (□) 0.4% [116]. Reproduced from *Biotechnol. Bioeng.* vol.33, no 1, pp.79-89, 1989 with the permission of John Wiley and sons.

CONCLUSIONS

The objective of this chapter was to give the main properties and characteristics of water soluble biopolymers, especially polysaccharides. We presented the polyelectrolyte characteristics and how we can use the polyelectrolyte behavior to determine the conformation of these polymers. It is especially important with stereoregular bacterial polysaccharides where helical conformations are frequent. Nuclear magnetic resonance is very important to get information on the chemical structure of biopolymers but also on the conformation. The techniques adopted to determine the macromolecular parameters have been described. It is clear that

SEC with 3 detectors on line is a very powerful technique to determine the molar mass, the molar mass distribution, the persistence length and the relationship parameters between intrinsic viscosity or radius of gyration and the molar mass. In addition, rheology is essential to test the behavior in connection with the applications as thickeners or gelling polymers.

Polysaccharides also give different types of physical gels stabilized by different cooperative interactions. These gels are often rigid and behave differently from synthetic cross-linked polymers. The enthalpic contribution is large in such physical gels but entropic for chemical gels. Ionic selectivity and stereoregularity in gel forming ionic polysaccharides are very important to control the ability to induce helix formation and consequently gelation and to impose their physical properties. The different mechanisms involved in gelling systems need to be examined to better understand their behavior and their domains of applications. The physical gelation is usually directly related with the chemical structure and concentration of the polymer involved; it depends strongly on the thermodynamic conditions (pH, temperature, nature and concentration of ions in solution).

CONFLICT OF INTEREST

The author (editor) declares no conflict of interest, financial or otherwise.

ACKNOWLEDGEMENTS

Declare none.

REFERENCES

[1] M. Rinaudo, "Advances in Characterization of Polysaccharides in Aqueous Solution and Gel State", In: *Polysaccharides: Structural Diversity and Functional Versatility,* S. Dimitriu, Ed., 2nd ed Marcel Dekker: New-York, USA, 2005, pp. 237-252.

[2] M. Rinaudo, "Relation between the Molecular Structure of Some Polysaccharides and Original Properties in Sol and Gel States", *Food Hydrocoll.,* vol. 15, pp. 433-440, 2001. [http://dx.doi.org/10.1016/S0268-005X(01)00041-8]

[3] M. Rinaudo, "Main properties and current applications of some polysaccharides as biomaterials", *Polym. Int.,* vol. 57, no. 3, pp. 397-430, 2008. [http://dx.doi.org/10.1002/pi.2378]

[4] M. Rinaudo, "Physical properties of chitosan and derivatives in sol and gel states", In: *Chitosan-Based Systems for Biopharmaceuticals: Delivery, Targeting and Polymer Therapeutics,* B. Sarmento, J. das Neves, Eds., J. Wiley and Sons: New-Jersey, USA, 2014, pp. 63-80. [http://dx.doi.org/10.1002/9781119962977.ch2]

[5] I. Younes, and M. Rinaudo, "Chitin and chitosan preparation from marine sources. Structure, properties and applications", *Mar. Drugs,* vol. 13, no. 3, pp. 1133-1174, 2015. [http://dx.doi.org/10.3390/md13031133] [PMID: 25738328]

[6] P.L. Nasatto, F. Pignon, J.L. Silveira, M.E. Duarte, M.D. Noseda, and M. Rinaudo, *Methylcellulose, a cellulose derivative with original physical properties and extended applications*vol. 7. Polymers, 2015,

pp. 777-803.

[7] M. Rinaudo, "Non-covalent interactions in polysaccharide systems", *Macromol. Biosci.*, vol. 6, no. 8, pp. 590-610, 2006. [http://dx.doi.org/10.1002/mabi.200600053] [PMID: 16881089]

[8] N. Berriaud, M. Milas, and M. Rinaudo, "Characterization and Properties of Hyaluronic Acid (Hyaluronan)", In: *Polysaccharides: Structural Diversity and Functional Versatility,* S. Dumitriu, Ed., Marcel Dekker: New-York, USA, 1998, pp. 313-334.

[9] M. Milas, and M. Rinaudo, "Characterization and Properties of Hyaluronic Acid (Hyaluronan)", In: *Polysaccharides: Structural Diversity and Functional Versatility, 2nd,* S. Dimitriu, Ed., Marcel Dekker: New-York, USA, 2005, pp. 535-549.

[10] M. Milas, and M. Rinaudo, "The Gellan Sol-Gel Transition", *Carbohydr. Polym.*, vol. 30, pp. 177-184, 1996. [http://dx.doi.org/10.1016/S0144-8617(96)00090-2]

[11] M. Milas, and M. Rinaudo, "On the Electrostatic Interactions of Ionic Polysaccharides in Solution", *Curr. Trends Polym. Sci.*, vol. 2, pp. 47-67, 1997.

[12] M. Rinaudo, "Polyelectrolyte Properties of a Plant and Animal Polysaccharide", *Struct. Chem.*, vol. 20, no. 2, pp. 277-289, 2009. [http://dx.doi.org/10.1007/s11224-009-9426-z]

[13] A. Katchalsky, Z. Alexandrowicz, and O. Kedem, "Polyelectrolyte solutions", In: *Chemical Physics of ionic solutions,* B.E. Conway, R.G. Barradas, Eds., Wiley & sons: New York, USA, 1966, pp. 295-346.

[14] G.S. Manning, "Limiting laws for equilibrium and transport properties of polyelectrolyte solutions", In: *Polyelectrolytes, E.; Selegny,* M. Mandel, U.P. Strauss, Eds., Reidel.Publishing Company: Dordrecht, Holland, 1974, pp. 9-37. [http://dx.doi.org/10.1007/978-94-010-2185-2_2]

[15] G.S. Manning, "Limiting laws and counterions condensation in polyelectrolyte solutions.I. Colligative properties", *J. Chem. Phys.*, vol. 51, pp. 924-933, 1969. [http://dx.doi.org/10.1063/1.1672157]

[16] M. Rinaudo, "Comparison between experimental results obtained with hydroxylated polyacids and some theoretical models", In: *Polyelectrolytes,* E. Selegny, M. Mandel, U.P. Strauss, Eds., Reidel publishing Company: Dordrecht, Holland, 1974, pp. 157-193. [http://dx.doi.org/10.1007/978-94-010-2185-2_10]

[17] M. Rinaudo, and A. Domard, "Preparation and characterization of I -L-glutamic acid oligomers", *Biopolymers,* vol. 14, pp. 2035-2048, 1975. [http://dx.doi.org/10.1002/bip.1975.360141005]

[18] G. Ravanat, and M. Rinaudo, "Investigation on oligo- and polygalacturonic acids by potentiometry and circular dichroism", *Biopolymers,* vol. 19, pp. 2209-2222, 1980. [http://dx.doi.org/10.1002/bip.1980.360191206]

[19] J.F. Thibault, and M. Rinaudo, "Gelation of Pectinic Acids in The Presence of Calcium Counterions", *Br. Polym. J.*, vol. 17, no. 2, pp. 181-184, 1985. [http://dx.doi.org/10.1002/pi.4980170217]

[20] M. Milas, X. Shi, and M. Rinaudo, "On the physicochemical properties of gellan gum", *Biopolymers,* vol. 30, no. 3-4, pp. 451-464, 1990. [http://dx.doi.org/10.1002/bip.360300322] [PMID: 2279074]

[21] C. Rochas, and M. Rinaudo, "Activity coefficients of counterions and conformation in κ-carrageenan systems", *Biopolymers,* vol. 19, pp. 1675-1687, 1980. [http://dx.doi.org/10.1002/bip.1980.360190911]

[22] R. Chandrasekaran, R.P. Millane, S. Arnott, and E.D. Atkins, "The crystal structure of gellan", *Carbohydr. Res.,* vol. 181, pp. 1-15, 1988. [http://dx.doi.org/10.1016/0008-6215(88)80151-4]

[23] M. Rinaudo, and M. Milas, "Polyelectrolyte behavior of bacterial polysaccharide from Xanthomonas campestris. Comparison with carboxymethylcellulose", *Biopolymers,* vol. 17, no. 11, pp. 2663-2678, 1978. [http://dx.doi.org/10.1002/bip.1978.360171113]

[24] C. Rochas, M. Rinaudo, and M. Vincendon, "Structural and conformational investigation of carrageenans", *Biopolymers,* vol. 19, pp. 2165-2175, 1980. [http://dx.doi.org/10.1002/bip.1980.360191203]

[25] M. Rinaudo, J. Mazet, and M. Milas, "PropriA(c)tA(c)s thermodynamiques des polyanions A densitA(c) de charge variable A", C.R. Acad. Sci", *Ser. C,* vol. 276, pp. 1401-1404, 1973.

[26] A. Katchalsky, and Z. Alexandrowicz, "On the additivity of osmotic properties of polyelectrolyte- salt solution", *J.Polym. Sci. Part A,* vol. 1, pp. 2093-2099, 1963.

[27] Z. Alexandrowicz, and A. Katchalsky, "Colligative properties of polyelectrolyte solution in excess of salt", *J.Polym. Sci. Part A,* vol. 1, pp. 3231-3260, 1963.

[28] E. Fouissac, *"Contribution à l'obtention d'acide hyaluronique par voie fermentaire et étude de ses propriétés physico-chimiques"*, PhD. thesis, The University of Grenoble, France, 1992.

[29] M. Rinaudo, G. Pavlov, and J. Desbrieres, "Solubilization of chitosan in strong acid medium", *Int. J. Polym. Anal. Charact.,* vol. 5, pp. 267-276, 1999. [http://dx.doi.org/10.1080/10236669908009742]

[30] G. Gravanis, M. Milas, M. Rinaudo, and A.J. Clarke-Sturman, "Conformational transition and polyelectrolyte behaviour of a succinoglycan polysaccharide", *Int. J. Biol. Macromol.,* vol. 12, no. 3, pp. 195-200, 1990. [http://dx.doi.org/10.1016/0141-8130(90)90032-6] [PMID: 2271490]

[31] M. Rinaudo, M. Milas, F. Lambert, and M. Vincendon, "Proton and carbon-13 NMR investigation of xanthan gum", *Macromolecules,* vol. 16, pp. 816-819, 1983. [http://dx.doi.org/10.1021/ma00239a018]

[32] C. Rochas, M. Rinaudo, and M. Vincendon, "Spectroscopic characterization and conformation of oligo I -carrageenans", *Int. J. Biol. Macromol.,* vol. 5, pp. 111-116, 1983. [http://dx.doi.org/10.1016/0141-8130(83)90021-1]

[33] K. Haxaire, E. Buhler, M. Milas, S. Perez, and M. Rinaudo, "Predictive and experimental behavior of hyaluronan in solution and solide state", In: *Hyaluronan, vol. 1: Chemical, biochemical and biological aspects,* J.F. Kennedy, G.O. Phillips, P.A. Williams, V.C. Hascall, Eds., Woodhead Pub.: Cambridge (England), 2002, pp. 37-46.

[34] H. Benoit, and P. Doty, "Light Scattering from Non-Gaussian Chains", *J. Phys. Chem.,* vol. 57, no. 9, pp. 958-963, 1953. [http://dx.doi.org/10.1021/j150510a025]

[35] T. Odijk, "Polyelectrolytes near the rod limit", *J. Polym. Sci., Polym. Phys. Ed,* vol. 15, pp. 477-483, 1977. [http://dx.doi.org/10.1002/pol.1977.180150307]

[36] T. Odijk, "On the ionic-strength dependence of the intrinsic viscosity of DNA", *Biopolymers,* vol. 18, no. 12, pp. 3111-3113, 1979. [http://dx.doi.org/10.1002/bip.1979.360181215] [PMID: 518964]

[37] W. Reed, "Light-scattering results on polyelectrolyte conformations, diffusion and interparticle interactions and correlations", In: *Macro-ion characterization from dilute solutions to complex fluids, ACS Symp. Series 548,* K.S. Schmitz, Ed., American Chemical Society: Washington, DC, USA, 1994,

pp. 297-314.

[38] K. Mazeau, "S. Perez M. Rinaudo. A" Predicted Influence of N-Acetyl Group Content On The Conformational Extension Of Chitin And Chitosan Chains", *J. Carbohydr. Chem,* vol. 19, pp. 1269-1284, 2000.
[http://dx.doi.org/10.1080/07328300008544150]

[39] K. Haxaire, I. Braccini, M. Milas, M. Rinaudo, and S. Perez, "Conformational behavior of hyaluronan in relation to its physical properties as probed by molecular modeling", *Glycobiology,* vol. 10, no. 6, pp. 587-594, 2000.
[http://dx.doi.org/10.1093/glycob/10.6.587] [PMID: 10814700]

[40] C.L. Petkowicz, M. Rinaudo, M. Milas, K. Mazeau, T. Bresolin, F. Reicher, and J.L. Ganter, "Conformation of Galactomanan, Experimental and Modeling Approaches", *Food Hydrocoll,* vol. 13, pp. 263-266, 1999.
[http://dx.doi.org/10.1016/S0268-005X(99)00008-9]

[41] C. Petkowicz, F. Reicher, and K. Mazeau, "Conformational analysis of galactomannans: from oligomeric segments to polymeric chains", *Carbohydr. Polym.,* vol. 37, pp. 25-39, 1998.
[http://dx.doi.org/10.1016/S0144-8617(98)00051-4]

[42] J. Brugnerotto, J. Desbrieres, G. Roberts, and M. Rinaudo, "Characterization of chitosan by steric exclusion chromatography", *Polymer (Guildf.),* vol. 42, pp. 9921-9927, 2001.
[http://dx.doi.org/10.1016/S0032-3861(01)00557-2]

[43] E. Fouissac, M. Milas, M. Rinaudo, and R. Borsali, "Influence of the ionic strength on the dimensions of sodium hyaluronate", *Macromolecules,* vol. 25, pp. 5613-5617, 1992.
[http://dx.doi.org/10.1021/ma00047a009]

[44] M. Rinaudo, "The relation between the chemical structure of polysaccharides and their physical properties", In: *Gums and Stabilizers for the Food Industry 6,* G.O. Phillips, D.J. Wedlock, P.A. Williams, Eds., IRL Press at Oxford University Press: Oxford, UK, 1992, pp. 51-61.

[45] E. Balnois, S. Stoll, K.J. Wilkinson, J. Buffle, M. Rinaudo, and M. Milas, "Conformations of succinoglycan as observed by atomic force microscopy", *Macromolecules,* vol. 33, no. 20, pp. 7440-7447, 2000.
[http://dx.doi.org/10.1021/ma0002951]

[46] I. Morfin, W.F. Reed, M. Rinaudo, and R. Borsali, "Further evidence of liquid-like correlations in polyelectrolyte solutions", *J. Phys. II,* vol. 4, no. 6, pp. 1001-1019, 1994.
[http://dx.doi.org/10.1051/jp2:1994180]

[47] L. Chazeau, M. Milas, and M. Rinaudo, "Conformations of xanthan in solution: analysis by steric exclusion chromatography", *Int. J. Polym. Anal. Charact.,* vol. 2, pp. 21-29, 1995.
[http://dx.doi.org/10.1080/10236669508233892]

[48] A. Malovikova, M. Milas, M. Rinaudo, and R. Borsali, "Viscosimetric behavior of Na-polygalacturonate in the presence of low salt content", In: *Macro-ion characterization from dilute solutions to complex fluids, ACS Symp. Series 548,* K.S. Schmitz, Ed., American Chemical Society: Washington, DC, USA, 1994, pp. 315-321.

[49] I. Roure, M. Rinaudo, and M. Milas, "Viscometric Behavior of Dilute Polyelectrolytes. Role of Electrostatic Interactions", *Ber. Bunsenges. Phys. Chem,* vol. 100, pp. 703-706, 1996.
[http://dx.doi.org/10.1002/bbpc.19961000604]

[50] R. Borsali, M. Rinaudo, and L. Noirez, "Light scattering and small angle neutron scattering from polyelectrolyte solutions: the succinoglycan", *Macromolecules,* vol. 28, pp. 1085-1088, 1995.
[http://dx.doi.org/10.1021/ma00108a040]

[51] M. Milas, M. Rinaudo, R. Duplessix, R. Borsali, and P. Lindner, "Small angle neutron scattering from polyelectrolyte solutions: from disordered to ordered xanthan chain conformation", *Macromolecules,* vol. 28, pp. 3119-3124, 1995.

[http://dx.doi.org/10.1021/ma00113a014]

[52] M. Rinaudo, C. Rochas, and B. Michels, "Etude par absorption ultrasonore de la fixation sélective du potassium sur le carraghenane", *J. Chim. Phys.,* vol. 80, no. 3, pp. 305-308, 1983.

[53] M. Rinaudo, and C. Rochas, "Investigations on aqueous solution properties of κ-carrageenans Solution properties of polysaccharides", In: *Solution Properties of Polysaccharides, ACS Symp. Series 150,* D. Brant, Ed., American Chemical Society: Washington,DC (USA), 1981, pp. 367-378.

[54] M. Rinaudo, and M. Milas, "Proprietes electrochimiques des polyelectrolytes a densite de charge elevéee", *J. Chim. Phys.,* vol. 66, no. 9, pp. 1489-1496, 1969.

[55] R. Zana, C. Tondre, M. Rinaudo, and M. Milas, "Etude ultrasonore de la fixation sur site des ions alcalins de densite de charge variable", *J. Chim. Phys.,* vol. 68, pp. 1258-1260, 1971.

[56] M. Rinaudo, R. Auzely, and K. Mazeau, "Polysaccharides and carbohydrate polymers", In: *Encyclopedia of Polymer Science and Technology,* H.F. Mark, Ed., vol. 11. Wiley-Interscience: New York, USA, 2004, pp. 200-261.

[57] M. Rinaudo, and M. Milas, "Gellan gum, a bacterial gelling polymer", In: *Novel Macromolecules in Food Systems,* G. Doxastakis, V. Kiosseoglou, Eds., Elsevier Science B.V: Amsterdam, The Netherlands, 2000, pp. 239-263.
[http://dx.doi.org/10.1016/S0167-4501(00)80012-6]

[58] M. Hirrien, J. Desbrieres, and M. Rinaudo, "Physical properties of methylcelluloses in relation with the conditions for cellulose modification", *Carbohydr. Polym.,* vol. 31, pp. 243-252, 1996.
[http://dx.doi.org/10.1016/S0144-8617(96)00118-X]

[59] M. Hirrien, C. Chevillard, J. Desbrieres, M.A. Axelos, and M. Rinaudo, "Thermogelation of methylcelluloses: new evidence for understanding the gelation mechanism", *Polymer (Guildf.),* vol. 39, no. 25, pp. 6251-6259, 1998.
[http://dx.doi.org/10.1016/S0032-3861(98)00142-6]

[60] M. Rinaudo, I. Roure, and M. Milas, "Use of steric exclusion chromatography to characterize hyaluronan, a semi-rigid polysaccharide", *Int. J. Polym. Anal. Charact,* vol. 5, pp. 277-287, 1999.
[http://dx.doi.org/10.1080/10236669908009743]

[61] M. Rinaudo, and B. Tinland, "Some problems in aqueous size exclusion chromatography of synthetic polymers and biopolymers characterization", *J. Appl. Polym. Sci. Symp.,* vol. 48, pp. 19-31, 1991.
[http://dx.doi.org/10.1002/app.1991.070480003]

[62] P. Mathieu, T. Conrozier, E. Vignon, Y. Rozand, and M. Rinaudo, "Rheologic behavior of osteoarthritic synovial fluid after addition of hyaluronic acid: a pilot study", *Clin. Orthop. Relat. Res.,* vol. 467, no. 11, pp. 3002-3009, 2009.
[http://dx.doi.org/10.1007/s11999-009-0867-x] [PMID: 19418104]

[63] T. Conrozier, P. Mathieu, E. Vignon, M. Piperno, and M. Rinaudo, "Differences in the osteoarthritic synovial fluid composition and rheology between patients with or without flare: a pilot study", *Clin. Exp. Rheumatol.,* vol. 30, no. 5, pp. 729-734, 2012.
[PMID: 22935334]

[64] N. Berriaud, *"Contribution à l'étude des propriétés du hyaluronane: comportement rhéologique, hydratation"*, PhD thesis, The University of Grenoble, France, 1994.

[65] I. Roure, *"Rhéologie et dimensions des chaînes de hyaluronane en solution"*, PhD thesis, The University of Grenoble, France, 1997.

[66] M. Milas, M. Rinaudo, and B. Tinland, "The viscosity dependence on concentration, molecular weight and shear rate of xanthan solutions", *Polym. Bull.,* vol. 14, pp. 157-164, 1985.
[http://dx.doi.org/10.1007/BF00708475]

[67] E. Fouissac, M. Milas, and M. Rinaudo, "Shear-rate, concentration, molecular weight, and temperature viscosity dependences of hyaluronates, a wormlike polyelectrolyte", *Macromolecules,* vol. 26, pp.

6945-6951, 1993.
[http://dx.doi.org/10.1021/ma00077a036]

[68] T.K. Kwei, M. Nakazawa, S. Matsuoka, M.K. Cowman, and Y. Okamoto, "Concentration dependence of solution viscosities of rigid rod polymers", *Macromolecules,* vol. 33, no. 2, pp. 235-236, 2000. [http://dx.doi.org/10.1021/ma9912002]

[69] V.P. Kapoor, M. Milas, F.R. Taravel, and M. Rinaudo, "Rheological Proprties of a Seed Galactomannan from Cassia Siamea Lamk", *Food Hydrocoll.,* vol. 10, pp. 167-172, 1996. [http://dx.doi.org/10.1016/S0268-005X(96)80031-2]

[70] V.P. Kapoor, F.R. Taravel, J.P. Joseleau, M. Milas, H. Chanzy, and M. Rinaudo, "Cassia spectabilis DC seed galactomannan: structural, crystallographical and rheological studies", *Carbohydr. Res.,* vol. 306, no. 1-2, pp. 231-241, 1998.
[http://dx.doi.org/10.1016/S0008-6215(97)00241-3] [PMID: 9691448]

[71] J. Ganter, M. Milas, and M. Rinaudo, "Study of Solution Properties of Galactomannan from the Seeds of Mimosa Scabrella", *Carbohydr. Polym.,* vol. 17, pp. 171-175, 1992.
[http://dx.doi.org/10.1016/0144-8617(92)90001-7]

[72] M. Milas, M. Rinaudo, M. Knipper, and J.L. Schuppiser, "Flow and viscoelastic properties of xanthan gum solutions", *Macromolecules,* vol. 23, pp. 2506-2511, 1990.
[http://dx.doi.org/10.1021/ma00211a018]

[73] W.P. Cox, and E.H. Merz, "Correlation of dynamic and steady flow viscosities", *J. Polym. Sci., Polym. Phys. Ed.,* vol. 28, pp. 619-622, 1958.

[74] M. Rinaudo, "Gelation of Polysaccharides", *J. Intell. Mater. Syst. Struct.,* vol. 4, pp. 210-215, 1993. [http://dx.doi.org/10.1177/1045389X9300400210]

[75] G.T. a) Grant, E.R. Morris, D.A. Rees, P.J. Smith, and D. Thom, "Biological interactions between polysaccharides and divalent cations: the egg-box model", *FEBS Lett.,* vol. 32, no. 1, pp. 195-198, 1973.
[http://dx.doi.org/10.1016/0014-5793(73)80770-7] I. b) Braccini, and S. Perez, "Molecular basis of C(2+)-induced gelation in alginates and pectins: the egg-box model revisited", *Biomacromolecules,* vol. 2, no. 4, pp. 1089-1096, 2001.
[http://dx.doi.org/10.1021/bm010008g] [PMID: 11777378]

[76] R. Kohn, "Ion binding on polyuronates-alginate and pectin", *Pure Appl. Chem.,* vol. 42, no. 3, pp. 371-397, 1975.
[http://dx.doi.org/10.1351/pac197542030371]

[77] M. a) Rinaudo, "Seaweed polysaccharides", In: *Comprehensive Glycoscience, 2.21: Polysaccharide Functional Properties,* J.P. Kamerling, Ed., Elsevier B.V., 2007, pp. 691-735.
[http://dx.doi.org/10.1016/B978-044451967-2/00140-9] J.F. b) Thibault, and M. Rinaudo, "Interactions of Mono-And Divalent Counterions with Alcali- and Enzyme-Deesterified Pectins in Salt-Free Solutions", *Biopolymers,* vol. 24, pp. 2131-2134, 1985.
[http://dx.doi.org/10.1002/bip.360241109]

[78] J.F. Thibault, and M. Rinaudo, "Chain association of pectic molecules during calcium-induced gelation", *Biopolymers,* vol. 25, pp. 455-468, 1986.
[http://dx.doi.org/10.1002/bip.360250306]

[79] J.F. Thibault, and M. Rinaudo, "Gelation of pectinic acids in the presence of calcium counterions", *Br. Polym. J.,* vol. 17, no. 2, pp. 181-184, 1985.
[http://dx.doi.org/10.1002/pi.4980170217]

[80] A. Ghadban, L. Albertin, M. Rinaudo, and A. Heyraud, "Biohybrid glycopolymer capable of ionotropic gelation", *Biomacromolecules,* vol. 13, no. 10, pp. 3108-3119, 2012.
[http://dx.doi.org/10.1021/bm300925j] [PMID: 22928921]

[81] A. Ghadban, E. Reynaud, M. Rinaudo, and L. Albertin, "RAFT copolymerization of alginate-derived

macromonomers ?" synthesis of a well-defined poly(HEMAm)-graft-(1 +'4)-I -L-guluronan copolymer capable of ionotropic gelation", *Polym. Chem.,* vol. 4, pp. 4578-4585, 2013. [http://dx.doi.org/10.1039/c3py00730h]

[82] F. Mazen, M. Milas, and M. Rinaudo, "Conformational transition of native and modified gellan", *Int. J. Biol. Macromol.,* vol. 26, no. 2-3, pp. 109-118, 1999. [http://dx.doi.org/10.1016/S0141-8130(99)00070-7] [PMID: 10517517]

[83] E.R. Morris, K. Nishinari, and M. Rinaudo, "Gelation of Gellan. A Review", *Food Hydrocoll.,* vol. 28, pp. 373-411, 2012. [http://dx.doi.org/10.1016/j.foodhyd.2012.01.004]

[84] C. Rochas, and M. Rinaudo, "Mechanism of gel formation in I -carrageenan", *Biopolymers,* vol. 23, pp. 735-745, 1984. [http://dx.doi.org/10.1002/bip.360230412]

[85] M. Rinaudo, and S. Landry, "On the volume change on non-covalent gels in solvent-non solvent mixtures", *Polym. Bull.,* vol. 17, pp. 563-565, 1987. [http://dx.doi.org/10.1007/BF00283094]

[86] C. Rochas, M. Rinaudo, and S. Landry, "Relation between the molecular structure and mechanical properties of carrageenan gels", *Carbohydr. Polym.,* vol. 10, pp. 115-127, 1989. [http://dx.doi.org/10.1016/0144-8617(89)90061-1]

[87] M. Rinaudo, "Physicochemical properties of pectins in solution and gel states", In: *Pectins and Pectinases, Progress in Biotechnology 14,* J. Visser, A.G. Voragen, Eds., Elsevier: Amsterdam, The Netherland, 1996, pp. 21-33. [http://dx.doi.org/10.1016/S0921-0423(96)80243-7]

[88] H. Andriamanantoanina, and M. Rinaudo, "Relationship between the molecular structure of alginates and their gelation in acidic conditions", *Polym. Int.,* vol. 59, no. 11, pp. 1531-1541, 2010. [http://dx.doi.org/10.1002/pi.2943]

[89] H.H. Winter, and F.J. Chambon, "Analysis of linear viscoelasticity of a crosslinking polymer at the gel point", *J. Rheol. (N.Y.N.Y.),* vol. 30, pp. 367-382, 1986. [http://dx.doi.org/10.1122/1.549853]

[90] M. Ciancia, M. Milas, and M. Rinaudo, "On the specific role of coions and counterions on I - carrageenan conformation", *Int. J. Biol. Macromol.,* vol. 20, no. 1, pp. 35-41, 1997. [http://dx.doi.org/10.1016/S0141-8130(97)01149-5] [PMID: 9110183]

[91] I.T. Norton, E.R. Morris, and D.A. Rees, "Lyotropic effects of simple anions on the conformation and interactions of kappa-carrageenan", *Carbohydr. Res.,* vol. 134, no. 1, pp. 89-101, 1984. [http://dx.doi.org/10.1016/0008-6215(84)85025-9]

[92] H. Grasdalen, and O. Smidsrod, "Iodide-specific formation of k-carrageenan single helices. 127 I NMR spectroscopic evidence for selective site binding of iodide anions in the ordered conformation", *Macromolecules,* vol. 14, pp. 1842-1845, 1981. [http://dx.doi.org/10.1021/ma50007a051]

[93] C. Chevillard, and M.A. Axelos, "Phase separation of aqueous solution of methylcellulose", *Colloid Polym. Sci.,* vol. 275, no. 6, pp. 537-545, 1997. [http://dx.doi.org/10.1007/s003960050116]

[94] P.L. Nasatto, F. Pignon, J.L. Silveira, M.E. Duarte, M.D. Noseda, and M. Rinaudo, "Influence of Molar Mass and Concentration on the Thermogelation of Methylcelluloses", *Inter. J. Polym. Anal. Ch.,* vol. 20, pp. 110-118, 2015. [http://dx.doi.org/10.1080/1023666X.2015.973155]

[95] M. Rinaudo, R. Auzely, C. Vallin, and I. Mullagaliev, "Specific interactions in modified chitosan systems", *Biomacromolecules,* vol. 6, no. 5, pp. 2396-2407, 2005. [http://dx.doi.org/10.1021/bm0580025] [PMID: 16153074]

[96] M. Rinaudo, R. Auzely, S. Kadi, A. Bresin, and E. Kubik, New derivatives of hyaluronic acid, their preparation process and their uses, 2007.WO2007059890

[97] E.R. Morris, "Mixed Polymer Gels", In: *Food Gels,* P. Harris, Ed., vol. 8. Elsevier Applied Science: London, UK, 1990, pp. 291-359. [http://dx.doi.org/10.1007/978-94-009-0755-3_8]

[98] F.M. Goycoolea, M. Milas, and M. Rinaudo, "Heterotypic interactions of deacetylated xanthan with a galactomannan of high galactose substitution during synergistic gelation", In: *Gums and Stabilisers for the Food Industry 10,* P.A. Williams, G.O. Phillips, Eds., Royal Society of Chemistry: Cambridge, UK, 2000, pp. 229-240. [http://dx.doi.org/10.1533/9781845698355.3.229]

[99] F.M. Goycoolea, M. Milas, and M. Rinaudo, "Associative phenomena in galactomannan-deacetylated xanthan systems", *Int. J. Biol. Macromol.,* vol. 29, no. 3, pp. 181-192, 2001. [http://dx.doi.org/10.1016/S0141-8130(01)00164-7] [PMID: 11589971]

[100] M. Rinaudo, M. Milas, T. Bresolin, and J. Ganter, "Physical properties of xanthan galactromannan and their mixtures in aqueous solutions", *Macromol. Chem. Phys.,* vol. 140, pp. 115-124, 1999.

[101] T.M. Bresolin, M. Milas, M. Rinaudo, and J.L. Ganter, "Xanthan-galactomannan interactions as related to xanthan conformations", *Int. J. Biol. Macromol.,* vol. 23, no. 4, pp. 263-275, 1998. [http://dx.doi.org/10.1016/S0141-8130(98)00061-0] [PMID: 9849624]

[102] P. Cairns, M.J. Miles, V.J. Morris, and G.J. Brownsey, "X-ray fibre-diffraction studies of synergistic, binary polysaccharide gels", *Carbohydr. Res.,* vol. 160, pp. 411-423, 1987. [http://dx.doi.org/10.1016/0008-6215(87)80327-0]

[103] R. Auzely-Velty, and M. Rinaudo, "New supramolecular assemblies of a cyclodextrin grafted chitosan through specific complexation", *Macromolecules,* vol. 35, no. 21, pp. 7955-7962, 2002. [http://dx.doi.org/10.1021/ma020664o]

[104] V.B. Tolstoguzov, "Functional properties of protein-polysaccharide mixtures", In: *Functional Properties of Food Macromolecules,* J.R. Mitchell, D.A. Ledward, Eds., Elsevier Applied Science Pub.: London, UK, 1986, pp. 385-415.

[105] J.C. Harrington, and E.R. Morris, "Conformational ordering and gelation of gelatin in mixtures with soluble polysaccharides", *Food Hydrocoll.,* vol. 23, pp. 327-336, 2009. [http://dx.doi.org/10.1016/j.foodhyd.2008.03.003]

[106] P. Erbacher, S. Zou, T. Bettinger, A.M. Steffan, and J.S. Remy, "Chitosan-based vector/DNA complexes for gene delivery: biophysical characteristics and transfection ability", *Pharm. Res.,* vol. 15, no. 9, pp. 1332-1339, 1998. [http://dx.doi.org/10.1023/A:1011981000671] [PMID: 9755882]

[107] J. Zhang, and C. Rochas, "Interactions between agarose and k-carrageenan in aqueous solution", *Carbohydr. Polym.,* vol. 13, pp. 257-271, 1990. [http://dx.doi.org/10.1016/0144-8617(90)90058-Z]

[108] D. Thom, I.C. Dea, E.R. Morris, and D.A. Powell, "Interchain associations of alginate and pectins", *Prog. Food Nutr. Sci.,* vol. 6, pp. 97-108, 1982.

[109] K. Toft, H. Grasdalen, and 0. Smidsrod, "Synergistic gelation of alginates and pectins", In: *Chemistry and Function of Pectins. ACS Symposium Series 310,* M.L. Fishman, J.J. Jen, Eds., American Chemical Society: Washington DC (USA), 1986.

[110] V.B. Tolstoguzov, V.P. Belkina, V.J. Gulov, V.J. Grinberg, E.F. Titova, and E.M. Belavzev, "State of phase, structure and mechanical properties of the gelatinous system water-gelatin-dextran", *Starke,* vol. 26, pp. 130-138, 1974. [http://dx.doi.org/10.1002/star.19740260407]

[111] F. Tanaka, *Polymer Physics. Applications to molecular association and thermoreversible gelation.*

Cambridge University Press: Cambridge, UK, 2011.
[http://dx.doi.org/10.1017/CBO9780511975691]

[112] M. Djabourov, K. Nishinari, and S.B. Ross-Murphy, *Physical gels from Biological and synthetic polymers.* Cambridge University Press: Cambridge, UK, 2013.
[http://dx.doi.org/10.1017/CBO9781139024136]

[113] G.M. Kavanagh, and S.B. Ross-Murphy, "Rheological characterization of polymer gels", *Prog. Polym. Sci.,* vol. 23, pp. 533-562, 1998.
[http://dx.doi.org/10.1016/S0079-6700(97)00047-6]

[114] K. Almdal, J. Dyre, S. Hvidt, and O. Kramer, "Towards a phenomenological definition of the term ?~gel ?(tm)", *Polym. Gels Netw.,* vol. 1, no. 1, pp. 5-17, 1993.
[http://dx.doi.org/10.1016/0966-7822(93)90020-I]

[115] C. Rochas, M. Rinaudo, and S. Landry, "Role of the molecular weight on the mechanical properties of I -carrageenan gels", *Carbohydr. Polym.,* vol. 12, pp. 255-266, 1990.
[http://dx.doi.org/10.1016/0144-8617(90)90067-3]

[116] A. Martinsen, G. Skjåk-Braek, and O. SmidsrA,d, "Alginate as immobilization material: I. Correlation between chemical and physical properties of alginate gel beads", *Biotechnol. Bioeng.,* vol. 33, no. 1, pp. 79-89, 1989.
[http://dx.doi.org/10.1002/bit.260330111] [PMID: 18587846]

CHAPTER 9

Biopolymers from Mesquite Tree (*Prosopis* spp.)

Yolanda L. López-Franco*, **Alma R. Toledo-Guillén** and **Jaime Lizardi-Mendoza**

Biopolymers Group, Research Center for Food and Development, (C.I.A.D., A.C.), Hermosillo, Sonora 83304, Mexico

Abstract: The mesquite tree (*Prosopis* spp.) is a native plant species in arid zones that has been documented as a potential source of biopolymers, including polysaccharides, and more recently, proteins. The native flora of the Sonoran Desert offers considerable potential for the recovery of such compounds. Mesquite trees produce a water-soluble exudate known as mesquite gum (MG). Pods contain seeds with high levels of protein, and the endosperm yields a novel gum of the galactomannan family. MG and galactomannan (GM) are not yet considered food additives in the international market due to lack of FDA approval. However, structural, physicochemical and functional studies have shown that MG and GM are potential substitutes for commercial Arabic and guar gums, respectively. In addition, it was recently documented that mesquite seeds are a rich source of protein, with considerable quantities of essential amino acids. This chapter presents information about the structural and chemical characterization and applications of the biopolymers that can be obtained from *P. velutina.*

Keywords: Biopolymers, Galactomannan, Mesquite gum, Physico-chemical properties, Protein, Structural characteristics.

INTRODUCTION

Mesquite is the common name given to the trees of the *Prosopis* genus that are widespread in arid and semiarid regions of the world. The genus *Prosopis* grows mostly in the North America, South America, northern Africa and eastern Asia [1]. Mesquite is a multipurpose tree with the capacity to provide fuel, timber, fodder and edible pods, as well as biopolymers with characteristics similar to those currently used commercially.

The bark of *Prosopis* produces a viscous gum in the cavities of the xylem and

* **Corresponding author Yolanda L. López-Franco**: Biopolymers Group, Research Center for Food and Development. Carretera a La Victoria Km 0.6, Ej. La Victoria, Hermosillo, Sonora, 83304 Mexico; Tel: +52 (662) 289 2400 x364; Fax: +52 (662) 280 0421; E-mail: lopezf@ciad.mx

Martin Masuelli & Denis Renard (Eds.)

phloem as a response to insect attack, wounding or weather conditions (temperature). Drought, age, and health of the tree can affect gum exudation. Mesquite gum is highly soluble in aqueous media (above 50% w/w) [2] and insoluble in alcohol or ether. The most important physical properties of the gum are its light color, high water solubility and viscosity, lack of odor and slightly astringent taste. The composition of gum collected in Sonora, Mexico, is 10% moisture, 2.6% ash, 0.35% tannin, 0.57% total nitrogen, 3.7% protein, and 96% total sugars, which include 71% arabinose (Ara), 26% galactose (Gal) and 3% uronic acids (UAc) [3, 4]. The intrinsic viscosity of the mesquite gum was calculated to be $[\eta] = 1.47 \times 10^{-2}\ Mw^{0.50}$ ($ml.g^{-1}$) [5]. This composition may vary with the species and variety of mesquite.

The mesquite tree produces copious pods. Its productivity depends on the age of the tree and soil characteristics. Pods are manually collected by rural families for use as feed for small livestock. Pods are composed of an external fibrous exocarp (56 wt%) that surrounds the mesocarp or pulp (composed mainly of sugars); the endocarp (35 wt%), is fibrous but much harder than the exocarp. Inside the endocarp are seeds (9 wt%) with 0.04 g average weight, 0.99 cm length, 0.56 cm in width and 0.22 cm in thickness. The endosperm of mesquite seeds develops alongside the germ (cotyledon) and completely envelops it. The endosperm itself is protected by a seed coat. The hard compact endosperm contains galactomannans that can reach 88% of its weight. The *P. pallida* seed cotyledon is 65% protein [6], while cotyledons from *P. velutina* are 32% protein. This protein fraction is well-balanced in essential amino acids [7]. Therefore, the use of mesquite biopolymers should be considered a potential source of valuable commodities and rural employment in arid regions.

Structural studies are a basic requirement for understanding the functional properties of biopolymers such as polysaccharides and proteins. The physicochemical characteristics of these macromolecules are defined by molecular weight; monosaccharide composition and sequence; conformation, configuration and the position of glycoside linkages; particle size; solubility; size distribution; rheological behavior; gelling properties; and the surface and interfacial properties that are directly related to the structure. On the other hand, the chemical composition and molecular structure of biopolymers often depend on the source, extraction methods and any further processing conditions. The physicochemical and structural features of biopolymers obtained from mesquite, as well as their applications, are discussed throughout this chapter.

Mesquite Gum (Exudate from *Prosopis*)

Mesquite gum is a polysaccharide exudate obtained from the bark of *Prosopis* spp. trees (Fig. **1**). In rural areas of Sonora, Mexico, mesquite gum is collected mostly from wild *P. velutina* trees and is used in small amounts in various domestic applications, folk medicine and as a chewable sweet [8].

Fig. (1). Exudate from a mesquite tree (*Prosopis velutina*).

Chemical Structure

The chemical composition of mesquite gum was reported many years ago [9]–[14]. Mesquite gum (from *P. juliflora*) is a highly branched polysaccharide with L-Ara, D-Gal and 4-O-methyl-D-glucuronic acid residues. More recently, it was demonstrated that after partial acid hydrolysis, the inner backbone chains of *P. juliflora* gum contain D-Gal residues linked by 1→3 and/or 1→6 linkages with β-D-configuration and residues of glucuronic acid and 4-O-methyl-D-glucuronic acid bound to D-Gal by α-D-(1→4) and β-D-(1→6) linkages as the end groups [15, 16].

In addition, it was found that the acid-labile peripheral chains are complex oligosaccharides containing L-Ara. Apart from the carbohydrate component, mesquite gum contains 4% protein, and tannins (0.35%) have also been detected [4, 17, 18]. An arabinogalactan (AG) from mesquite gum (*P. velutina*) contains a galactan-core polysaccharide composed of glucuronic acid and D-Gal. Branching occurs by the attachment of L-Ara oligomers of as many as ten sugar residues to glucuronic acid, as revealed by the identification of GlcUA-Ara1-10 by MALDI-TOF MS. Small amounts of other sugars, namely D-mannose (Man), D-glucose (Glc) and D-xylose (Xyl), are present.

Hydrophobic Interaction Chromatography (HIC) showed that the gum consists of three main molecular mass fractions with varying protein content [3, 4]. The main component (Fraction I) is the arabinogalactan fraction (AG), which represents 94.77% of the gum and contains 2.79% protein. It has a molecular mass of 3.89×10^5, a radius of gyration (*Rg*) ~50 nm and a hydrodynamic radius (*Rh*) ~10 nm determined by static and dynamic light scattering. A second component is split into two sub-fractions (Fraction IIa and IIb) that are referred to as the constituents of an arabinogalactan-protein fraction (AGP). These sub-fractions represent ~2.5% and 1.17% of the total, contain 11% and 29% protein, have a molecular mass of 4.8 and 3.8×10^5, *Rg* values of 55 and 49 nm, and *Rh* values of 13 and 16 nm, respectively. The factor form was calculated for the AGP fraction after inspection of the average molecular weight and architecture (radius of gyration) data using a different model setting. The results showed that the AGP fraction was best fit by a polydisperse coil structure with a polydispersity index of 2.

These experimental data deviate from a globular structure, such as the wattle blossom-type structure typical of AGP, but are in agreement with the twisted hairy rope form of AGP from gum Arabic [3, 19, 20]. The third component (Fraction III), referred to as glycoprotein (GP), represents only ~1.6% of the total and contains 24% protein. The GP fraction has a molecular mass of 8.06×10^5, *Rg* 71 nm and *Rh* 24 nm. The neutral sugar content of the three components is similar (Ara, Gal, GlcUA, Glc). However, the AGP and GP fractions contain Man and Xyl [4].

The presence of AGP was determined by diffusion gel with the β-glucosyl Yariv reagent (1, 3, 5-tris [4-β-D-glucopyranosyloxyphenylazo]-2, 4, 6-trihydroxy-benzene) in fractions obtained by HIC [21]. Fraction IIa and Fraction IIb showed halos corresponding to AGP with concentrations of 2.63 and 0.53 mg/mL, respectively. Quantification was made with a standard curve of gum Arabic and in accordance with previous studies [3]. Amino acid profiles and lipid compounds in mesquite gum were determined. The amino acid profile indicated that mesquite gum has a high amount of hydroxyproline, serine, glycine and valine (values of 329.79, 245.30, 132.14 and 112.64 residues/1000 residues, respectively). Hydroxyproline was particularly noted because it is part of the protein fraction of the AGP. This amino acid was the most abundant amino acid in the gum (27.28%). With this amino acid composition information, *P. velutina* AGP can be considered a classical AGP [21]. Further structural investigation would be necessary to deepen understanding of the secondary structure of the AGP polypeptide component of mesquite gum, as has been reported for gum Arabic [20].

There is evidence that the AGPs from *Pyrus communis* and *Nicotiana alata* contain a glycosylphosphatidylinositol (GPI) lipid anchor [22, 23]. GPI components such as mannose, inositol, GlcN and lipids were reported in gum Arabic, but after extensive analysis and experimental argument, the authors were not able to confirm the existence of a GPI anchor in gum Arabic [24]. Mesquite gum contains mannose, glucose, xylose, short-chain fatty acids (C16:0, C18:0, C20:0), and a long-chain fatty acid (C22:0). More evidence is needed to confirm the hypothesis that some AGP in mesquite gum contain GPI lipid groups.

Mesquite gum also contains phenolic compounds (0.3-2%). These compounds have reported bioactive effects related to their antioxidant properties [25, 26]. Flavonoids and phenolic acids have numerous biological and pharmacological properties including analgesic, anti-inflammatory, anti-hypertensive, antibacterial and anticancer properties [27, 28]. Previous studies of polyphenol composition in mesquite gum detected flavonoids (apigenin and quercetin) and phenolic acid (ρ-coumaric acid); further confirmation of the presence and role of these compounds is underway.

Surface Properties and Microstructure

Surface property studies have been conducted to determine the surface activity of mesquite gum and its fractions [29] by subjecting Langmuir monolayers at the air-water interface to different surface pressures. These monolayers were transferred from the water surface to a solid substrate (*e.g.*, mica) to obtain Langmuir-Blodgett (LB) films for direct visualization by atomic force microscopy (AFM). Inspection of the LB films showed the formation of domains of mesquite molecules at 5, 10, 15 and 20 mN/m surface pressures, with different roughness and height for all fractions. AFM images of LB films of the AGP fraction showed the formation of molecular aggregates at different surface pressures. The presence of large domains in the 10 mN/m LB film could be caused by an incomplete molecular arrangement. In contrast, at higher surface pressures, homogeneous spherical structures (1.32 nm tall) appeared throughout the topographical image (Fig. **2**). In that case, amphiphilic gum molecule domains have already formed a monolayer at the air-water interface at the initial starting pressure, and along with protein penetration, these domains grow, coalesce and finally form a homogeneous layer.

As a consequence of its chemical structure, mesquite gum has interesting functional properties that have allowed its application in various fields. The functionality of mesquite gum is mainly due to its emulsifying and stabilizing emulsion capacity (o/w) [30 - 33] and its flavor encapsulation [8, 17, 34]. These functional properties have been detailed [35].

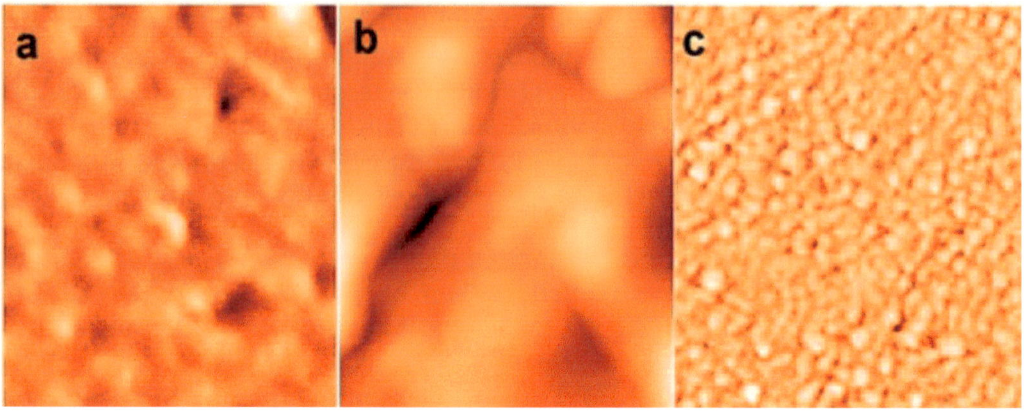

Fig. (2). AFM images from AGP fraction Langmuir-Blodgett films extracted at: a) 5 mN/m; b) 10 mN/m; and c) 20 mN/m surface pressure.

Applications

Mesquite gum has chemical, structural and functional properties similar to gum Arabic, the most studied polysaccharide exudate, which has many industrial applications. Comparison of the physicochemical characteristics and functional properties of both exudates determines that mesquite gum could be a substitute for gum Arabic in many applications [30 - 34]. However, regulatory agencies, such as the US Food and Drug Administration (FDA) or the Codex Committee on Food Additives (CCFA) in Europe, have not yet granted the "Generally Recognized As Safe" (GRAS) status to mesquite gum. In Mexico, its use has been approved as an adjuvant in processed foods [36, 37]. An adjuvant is defined as a *"substance or matter that is not consumed as a food ingredient by itself, and is intentionally used in the processing of raw materials, products or their ingredients to fulfill a technological purpose during treatment or processing"*. The approval of mesquite gum as an adjuvant in foods was given after conducting a multigenerational toxicological study in rats fed with a diet containing mesquite gum compared to a control diet with cellulose. The study showed no differences in the histopathological analysis of rats fed with either of the two diets [38].

The main objection to the use of mesquite gum in food is the presence of tannins (polyphenols) due to their potential toxicity, even though gum Arabic may have higher tannin content [18]. The tannin content of mesquite gum could be limited by simple drop classification or ultrafiltration processing. This technology removes approximately 70% of polyphenols and the obtained product has similar analytical and functional properties to gum Arabic [17, 18]. In addition, mesquite gum has been traditionally consumed as hard candy by generations of natives of the Sonora region (Mexico) without acute or chronic toxicity [39].

Several industrial applications of mesquite gum have been proposed due to its excellent functional properties as emulsifying, stabilizing or encapsulation agents in a wide variety of applications [35, 40, 41]. However, the mesquite gum supply is not constant, and gum yield and quality varies considerably by year due to variable mesquite growing conditions (*i.e.*, climate, age of trees, place of exudation and species). Another constraint is the lack of commercial plantations and efficient collection systems compared to those for gum Arabic in Sudan.

Galactomannans (Mesquite Seed Gum)

Mesquite trees produce elongated pods that are straight or curved and have a tan-brown color when mature. The pod weighs ca. 1.37 g, are 11.71 cm in length, 0.79 cm in width, 0.64 cm in thickness and there are 12 seeds per pod (n= 100) [42]. Pods contain protein (9-17%), fat (1.2-4.3%), sugar (30-48%), calcium (0.5%), phosphorus (0.44%) and dietary fiber (16-34%) [1, 6]. Pods are manually collected by rural families and used to feed small livestock.

Physical Characteristics of Mesquite Pods

Pods have three components: the fibrous exocarp (56 wt%), that surrounds the mesocarp or pulp, the main constituents of which are sugar (48%), crude fiber (16%), fat (2.0%) and protein (9.5%). The endocarp (35 wt%), which is also fibrous but much harder than the exocarp, contains sugar (6%), protein (6.5%), fat (1.3%), and crude fiber (45%); and the seed (9% wt%), with protein (32%), sugar (53%), fat (7.9%) and crude fiber (11%) [1, 6]. These fractions of the total pod weight can vary by species.

Physical Characteristics from Mesquite Seeds

The seeds are released from endocarp with a mill with a particular geometry, and the freed seeds are separated from the rest of the pods using sieves. Table **1** shows the chemical composition and physical characteristic of mesquite seeds, which are similar to those of other *Prosopis* seeds. Protein content (32%) [42] was similar to *P. ruscifolia* (33.81%) [43] but lower than other legumes, such as soybean (*Glycine max*) (36.7%) [44].

Table 1. Chemical composition and physical characteristics of mesquite seeds (*P. velutina*).

Chemical composition		Dimensions		Color	
Moisture (%)	3.37 ± 0.20	Weight** (g)	4.11 ± 0.03	*L*	21.46 ± 1.63

(Table 1) contd.....

Chemical composition		Dimensions		Color	
Ash (%)	3.16 ± 0.04	Length (cm)	0.99 ± 0.15	*a*	4.34 ± 0.61
Fat (%)	7.94 ± 0.29	Width (cm)	0.56 ± 0.10	*b*	17.80 ± 1.15
Protein (%)	32.12 ± 0.52	Thickness (cm)	0.22 ± 0.07	*c*	18.51 ± 1.25
Carbohydrates* (%)	53.41			*h*	75.96 ± 1.72

All measurements are on a dry-weight basis except for moisture.
Data are the means of triplicate analysis ± SD
* Determined by difference
**Weight from 100 seeds

The *L* value (21.46) indicates low brightness or a dark appearance. The values of *a* (4.34) and *b* (17.80) were positive, showing a tendency toward red and yellow, respectively. According to the hue angle (*h*), the seed was brown, with partial saturation according to the *c* parameter.

Seeds are composed of a hull, cotyledon and endosperm (Fig. **3**). The endosperm contains water-soluble galactomannan polysaccharides, which are among the most important plant-derived gums used in the food industry. They can be used as emulsifiers, texture modifiers or stabilizers in food products. The chemical composition, fine structure and physical and functional properties of these polysaccharides can vary significantly with the plant source, environment, and method of production.

There are four major sources of galactomannans of commercial interest: guar plant (*Cyamopsis tetragonoloba)*, carob tree (*Ceratonia siliqua*), tara shrub (*Caesalpinia spinose*) and fenugreek plant (*Trigonella foenum-graecum*). Mesquite (*Prosopis velutina*) is another legume that supplies galactomannan and has commercial potential. The endosperm of guar, carob, tara and fenugreek seeds is a source of galactomannan gums. Guar and carob (locust bean) gums are widely used, but tara and fenugreek are limited by availability and price. Information about these gums is reported by Weilinga [45]. Use of mesquite galactomannan is still in the research phase, especially in countries where *Prosopis* grows widely, to find similar routes of exploitation of galactomannans for use in the food industry.

Galactomannan Extraction Process

Extraction of galactomannans requires removal of seeds using a mill that allows separation of the seeds from the pods. The mill breaks the pods, and by means of vibrating sifters of different sizes, pod fractions (mesocarp, endocarp and seeds) are sorted. Air separation is required to obtain clean seeds.

Galactomannans can be extracted from whole seed or endosperm. Seeds are first roasted, grounded and defatted. Galactomannan is obtained by alcoholic precipitation. The disadvantage of this simple process is that the gum has somewhat lower quality and lower viscosity, as well as a light yellow color due to small amounts of residual germ particles. Alternatively, the endosperm can be extracted from the seeds after removing the hull through a process called "*peeling*" [45]. Briefly, the seed hull is removed with concentrated sulfuric acid at high temperature. This facilitates the separation of the germ from the endosperm, and thus galactomannans with high viscosities can be obtained. The disadvantage of this process is the acid effluent problem [45]. A similar technic using alkali (1 M NaOH) was developed to separate the endosperm from the seeds from mesquite (*P. pallida*) [46]. Cerqueira and co-workers [47] have reported other methodologies for the extraction of galactomannans from *Adenanthera pavonina, Caesalpinia pulcherrima and Gleditsia triacanthos*. Seeds were mechanically broken and endosperm was manually separated from the germ and the hull. The endosperm was suspended in ethanol at 70 °C for 15 min, the ethanol was decanted, water was added and the suspension was left to rest for 24 hours. Subsequently, water was added and the suspension was mixed for 5 min. This suspension was filtered through a nylon net and centrifuged for 20 min at 20 °C. The hydrocolloid was precipitated with absolute ethanol, the ethanol was decanted and the extracted polysaccharide was freeze dried.

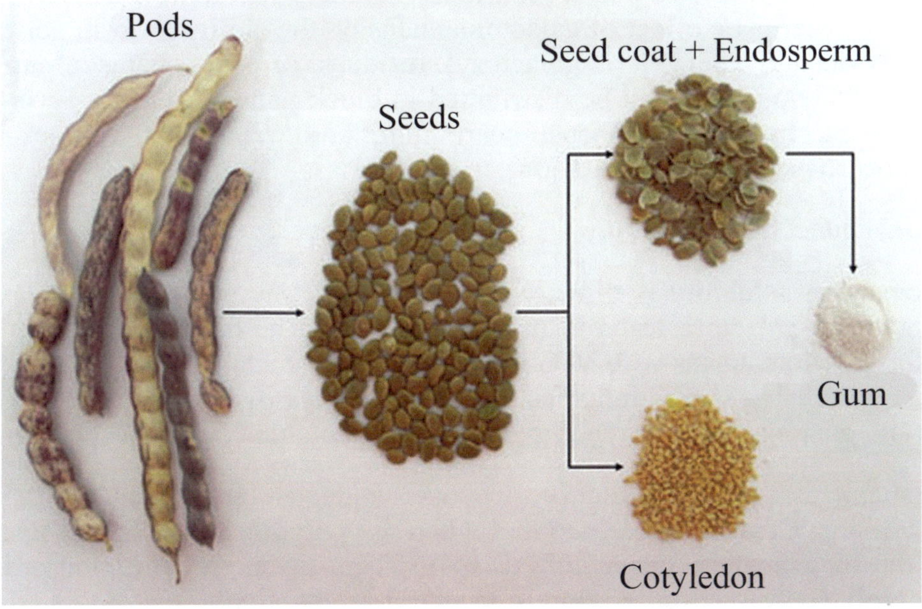

Fig. (3). Pods, seeds, seed components and galactomannan from mesquite (*Prosopis* spp.).

In the literature, many studies report the separation of endosperm and extraction of polysaccharides from numerous legume seeds. In the case of mesquite seed, the procedure is as follows. Mesquite seeds are broken using a household coffee mill. The endosperm is manually separated from the germ and the hull and suspended in double-distilled water at 25 °C for 24 h. After filtration through layers of gauze, polysaccharides are recovered by precipitation with absolute ethanol, freeze dried and ground. This product is suspended in distilled water at 25 °C, centrifuged and freeze dried [48]. The extraction yields and physicochemical characteristics of galactomannans depend on the extraction and purification method.

Chemical Structure

To better understand the physicochemical and functional properties of galactomannans, their chemical structure should be determined. Galactomannans are polysaccharides composed of a β-(1-4)-D-mannan backbone with a single D-Gal branch with a α-(1-6) linkage. The mannose-galactose (M/G) ratio depends on the source; for locust bean, guar, tara and mesquite seed gums, M/G ratios of 3.5, 1.5, 3, and 1.1-1.5, respectively, have been reported [48 - 51]. The distribution of D-galactosyl residues over the mannan backbone is important for the functional properties of polysaccharides, including rheological behavior [52, 53] and the synergistic effect of galactomannans on the ability to form gels with other polysaccharides [49]. D-galactosyl residues or side chains along the mannose backbone chain can be distributed in block patterns, as in guar and *P. juliflora* gum. In tara and locust bean gum, both random and block-wise distributions have been observed [54].

Physicochemical Characteristics

Galactomannan solubility is directly related to M/G ratios, and an increase in substitution by galactose leads to higher solubility. The highest solubility was found for galactomannans with M/G ratios of 1:1, in which high substitution with galactose establishes steric hindrance that helps it to dissolve easily in cold water. As a result, guar and mesquite seed gums are more soluble than locust bean gum.

Galactomannans produce highly viscous aqueous solutions at 0.5-2% concentrations. Cruz [6] explained that when the polysaccharide is dissolved in water, the long mannan chain unfolds to form an open, flexible, non-ordered random coil conformation. At low concentration, the coils are separated and viscosity increases linearly with concentration. This behavior was also reported by Estévez and co-workers [55] in galactomannans from *P. chilensis* (Mol) Stuntz. Furthermore, when the effect of pH on viscosity was studied, no differences were detected at pH 3.0, 4.0, or 5.0. Table **2** compares the main physicochemical characteristics of galactomannans from *Prosopis* and commercial seeds.

Galactomannans from *Prosopis* are more substituted over the mannan backbone in comparison to guar gum. There are also differences in intrinsic viscosity and molecular weight between polysaccharides, but they have polydispersities similar to macromolecules of botanical origin.

One of the most important functional features of galactomannans is the control of solution and dispersion rheology, which is attributed to their high molecular weight, chain entanglement and polysaccharide solvent interactions. Galactomannans have the ability to form very viscous solutions at relatively low concentrations that are only slightly affected by pH, ionic strength and heat processing. Viscosity of galactomannans tends to remain constant over a broad pH range (1-10.5) and ionic strength, mainly due the neutral character of their molecules. Galactomannan solutions usually exhibit non-Newtonian behavior, in which viscosity decreases with an increase in shear rate [59].

Table 2. Physicochemical characteristics of galactomannans from *Prosopis* and commercial seeds.

Sample	M/G ratio	[η] (dL/g)	Mv (g/mol)	Reference
P. pallida	1.4	10.20	600,000-1,800,000	[6]
	1.1	--	900,000-1,000,000	[46]
P. juliflora	1.7	11.78	1,140,000	[56]
P. velutina	1.5	12.36	1,200,000	[48]
P. glandulosa	2.0	11.0	1,500,000	[57]
C. siliqua	3.2	15.20	2,030,000	[58]
C. tetragonoloba	1.4	13.31	1,400,000	[48]

Synergistic Gelification

The ability of galactomannans to form synergistic gels with other polysaccharides (*e.g.*, xanthan, carrageenan) has been studied. This characteristic has been attributed only to low-substitution galactomannans (*e.g.*, locust bean gum). The mechanism in these inter-polymer systems is the direct association of unsubstituted mannan chains with the surfaces of ordered xanthan helices [51]. In these systems, gel formation occurs at 50-60 °C, independent of salt concentration and xanthan conformation. In contrast, galactomannans with substitutions similar to guar gum have the capacity to form gels with xanthan under more favorable conditions, as the disordered conformation occurs at 25 °C [49, 51]. The synergistic gelification of deacetylated xanthan-locust bean gum mixtures and xanthan-mesquite (*P. pallida and P. juliflora*) seed gum mixtures was studied [51]. The results confirm the interaction between xanthan and galactomannans with low M/G ratios using a mechanism with three hallmarks: a transition with a lower thermal stability close to room temperature; high cooperativity and large enthalpy; and an optimum stoichiometry (w/w) for galactomannan/xanthan.

The possible interaction between galactomannan from other *Prosopis* species with xanthan, agar or carrageenan to form gels is still a challenge. More research is necessary to increase understanding of the role of galactose substitution on the synergistic gelification of these polysaccharides.

Applications

Galactomannans have a broad range of applications as thickeners, stabilizers in emulsion and suspension systems, or emulsifiers and gelling agents in gel-based products [60]. They have also been used in food, cosmetic and pharmaceutical products as a matrix for tablets [61, 62], edible coating materials [63], and microsphere-forming agents for target-specific drug release to the colon [64]. Galactomannans are also an important source of dietary fiber, which may exhibit bioactivities such as controlling blood glucose and insulin levels and reducing the risks of cardiovascular disease and colon cancer [65]. Other applications of galactomannans are found in remediation processes to remove petroleum hydrocarbon diesel fractions (TPH-diesel) from oil-contaminated volcanic soils [66]. In this study, guar gum at 2 ppm was more effective than locust bean gum in removing up to 82% of TPH-diesel. The possibility of producing chemically modified derivatives makes these polysaccharides versatile biomaterials that can be employed in many biomedical applications, both alone and in composites and blends. This opens new research opportunities for mesquite seed gum, which has not been used in industrial applications.

Mesquite Seed Protein

Proteins are complex organic nitrogen substances found in animal and plant cells. They are linear polymers of amino acids that fold into a remarkable diversity of three-dimensional forms that provide a corresponding variety of functions [67]. Based on their functions in the cell, plant proteins can be classified into three main groups: structural proteins, proteins with biological activity and reserve storage proteins [68]. Storage proteins, present only in seeds, accumulate in large amounts and are quickly hydrolyzed to release amino acids. These proteins are a nitrogen source for the early stages of plant germination [69]. In legume seeds, storage proteins are dominant [70].

Proteins are important food ingredients due to their functional properties, such as emulsification, solubility, foaming properties, water- and oil-absorption capacities and gelling properties. These properties give proteins a wide range of applications. The most frequently used seed proteins are from sunflower, pea and soybean due to their nutritional value and functional properties [71]. However, their production, especially in the case of soy, is expensive. There is a constant search for alternative protein sources for use in different food systems, primarily to increase nutritional quality and provide functional properties such as texture, structural features or flavor. One notable alternative protein source is the legume seed.

Legume seeds are considered a good source of protein and are usually low-cost and widely accepted. They constitute the second most important food group for humans and animals after cereals, with twice the level of protein found in cereal grains [72]. Despite the relatively high cost of protein from meat, eggs and dairy products, their intake is 10-fold greater than that of legume proteins, even though legumes have more abundant protein (20-40%). Some species of *Prosopis* seeds have up to 60% protein content [73]. *P. velutina* seeds could be considered an unconventional protein source, as this species has high protein content that is well-balanced in essential amino acids [7]. This represents an opportunity to study mesquite seed protein to enhance its utilization for human food consumption and increase the market value of the seed.

Chemical Composition

Proteins from legumes differ in physicochemical and structural characteristics that affect their functional properties and applications in the food industry. Fractionation and characterization of mesquite seed proteins have been reported [7]. Mesquite-seed protein content was 32.12%, higher than of *P. farcta* (18.45%) [74]; *P. africana* (20.54%) [75]; and *P. chilensis* (22.39%) [76]. The protein from mesquite gum was fractionated according to the Osborne method, which involves sequential extraction of proteins using water (albumin), salt (globulin), alkaline (glutelin) and alcoholic solutions (prolamin) [77]. The data showed that globulin constituted the major portion of mesquite seed protein, at 56.47%, followed by glutelin (51.62%), prolamin (17.42%) and albumin (13%) [7].

Nutritional values or quality of protein is governed by amino acid composition, particularly the ratio of essential amino acids to non-essential amino acids. Recently, hydrophobic amino acids such as glycine, alanine, valine, phenylalanine, isoleucine and leucine, as well as glutamic and aspartic acids, were reported to be present in large amounts in whole seeds and protein fractions from mesquite. Methionine was the limiting amino acid in all fractions (Table **3**). These results are consistent with those reported for other legume seed proteins [78].

Essential amino acids, such as histidine, threonine, valine, phenylalanine, isoleucine, leucine and lysine, were found in considerable amounts in the seed, globulin, glutelin and prolamin fractions. In general, the amino acid profile of mesquite seeds seems to be acceptable relative to the recommended amino acid requirements for an adult of 70 kg [79].

Mesquite seed proteins have higher levels of histidine, lysine and threonine than commercial legumes such as soy, chickpeas and peanuts. In cereals, lysine is the limiting amino acid [80]. The combination of cereals with mesquite seed proteins could produce proteins with better nutritional value for incorporation into human food manufacturing.

Table 3. Amino acid profile of seed and protein fractions from *P. velutina*.

Amino acid	(g /100 g of protein)					Requirement, 70 kg adult
	Seed	Albumin	Globulin	Glutelin	Prolamin	Joint W. H. O. (2007) [79]
Aspartic acid	8.00 ± 0.22	4.23 ± 0.04	5.03 ± 0.16	4.63 ± 0.20	9.06 ± 0.12	
Glutamic acid	19.20 ± 0.48	6.48 ± 0.07	8.17 ± 0.26	8.58 ± 0.40	15.10 ± 0.22	
Serine	1.27 ± 0.04	0.97 ± 0.01	0.47 ± 0.02	0.90 ± 0.03	3.15 ± 0.03	
Histidine	3.32 ± 0.10	0.74 ± 0.00	1.83 ± 0.06	2.88 ± 0.11	3.15 ± 0.01	
Glycine	16.58 ± 0.36	3.94 ± 0.03	9.05 ± 0.36	8.22 ± 0.19	12.47 ± 0.16	
Threonine	5.05 ± 0.19	2.33 ± 0.01	2.91 ± 0.15	2.94 ± 0.16	5.37 ± 0.06	1.05
Arginine	2.24 ± 0.08	1.46 ± 0.00	1.17 ± 0.02	1.86 ± 0.08	4.13 ± 0.04	
Alanine	3.22 ± 0.10	2.07 ± 0.01	1.97 ± 0.08	2.41 ± 0.12	4.15 ± 0.05	
Tyrosine	2.66 ± 0.08	1.42 ± 0.02	1.85 ± 0.99	2.19 ± 0.10	3.73 ± 0.06	
Methionine	0.49 ± 0.02	0.24 ± 0.00	0.68 ± 0.01	0.88 ± 0.02	1.07 ± 0.00	1.05
Valine	3.86 ± 0.14	1.85 ± 0.01	2.72 ± 0.12	3.13 ± 0.14	4.70 ± 0.05	1.82
Phenylalanine	3.82 ± 0.13	0.88 ± 0.01	2.16 ± 0.10	2.88 ± 0.12	3.36 ± 0.02	1.75
Isoleucine	3.19 ± 0.11	1.01 ± 0.01	2.24 ± 0.10	2.47 ± 0.10	3.21 ± 0.03	1.40
Leucine	6.47 ± 0.23	1.90 ± 0.02	4.58 ± 0.10	4.33 ± 0.19	4.89 ± 0.05	2.73
Lysine	5.64 ± 0.17	4.21 ± 0.03	4.49 ± 0.08	3.55 ± 0.17	8.94 ± 0.09	2.10

Values are expressed as the mean ± standard deviation (n=3).

Physicochemical Characteristics

Molecular weights and thermal parameters of the mesquite protein fractions were determined using SDS-PAGE and differential scanning calorimetry techniques. All fractions showed protein patterns with a significant number of polypeptide subunits, with molecular weights ranging from 6.5 to 66.2 kDa. Denaturation temperatures between 37 and 70 °C and enthalpies from 0.0071 to 0.4444 J/g were obtained from the protein fractions. The thermal parameters were very low, possibly because most protein was denatured before the thermal analysis. Determination of functional properties could increase the value of mesquite seed protein as a food ingredient, thus expanding its market value.

Applications

Proteins are used in the food industry for their functional properties, such as solubility, water- and fat-binding capacities, gel formation, and emulsifying and foaming abilities [81]. The functional properties of proteins are key for their application in pharmaceuticals, cosmetics and food products. These properties are related to their physicochemical characteristics and molecular structure, which determines intra- and intermolecular interactions such as the formation of covalent or non-covalent bonds. Studies on the physicochemical characteristics and functional properties of mesquite (*Prosopis* spp.) proteins are scarce. However, the results described in this section are indicative of the potential uses of these biopolymers.

CONCLUSION

Native plant species of arid regions can be considered potential sources of useful hydrocolloids. In the Sonoran desert and other arid regions, the native flora, particularly mesquite trees, offers considerable potential for the recovery of such compounds. Mesquite bark exudes a water-soluble gum (mesquite gum) with similar properties to gum Arabic, and its seeds contain galactomannans (comparable to guar gum) and high-quality protein.

Physicochemical characteristics and functional properties of these biopolymers evidence the great potential of mesquite tree products for food and other industrial applications. Furthermore, these macromolecules also offer the possibility of manufacturing novel materials or food products with improved properties.

Globally, there are several research efforts to expand the knowledge about mesquite hydrocolloids. For example, the bioactive properties associated with mesquite hydrocolloids, such as antioxidant and hypocholesterolemic capacity, among others, have not been completely elucidated. Modification of chemical structures is being explored to develop novel food additives and controlled drug delivery systems. Similarly, soft matter materials based on mesquite gums are in development for applications in the food industry. Several technological improvements are proposed to make use of the seed protein and other mesquite tree coproducts.

One main challenge for the use of mesquite hydrocolloids is building a sustainable production chain that takes into account economic, ecological and social aspects. Currently, *Prosopis* biomass is mainly used for charcoal production, and its fruit is used for feed. Mesquite hydrocolloids could be an alternative added-value product that enables development of arid and semi-arid regions.

CONFLICT OF INTEREST

The author (editor) declares no conflict of interest, financial or otherwise.

ACKNOWLEDGEMENTS

Declared none.

REFERENCES

[1] P. Felker, G. Takeoka, and L. Dao, "Pod mesocarp flour of north and south american species of Leguminous tree Prosopis (mesquite): Composition and food applications", *Food Rev. Int.,* vol. 29, no. 1, pp. 49-66, 2013. [http://dx.doi.org/10.1080/87559129.2012.692139]

[2] F.M. Goycoolea, E.R. Morris, R.K. Richardson, and A.E. Bell, "Solution rheology of mesquite gum in comparison with gum arabic", *Carbohydr. Polym.,* vol. 27, no. 1, pp. 37-45, 1995. [http://dx.doi.org/10.1016/0144-8617(95)00031-2]

[3] Y.L. López-Franco, M.A. Valdez, J. Hernández, A.M. Calderón de la Barca, M. Rinaudo, and F.M. Goycoolea, "Macromolecular dimensions and mechanical properties of monolayer films of Sonorean mesquite gum", *Macromol. Biosci.,* vol. 4, no. 9, pp. 865-874, 2004. [http://dx.doi.org/10.1002/mabi.200400055] [PMID: 15468295]

[4] Y.L. López-Franco, A.M. de la Barca, M.A. Valdez, M.G. Peter, M. Rinaudo, G. Chambat, and F.M. Goycoolea, "Structural characterization of mesquite (Prosopis velutina) gum and its fractions", *Macromol. Biosci.,* vol. 8, no. 8, pp. 749-757, 2008. [http://dx.doi.org/10.1002/mabi.200700285] [PMID: 18446807]

[5] M. Rinaudo, F.M. Goycoolea, and M.A. Valdez, "Emulsifying properties of mesquite gum," Foods Food Ingred", *J Jpn,* vol. 213, no. 3, pp. 239-248, 2008.

[6] A.G. Cruz, *"Production and Characterization of Prosopis Seed Galactomannan"*, PhD Dissertation ETH no. 13153, Swiss Federal Institute of Technology Zurich, Zurich, Germany, 1999.

[7] Y.L. López-Franco, L. García-Beltrán, J. Lizardi-Mendoza, and A.R. Toledo-Guillén, "Seeds from Mesquite (Prosopis spp) As Source of Protein", In: *Agricultural Research Updates.* vol. 11. Nova Science Publishers Inc: New York, 2015, pp. 145-155.

[8] F.M. Goycoolea, A.M. Calderón de la Barca, J.R. Balderrama, and J.R. Valenzuela, "Immunological and functional properties of the exudate gum from northwestern Mexican mesquite (Prosopis spp.) in comparison with gum arabic", *Int. J. Biol. Macromol.,* vol. 21, no. 1-2, pp. 29-36, 1997. [http://dx.doi.org/10.1016/S0141-8130(97)00037-8] [PMID: 9283013]

[9] E.V. White, "The Constitution of Mesquite Gum. I. The methanolysis products of methylated mesquite gum", *J. Am. Chem. Soc.,* vol. 68, no. 2, pp. 272-275, 1946. [http://dx.doi.org/10.1021/ja01206a037]

[10] E.V. White, "The constitution of mesquite gum; partial hydrolysis of mesquite gum", *J. Am. Chem. Soc.,* vol. 69, no. 3, pp. 622-623, 1947.

[http://dx.doi.org/10.1021/ja01195a046] [PMID: 20289449]

[11] E.V. White, "The constitution of mesquite gum: hexamethyl-3-glucuronosido-methyl-galactoside methyl ester", *J. Am. Chem. Soc.,* vol. 69, no. 10, pp. 2264-2266, 1947. [http://dx.doi.org/10.1021/ja01202a007] [PMID: 20268277]

[12] J. I. Cunneen, and F. Smith, "The constitution of mesquite gum. Part I. Isolation of 6- and 4-glucuronosidogalactose", *J. Chem. Soc. Resumed,* no. 227, pp. 1141-1146, 1948. [http://dx.doi.org/10.1039/jr9480001141]

[13] J. I. Cunneen, and F. Smith, "The constitution of mesquite gum. Part II. Methylated mesquite gum", *J. Chem. Soc. Resumed,* no. 228, pp. 1146-1157, 1948. [http://dx.doi.org/10.1039/jr9480001146]

[14] M. A. Akher, F. Smith, and D. Spriestersbach, *The constitution of mesquite gum. Part IV. Determination of the structure of the amide of 6-β-(4-methyl D-glucopyruronosyl)α-methy--D-galactopyranoside,* no. 697, pp. 3637-3640, 1952.*J. Chem. Soc. Resumed,* no. 697, pp. 3637-3640, 1952. [http://dx.doi.org/10.1039/JR9520003637]

[15] G.O. Aspinall, and C.C. Whitehead, "Mesquite gum. I. The 4-O-methylglucuronogalactan core", *Can. J. Chem.,* vol. 48, no. 24, pp. 3840-3849, 1970. [http://dx.doi.org/10.1139/v70-647]

[16] G.O. Aspinall, and C.C. Whitehead, "Mesquite gum. II. The arabinan peripheral chains", *Can. J. Chem.,* vol. 48, no. 24, pp. 3850-3855, 1970. [http://dx.doi.org/10.1139/v70-648]

[17] F.M. Goycoolea, A.M. Calderón de la Barca, J.R. Balderrama, J.R. Valenzuela, and G. Hernández, "Processing and Functional Behaviour of Low-Tannin Mesquite Gum", In: *Gums and Stabilisers for the Food Industry.* vol. 9. The Royal Society of Chemistry: Cambridge, UK, 1998, pp. 305-313. [http://dx.doi.org/10.1533/9781845698362.4.305]

[18] Y.L. López-Franco, R.E. Cordova-Moreno, F.M. Goycoolea, M.A. Valdez, J. Juarez-Onofre, and J. Lizardi-Mendoza, "Classification and physicochemical characterization of mesquite gum (Prosopis spp.)", *Food Hydrocoll.,* vol. 26, no. 1, pp. 159-166, 2012. [http://dx.doi.org/10.1016/j.foodhyd.2011.05.006]

[19] R.C. Randall, G.O. Phillips, and P.A. Williams, "The role of the proteinaceous component on the emulsifying properties of gum arabic", *Food Hydrocoll.,* vol. 2, no. 2, pp. 131-140, 1988. [http://dx.doi.org/10.1016/S0268-005X(88)80011-0]

[20] W. Qi, C. Fong, and D.T. Lamport, "Gum arabic glycoprotein is a twisted hairy rope : a new model based on o-galactosylhydroxyproline as the polysaccharide attachment site", *Plant Physiol.,* vol. 96, no. 3, pp. 848-855, 1991. [http://dx.doi.org/10.1104/pp.96.3.848] [PMID: 16668264]

[21] Y.L. López-Franco, F. Bainori-Vasquez, A.K. Sánchez-Castañeda, K.G. Martínez-Robinson, T. Madera-Santana, A.R. Toledo-Guillén, and J. Lizardi-Mendoza, "Chemical and Structural Composition of Arabinogalactan Proteins (AGP) of the Mesquite (Prosopis spp)", In: *Legumes: Types, Nutritional Composition and Health Benefits.* Nova Science Publishers Inc: New York, 2013, pp. 257-270.

[22] J.J. Youl, A. Bacic, and D. Oxley, "Arabinogalactan-proteins from Nicotiana alata and Pyrus communis contain glycosylphosphatidylinositol membrane anchors", *Proc. Natl. Acad. Sci. USA,* vol. 95, no. 14, pp. 7921-7926, 1998. [http://dx.doi.org/10.1073/pnas.95.14.7921] [PMID: 9653116]

[23] J. Svetek, M.P. Yadav, and E.A. Nothnagel, "Presence of a glycosylphosphatidylinositol lipid anchor on rose arabinogalactan proteins", *J. Biol. Chem.,* vol. 274, no. 21, pp. 14724-14733, 1999. [http://dx.doi.org/10.1074/jbc.274.21.14724] [PMID: 10329668]

[24] M.P. Yadav, J.M. Igartuburu, Y. Yan, and E.A. Nothnagel, "Chemical investigation of the structural basis of the emulsifying activity of gum arabic", *Food Hydrocoll.,* vol. 21, no. 2, pp. 297-308, 2007. [http://dx.doi.org/10.1016/j.foodhyd.2006.05.001]

[25] D.M. Anderson, and N.A. Morrison, "The characterization of four proteinaceous Acacia gums which are not permitted food additives", *Food Hydrocoll.,* vol. 3, no. 1, pp. 57-63, 1989. [http://dx.doi.org/10.1016/S0268-005X(89)80033-5]

[26] K.T. Chung, C.I. Wei, and M.G. Johnson, "Are tannins a double-edged sword in biology and health?", *Trends Food Sci. Technol.,* vol. 9, no. 4, pp. 168-175, 1998. [http://dx.doi.org/10.1016/S0924-2244(98)00028-4]

[27] M.I. Abdelhady, A.M. Kamal, and M. Youns, "Biological activity and total phenolic contents of ethanolic extracts of three species of Acacia leaves", *J. Pharm. Res. Vol,* vol. 5, no. 1, pp. 691-695, 2012.

[28] A. Plazonić, F. Bucar, Z. Males, A. Mornar, B. Nigović, and N. Kujundzić, "Identification and quantification of flavonoids and phenolic acids in burr parsley (Caucalis platycarpos L.), using high-performance liquid chromatography with diode array detection and electrospray ionization mass spectrometry", *Molecules,* vol. 14, no. 7, pp. 2466-2490, 2009. [http://dx.doi.org/10.3390/molecules14072466] [PMID: 19633617]

[29] Y.L. López-Franco, *"Estructura y Propiedades de Superficie de las Proteoglicanas de la Goma de Mezquite (Prosopis velutina)",* PhD Thesis, Centro de Investigación en Alimentación y Desarrollo, A.C., Sonora, México, 2005.

[30] J.I. Acedo-Carrillo, A. Rosas-Durazo, R. Herrera-Urbina, M. Rinaudo, F.M. Goycoolea, and M.A. Valdez, "Zeta potential and drop growth of oil in water emulsions stabilized with mesquite gum", *Carbohydr. Polym.,* vol. 65, no. 3, pp. 327-336, 2006. [http://dx.doi.org/10.1016/j.carbpol.2006.01.016]

[31] M.A. Valdez, J.I. Acedo-Carrillo, A. Rosas-Durazo, J. Lizardi, M. Rinaudo, and F.M. Goycoolea, "Small-deformation rheology of mesquite gum stabilized oil in water emulsions", *Carbohydr. Polym.,* vol. 64, no. 2, pp. 205-211, 2006. [http://dx.doi.org/10.1016/j.carbpol.2005.11.028]

[32] E.J. Vernon-Carter, S.A. Gómez, C.I. Beristaín, G. Mosqueira, R. Pedroza-Islas, and R.C. Moreno-Terrazas, "Color degradation and coalescence kinetics of Aztec Marigold oleoresin-in-water emulsions stabilized by mesquite or Arabic gums and their blends", *J. Texture Stud.,* vol. 27, no. 6, pp. 625-641, 1996. [http://dx.doi.org/10.1111/j.1745-4603.1996.tb00997.x]

[33] E.J. Vernon-Carter, R. Pedroza-Islas, and C.I. Beristain, "Stability of Capsicum annuum oleoresin-i--water emulsions containing Prosopis and Acacia gums", *J. Texture Stud.,* vol. 29, no. 5, pp. 553-567, 1998. [http://dx.doi.org/10.1111/j.1745-4603.1998.tb00183.x]

[34] C.I. Beristain, and E.J. Vernon-Carter, "Utilization of mesquite (Prosopis-juliflora) gum as emulsion stabilizing agent for spray-dried encapsulated orange peel oil", *Dry. Technol.,* vol. 12, no. 7, pp. 1727-1733, 1994. [http://dx.doi.org/10.1080/07373939408962195]

[35] Y.L. López-Franco, I. Higuera-Ciapara, F.M. Goycoolea, and W. Wang, "Other exudates: Tragancanth, Karaya, Mesquite Gum and Larchwood Arabinogalactan", In: *Handbook of Hydrocolloids.* 2nd ed Woodhead International Limited: Cambridge, 2009, pp. 495-534. [http://dx.doi.org/10.1533/9781845695873.495]

[36] E. J. Vernon-Carter, Z. Nieto-Villalobos, O. Velasquez-Madrazo, P. Valle-Vega, A. S. de Aluja, F. Constantino, C. Zamora, R. Pedroza-Islas, M. Janczur, H. Bourges-Rodríguez, J. Espinoza, and L. de la Fuente-Dorado, "Evaluation of Mesquite Gum Toxicity", *Multidiciplinary Committee Report submitted to the Mexican Welfare Secretariat,* .

[37] "ACUERDO por el que se determinan los aditivos y coadyuvantes en alimentos, bebidas y suplementos alimenticios, su uso y disposiciones sanitarias", *Secretaría de Salud, México. DIARIO OFICIAL (Tercera Sección), Lunes 16 de julio de 2012,* 2012. Available: http://www.salud.gob.mx/cdi/nom/compi/Acuerdo_aditivos_160712.pdf [Accessed: 3rd Nov. 2015]

[38] A. Flores-Domínguez, Z. Nieto-Villalobos, F. Constantino-Casas, E.J. Vernon-Carter, R. Pedroza-Islas, and M. Janczur, "Estudio Toxicológico Multigeneracional de la Goma de Mezquite Utilizando Ratas Wistar", In: *El Mezquite Árbol de Usos Múltiples. Estado Actual del Conocimiento en México.* 1st ed Universidad de Guanajuato: México, 2000, pp. 227-247.

[39] J.R. Balderrama, *"Caracterización Fisicoquímica y Análisis del Aprovechamiento de la Goma Chúcata y Galactomanana del Mezquite (Prosopis spp.) Como Posibles Hidrocoloides Alimentarios"*, Bch Thesis, Universidad de Sonora, México, 1998.

[40] E.J. Vernon-Carter, C.I. Beristain, and R. Pedroza-Islas, "Mesquite Gum (Prosopis Gum)", In: *Developments in Food Science. Novel Macromolecules in Food Systems.* Elsevier: Amsterdam, 2000, pp. 217-238.
[http://dx.doi.org/10.1016/S0167-4501(00)80011-4]

[41] Y.L. López-Franco, F.M. Gooycolea, and J. Lizardi-Mendoza, "Gum of Prosopis/Acacia Species", In: *Polysaccharides. Bioactivity and Biotechnology.* Springer International Publishing: Switzerland, 2015, pp. 641-662.

[42] L. García-Beltrán, *"Extracción y Caracterización Fisicoquímica de la Proteína de la Semilla de Mezquite (Prosopis spp)"*, Bch Thesis, Instituto Tecnológico de Sonora, México, 2015.

[43] M. Freyre, E. Astrada, C. Blasco, C. Baigorria, V. Rozycki, and C. Bernardi, "Valores nutricionales de frutos de vinal (Prosopis ruscifolia): consumo humano y animal", *Cienc. Tecnol. Aliment.,* vol. 4, no. 1, pp. 41-46, 2003.
[http://dx.doi.org/10.1080/11358120309487617]

[44] A. Fernández-Quintela, M.T. Macarulla, A.S. Del Barrio, and J.A. Martínez, "Composition and functional properties of protein isolates obtained from commercial legumes grown in northern Spain", *Plant Foods Hum. Nutr.,* vol. 51, no. 4, pp. 331-342, 1997.
[http://dx.doi.org/10.1023/A:1007936930354] [PMID: 9650726]

[45] W.C. Wielinga, "Galactomannan", In: *Handbook of Hydrocolloids.* 2nd ed Woodhead Publishing Limited: Cambridge, UK, 2009, pp. 228-251.
[http://dx.doi.org/10.1533/9781845695873.228]

[46] L. Chaires-Martinez, J.A. Salazar-Montoya, and E.G. Ramos-Ramirez, "Physicochemical and functional characterization of the galactomannan obtained from mesquite seeds (Prosopis pallida)", *Eur. Food Res. Technol.,* vol. 227, no. 6, pp. 1669-1676, 2008.
[http://dx.doi.org/10.1007/s00217-008-0892-0]

[47] M.A. Cerqueira, A.C. Pinheiro, B.W. Souza, Á.M. Lima, C. Ribeiro, C. Miranda, J.A. Teixeira, R.A. Moreira, M.A. Coimbra, M.P. Gonçalves, and A.A. Vicente, "Extraction, purification and characterization of galactomannans from non-traditional sources", *Carbohydr. Polym.,* vol. 75, no. 3, pp. 408-414, 2009.
[http://dx.doi.org/10.1016/j.carbpol.2008.07.036]

[48] Y.L. López-Franco, C.I. Cervantes-Montaño, K.G. Martínez-Robinson, J. Lizardi-Mendoza, and L.E. Robles-Ozuna, "Physicochemical characterization and functional properties of galactomannans from mesquite seeds (Prosopis spp.)", *Food Hydrocoll.,* vol. 30, no. 2, pp. 656-660, 2013.
[http://dx.doi.org/10.1016/j.foodhyd.2012.08.012]

[49] S. Barak, and D. Mudgil, "Locust bean gum: processing, properties and food applicationsa review", *Int. J. Biol. Macromol.,* vol. 66, pp. 74-80, 2014.
[http://dx.doi.org/10.1016/j.ijbiomac.2014.02.017] [PMID: 24548746]

[50] V.D. Prajapati, G.K. Jani, N.G. Moradiya, N.P. Randeria, B.J. Nagar, N.N. Naikwadi, and B.C.

Variya, "Galactomannan: a versatile biodegradable seed polysaccharide", *Int. J. Biol. Macromol.,* vol. 60, pp. 83-92, 2013.
[http://dx.doi.org/10.1016/j.ijbiomac.2013.05.017] [PMID: 23707734]

[51] F.M. Goycoolea, M. Milas, and M. Rinaudo, "Associative phenomena in galactomannan-deacetylated xanthan systems", *Int. J. Biol. Macromol.,* vol. 29, no. 3, pp. 181-192, 2001.
[http://dx.doi.org/10.1016/S0141-8130(01)00164-7] [PMID: 11589971]

[52] R. Chandrasekaran, A. Radha, and K. Okuyama, "Morphology of galactomannans: an X-ray structure analysis of guaran", *Carbohydr. Res.,* vol. 306, no. 1–2, pp. 243-255, 1998.
[http://dx.doi.org/10.1016/S0008-6215(97)00274-7]

[53] T.M. Bresolin, M. Milas, M. Rinaudo, and J.L. Ganter, "Xanthan-galactomannan interactions as related to xanthan conformations", *Int. J. Biol. Macromol.,* vol. 23, no. 4, pp. 263-275, 1998.
[http://dx.doi.org/10.1016/S0141-8130(98)00061-0] [PMID: 9849624]

[54] P.J. Daas, H.A. Schols, and H.H. de Jongh, "On the galactosyl distribution of commercial galactomannans", *Carbohydr. Res.,* vol. 329, no. 3, pp. 609-619, 2000.
[http://dx.doi.org/10.1016/S0008-6215(00)00209-3] [PMID: 11128589]

[55] A.M. Estévez, C. Saenz, M.L. Hurtado, B. Escobar, S. Espinoza, and C. Suarez, "Extraction methods and some physical properties of mesquite (Prosopis chilensis (Mol) Stuntz) seed gum", *J. Sci. Food Agric.,* vol. 84, no. 12, pp. 1487-1492, 2004.
[http://dx.doi.org/10.1002/jsfa.1795]

[56] F. Rincón, J. Muñoz, P. Ramirez, H. Galan, and M. Carmen Alfaro, "Physicochemical and rheological characterization of Prosopis juliflora seed gum aqueous dispersions", *Food Hydrocoll.,* vol. 35, pp. 348-357, 2014.
[http://dx.doi.org/10.1016/j.foodhyd.2013.06.013]

[57] G.C. Martínez-Ávila, A.Y. Hernández-Almanza, F.D. Sousa, R. Moreira, G. Gutierrez-Sanchez, and C.N. Aguilar, "Macromolecular and functional properties of galactomannan from mesquite seed (Prosopis glandulosa)", *Carbohydr. Polym.,* vol. 102, pp. 928-931, 2014.
[http://dx.doi.org/10.1016/j.carbpol.2013.10.051] [PMID: 24507365]

[58] C.T. Andrade, E.G. Azero, L. Luciano, and M.P. Gonçalves, "Solution properties of the galactomannans extracted from the seeds of Caesalpinia pulcherrima and Cassia javanica: comparison with locust bean gum", *Int. J. Biol. Macromol.,* vol. 26, no. 2-3, pp. 181-185, 1999.
[http://dx.doi.org/10.1016/S0141-8130(99)00075-6] [PMID: 10517527]

[59] N. Garti, Z. Madar, A. Aserin, and B. Sternheim, ""Fenugreek galactomannans as food emulsifiers," Food Sci. Technol.-Lebensm.-", *Wiss. Technol.,* vol. 30, no. 3, pp. 305-311, 1997.
[http://dx.doi.org/10.1006/fstl.1996.0179]

[60] V.P. Kapoor, K. Pandey, M. Khanna, A.K. Dwiredi, and S. Singh, "Pharmaceutical applications of the galactomannan from the seeds of Cassia javanica Linn", *Trends Carbohydr. Chem.,* vol. 5, pp. 61-69, 1999.

[61] S. Baveja, K. Rao, J. Arora, N. Mathur, and V. Vinayak, ""Chemical investigations of some galactomannan gums as matrix tablets for sustained drug delivery," Indian J. Chem. Sect. B-Org", *ChemMedChem,* vol. 30, no. 2, pp. 133-137, 1991.

[62] M. Pauly, O. Freis, and G. Pauly, "Galactomannan and xyloglucan: Bio-active polysaccharides", *Cosmet. Toilet.,* vol. 114, no. 7, pp. 65-78, 1999.

[63] A.M. Lima, M.A. Cerqueira, B.W. Souza, E.C. Santos, J.A. Teixeira, R.A. Moreira, and A.A. Vicente, "New edible coatings composed of galactomannans and collagen blends to improve the postharvest quality of fruits - Influence on fruits gas transfer rate", *J. Food Eng.,* vol. 97, no. 1, pp. 101-109, 2010.
[http://dx.doi.org/10.1016/j.jfoodeng.2009.09.021]

[64] M.K. Chourasia, and S.K. Jain, "Potential of guar gum microspheres for target specific drug release to colon", *J. Drug Target.,* vol. 12, no. 7, pp. 435-442, 2004.

[http://dx.doi.org/10.1080/10611860400006604] [PMID: 15621668]

[65] E. Theuwissen, and R.P. Mensink, "Water-soluble dietary fibers and cardiovascular disease", *Physiol. Behav.,* vol. 94, no. 2, pp. 285-292, 2008. [http://dx.doi.org/10.1016/j.physbeh.2008.01.001] [PMID: 18302966]

[66] A. Hernández-Espriú, E. Sanchez-Leon, P. Martinez-Santos, and L.G. Torres, "Remediation of a diesel-contaminated soil from a pipeline accidental spill: enhanced biodegradation and soil washing processes using natural gums and surfactants", *J. Soils Sediments,* vol. 13, no. 1, pp. 152-165, 2013. [http://dx.doi.org/10.1007/s11368-012-0599-5]

[67] J.M. Teijón, *Bioquímica Estructural.* España: Tébar, 2001.1st

[68] S. Mandal, and R.K. Mandal, "Seed storage proteins and approaches for improvement of their nutritional quality by genetic engineering", *Curr. Sci.,* vol. 79, no. 5, pp. 576-589, 2000.

[69] D. Spencer, "The physiological-role of storage proteins in seeds", *Philos. Trans. R. Soc. Lond. B Biol. Sci.,* vol. 304, no. 1120, pp. 275-285, 1984. [http://dx.doi.org/10.1098/rstb.1984.0023]

[70] R.C. Foley, L-L. Gao, A. Spriggs, L.Y. Soo, D.E. Goggin, P.M. Smith, C.A. Atkins, and K.B. Singh, "Identification and characterisation of seed storage protein transcripts from Lupinus angustifolius", *BMC Plant Biol.,* vol. 11, p. 59, 2011. [http://dx.doi.org/10.1186/1471-2229-11-59] [PMID: 21457583]

[71] J.E. Kinsella, and L. Phillips, "Structure: Function Relationships in Food Proteins, Film, and Foaming Behavior", In: *Food Proteins.* The American Oil Chemistry Society: Champaign, IL, 1989, pp. 52-77.

[72] M.A. Olvera-Novoa, and L. Olivera-Castillo, "Potencialidad del uso de las leguminosas como fuente proteica en alimentos para peces", *Memorias del IV Simposium Internacional de Nutrición Acuícola,* 2000pp. 327-348

[73] J.R. Juárez-Argumedo, M. Alvarado, and R.D. Valdez, "Escarificación de semillas de mezquite (Prosopis laevigata) para aumentar la eficiencia en la germinación", In: *5tas. Jornadas de Investigación.* Universidad Autónoma de Zacatecas: México, 2001.

[74] H.B. Lajnef, H. Mejri, A. Feriani, S. Khemiri, E. Saadaoui, N. Nasri, and N. Tlili, "Prosopis farcta seeds: Potential source of protein and unsaturated fatty acids?", *J. Am. Oil Chem. Soc.,* vol. 92, no. 7, pp. 1043-1050, 2015. [http://dx.doi.org/10.1007/s11746-015-2660-1]

[75] J.T. Barminas, H.M. Maina, and J. Ali, "Nutrient content of Prosopis africana seeds", *Plant Foods Hum. Nutr.,* vol. 52, no. 4, pp. 325-328, 1998. [http://dx.doi.org/10.1023/A:1008045218320] [PMID: 10426119]

[76] V. Briones-Labarca, C. Muñoz, and H. Maureira, "Effect of high hydrostatic pressure on antioxidant capacity, mineral and starch bioaccessibility of a non-conventional food: Prosopis chilensis seed", *Food Res. Int.,* vol. 44, no. 4, pp. 875-883, 2011. [http://dx.doi.org/10.1016/j.foodres.2011.01.013]

[77] T.B. Osborne, *The Vegetable Proteins.* 2nd ed Longmans, Green & Co.: London, 1924.

[78] D. Prakash, and P.S. Misra, "Protein content and amino acid profile of some wild leguminous seeds", *Plant Foods Hum. Nutr.,* vol. 38, no. 1, pp. 61-65, 1988. [http://dx.doi.org/10.1007/BF01092311] [PMID: 3231595]

[79] W.H.O. Joint, "Protein and Amino Acid Requirements in Human Nutrition", *World Health Organization Technical Report Series: Singapore,* .

[80] D. Pérez-Conesa, G. Ros, and M.J. Periago, "Protein nutritional quality of infant cereals during processing", *J. Cereal Sci.,* vol. 36, no. 2, pp. 125-133, 2002. [http://dx.doi.org/10.1006/jcrs.2002.0463]

[81] L. Day, "Proteins from land plants – Potential resources for human nutrition and food security",

Trends Food Sci. Technol., vol. 32, no. 1, pp. 25-42, 2013. [http://dx.doi.org/10.1016/j.tifs.2013.05.005]

CHAPTER 10

Direct Measurement of Free Volume Properties in Polymeric Materials

Asmita Sen Gupta*

Physics Department, Visva-Bharati Central University, Santiniketan-731235, W.B, India

Abstract: There has been a long history of theoretical and experimental papers which are concerned with the development of the free volume concept to explain transport and diffusion in polymer systems. Several theoretical efforts also took place to clinch the mechanism of the glass transition on the basis of free volume concept. In spite of the strong presence of free volume concept in scientific arena, it carries ambiguity in terms of definition. Many consider free volume concept as nothing but a guide to discussion since it has no physical existence. Past papers on theory and experiments dealt with the development of free volume concept to explain transport and diffusion in polymers. Many models based on free volume concept, show a number of unsatisfactory features as they stand at present. However, a number of theoretical approaches emerged to achieve a unanimous definition of free volume concept. Since the last few decades, Positron Annihilation Lifetime Spectroscopy (PALS) has become a well-established experimental technique for gathering information on the microstructure of free volume in polymeric materials. A brief discussion is presented in this chapter on free volume models along with PALS effectiveness in investigating free volume properties.

Keywords: Free volume, Models, PALS technique, Polymer and Biopolymer.

INTRODUCTION

Free volume is practically related to the motion and thus is a few centuries old concept. The great Greek philosopher Epicurus (341-270BC), offered 'free space' [1] concept to explain movement of atoms and an acknowledgement is due to Epicurus' school, for this great service. Though it is known that the equilibrium and transport properties of fluids depend on the space available for molecular motion, but the lack of methods to measure free volume has limited the development of this perspective [2].

* **Corresponding author Asmita Sen Gupta:** Physics Department, Visva-Bharati Central University, Santiniketan, India; E-mail: asmita_sengupta@hotmail.com

Martin Masuelli & Denis Renard (Eds.)

EYRING [3] in '36-'37 defined liquid by stating that they treated it as liquid which was constituted with 'individual molecules' and 'each moving in an average potential field' because of the presence of respective neighbors. He called the 'partition function' equation of the liquid as the definition of free volume, V_f:

$$F_l = V_f \frac{(2\pi mkT)^{\frac{3}{2}}}{h^3} R_l V_i b_l \exp\left(\frac{\Delta E}{RT}\right) \quad \textbf{(1)}$$

Here ΔE is the energy of vaporization, R_l and $V_i b_l$ are the rotational and vibrational parts of the partition function of a molecule in the liquid respectively, m, k, h and T are the molecule mass, Boltzmann constant, Planck constant and absolute temperature respectively.

Yang Yu in his thesis defined that 'Free volume' is more a geometric and descriptive concept than energetic [4].

To define free volume, Vrentas & Vrentas [5] stated that free volume theory depends on the specific volume of a single component or a mixture, which is constituted with three different volumetric components, *i.e.* occupied volume, interstitial free volume and hole free volume. The free volume theory is popular among the researchers dealing with polymers. It is easy to understand many properties of polymer at a molecular level with the help of this theory when clubbed with quantum and statistical mechanical calculations. Inspite of development of several free volume theories, a simple expression of free volume (V_f) can be more suitable for applications, such as [6], total volume (V_t) minus the occupied Volume (V_0): $V_f = V_t - V_0$.

'Free volume' is related to the microscopic empty space existing among the molecules and is represented by V_f. The specific free volume is given by the relation $V_f = V - V_{occ}$ where V is the specific total volume in cm^3/g and V_{occ} is the specific occupied volume which can be defined as the van der Waals volume V_W *i.e.* the volume of the space occupied by the molecule. In normal thermal energies this molecular space remains impenetrable to other molecules and V_f stands for the total free volume. It can be classified into two groups, namely, 'excess free volume' which appears due to the zigzag orientation of free volume chains forming irregular shape in sub nanometer scale and mostly appears in the amorphous region of polymers causing a decrease of about 10% in the density of the material in comparison to the corresponding crystal and the 'interstitial free volume' which is due to the unoccupied spaces appearing between two regularly packed chains and exists in both amorphous and crystalline region of polymer material. The following Figs. (**1** and **2**) show the two kinds of free volumes in the material:

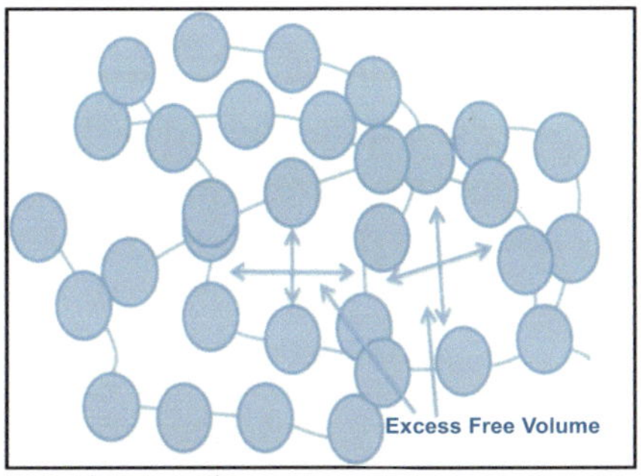

Fig. (1). Excess free volumes in amorphous regions.

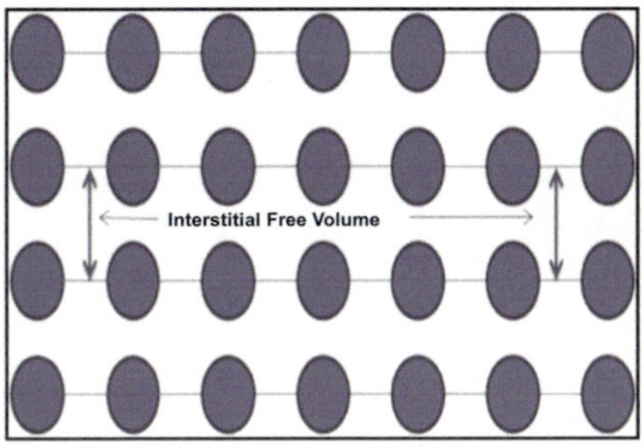

Fig. (2). Interstitial free volume in crystalline region.

These local free volumes are generally termed as "holes", thus the relation above can be rewritten in terms of hole free volume as $V_{fh} = V_f = V - V_{occ}$ and this V_{occ} involves both van der Waals volume and the interstitial free volume *i.e.* $V_{occ} = V_W + V_{fi.}$ The presence of hole free volume controls the diffusion of small molecules through molecular materials and also the molecular mobility [7, 8].

The occupied volume resembles approximately the crystalline volume and can be expressed as $V_{occ} \approx V_c \approx 1.45V_W$ [9]. The local free volumes or "holes" in the sub nanometer scale can be defined in terms of the average hole volume V_h and the specific hole number density, N_h' (in g^{-1}) by the expression, $V_f = V_h \times N_h'$.

The other important parameter for free volume concept is the free volume fraction which is defined as the ratio of the free volume to the total volume, *i.e.* $f = V_f / V$ and $y = 1\text{-}f$ = the corresponding occupied fraction.

Since there are several ways to define the free volume in polymer materials, the values reported for the free volume fraction, f, show a wide variation in the range from few % (percent) to about 20% [6].

However, Fujita [10] asserted that experimentally the specific volume of a given liquid, V (= Nv ; where v is the liquid volume per molecule), is measurable where N is the total number of the molecules per gram of liquid, then Nv_c (where v_c is the volume occupied by entity of one molecule) may be estimated from suitable volume data. Thus if the hole volume assumption is accepted, $V_f = V - Nv_c$ can be calculated and subsequently *f* becomes a measurable parameter. It means that hole volume assumption, the temperature dependence of molecular mobility, *m* for pure liquids can be predicted from experimental information on V and v_c .

FREE VOLUME MODELS

The free volume in polymers arises because of the inefficient packing of the macromolecules. It seems that the free volume theory is broadly appreciated by scientists dealing with polymer materials [6]. It is to be noted that irregular packing of the macromolecules generates free volume from three principal sources including the motion of chain ends, side chains and the main chain.

Polymers are of big macromolecular structure constructed with small monomer units and this proposition proposed number of theoretical models, mainly to deal with molecular mobility and adsorption.

In 1913 A.J. Batschinski [11] proposed a linear relation between the fluidity of a simple liquid and the free volume. He argued to define free volume as the difference between the specific volume and the van der Waals volume. His proposed simple model was:

$$\eta = \frac{1}{V_f} - \frac{1}{v - v_0} \qquad \textbf{(2)}$$

Here, η is the viscosity, V_f is the free volume, v_0 is the occupied volume, v is the total volume and the equation qualitatively explains the temperature dependence on viscosity but in the absence of a good model to describe the structure and property relationship of liquids and glasses, a century old Batschinski (Eq-2) model is still a considered option for these mediums [12]. A. K. Doolittle [13], in the year 1951 after having an experimental observation in hydrocarbon liquids proposed a model by connecting free volume with viscosity of molecular liquids. His Semi-empirical equation for viscosity of liquids was:

$$\eta = A\exp\left(B\frac{v_o}{v_f}\right) \qquad \textbf{(3)}$$

where η is the coefficient of viscosity, v_0 and v_f are occupied and free volumes, respectively. A and B are numerical constants. Viscosity of a polymer, diluted with a solvent, when expressed in relation to the free volume, was proposed by Kelley and Bueche as shown below [14]:

$$C_P^{\frac{4}{3}}\exp\frac{1}{f} \tag{4}$$

where B is a constant, C_p is the polymer solution and fractional free volume was given by:

$$f=\frac{V-V_0}{V} \tag{5}$$

where v is the specific volume and v_0 is the occupied specific volume. They developed model further by evaluating 'f' for the solution from the free volume of the mixture. The mixing rule as developed by them gave satisfactory viscosity predictions for liquid solvents [15].

Many classical free-volume concepts of diffusion of small molecules in liquids initiated with the theory of Cohen and Turnbull [7] which was an extended model of Doolittle's model [13]. Many other models are actually modified versions of Cohen and Turnbull model. Their model, also applied to polymers, assumes that molecular transport may occur only when voids having a volume greater than some critical value v^* are formed by the statistical redistribution of the free volume and no energy is required for free volume redistribution. Adopting a simple consideration to define relation between diffusion constant D in a liquid of hard spheres and the free volume, their model based on the above mentioned concepts, proposed the relation as:

$$D=A\ \exp[-\gamma^{v^*}/v_f] \tag{6}$$

where v * is the minimum required volume of the void and A and γ are constants. Cohen and Turnbull also concluded that liquid and glassy states in combination constitute a single thermodynamically well-defined phase and glass transition results from the free volume [16]. To predict the diffusion coefficients by free volume theory, Vrentas and Vrentas [17] basically formulated the following equation:

$$D_1= \ln \bar{D}_0 - \frac{E^*}{RT} - \left\{\frac{(1-W_2)\,\hat{V}_1^* + \xi w_2 \hat{V}_2^*}{\frac{\hat{V}_{FH}}{\gamma}}\right\} \tag{7}$$

where D_1 is the diffusion coefficient of the penetrant, D_0 is a constant, E* is the effective energy per mole that a molecule needs to overcome attractive forces, R the gas constant, T the temperature, ξ the 'size parameter', $\tilde{V}_1^*$ and $\tilde{V}_2^*$ are the specific volumes of penetrant and polymer, $\tilde{V}_{FH}$ the average hole free volume per unit mass of mixture, γ an average overlap value in the mixture, and w_i is the mass fraction of the component I, where 1 is for penetrant and 2 is for polymer. Vrentas and Duda [18 - 20] modified the theories of Cohen and Turnbull to propose a theory based on the utilization of the average free volume per jumping unit. It introduced the concept of specific shape of the penetrant molecule, a concept not considered in other earlier models. Vrentas and Duda [18] have also reported the deficiencies of the model in describing diffusion in glassy polymers, relaxing polymers and cross linked polymers. From their theory, the following equations were derived by Vrentas, Vrentas and Duda [17, 20]:

$$\text{In}\left[\frac{D_1}{D_{1(O)}}\right] = \phi_1 F / [\frac{f_2^2}{\gamma} + \phi_1 \frac{f_2}{\gamma}\left(\frac{f_1}{\gamma} - \frac{f_2}{\gamma}\right)] \quad \textbf{(8)}$$

$$F = \xi \tilde{V}_2^* f_1 / \tilde{V}_2^0 \gamma - \tilde{V}_1^* f_2 / \tilde{V}_1^0 \gamma \quad \textbf{(9)}$$

Where D_1 is the solvent self diffusion coefficient, $D_1(o)$ refers to the value of D_1 evaluated at zero solvent concentration, ϕ_I is the volume fraction of component *I,* and f_I is the fractional hole free volume of pure component *I* at the concern temperature. Moreover, $\tilde{V}_I^*$ is the specific critical hole free volume of *I*, necessary for a jump, ξ is the ratio of the molar volume of the solvent jumping unit to that of the polymer, and $\tilde{V}_I^*$, is the specific volume of component *I* of a temperature of concern. Their above mentioned equation was based on the assumtion that the polymer and solvent mixing does not introduce any change in the volume and thus partial specific volumes of the both are independent of composition. A notable and considered as advance free volume diffusion coefficient model was proposed by Fujita [21] and Fujita *et al.* [22]. It was, though formed in earlier days but considered as advanced because few parameters are necessary to describe the concentration dependence of the diffusion-coefficient can be collected from physical- chemical properties of the polymer and penetrant. To calculate the thermodynamic diffusion coefficient, D_T, a relation proposed by Fujita was like:

$$D_T = A_d RT \exp\left(-\frac{B_d}{V_f}\right) \quad \textbf{(10)}$$

Where V_f is the average fractional free volume. A_d, proportional coefficient is dependent on the size and shape of the penetrant and B_d is used as penetrant concentration but temperature independent. However, two parameters, A_d and B_d are adjustable and can be used by fitting experimental diffusion data. Fujita [10] and Vrentas, Vrentas and Duda [23] discussed in detail, the differences of their respective models as well as the utility of each models, but others considered that both the theoretical models are useful even in a single experimentation. Mafi *et al.* [24] found that Fujita and Vrentas-Duda's theories are much more accurate especially for dilute aqueous feed solutions. It is known that free volume depends on temperature as well as on the number of polymer chain ends present in the system. End chain units are less dense than the covalently bonded inter chain units and thus chain ends can be considered as an impurity when considering the packing of chains, and more impurity results in a lower T_g. Hence, glass transition temperature is dependent on free volume, which in turn is dependent on the average molecular weight of the polymer sample. This relationship is described by the Flory-Fox equation. Their equation has been used for over 50 years and is still popular. The dependence of Tg on the molecular number follows the equation developed by Fox-Flory [25, 26]:

$$T_g = T_g^{\infty} - \frac{K}{(\alpha_{r} - \alpha_g)M_n} \quad \textbf{(11)}$$

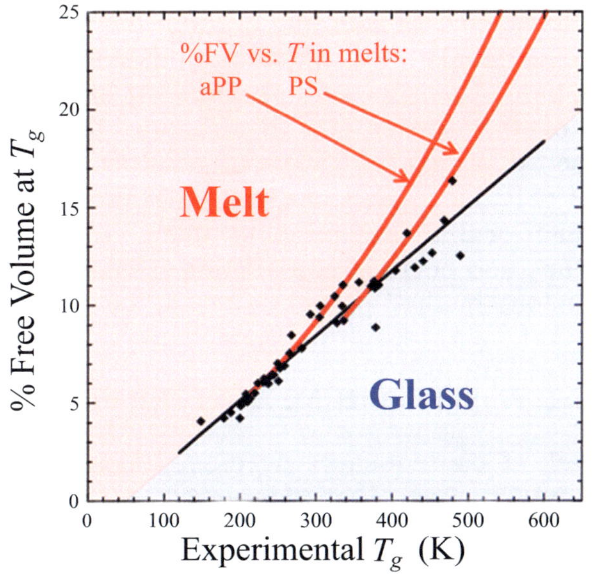

Fig. (3). Variation of Free volume (%) with T_g (Reprinted with permission from ACS Macro Letters, (c) 2015, ACS.)

where Tg$^{\infty}$ is the glass transition temperature at infinite molecular weight, M_n is the molecular weight, K is a fitting constant, α_r and α_g are the cubic (volume) expansion coefficients in the rubbery and glassy states [27], respectively. Whether free volume can serve as a tenable basis for a theory of the glass transition, is an issue having a long history, see (Fig. **3**) [28]. Present day research is taking interest to find relations between polymeric glass transition and bulk properties. However, two broad connections was reported in a 'Letter'. At first, theoretically determined polymeric free volume in the melt was correlated with their experimentally determined T_g values. Secondly, they showed a linear correlation between the experimental T_g and the Locally Correlated Lattice (LCL) model calculation for the free volume percent expected at T_g which has a trend of prediction about boundary of T dependent minimum necessity of free volume to separate between melt and glassy regimes. Moreover, their theoretical estimation of free volume values at T_g ranges in between 4 to 16% and their dependence on temperature signifies the role of temperature in glass formation.

Free volume distributions have been studied theoretically by using kinetic theories and molecular dynamics. The free volume theory implies that the void fraction is the dominant content as D is proportional to both the kinetic velocity and the probability of finding enough free volume for the hopping of the penetrant [7, 8]. More recent models have tried to take into account the distribution of the free volume and even its dynamical redistribution as the polymer evolves performing thermal motion with time. Takeuchi and Okazak [30] have suggested a simple integral expression of the cluster size distribution that shows very good correlation with the self-diffusion coefficient of penetrant particles of He and O_2 for their hypothetical polymer models. Experimental probes for free volume at molecular and atomic scales are possible using small-angle diffraction, fluorescent spectroscopy and positron annihilation lifetime spectroscopy, which has emerged as the unique method providing resolution of a few Angstroms in size directly.

FREE VOLUME AND POSITRONIUM: AN OVERVIEW

It is important to look at consequences developed after the discovery of the Theory of relativity by Einstein. That theory helped to interpret universe as structured with matter and anti-matter. A particle has an anti-particle and there is no bar to say that there is anti-universe and can go further. Dirac in 1931, offered both positive and negative solution for his relativistic wave equation for the energy of electron and thus negative energy solution had a positive energy solution, which means electron has a charge of + e, when commoner would love to search –e [31]. From this, Dirac introduced 'hole theory' which speaks by just assuming that all negative energy states normally has an occupancy of electrons. A positron shows as 'holes' in these states if an electron switches over to a positive energy level.

In 1933, Anderson discovered positrons in his cloud chamber experimentation

[32]. Photographs of cosmic ray shower inside cloud chamber showed few tracks deflected by a magnetic field in a direction which was unexpected from an electron. The mass of that particle denoted by length and density of that track was almost same of an electron. This discovery was further authenticated by Blackett and Occhialini in 1933 [33, 34]. A year later Mohorovicic predicted that the positron and the electron could form a bound state [35]. In 1945, Ruark suggested a name, 'POSITRONIUM' for this bond system and also pointed out some of the important characteristics of that system [36]. In 1949, Ore proposed a theoretical model based on energy considerations for evaluation of the probability of positronium formation in gases [37]. Energy range required for formation of positronium is called Ore gap and the Norwegian Scientist Ore developed the theory for gasses and showed that with a minimum energy, E_{min}, Positronium formation is possible and that is E_{min} = V- 6.8eV, where V is the ionization potential of the gas and the value 6.8eV is the binding energy of the positronium. Within the gap of few years, in 1951 Martin Deutsch experimentally identified for the first time the Ps atom by forming it in Freon gas. Two months later he published his paper about measurement of lifetime of the Ortho-Ps [38]. This successful experimental evidence of theoretical findings of earlier days created a stimulus agenda of research with Positronium by executing number of innovative measurement protocols. Ps physics progressed rapidly in collaboration with variety of different experiments focusing on material science. In between Dirac and Deutsch and many other scholars engaged themselves to work on this astonishing problem of matter and antimatter and those discoveries tutored many countries, generations, individuals, groups and institutions in last few decades and is still continuing. Their scholastic signature is more important than any awards as because their discoveries satisfied, engaged everyone and will do so in near future also. Free volume along with hole free volume (FVH) characteristics are very important parameters of materials to determine properties like deformability, permeability, elasticity and aging of polymers. The size and distribution of FVH are also very important topic of research, several theoretical models of free volume are developed by Positron scientists to correlate this concept with positron parameters. The free volume theory is popular among the researchers dealing with polymers. It is easy to understand many properties of polymer at a molecular level with the help of this theory when clubbed with quantum and statistical mechanical calculations [6].

POSITRONIUM

Positronium is a hydrogen-like bound state composed of a positron and an electron. They circulate a common center. The atomic mass is 0.0011 amu. The diameter (distance of the two particles) is $3a_0$ (a_0 - classical Bohr radius).

Table 1. Properties of the positronium atoms [39].

Property	Types of State Singlet $(p\text{-}P_S)$	Triplet $(O\text{-}P_S)$
Symbol	${}^{S}P_{s}$	${}^{T}P_{s}$
Spin State	e ↑↓ e^{+}	e ↑↑ e^{+}
Quantum number	J=0	J=1
Share %	25	75
Proper lifetime,s	1.25×10^{-10}	1.4×10^{-7}
Annihilation type	2γ	3γ

Brandt [40] proposed a model to explain the increase in the o-Ps lifetime in polymers. Positronium atoms are formed when the free volume is equal to the volume of Ps atom or larger than a certain critical value. It is termed as free volume model. In a lattice L of electron density distribution $\Psi L(r)^2$ composed of neutral atoms or molecules with closed shells, the electron pick up rate, γ_p of o-Ps in the lattice L is:

$$\gamma_P = \pi c r_0^2 \int_L \Psi_L^*(r)\ \Psi_{\beta+}^*\ \Psi_{\beta+}(r)\ \Psi_L(r)\ dr^3 \qquad \textbf{(12)}$$

where Ψ_β is the wave function of the positron in the field of the electron to which it is bound as o-Ps and the field of the lattice, r_o is the classical electron radius and c is the velocity of light [40].

Ignoring the internal structure of Ps and approximating the lattice interaction by a square well potential and considering the electron density (ρ_o) as a constant between the walls, equation (12) after modification by Brandt *et al.*, becomes [40]:

$$\gamma_\rho = \pi c r_0^2\, \rho_o \int_{Vo} \Psi_{Ps}^*(r)\ \ \Psi_{Ps}(r)\ dr^3 \qquad \textbf{(13)}$$

Eldrup *et al.* [41] and Tao [42] suggest a relation between the annihilation rate γ_P and the hole size by using a spherical potential well of radius Ro with a virtual electron layer of thickness ΔR.

In positron annihilation spectroscopy, the observed lifetime is the inverse of the integral of the positron (ρ+) and the electron (ρ-) densities at site where the annihilation takes place [43],

$$\tau = \text{constant}^{*} \int (p + p - dr)^{-1} \quad \textbf{(14)}$$

According to eq.(14) a large hole, which has a lower average electron density, is expected to have a longer Ps lifetime. The correlation between the free volume in molecular systems and the observed o-Ps lifetime is expressed in semi empirical equation [44] between the mean o-Ps lifetime τ_3 and mean radius of holes R as:

$$\tau 3 = \frac{1}{2}\left[1 - \frac{R}{R_0} + \frac{1}{2\pi} Sin(2\pi \frac{R}{R_0})\right]^{-1} \quad \textbf{(15)}$$

where τ_3 is the positronium lifetime in free volume, R the hole radius where R_0=R+ΔR and the free volume is given in the spherical geometry approximation as:

$$V_h = 4/3\ \pi\ R^3 h \quad \textbf{(16)}$$

The use of positronium lifetimes to determine the size of voids faces severe difficulties [45]. But lifetime system can be engaged to determine the correlation between lifetime and free-volume with an approach of approximation.

In general, the simplest expression of free-volume (V_h) can be defined as the difference between the total volume and the occupied volume ($V_h = V_t - V_o$), where the free volume [46, 47] is an open space that is freely moving in media. Kobayshi *et al.,* [48]. Proposed a semi empirical relation which may be used to evaluate the fractional free volume F_h given by:

$$F_h = A\ I_3 V_h \quad \textbf{(17)}$$

Where A is normalization constant, its value is 1-2 nm^3 for polymers [49]. I_3 is the intensity of τ_3, given in a percentage form and V_h is given in nm^3.

Simha-Somcynsky Lattice-Hole Theory [50, 51] proposed that lattice sites are occupied by polymer molecule and holes. The combinatorial factor is stated as polymer molecules and holes representing disorder in the polymer. The ratio of the occupied sites on all of lattice sites is y, which is called the volume fraction:

$$y = \frac{sN}{sN + N_b} \quad \textbf{(18)}$$

where s is number of segments, N is the no. of macromolecules and N_b is the no. of unoccupied lattice sites in S-S theory, $sN + N_b$ is the number of total cell sites that is available.

The configurational partition function is represented with the combinatorial factor, free volume and partition function:

$$Z_{conf} = g(N, y)\,[\,v_f\,(\tilde{V}, y)]^{cN}\ \exp\left[-\frac{E_0(\tilde{V} \neg, y)}{kT}\right] \quad \textbf{(19)}$$

The combinatorial factor denoted by g (N, y) and representing the occupation possibilities of the polymer molecules into the lattice sites, is obtained using the number of molecules, occupied site fraction and hole fraction.

$$g(N, y) = y\text{-}N\,(1\text{-}y)\text{-}sN\left(\frac{1-y}{y}\right) \quad \textbf{(20)}$$

is factor in the SS theory is derived from Flory-Huggins Mean Field.

Theory [52 - 55]. Gunter Dlubek *et al.* [56 - 58] observed a common hitch in free volume theory when applied to define free volume of a specific character which can be characterized as capable in deciding the structural dynamics and can be measureable.

However, Simha-Somcynsky Lattice-Hole Theory based equation is very successful in the estimation of the fraction of the excess through the results of Pressure-Volume- Temperature (PVT) experiments [59, 60].

The S-S theory describes structure of a liquid by a cell and can explain disorder by assuming statistical mixture of occupied and unoccupied cells of same size.

In 1950 Fox- Flory [61] studied free volume and glass transition in polystyrene as a function of molecular weight and relaxation of time. They found that in case of infinite molecular weight, the specific free volume, vf, could be described above T_g as:

$$V_f = K + (\alpha_R - \alpha_G)T \quad \textbf{(21)}$$

Where K was related to the free volume at 0^0K and $\alpha_{R,}$ α_G respectively represented cubic volume expansion coefficients in the rubbery and glassy states. They also found that below T_g the same specific volume-temperature relationships held for all the polystyrenes, independent of molecular weight. Fox-Flory concluded that the local conformational arrangement of the polymer segments was independent of molecular weight and temperature as well as glass transition temperature was an iso-free-volume state.

Simha-Boyer [62] claimed that the free volume at T= T_g should be defined as:

$$V - (V_{0.R} + \alpha_G T) = V_f \tag{22}$$

Substitution of the quantity:

$$V = V_{0,R} + \alpha_R T\ , \tag{23}$$

which again leads to a relation:

$$(\alpha_R - \alpha_G)T_g = K_1 \tag{24}$$

From the Fig. (**4**), V is the specific volume, $V_{0,R}$ and $V_{0,G}$ are the volumes extrapolated to 0 Kelvin using α_R, α_G as the coefficients of expansions. Simha and Boyer concluded that:

$$K_1 = (\ \alpha_R - \alpha_G)\ Tg = 0.113 \tag{25}$$

Modifying the Tao-Eldrup model for the void dimension in several nanometers, Goworek *et al.* [63, 64] introduced the pick-off annihilation model with finite spherical and cylindrical wells in which the pickoff annihilation rate depend on the o-Ps states. Professor T. Goworek summarized the status of Tao-Eldrup model in present research context which is very important and also explains the necessity of introducing an extended model to describe free volume related to pick off process. The routine applications of positron technique for determination of free volume holes, the intrinsic decay rate is usually ignored as a small contribution. But in large voids, the pick-off rate is smaller than intrinsic decay one. The pick-off process and similar models which are main content of Tao-Eldrup model is "no longer the dominant one" [65]. Many factors which are treated with less importance for annihilation rate, now can show their essentiality, such as, chemical effects, sorption processes. The lifetime is inverse of annihilation rate, and if it decreases, the lifetime should lengthen [66].

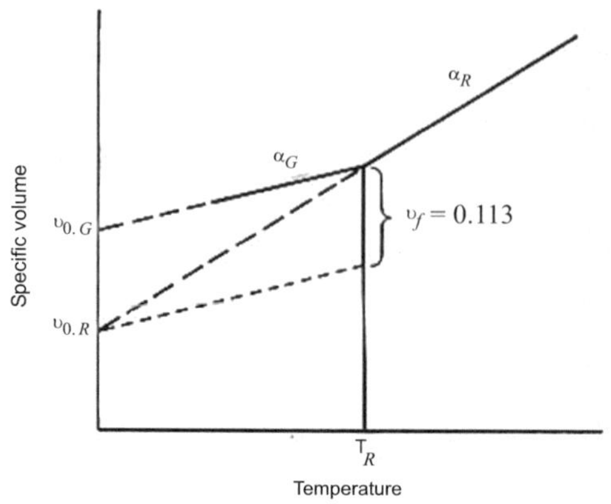

Fig. (4). A schematic diagram illustrating free volume as calculated by Simha- Boyer.

However, Tao- Eldrup model is often used by the polymer researchers due to model's demand of easy parameters and its effectiveness to estimate a mean value of free-volume size at the atomic molecular level for most polymers. Polyimides [67] and carbon contained materials shows no orthopostronium or a fraction of <2% o-Ps intensity. This chemical process create hindrance to use Tao-Eldrup model to gather the correct free volume data from Positron Annihilation Lifetime System [67]. To deal with such scenarios, Kao and Jean *et al.*, rewrote [67]. Tao-Eldrup model in a general form as,

$$\tau_2 = c\,[1 - \frac{R}{R+\Delta R} + \frac{1}{2\pi}\sin\left(\frac{2\pi R}{R+\Delta R}\right)]^{-1} \quad \textbf{(26)}$$

and presented in an explicit figure in their august publication (see Fig. **5**).

Biopolymers

Human civilization is blessed with biopolymers or natural polymers. Increasing demand of various industrial manufacturer for quality biopolymers production though processing is dependent mainly on trial and error methodology. Hence, an all-out efforts are on research agenda to bring rational approach to control and stabilize physical properties of this class of polymers.

Polymers produced by living organisms are polymeric bio molecules or biopolymers. They also contain monomeric units which are covalently bonded to form larger structures. Unlike synthetic polymers the interactions of biopolymers are mediated by water and have charges giving rise to the electrostatic interactions by Coulomb's law *i.e.* $F=q_1q_2 / r^2D$, D being the dielectric constant of the medium.

Three main classes of biopolymers are classified according to the monomeric units used and the structure of the biopolymer formed, they are Poly nucleotides (DNA and RNA), which are long polymers composed of 13 or more, nucleotide monomers, polypeptides, which are short polymers of amino acids; and polysaccharides, which are often linear bonded polymeric carbohydrate structures [68 - 71]. Besides, cellulose is the most common organic compound and biopolymer on Earth. All plant matter contains around 33% is cellulose.

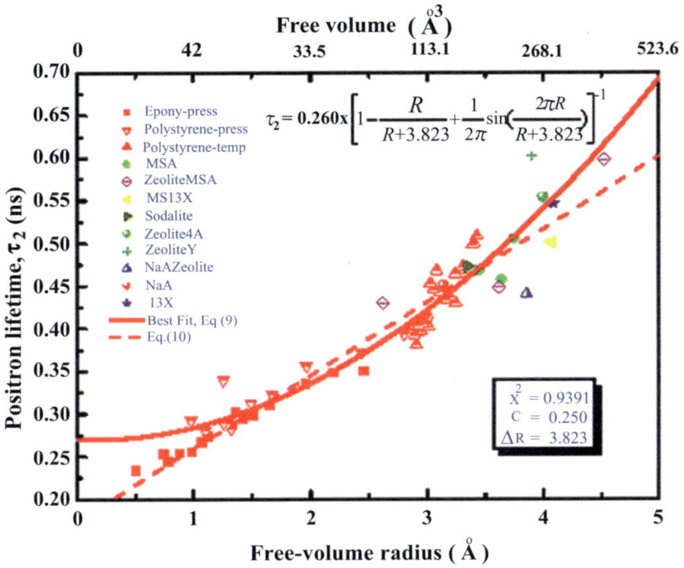

Fig. (5). Variation in Positron lifetime with Free volume radius (Reprinted with permission from Ref N°: 67, 2011,© ACS).

Biopolymers mainly known in three separate classes though there are exceptions. Nucleic acids, proteins and carbohydrates originated biopolymers are at use. Otherwise, bio-based monomers are polymerized to traditional or novel polymers. The proteins are naturally occurring polyamides from plant and animal origin that are highly suitable as raw material for the production of biopolymers.

Molecular mobility and diffusion mechanisms contribute to failure of biopolymers used in food, drug, minerals *etc.* [72]. However, governing principals of diffusion mechanism in biopolymers is not yet clear though it plays an important role.

There is broad unanimity that their structure and dynamics are inseparable with free volume which exists in between molecules of all individual matrix in these category of materials due to irregular packing, fluctuations of density and internal environmental constraints. Numerous sub-nanometer sized voids constitute their free volume for which many theories have been developed to explain self-diffusion and the diffusion of intruder molecules in terms of the size and size distributions of these free volume elements [19, 73].

The importance of characterization of the free volume in these systems, is of immense importance, as that can help to enhance our understanding for progress in designing of bio-encapsulants [72].

In an experimentation on vitrification properties of mixtures of several biopolymers with sugars, it was observed by Stefan Kasapis [74] that the free volume followed a mutual relationship with the number average molecular weight of the biopolymer, and in association with imperfect intermolecular around the ends of the molecules. The effect of free volume on the kinetics of vetrification is quantified by the friction coefficient (ζo) per monomer unit of the biopolymer.

The Flory-Huggins free volume (FHFV) theory has reasonable success in predicting the behaviour of synthetic polymer systems and some biopolymer systems like dextran and keratin solutions [75, 76]. Recently, it also has been used to model the water activity of food-borne carbohydrates like low molecular weight sugars [77], and mixtures of maltodextrin and maltose [78].

The Flory–Huggins theory [79] states that:

$$\frac{\mu_W}{RT} = \ln(1-\Phi) + (1-\frac{1}{N})\Phi + \chi\Phi^2 \tag{27}$$

$$\frac{\mu_S}{RT} = N\ [\ \ln\Phi / N - (1 - 1/N)(1 - \Phi) + \chi(1-\Phi)^2 \tag{28}$$

with Φ as the volume fraction of dissolved biopolymer, R the gas constant, T the absolute temperature, and N ¼ vS/vW is the ratio of the molar volume of solute and water, μ_s and $\mu_{w,}$ respectively. For the starch-water mixture N is taken to be infinite. In comparatively moist polysaccharides, Flory-Huggins (FH) theory is applicable [80]. This theory is also useful to explain the melting point depression of the starch crystalline phase [81 - 83]. The FH theory has been used frequently to narrate the moisture sorption behavior of starch [84], maltodextrins [85, 86], dextrans and pullulans [87]. The materials of biological origin as like food proteins, cheese, dried fruits, honey and candy are showing interest in the application of WLF (Williams-Landel-Ferry) equation [88 - 90], which is an empirical equation that describes the strong temperature dependence of viscosity and relaxation time for polymers. The temperature dependence of several physical/chemical phenomena that takes place during processing and storage has been studied using the WLF equation it is possible to study the temperature dependence of many physical and chemical phenomena which happens during processing and storage. The WLF equation has found extensive applications in synthetic polymer research where it is used in many areas of food polymers where temperature is critical, such as crystallization, recrystallization, and non-enzymatic browning reactions. Most importantly, literature suggests that studies on biomaterials have shown that the constants are material specific and should be obtained experimentally [91]. To describe the behavior observed in fragile systems, the VFT equation has often been employed [92]. The VFT expression of the viscosity was given by:

$$\eta = \eta_0 \exp\left(\frac{B_{VFT}}{T - T_0}\right) \quad \textbf{(29)}$$

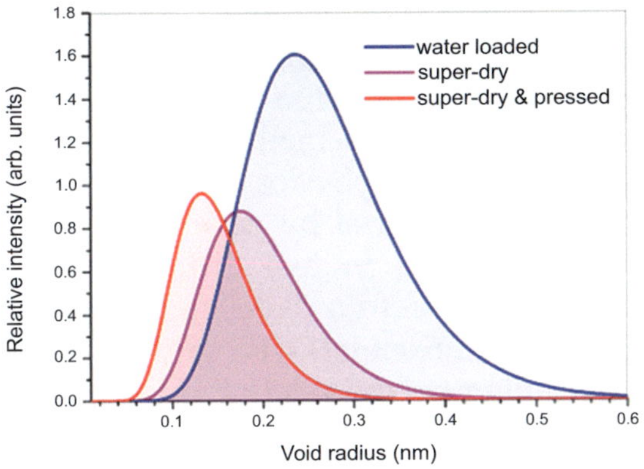

Fig. (6). Relative intensity for different void radius ("Reprinted with permission from Ref. No: 97 Copyright {2014} ACS.").

where T is the absolute temperature, η_0 is the value of the viscosity at the high temperature limit, and B_{VFT} is a quantity related with the activation barrier. T_0 is the so called ideal glass transition temperature at which the viscosity or the relaxation time diverges, or the configurational entropy of the supercooled liquid vanishes [88 - 90]. Bio-polymers are distinguished for their inclination to absorb moisture as they are hygroscopic. Water works as plasticizer for biopolymer matrices and the glass transition temperature of these systems decreases strongly as function of increasing water content [93].Diffusional mobility of small molecules in biopolymer matrices, diffusional mobility of tiny molecules of water, gases, organic compounds increase rapidly when water content increases. It happens in the glassy as well in rubbery state also [94]. Literature suggests that PALS studies shows that the sorption of water molecules introduces complex changes in the molecular arrangements of biopolymer matrices [95]. An important observation was made by M. Roussenova *et al.* [96] that in all cases water has a strong plasticizing effect on the biopolymer matrices, the observed changes in v_h (hole volume) as probed by Ps are system-specific (see Fig. **6**). Moreover, they suggested that to engage Positron Annihilation lifetime spectroscopy to investigate free volume properties in a broad range of biopolymers is fruitful for Pharmaceutical industry. However, to study the free volume characteristic behavior in glucose-gelatin compounds, on dependence of water loading, drying, was investigated with PALS by Christoph Hugenschmidt *et al.* [97].

Positron Annihilation Spectroscopy (PAS): Basics and Instrumentation

The study of positron-matter interaction has a history over last several decades. The state of the positron-electron system in the matter has been found to be only responsible for the characteristics of quantum electrodynamics annihilation process. The positron annihilation method has been vastly explored and has been established as a purely non-destructive method of obtaining information on various problems in condensed matter physics, it is non-destructive because all information is carried out of the material by penetrating annihilation radiation. Positron-electron annihilation occurs in two different ways, namely, the 2γ –annihilation from free positron-electron state and the annihilation from the positron –electron bound state (positronium) with subsequent emission of 2γ or 3γ annihilation radiations depending on the relative orientation of their spins The interaction of positrons with electrons in condensed matter has made it possible to use positrons as the probe for the determination of the electronic structure of materials. Also it has been realized that positrons have strong affinity to structural defects in materials and the interactions of positrons with surrounding medium in presence of defects play an important role in providing information about the nature, concentration and size of the defects. Positron-electron annihilation process: Energetic positrons emitted from a suitable radioactive isotope enter the material and lose their energies due to collisions, electron-hole pair creation [98] and finally by phonon emission [99]. In the process their energies drop from several hundred keV to thermal energy *i.e.* ~kT, k being the Boltzmann constant and T is the material temperature within 10^{-12} seconds and thus reach in thermal equilibrium with the host material. Depending on the initial energy the penetration depth of positron in material varies from 10 to 1000μm ensuring that the annihilation event occurs at the bulk rather than on the surface of the material. The positron can attain such a low energy in contrast to electron because there is only one positron at a time in the material and thus the exclusion principle does not prevent it from going to its low energy state. Positron-electron annihilation process converts the particle masses into electromagnetic energy and is a relativistic process. The probability of the various modes of slow positron –electron annihilation process to occur is determined by several selection rules as derived from the invariance properties of quantum electrodynamics. The released amount of energy in the annihilation process is same as the total rest mass energy of the annihilating pair *i.e.* ~2 m_oC^2. The conservation of momentum for slowly moving annihilating pair demands that there should be at least one or two bodies present (particles or quanta) in order to absorb the total momentum of the annihilating pair to complete the annihilation process. One gamma annihilation is possible only when a third body, an electron or a nucleus is present to absorb the recoil momentum. Fig. (7) shows the Feynman diagram for two and three photon annihilations [100].

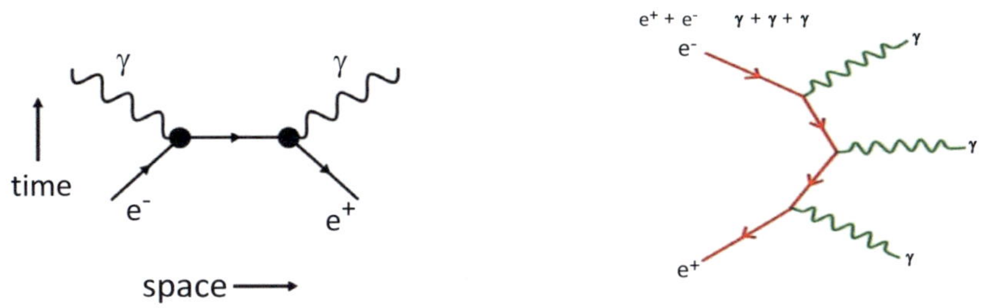

Fig. (7). Feynman diagram.

The cross section for any particular process is multiplied by a factor of the order of fine structure constant $\alpha = 1/137$ due to the introduction of an additional vertex for that particular process and thus the cross section for three photon annihilation is more than two orders of magnitude smaller than that for the two photon process. Single photon annihilation cross section is further reduced by an additional factor of $\lambda_c^3\,\rho$ where λ_c is the Compton wavelength of electrons and ρ is the density of the electrons or atoms. In any physical situation the largest value that ρ can have is such that it can make the additional factor of the order of α^3. Hence, the probability of two photon annihilation is considerably greater than that for one or three photon annihilation, the ratios of the cross sections for respective processes are $\sigma_{(3)} / \sigma_{(2)} \approx \alpha$ and $\sigma_{(1)} / \sigma_{(2)} \approx \alpha^4$ where σ is the cross section for the process and α is the fine structure constant. Only in case of a spin correlated state like ortho positronium where the selection rule forbids two photon processes, three gamma annihilation can occur. The cross- section of the 2γ – annihilation for a free positron and a stationary electron as obtained by Dirac [31] is given by,

$$\sigma(2) = \frac{\pi r_o^2}{\gamma+1}\left[\frac{\gamma^2+4\gamma+1}{\gamma^2-1}\ln\left(\gamma+\sqrt{\gamma^2-1}\right)-\frac{\gamma+3}{\sqrt{(\gamma^2-1)}}\right] \tag{30}$$

where $\gamma = (1 - v^2/c^2)^{-1/2}$ and r_o = the classical electron radius = e^2/m_oc^2

In the limit of non-relativistic velocities *i.e.* at low positron energies, the cross section reduces to

$$\sigma_{(2)} = \pi r_o^2 c / v \quad \textbf{(31)}$$

where v is the electron - positron relative velocity, c is the velocity of light and r_o is the classical electron radius [101]. It is found from Dirac's result that for large v, $\sigma_{(2)}$ becomes negligible in comparison to the cross-section of the processes of ionization and excitation, thus these processes are predominant as long as they are energetically possible. Hence, Dirac's free annihilation probability or the annihilation rate can be expressed as,

$$\lambda = \sigma_{(2)} v n_e = \pi r_o^2 c n_e \quad \textbf{(32)}$$

i.e. independent of positron velocity and is directly proportional to the electron density at positron site. The inverse of the annihilation rate gives the lifetime of positron. Thus the annihilation rate can provide a direct determination of the electron concentration encountered by a positron. It is therefore possible to use positron as probe for the determination of electron density in the material. The mean lifetime of positron in condensed matter generally varies from 100 pico-seconds to 10 micro-seconds depending on the electron density at the annihilation site.

Due to the conservation principles of momentum and energy two other characteristics are observed for two- photon annihilation process in view of their relative emission directions and energies. It can, thus, provide useful information about the initial state of the positron – electron annihilating pair. The kinetic energy of the annihilating pair is typically few electron volts and in center of mass frame the two photons emit strictly in the opposite directions each with an energy equals to the rest mass energy *i.e.* m_0c^2 (= 511 keV) whereas in laboratory frame the emission of two photons deviate from colinearity by an angle, very small (<1°) because of the non-zero component of the momentum of the annihilating pair. The thermalized positrons have zero momentum, therefore the observed angular distribution for varying angles represent the momentum distribution of the annihilated electrons in the material. The conservation principles provide a similar characteristics of the two photon annihilation process, that is, the non-zero component of the momentum of the annihilating pair along the emission line of the photons causes a Doppler shift in the energy of the annihilation radiation Thus the line shape of the annihilation radiation also effectively represents the momentum distribution of the annihilating electrons in the material.

On the basis of the three different characteristics of the positron-electron annihilation process three different measurements are evolved, they are, positron lifetime, Doppler broadening annihilation line shape and angular correlation of annihilation radiation measurements respectively. Among the three, positron annihilation lifetime spectroscopy (PALS) provides a quantitative description about the micro structural configuration of the material and the Doppler broadening annihilation radiation (DBAR) line shape measurement that provides the electron momentum distribution are most commonly used techniques and will be discussed in this chapter.

Positron Source

The most widely used positron emitter is ^{22}Na radioactive isotope which emits positron following the decay reaction:

$$^{22}_{11}Na \rightarrow ^{22}_{10}Ne + \beta^{+} + \nu_e + \gamma \quad \textbf{(33)}$$

^{22}Na isotope provides a relatively high positron yield, 90.4%, has a half-life of about 2.6 years, and emits 1.27 MeV γ simultaneously with the emission of positron indicating the birth of positron in its lifetime, generally available as an aqueous solution NaCl for laboratory use. Depending on the necessity of the source activity, few drops of source solution are evaporated on a thin Ni, Al or Kapton foil and is covered by another similar foil. The edges are sealed and the dimension of the drops are extremely minimum so that the source becomes a point source (see Fig. **8**). The source is sandwiched between two identical pieces of the target sample. The decay scheme of ^{22}Na is shown as following:

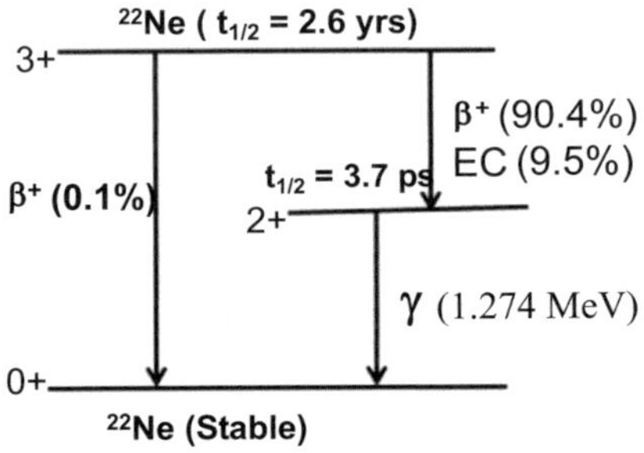

Fig. (8). The decay scheme of ^{22}Na.

Positron Annihilation Lifetime Spectroscopy

Basic principle for the positron lifetime measurement is that when an energetic positrons emitted from a suitable radioactive source enter the material it lose their energy due to ionization and excitation of atoms, excitation of plasmons and finally by creation of electron –hole pairs and phonons. Positrons are then thermalized within the material in a few picoseconds time and during the span of life of 100-400ps positrons migrate a mean distance of typically 100 nm before getting annihilated with electrons. The energy, momentum and spin are conserved during the annihilation process and it leads to the emission of two photons, each with the energy of $mc^2 = 0.511$ MeV (where m is the mass of electron and c is the vacuum velocity of light) in the opposite directions almost collinearly. Because of the conservation of energy, momentum and spin, the emitted annihilation photons carry information about the properties of electrons at the positron site. Thus the analysis of the spectroscopic properties of annihilation photons with respect to their energy, angular distribution and moment of emission can provide detail information regarding the properties of the electrons participating in the annihilation process.

Fig. **9** shows the principle of positron annihilation measurements. ^{22}Na radio isotope emits a 1.28MeV photon almost simultaneously with the emission of positron and that is considered as the 'birth signal' of positron. When positron – electron annihilates, 0.511 MeV photons appear as the 'death signal' of positron, Thus the time difference between the 1.28 MeV 'birth' photon and 0.511 MeV 'death' photon provides the lifetime of positron. The mean lifetime of an ensemble of positrons varies inversely to the concentration of electrons at the annihilating positron site. Fig. **10** is a schematic presentation of Positron Lifetime Spectrometer.

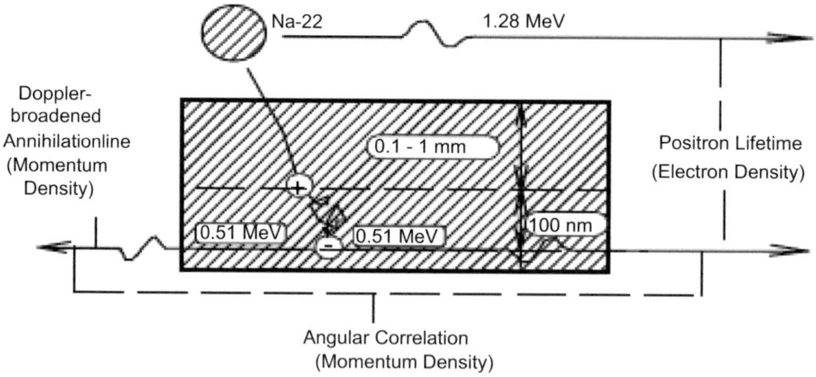

Fig. (9). Principle of Positron Annihilation Measurement.

It is a conventional Fast-Fast coincidence system for nuclear spectroscopy where the time is measured as well as energy selection is done by using the fast channel comprising with the ' Time to Amplitude Converter' and Fast Differential Discriminator. With the help of positron trapping model the experimental lifetime data are analyzed. The positron lifetime spectrum is convoluted with the resolution function of the spectrometer which is assumed to be a single or a sum of two or three Gaussian functions and the experimental lifetime spectrum is fitted with the convoluted spectrum using standard nonlinear fitting procedures. Usually the standard PATFIT and LT9.0 computer programs are used to obtain different characteristic lifetime values with corresponding intensities.

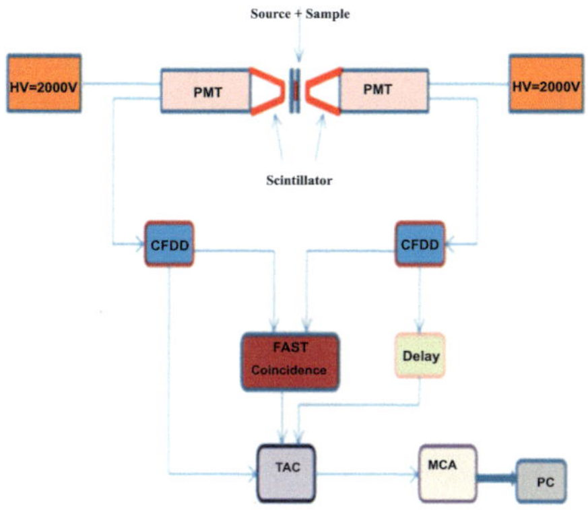

Fig. (10). Schematic diagram of PALS.

There are two other observable effects arising out of the momentum, **p** of the electron-positron annihilating pair, the first effect is due to the deviation of the angle of emission of two annihilating photons from the collinearity by a value of $\theta \sim \tan \theta = p_x / m_o c$ where p_x is the momentum component perpendicular to the original emission direction of photons. Angular Correlation of Annihilation Radiation (ACAR) is measured using twi scintillation detectors in coincidence and the angular correlation curve provides the momentum distribution of the annihilating pair. Since positrons are thermalized before annihilation, the curve effectively gives the momentum distribution of electrons. The second effect occurs due to the deviation of energy of the annihilation photons from $E_o = m_o c^2$ (m_o is the rest mass of an electron) by $+\delta E$ and $-\delta E$ where $\delta E = p_y c /2$, p_y being the momentum component of the annihilating pair parallel to the original emission direction of the annihilation photons, it also effectively reflects the momentum distribution of electrons.

Application of Positron Annihilation Technique in Natural Rubber

Natural rubber is produced from a milky white liquid called latex that oozes from certain plants when they are cut. The latex is having about one third water and one third rubber particles held in the form of a colloidal suspension, it is a polymer of isoprene (also known as 2-methylbuta-1,3-diene) with the chemical formula (C5H8)n. In fact it is made of many thousands of basic C5H8 units *i.e.* the monomer of isoprene, they are loosely joined to make long, tangled chains of molecules which can be untangled if they are pulled apart but they also spring back straight together while released and thus rubber is elastic. However, natural rubber is an important renewable material, owing to its different properties such as its high mechanical performance, good elasticity along with other unique properties it can produce various industrial products [102] though the raw or crude rubber is very little useful as it possess undesirable properties like low tensile strength, elasticity only over a limited range of temperature *etc.* also it becomes softer, more plastic and sticky on heating and brittle on cooling (see Fig. **11**).

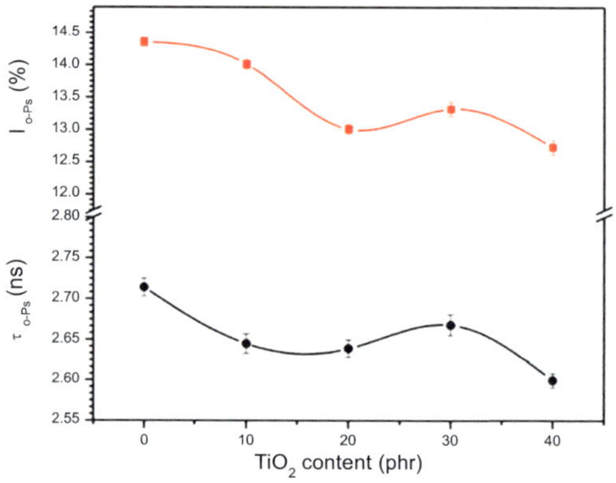

Fig. (11). Variations in positron parameters.

Amorphous polymers like natural rubber contains excess free volume in the amorphous region which is responsible for a decrease in the density by about 10% compared to its crystalline state. Addition of fillers or producing free radicals by irradiation with high energy e- / γ rays are proved to be useful for creating cross linking between polymer chains as well as decreasing the amount of free volumes within the polymers (see Fig. **12**) [103 - 105].

A series of investigations are performed to observe the changes in thermal properties and free volume microstructures in natural rubber while filled with three different fillers (TiO_2, nanoclay, nanosilica) at different concentrations and also after being subjected to high energy electron irradiation.

The results of the investigations show the changes in the positron lifetime parameters with filler concentration which reflect that the size and concentration of free volume holes decrease with the increase of filler concentration. Also the probable creation of new positronium trapping sites at the interface of filler-an--natural rubber can be predicted. In case of high energy electron irradiated natural rubber with the same filler contents, the values of the ortho positronium lifetime,

$\tau_{o\text{-}Ps}$ and the hole free volume Vh are found to be significantly lesser than that of un irradiated samples with same filler concentration [103]. The reason is, irradiation with high energy electron beam sets in two simultaneous processes within the polymer, primarily, it produces free radicals and as a result forms new bonds or cross-linking which decreases o-Ps lifetime as well as free volume. Secondly, it ruptures molecular chains of natural rubber as well as creates new unoccupied spaces (*i.e.* free volume in polymer) and increases o-Ps lifetime as well as the size of the free volumes. The high energy electron irradiation on natural rubber filled with TiO_2 and nano clay filler produces free radicals within the samples which causes cross-linking between natural rubber chains and decrease the o-Ps lifetime as well as the size of free volumes. The difference in the values of positron parameters (*i.e.* ortho positronium lifetime and corresponding intensity) and the calculated mean free volumes and free volume fractions for an irradiated and irradiated natural rubber with TiO_2 filler at different concentrations can be observed from the following figures (see Fig. **13**).

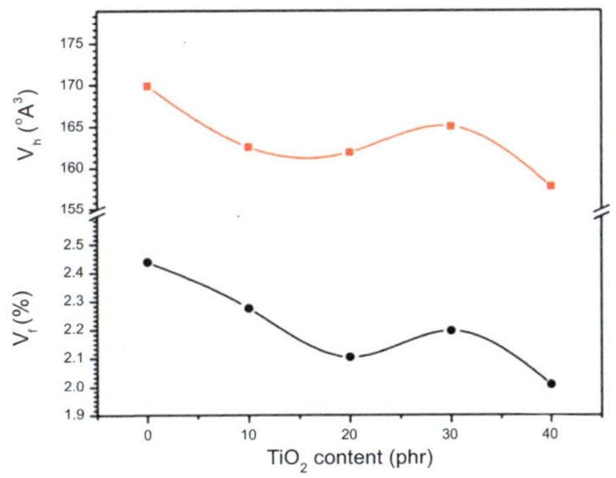

Fig. (12). Variations in V_h and V_f with TiO_2 filler content in un irradiated with TiO_2 filler content in natural rubber. an irradiated natural rubber.

Thus, the above observations can be discussed in the light that since the filler particles have a diameter of sub-nanometer order they fill up the free volume holes within the natural rubber and produce cross linking between the natural rubber chains which decrease the phase separations between different layers of rubber chain resulting to a decrease the amount of free volume in natural rubber (see Fig. **14**). High energy electron irradiation produces free radicals which also increase the cross linking process to a large extent. So besides addition of TiO_2 fillers high energy electron irradiation also causes cross linking between the natural rubber chains and decrease the phase separations between different layers of rubber chain which effectively improve the quality of natural rubber.

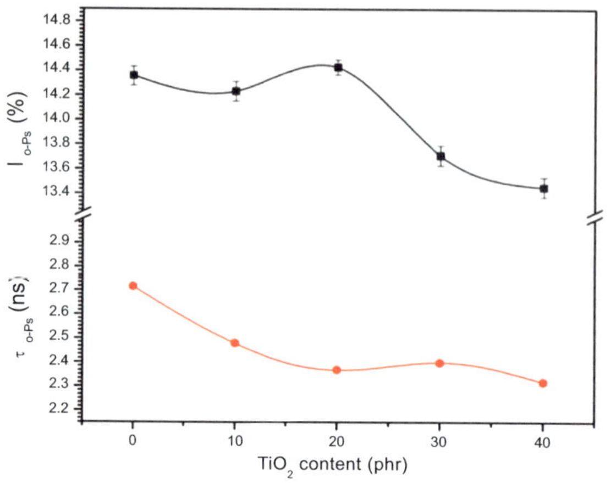

Fig. (13). Variations in positron parameters

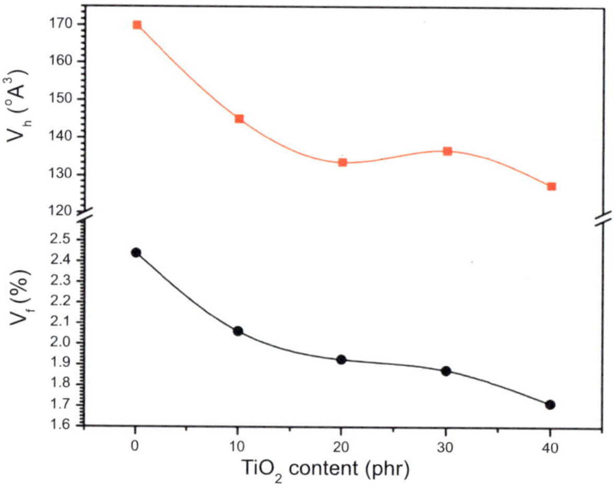

Fig. (14). Variations in V_h and V_f with TiO_2 filler content in un irradiated with TiO_2 filler content in irradiated Natural rubber. natural rubber.

CONCLUSION

This review is an attempt to give an introduction to the application of PALS in the study of polymers, biopolymers and natural rubber. The betterment of time resolution of positron lifetime spectroscopy and due to its sensitivity in measuring defect concentration in the range of 10^{-7}-10^{-4} ppm it has been possible to measure life time even of the order of 100-200ps. Thus PALS has emerged as a sophisticated technique for characterizing defects, micro voids or free volume holes of smaller dimension in polymeric materials. It is known that the quest for a relationship between the motion of molecules and free space in condensed matter has been an important topic for the scientists since the beginning. In the literature different theoretical models have clearly established the fact that the free volume measurements are correlated well with the physical, mechanical and transport properties of polymeric materials as well as bio polymers also. It is important to note that PALS is probably the only direct method of measuring free volume holes and their related content for understanding of the microstructure of polymeric system even in the sub nanometer scale compared to any other technique.

CONFLICT OF INTEREST

The author (editor) declares no conflict of interest, financial or otherwise.

ACKNOWLEDGEMENTS

The author is pleased to acknowledge Dr. A.K Saha, Dr. A. Mandal, Dr.S. Pan, Mr. Amir Sohel all her research scholars for their help and discussion. The author has not receive any fund from any agency to cover any costs for preparation or publication of this article.

REFERENCES

[1] K. David, "Epicurus", *The Stanford Encyclopedia of Philosophy, Summer Edition,* 2014. [Online] Available: http://plato.stanford.edu/archives/sum2014/entries/epicurus/

[2] B.A. Pazmiño Betancourt, P.Z. Hanakata, F.W. Starr, and J.F. Douglas, "Quantitative relations between cooperative motion, emergent elasticity, and free volume in model glass-forming polymer materials", *Proc. Natl. Acad. Sci. USA.* , vol. 112, no. 10, pp. 2966-2971, 2015. [http://dx.doi.org/10.1073/pnas.1418654112] [PMID: 25713371]

[3] H. Eyring, and J. Hirschfelder, "The theory of the liquid state", *J. Phys. Chem.* , vol. 41, no. 2, pp. 249-257, 1937. [http://dx.doi.org/10.1021/j150380a007]

[4] Y. Yu, Positron Annihilation Lifetime Spectroscopy: Studies of amorphous and crystalline molecular material. , PhD Thesis, University of Halle: Germany, 2011.

[5] J.S. Vrentas, and C.M. Vrentas, *Diffusion and Mass Transfer.* CRC Press, 2013, p. 134.

[6] Y.C. Jean, "Advances with Positron Spectroscopy of Solids and surfaces", In: *NATO Advanced Research workshop.* Italy, July 1993.

[7] M.H. Cohen, and D. Turnbull, "Molecular transport in liquids and glasses", *J. Chem. Phys.* , vol. 31, no. 5, pp. 1164-1169, 1959. [http://dx.doi.org/10.1063/1.1730566]

[8] D. Turnbull, and M.H. Cohen, "Free-volume model of the amorphous phase: glass transition", *J. Chem. Phys.* , vol. 34, no. 1, pp. 120-125, 1961. [http://dx.doi.org/10.1063/1.1731549]

[9] D.V. Krevelen, *Properties of polymers, their estimation and correlation with chemical structure.* 2nd ed Elsevier, 1976.

[10] H. Fujita, "Notes on free volume theories", *Polym. J.* , vol. 23, no. 12, pp. 1499-1506, 1991. [http://dx.doi.org/10.1295/polymj.23.1499]

[11] A.J. Batschinski, "Untersuchungen Aber die innere Reibnng der Flüssigkeiten. I", *Z. Phys. Chem.* ,

vol. 84, no. 1, pp. 643-706, 1913.
[http://dx.doi.org/10.1515/zpch-1913-0142]

[12] T. Egami, V. Levashov, R.S. Aga, and J.R. Morris, "Atomic dynamics in metallic liquids and glasses", *Mater. Trans.* , vol. 48, no. 7, p. 1729, 2007.
[http://dx.doi.org/10.2320/matertrans.MJ200763]

[13] A.K. Doolittle, "Studies in Newtonian flow. II. The dependence of the viscosity of liquids on free-space", *J. Appl. Phys.* , vol. 22, no. 12, pp. 1471-1475, 1951.
[http://dx.doi.org/10.1063/1.1699894]

[14] F.N. Kelley, and F. Bueche, "Viscosity and glass temperature relations for polymer-diluent systems", *J. Polym. Sci., Polym. Phys. Ed.* , vol. 50, no. 154, pp. 549-556, 1961.

[15] L.J. Gerhardt, A. Garg, Y.C. Bae, C.W. Manke, and Es. Gulari, "Effect of dissolved Carbon Dioxide on the viscosity of liquid PDMS", *Proc. of the XIth congr. on Rheology,* 1992pp. 348-350 Brussels
[http://dx.doi.org/10.1016/B978-0-444-89007-8.50141-6]

[16] A. Triolo, A. Mandanici, O. Russina, V. Rodriguez-Mora, M. Cutroni, C. Hardacre, M. Nieuwenhuyzen, H.J. Bleif, L. Keller, and M.A. Ramos, "Thermodynamics, structure, and dynamics in room temperature ionic liquids: the case of 1-butyl-3-methyl imidazolium hexafluorophosphate ([bmim][PF6])", *J. Phys. Chem. B.* , vol. 110, no. 42, pp. 21357-21364, 2006.
[http://dx.doi.org/10.1021/jp062895t] [PMID: 17048965]

[17] J.S. Vrentas, and C.M. Vrentas, "Predictive methods for self-diffusion and mutual diffusion coefficients in polymer–solvent systems", *Eur. Polym. J.* , vol. 34, p. 797, 1998.
[http://dx.doi.org/10.1016/S0014-3057(97)00205-X]

[18] J.S. Vrentas, and J.L. Duda, "Diffusion of small molecules in amorphous polymers", *Macromol.* , vol. 9, no. 5, pp. 785-790, 1976.
[http://dx.doi.org/10.1021/ma60053a019]

[19] J.S. Vrentas, and J.L. Duda, "Diffusion in polymer—solvent systems. I. Reexamination of the free-volume theory", *J. Polym. Sci., Polym. Phys. Ed.* , vol. 15, pp. 403-416, 1977.
[http://dx.doi.org/10.1002/pol.1977.180150302]

[20] J.L. Duda, J.S. Vrentas, S.T. Ju, and H.T. Liu, "Prediction of diffusion coefficients for polymer-solvent systems", *AIChE J.* , vol. 28, no. 2, pp. 279-285, 1982.
[http://dx.doi.org/10.1002/aic.690280217]

[21] H. Fujita, "Diffusion in polymer-diluent systems", In: *Fortschr. Hochpolym, Forsch.* vol. 3. 1961no. 1, .

[22] H. Fujita, A. Kishimoto, and K. Matsumoto, "Concentration and temperature dependence of diffusion coefficients for systems polymethyl acrylate and n-alkyl acetates", *Trans. Faraday Soc.* , vol. 56, pp. 424-437, 1960.
[http://dx.doi.org/10.1039/tf9605600424]

[23] J.S. Vrentas, C.M. Vrentas, and J.L. Duda, "Comparison of free-volume theories", *Polym. J.* , vol. 25, no. 1, pp. 99-101, 1993.
[http://dx.doi.org/10.1295/polymj.25.99]

[24] A. Mafi, A. Raisi, M. Hatam, and A. Aroujalian, "A comparative study on the free volume theories for diffusivity through polymeric membrane in pervaporation process", *J. Appl. Polym. Sci.* , vol. 131, no. 15, p. 40581, 2014.
[http://dx.doi.org/10.1002/app.40581]

[25] T.G. Fox, and P.J. Flory, "Second-order transition temperatures and related properties of polystyrene. I. Influence of molecular weight", *J. Appl. Phys.* , vol. 21, p. 581, 1950.
[http://dx.doi.org/10.1063/1.1699711]

[26] T.G. Fox, and P.J. Flory, "The glass temperature and related properties of polystyrene. Influence of molecular weight", *J. Polym. Sci., Polym. Phys. Ed.* , vol. 14, p. 315, 1954.

[27] L.H. Sperling, *Introduction to Physical Polymer Science.* 2nd John Wiley & Sons: Inc.: New York, 1992.

[28] D. Kivelson, and G. Tarjus, *Physics of Glasses,* P. Jund, R. Jullien ~AIP, Eds., Melville, NY, 1999.

[29] R.P. White, and J.E. Lipson, "Free Volume in the Melt and How It Correlates with Experimental Glass Transition Temperatures: Results for a Large Set of Polymers", *ACS Macro Lett.* , vol. 4, no. 5, pp. 588-592, 2015. [http://dx.doi.org/10.1021/acsmacrolett.5b00217]

[30] H. Takeuchi, and K. Okazaki, "Molecular dynamics simulation of diffusion of simple gas molecules in a short chain polymer", *J. Chem. Phys.* , vol. 92, p. 5643, 1990. [http://dx.doi.org/10.1063/1.458496]

[31] P.A. Dirac, "Quantized singularities in the electromagnetic field", *Proc. R. Soc. Lond. A.* , vol. 133, pp. 60-72, 1931. [http://dx.doi.org/10.1098/rspa.1931.0130]

[32] C.D. Anderson, "Cosmic-ray positive and negative electrons", *Phys. Rev.* , vol. 44, p. 406, 1933. [http://dx.doi.org/10.1103/PhysRev.44.406]

[33] P. M.S. Blackett, *Nobel Lecture.* , December 13, 1948.

[34] P.M. Blackett, and G.P. Occhialini, "The relationship between viscosity, elasticity and plastic strength of a soft material as illustrated by some mechanical properties of flour dough", *III. Proc. Royal Society of London. Series A.* , vol. 139, 1933no. 839, pp. 699-720+722+724+726. Published by: [online] Available: http://www.jstor.org/stable/96057

[35] S. Mohorovicic, "Möglichkeit neuer elemente und ihre bedeutung für die astrophysik", *Astron. Nachr.* , vol. 253, p. 94, 1934. [http://dx.doi.org/10.1002/asna.19342530402]

[36] A.E. Ruark, "Positronium", *Phys. Rev.* , vol. 68, p. 278, 1945. [http://dx.doi.org/10.1103/PhysRev.68.278]

[37] A. Ore, *Univ. i. Bergen Arobok, Naturvitenskap. Rekke, No. 9.*

[38] M. Deutsch, "Evidence for the formation of positronium in gases", *Phys. Rev.* , vol. 82, p. 455, 1951. [http://dx.doi.org/10.1103/PhysRev.82.455]

[39] V.P. Shantarovich, "Positronium Atom in Solids–Peculiarities of Formation and Interconnection with Free Volume Nanostructure", *J. Nucl. and Radiochem. Sci.* , vol. 7, no. 1, pp. R-37-R52, 2006. [http://dx.doi.org/10.14494/jnrs2000.7.R37]

[40] W. Brandt, S. Berko, and W.W. Walker, "Positronium decay in molecular substances", *Phys. Rev.* , vol. 120, p. 1289, 1960. [http://dx.doi.org/10.1103/PhysRev.120.1289]

[41] M. Eldrup, D. Lightbody, and J.N. Sherwood, "The temperature dependence of positron lifetimes in solid pivalic acid", *Chem. Phys.* , vol. 63, no. 1–2, pp. 51-58, 1981. [http://dx.doi.org/10.1016/0301-0104(81)80307-2]

[42] S.J. Tao, "Positronium annihilation in molecular substances", *J. Phys. Chem.* , vol. 56, p. 5499, 1972. [http://dx.doi.org/10.1063/1.1677067]

[43] D.M. Schrader, Y.C. Jean, Ed., *Positron and positronium chemistry*Elsevier Sci. Pub.: Amsterdam, 1988.

[44] H. Nakanishi, S.J. Wang, and Y.C. Jean, *In positron annihilation studies of fluids; S.C. Sharma,* W. Science, Ed., Singapore, 1988, p. 292.

[45] Y.C. Jean, and H. Shi, *Non Crist. Sol.* , vol. 172–174, p. 806, 1994. [http://dx.doi.org/10.1016/0022-3093(94)90582-7]

[46] M.H. Cohen, and D. Turnbull, "Molecular transport in liquids and glasses", *J. Chem. Phys.* , vol. 31, p. 1164, 1959.
[http://dx.doi.org/10.1063/1.1730566]

[47] G.S. Grest, and M.H. Cohen, "Liquids, glasses, and the glass transition: A free-volume approach", *Adv. Chem. Phys.* , vol. 48, p. 455, 1981.
[http://dx.doi.org/10.1002/9780470142684.ch6]

[48] Y. Kobayashi, W. Zheng, E.F. Meyer, J.D. McGrvey, A.M. Jamieson, and R. Simha, "Free volume and physical aging of poly (vinyl acetate) studied by positron annihilation", *Macromol.* , vol. 22, no. 5, pp. 2302-2306, 1989.
[http://dx.doi.org/10.1021/ma00195a052]

[49] J. Liu, Q. Deng, and Y.C. Jean, "Free-Volume Distributions of Polystyrene Probed by Positron-Annihilation-Comparison with Free-Volume Theories", *Macromol.* , vol. 26, p. 7149, 1993.
[http://dx.doi.org/10.1021/ma00078a006]

[50] R. Simha, and T. Somcyns, "On the statistical thermodynamics of spherical and chain molecule fluids", *Macromol.* , vol. 2, no. 4, pp. 342-350, 1969.
[http://dx.doi.org/10.1021/ma60010a005]

[51] R. Simha, "Configurational thermodynamics of the liquid and glassy polymeric states", In: *Macromol.* vol. 10. 1977, no. 5, pp. 1025-1030.
[http://dx.doi.org/10.1021/ma60059a028]

[52] P.J. Flory, "Molecular theory of rubber elasticity", *Polym. J.* , vol. 18, p. 18, 1985.

[53] M.L. Huggins, "Solutions of long chain compounds", *J. Chem. Phys.* , vol. 9, p. 440, 1941.
[http://dx.doi.org/10.1063/1.1750930]

[54] P.J. Flory, "Thermodynamics of high polymer solutions", *J. Chem. Phys.* , vol. 9, pp. 660-661, 1941.
[http://dx.doi.org/10.1063/1.1750971]

[55] P.J. Flory, "Thermodynamics of high polymer solutions", *J. Chem. Phys.* , vol. 10, pp. 51-61, 1942.
[http://dx.doi.org/10.1063/1.1723621]

[56] G. Dlubek, J. Pionteck, and D. Kilburn, "The Structure of the Free Volume in Poly (styrene-c--acrylonitrile) from Positron Lifetime and Pressure Volume Temperature (PVT) Experiments", *Macromol. Chem. Phys.* , vol. 205, pp. 500-511, 2004.
[http://dx.doi.org/10.1002/macp.200300103]

[57] G. Dlubek, A. Sengupta, J. Wawryszczuk, T. Goworek, R. Krause-Rehberg, D. Kilburn, M.A. Alam, J. Pionteck, H. Kaspar, and K.H. Lochhaas, *Acta Phys. Pol. A.* , vol. 107, no. 4, pp. 685-689, 2005.
[http://dx.doi.org/10.12693/APhysPolA.107.685]

[58] G. Dlubek, A. SenGupta, J. Pionteck, R. Krause-Rehberg, H. Kaspar, and K.H. Lochhaas, "Temperature dependence of the free volume in fluoroelastomers from positron lifetime and PVT experiments", *Macromol.* , vol. 37, no. 17, pp. 6606-6618, 2004.
[http://dx.doi.org/10.1021/ma049067n]

[59] G. Dlubek, D. Kilburn, and M.A. Alam, "Comments to the paper "The need to reconsider tradional free volume theory for polymer electrolytes"", *Electrochim. Acta.* , vol. 49, no. 28, pp. 5241-5247, 2004.
[http://dx.doi.org/10.1016/j.electacta.2004.06.032]

[60] G. Dlubek, M.Q. Shaikh, R. Krause-Rehberg, and M. Paluch, "Effect of free volume and temperature on the structural relaxation in polymethylphenylsiloxane: a positron lifetime and pressure-volum--temperature study", *J. Chem. Phys.* , vol. 126, no. 2, p. 024906, 2007.
[http://dx.doi.org/10.1063/1.2406077] [PMID: 17228972]

[61] T.G. Fox, and P.J. Flory, "Second-order transition temperatures and related properties of polystyrene. I. Influence of molecular weight", *J. Appl. Phys.* , vol. 21, pp. 581-591, 1950.

[http://dx.doi.org/10.1063/1.1699711]

[62] R. Simha, and R.F. Boyer, "On a general relation involving the glass temperature and coefficients of expansion of polymers", *J. Chem. Phys.*, vol. 37, pp. 1003-1007, 1962. [http://dx.doi.org/10.1063/1.1733201]

[63] T. Goworek, "Comments on the relation: positronium lifetime–free volume size parameters of the Tao–Eldrup model", *Chem. Phys. Lett.*, vol. 366, no. 1-2, pp. 184-187, 2002. [http://dx.doi.org/10.1016/S0009-2614(02)01569-5]

[64] T. Goworek, K. Ciecielski, B. Jansiniska, and J. Wawryszczuk, "Positronium in large voids. Silicagel", *Chem. Phys. Lett.*, vol. 272, pp. 91-95, 1997. [http://dx.doi.org/10.1016/S0009-2614(97)00504-6]

[65] T. Goworek, K. Ciesielski, B. Jasinska, and J. Wawryszczuk, "Temperature variations of average o-Ps lifetime in porous media", *Radiat. Phys. Chem.*, vol. 58, no. 5-6, pp. 719-722, 2000. [http://dx.doi.org/10.1016/S0969-806X(00)00246-2]

[66] T. Goworek, "Positronium as a probe of small free volumes", *J. Nucl. and Radiochem. Sc.*, vol. 1, no. 1, pp. 11-13, 2000. [http://dx.doi.org/10.14494/jnrs2000.1.11]

[67] K-S. Liao, H. Chen, S. Awad, J-P. Yuan, W-S. Hung, K-R. Lee, J-Y. Lai, C-C. Hu, and Y.C. Jean, "Determination of free-volume properties in polymers without orthopositronium components in positron annihilation lifetime spectroscopy", *Macromol.*, vol. 44, pp. 6818-6826, 2011. [http://dx.doi.org/10.1021/ma201324k]

[68] A.K. Mohanty, M. Misra, L.T. Drzal, Ed., *Natural Fibers, Biopolymers and Biocomposites*CRC Press, 2005. [http://dx.doi.org/10.1201/9780203508206]

[69] R. Chandra, and R. Rustgi, "Biodegradable polymers", *Prog. Polym. Sci.*, vol. 23, p. 1273, 1998. [http://dx.doi.org/10.1016/S0079-6700(97)00039-7]

[70] M.A. Meyers, P.Y. Chen, A.Y. Lin, and Y. Seki, "Biological materials: structure and mechanical properties", *Prog. Mater. Sci.*, vol. 53, p. 1, 2008. [http://dx.doi.org/10.1016/j.pmatsci.2007.05.002]

[71] A. Kumar, "Akshay Srivastava, Igor Yu Galaev, Bo Mattiasson. Smart polymers: Physical forms and bioengineering applications", *Prog. Polym. Sci.*, vol. 32, p. 1205, 2007. [http://dx.doi.org/10.1016/j.progpolymsci.2007.05.003]

[72] D.J. Mcclements, *Comprehensive Reviews in Food Science and Food Safety.* vol. 9. (c) John Wiley & Sons, 2010, pp. 374-397.

[73] M.H. Cohen, and G.S. Grest, "Liquid-glass transition, a free-volume approach", *Phys. Rev. B.*, vol. 20, p. 1077, 1979. [http://dx.doi.org/10.1103/PhysRevB.20.1077]

[74] S. Kasapis, "Advanced topics in the application of the WLF/free volume theory to high sugar/biopolymer mixtures: a review", *Food Hydrocoll.*, vol. 15, no. 4–6, pp. 631-641, 2001. [http://dx.doi.org/10.1016/S0268-005X(01)00048-0]

[75] J. Zhang, and G. Zografi, "The relationship between BET and free volume-derived parameters for water vapor absorption into amorphous solids", *J. Pharm. Sci.*, vol. 89, no. 8, pp. 1063-1072, 2000. [http://dx.doi.org/10.1002/1520-6017(200008)89:8<1063::AID-JPS11>3.0.CO;2-0] [PMID: 10906730]

[76] A.P. Pierlot, "Water in wool", *Text. Res. J.*, vol. 69, no. 2, p. 97, 1999. [http://dx.doi.org/10.1177/004051759906900204]

[77] X. He, A. Fowler, and M. Toner, "Water activity and mobility in solutions of glycerol and small molecular weight sugars: Implication for cryo-and lyopreservation", *J. Appl. Phys.*, vol. 100, p.

074702, 2006.
[http://dx.doi.org/10.1063/1.2336304]

[78] J. Ubbink, M.I. Giardiello, and H.J. Limbach, "Sorption of water by bidisperse mixtures of carbohydrates in glassy and rubbery states", *Biomacromolecules.* , vol. 8, no. 9, pp. 2862-2873, 2007. [http://dx.doi.org/10.1021/bm0701898] [PMID: 17691841]

[79] P.J. Flory, *Principles of Polymer Chemistry.* Cornell University Press, 1953.

[80] P.J. Flory, "Thermodynamics of crystallization in high polymers. IV. A theory of crystalline states and fusion in polymers, copolymers, and their mixtures with diluents", *J. Chem. Phys.* , vol. 17, no. 3, pp. 223-240, 1949.
[http://dx.doi.org/10.1063/1.1747230]

[81] R. Parker, and S.G. Ring, "Aspects of the physical chemistry of starch", *J. Cereal Sci.* , vol. 34, no. 1, pp. 1-17, 2001.
[http://dx.doi.org/10.1006/jcrs.2000.0402]

[82] I.A. Farhat, and J.M. Blanshard, "Aspects of the Physical Chemistry of Starch", *Carbohydr. Polym.* , vol. 34, no. 4, pp. 263-265, 1997.
[http://dx.doi.org/10.1016/S0144-8617(97)00086-6]

[83] J.W. Donovan, "Phase transitions of the starch–water system", *Biopolymers.* , vol. 18, no. 2, pp. 263-275, 1979.
[http://dx.doi.org/10.1002/bip.1979.360180204]

[84] N.W. Taylor, J.E. Cluskey, and F.R. Senti, "Water sorption by dextrans and wheat starch at high humidities", *J. Phys. Chem.* , vol. 65, no. 10, pp. 1810-1816, 1961.
[http://dx.doi.org/10.1021/j100827a029]

[85] G.K. Moates, T.R. Noel, R. Parker, and S.G. Ring, "The effect of chain length and solvent interactions on the dissolution of the B-type crystalline polymorph of amylose in water", *Carbohydr. Res.* , vol. 298, no. 4, pp. 327-333, 1997.
[http://dx.doi.org/10.1016/S0008-6215(96)00316-3]

[86] J. Ubbink, M-I. Giardiello, and H-J. Limbach, "Sorption of water by bidisperse mixtures of carbohydrates in glassy and rubbery states", *Biomacromolecules.* , vol. 8, no. 9, pp. 2862-2873, 2007. [http://dx.doi.org/10.1021/bm0701898] [PMID: 17691841]

[87] J. Eckelt, R. Sugaya, and B.A. Wolf, "Pullulan and dextran: uncommon composition dependent Flory-Huggins interaction parameters of their aqueous solutions", *Biomacromolecules.* , vol. 9, no. 6, pp. 1691-1697, 2008.
[http://dx.doi.org/10.1021/bm800217y] [PMID: 18471012]

[88] H. Vogel, "The law of the relation between the viscosity of liquids and the temperature", *Phys. Z.* , vol. 22, pp. 645-646, 1921.

[89] G.S. Fulcher, "Analysis of recent measurements of the viscosity of glasses", *J. Am. Ceram. Soc.* , vol. 8, pp. 339-355, 1925.
[http://dx.doi.org/10.1111/j.1151-2916.1925.tb16731.x]

[90] G. Tammann, W. Hesse, and Z. Anorg, "Die Abhängigkeit der Viscosität von der Temperatur bie unterkühlten Flüssigkeiten", *Allg. Chem.* , vol. 156, pp. 245-257, 1926.
[http://dx.doi.org/10.1002/zaac.19261560121]

[91] M. Erhan Yildiz, Sozer Nesli, and Jozef L. Kokini, ""Encyclopedia of Agricultural, Food, and Biological Engineering"", In: *Williams–Landel–Ferry (WLF) Equation.* Second Edition Taylor & Francis, 2011.

[92] H. Tokuda, K. Hayamizu, K. Ishii, M.A. Susan, and M. Watanabe, "Physicochemical properties and structures of room temperature ionic liquids. 1. Variation of anionic species", *J. Phys. Chem. B.* , vol. 108, pp. 16593-16600, 2004.
[http://dx.doi.org/10.1021/jp047480r]

[93] S. Kasapis, I.T. Norton, and J.B. Ubbink, *Modern biopolymer science bridging the divide between fundamental treatise and industrial application.* , Elsevier/Academic Press, 2009. [online] Available: http://www.123library.org/book_details/?id=45908

[94] R. Tromp, R. Parker, and S. Ring, "Water diffusion in glasses of carbohydrates", *Carbohydr. Res.* , vol. 303, pp. 199-205, 1997. [http://dx.doi.org/10.1016/S0008-6215(97)00160-2]

[95] D. Kilburn, J. Claude, T. Schweizer, A. Alam, and J. Ubbink, "Carbohydrate polymers in amorphous states: an integrated thermodynamic and nanostructural investigation", *Biomacromolecules.* , vol. 6, no. 2, pp. 864-879, 2005. [http://dx.doi.org/10.1021/bm049355r] [PMID: 15762653]

[96] M. Roussenova, and M.A. Alam, "PALS: A unique probe for the molecular organisation of biopolymer matrices", *J. Phys. Conf. Ser.* , vol. 443, p. 012044, 2013. [http://dx.doi.org/10.1088/1742-6596/443/1/012044]

[97] C. Hugenschmidt, and H. Ceeh, "The free volume in dried and H2O-loaded biopolymers studied by positron lifetime measurements", *J. Phys. Chem. B.* , vol. 118, no. 31, pp. 9356-9360, 2014. [http://dx.doi.org/10.1021/jp504504p] [PMID: 25046083]

[98] J.P. Carbotte, and H.L. Arora, "Thermalization time of positrons in metals", *Can. J. Phys.* , vol. 45, p. 387, 1967. [http://dx.doi.org/10.1139/p67-036]

[99] H.J. Mikeska, *Zeit. F. Phys.* , vol. 232, p. 159, 1970.

[100] A. SenGupta, Defect Studies in III-V Compound semiconductors, Alkali Halides and Other Application Oriented Materials Using The Positron Annihilation Technique. , Ph.D Thesis, The University of Calcutta: India, 1987.

[101] R.N. West, "Positron studies of condensed matter", *Adv. Phys.* , vol. 22, p. 263, 1973. [http://dx.doi.org/10.1080/00018737300101299]

[102] M. Ferreira, R.M. Moreno, P.S. Goncalves, and L.H. Mattoso, *Rubber Chem. Technol.* , vol. 75, no. 1, p. 171, 2002. [http://dx.doi.org/10.5254/1.3547668]

[103] A. Mandal, S. Mukherjee, S. Pan, and A. Sengupta, *Int. J. Mod. Phys.* , vol. 22, pp. 112-117, 2013.

[104] A. Mandal, S. Mukherjee, S. Pan, A.K. Saha, and A. Sengupta, "Int. Conf. on Recent Trends in Applied Physics and Material Science", *AIP Conf. Proc.* , vol. 1536, 2013p. 839

[105] A. Mandal, S. Pan, S. Mukherjee, A.K. Saha, S. Thomas, and A. Sengupta, "High energy electron irradiated polystyrene: free volume and thermal properties studied by PALS and DSC", *J. Polym. and Biopolym. Phys. Chem.* , vol. 2, no. 1, p. 25, 2014.

CHAPTER 11

Structural Analysis of Sulfated Polysaccharides

María C. Matulewicz[a,b,*] and **Marina Ciancia**[b,c]

[a] *Universidad de Buenos Aires. Facultad de Ciencias Exactas y Naturales. Departamento de Química Orgánica. Pabellón II, 3er P., Ciudad Universitaria, C1428EGA Buenos Aires, Argentina*

[b] *Universidad de Buenos Aires. Consejo Nacional de Investigaciones Científicas y Técnicas. Centro de Investigación de Hidratos de Carbono (CIHIDECAR). Facultad de Ciencias Exactas y Naturales Pabellón II, 3er P. Ciudad Universitaria, C1428EGA. Buenos Aires, Argentina*

[c] *Universidad de Buenos Aires. Facultad de Agronomía. Departamento de Biología Aplicada y Alimentos, Cátedra de Química de Biomoléculas, Av. San Martín 4453, C1417DSE Buenos Aires, Argentina*

Abstract: Sulfated polysaccharides are widely distributed in nature. The aim of this chapter is to give a brief description of methods of structural characterization of sulfated galactans, fucans, mannans, and arabinans from seaweeds, and sulfated polysaccharides rich in uronic acids (glycosaminoglycans and polysaccharides from green seaweeds), among other sulfated biopolymers.

These polysaccharides are heterogeneous with respect to chain length and sulfate content and must be purified to homogeneity before structural analysis is carried out.

Structural analysis of sulfated polysaccharides may be carried out by chemical methods: carbohydrate and sulfate content, monosaccharide composition, methylation/ ethylation and desulfation-methylation/ethylation, Smith degradation, *etc.*

Herein, the application of matrix-assisted laser-desorption ionization (MALDI) and electrospray ionization (ESI) mass spectrometry (MS), and nuclear magnetic resonance spectroscopy (NMR) in some of their wide variety of experimental methods is described.

^{1}H and ^{13}C NMR spectroscopy, together with 2D NMR techniques, provide very important information about sequence, interresidue linkage position and substitution pattern. MALDI-MS and ESI-MS have become important tools in the last years, however, they have limitations due to the labile nature of the sulfate group. In MALDI-MS, desorption difficulties with increasing molecular weight were also found. Thus, before application of mass spectrometry, depolymerization in controlled conditions is often required.

* **Corresponding author María C. Matulewicz**: CIHIDECAR-CONICET-UBA, Departamento de Química Orgánica, Facultad de Ciencias Exactas y Naturales, Universidad de Buenos Aires, Buenos Aires, Argentina; E-mail: cristina@qo.fcen-uba.ar

Martin Masuelli & Denis Renard (Eds.)

Keywords: ESI-MS, Glycosaminoglycans, MALDI-MS, NMR spectroscopy, Sulfated arabinans, Sulfated fucans, Sulfated galactans, Sulfated polysaccharides.

INTRODUCTION

Sulfated polysaccharides are extensively distributed in nature organisms from seaweeds to mammals. Their presence in marine organisms has been associated with their adaptation to the high ionic strength characteristic of marine environment. This is thought to be the reason why they are present in seaweeds and marine plants, but they are not found in terrestrial plants and fresh water algae. They have also been found in marine invertebrates, like sea urchins, sea cucumbers, and ascidians. The high content of sulfate groups on their chains makes them polyanionic compounds of electrostatic-dependent intermolecular binding properties with relevant health/disease-related basic proteins or positively charged assemblies like basic amino acid clusters in virus particles. Overall, the specific bindings of sulfated polysaccharides with basic proteins (or domains in viral particles) and thus binding affinities in molecular interactions are found in accordance with the resultant pharmacological actions showing effectiveness of these glycans. The nature of these interactions is not a mere consequence of the net charge due to the degree of sulfation, but it is mostly driven by stereospecific features like monosaccharide composition, sulfation and glycosylation sites, anomericity, and conformational preference in their backbones.

In terms of total biomass, sulfated polysaccharides from marine organisms are more abundant than glycosaminoglycans. However, as the latter are present in the extracellular matrix of all animal cell surfaces, and they are important for human health, they have driven the highest research efforts and human and economic resources, resulting in very detailed knowledge of their complex structures. This was possible through the development of several advanced analytical methodologies [1 - 3].

The chemical structure of sulfated polysaccharides is very diverse, but they can be divided in the following major groups according to characteristics of their backbones.

1. Sulfated Galactans

Sulfated galactans from red seaweeds have linear chains of alternating 3-linked β-galactopyranosyl residues (A-units) and 4-linked α-galactopyranosyl residues (B-units). The A-units always belong to the D-series, whereas the B-units include residues of the D- or L-series, many times occurring as 3,6-anhydrogalacto-pyranosyl moieties [4]. They are classified in carrageenans and agarans according to whether the 4-linked residues belong to the D- or L-series, respectively. The neutral, completely cyclized agaran is called agarose, commercially known as agar. Most of the other galactans have varied degrees of sulfation, as well as other substituents, as pyruvic acid ketal (occurring as 3-linked 4,6-*O*-(1′-carboxy) ethylidene-D-galactopyranose units), methoxyl groups and, sometimes, side chains, usually comprising single stubs of β-D-xylose. Fig. (**1A** and **1B**) show the structure of the commercially most important polysaccharides of this group.

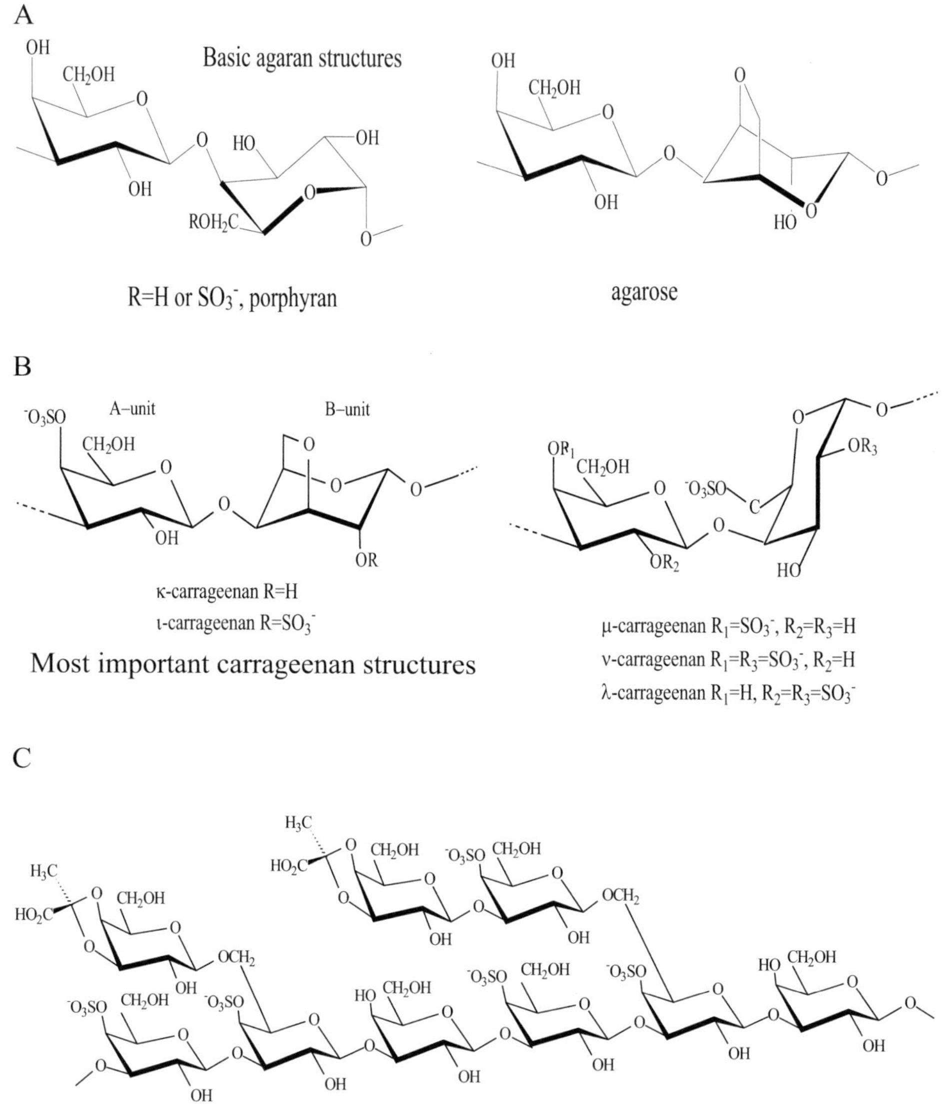

Fig. (1). Some sulfated galactan structures. A. Most important agaran structures. B. Major carrageenan structures. C. Galactans from *Codium* spp.

A sulfated galactan was isolated from the marine sea grass, *Rupia maritima* (Angiospermae, Spermatophyta) [5]. It was composed of regular tetrasaccharide repeating units containing both 4-linked α- and 3-linked β-D-galactose residues; however, these units were not distributed in an alternating order, but in the following regular tetrasaccharide repeating unit: [→3)-β-D-Gal 2-sulfate-(1→4) - α-D-Gal-(1→4)-α-D-Gal-(1→3)-β-D-Gal 4-sulfate-(1→]. It was localized in the plant cell walls, mostly in rhizomes and roots.

On the other hand, sulfated 3-linked β-D-galactans were found in green seaweeds from the Bryopsidales. These galactans showed complex structures with different sulfation patterns and a high amount of pyruvic acid ketals (Fig. **1C**). Some of these galactans were linear, while others had a high degree of branching on C-6, mostly with single stubs of 3,4-*O*-(1′-carboxy)ethylidene-β-D-galactopyranose units (*S*-isomer). Six-membered cyclic ketals linked to *O*-4 and *O*-6 of backbone units were also present in some cases. Different sulfation patterns were found for seaweeds of the suborders Bryopsidineae and Halimedineae [6, 7].

In the marine environment, the sulfated galactans were also described in some species of invertebrates. The discovery of sulfated galactans in tunicates (ascidians) was the first report of the existence of a polysaccharide composed exclusively of L-galactose residues [8]. These units constitute a 4-linked linear structure with sulfate groups on C-3. Heterogeneity and variation among different ascidian sulfated galactans species arise from the occurrence of non-sulfated L-galactose or D-glucose branches. A linear 3-linked α-L-galactan, sulfated on C-2 was isolated from the egg jelly layer of the sea urchin *Echinometra lucunter,* however, other urchins produce sulfated fucans (see later) [9].

2. Sulfated Fucans

Brown seaweeds synthesize fucoidans as sulfated polysaccharides. These sulfated polysaccharides gained much attention due to the fact that many of them have high and interesting biological activities [10, 11]. It is very difficult to systematize the fine structure of the fucoidans due to their complexity and variability. However, four major groups have been defined [12] within fucoidan polymers: 1) those containing high proportions of fucose, galactose and sulfate groups (galactofucans), which comprise a core of 6-linked β-D-galactopyranose units, 2) another group containing just fucose and sulfate (fucans), 3) fucoglucuronans, containing different monosaccharide moieties, less sulfate, and important proportions of uronic acids, and 4) fucomannoglucuronans, similar to those of the latter group, but rich in mannose (the latter two groups could also be included in only one group). Two main types of fucans have been defined: one carrying alternating 3- and 4-linked α-L-fucopyranosyl units and another one carrying mainly 3-linked α-L-fucopyranosyl units. These backbones are highly sulfated, mainly on C-2 and/or C-4, and sometimes also branched. For the galactofucan, a core containing 6-linked β-D-galactopyranosyl units has been postulated, whereas for the glucuronic acid-containing polymers, four different motifs were encountered. Xylose was found to appear mainly as branching points. Some terminal fucofuranose units have also been found [10].

In contrast to the algal polysaccharides, sulfated fucans from sea urchins and sea cucumbers (Echinodermata, Holothuroidea) have simple structures of linear chains of α-L-fucose in well-defined repetitive patterns [13]. Their specific sulfation patterns and the positioning of glycosidic bonds vary with the species. The simplest sea urchin sulfated fucan, composed of 3-linked α-L-fucopyranose 2-sulfate units, was found in the egg jelly of *Strongylocentrotus franciscanus* [14]. Other sea urchins so far studied possess sulfated fucans with different numbers of residues in the repeating units, which vary among the different species according to the position of the glycosidic linkages [α-(1→3) or α-(1→4)] and the sulfation on C-2 or C-4.

3. Sulfated Mannans

Red seaweeds from the order Nemaliales produce 3-linked α-D-mannans sulfated on C-2 and C-6, as those from *Nothogenia fastigiata* [15] or C-4 and C-6, as those from *Nemalion helminthoides* [16]; and in both cases, containing single stubs of β-D-xylopyranose on C-2. Moreover, for *Scinaia hatei* sulfation at C-4 and single stubs of xylose in C-2, C-4 and C-6 was also reported [17].

In addition, sulfated xylomannans were isolated from some species of the genus *Chondrophycus* (Ceramiales), but in this case, they were 4-linked β-D-mannans with sulfate on C-2 and single stus of β-D-xylopyranose or β-D-mannopyranose 2-sulfate linked to C-6 [18]. Recently, mannans with similar backbone and sulfation pattern were detected in green seaweeds from the genus *Codium* (Bryopsidales) [6, 19].

4. Sulfated Arabinans

Recently, a highly sulfated 3-linked pyranosic arabinan was isolated from green seaweed *Codium vermilara* [20] and it has been shown that sulfated pyranosic arabinans are usual components of the sulfated polysaccharide system from many green seaweeds [6], with some structural variants that are emerging.

5. Sulfated Polysaccharides Rich in Uronic Acids

Glycosaminoglycans (GAGs) chains from mammals are composed of disaccharide repeating units (Table **1**) comprising usually one uronic acid (D-glucuronic acid or L-iduronic acid) and one amino sugar (D-galactosamine or D-glucosamine). Hence, GAGs differ according to the type of hexosamine or hexuronic acid unit that they contain, as well as the configuration of the glycosidic linkage between these units. In addition, in some GAGs, uronic acid is replaced by an hexose, galactose in the case of keratan sulfate. Chondroitin sulfate and dermatan sulfate contain galactosamine (galactosaminoglycans), whereas heparin and heparan sulfate contain glucosamine (glucosaminoglycans). The amino sugar may be sulfated on C-4 or C-6 or on the non-acetylated nitrogen. Moreover, the remaining monosaccharides of the backbone may be sulfated at various positions. As a result, a simple octasaccharide could have over 1 000 000 different sulfation sequences. GAGs also vary in the geometry of the glycosidic linkage (α or β). At physiological pH, all carboxylic acid and sulfate groups are deprotonated, giving GAGs very high negatively charged densities, having heparin the highest negatively charged density of any known biomolecule. The only non-sulfated GAG is called hyaluronic acid. In nature, all GAG chains with the exception of the latter are covalently linked to a core protein to give a proteoglycan [21].

Table 1. Major repeating disaccharide units of some GAGs.

Glycosaminoglycan	Disaccharide unit	Features
Hyaluronic acid (HA)	D-GlcA-β(1→4)-D-GlcNAc-α(1→4)	Molecular weight 4-8000 kDa Non-sulfated, non-covalently attached to proteins in ECM Usually found in synovial fluid, vitreous humor, ECM of loose connective tissue. Excellent lubricators and shock absorbers.
Chondroitin sulfate (CS)	D-GlcA-β(1→3)-D-GalNAc4S-β(1→4)	Molecular weight 5-50 kDa Most abundant GAG in the body Found in cartilage, tendon, ligament, aorta Bind to proteins (like collagen) to form proteoglycan aggregates
Dermatan sulfate (DS)	L-IdoA-α(1→3)-D-GalNAc4S-β(1→4)	Molecular weight 15-40 kDa Found in skin, blood vessels, heart valves
Keratan sulfates I and II (KS)	D-Gal-β(1→4)-DGalNAc6S-β(1→3)	Molecular weight 4-19 kDa Most heterogeneous GAG KS I is found in conea KS II is found in cartilage aggregated with CS
Heparan sulfate (HS)	D-GlcA2S-α(1→4)-D-GlcNAc-α(1→4)	Very variable structure, related to that of heparin (which is much more sulfated and has important amounts of IdoA). Present in most of the tissues in mammals.

GAGs are present on all animal cell surfaces in the extracellular matrix (ECM), and some are known to bind and regulate a number of distinct proteins, including chemokines, cytokines, growth factors, morphogens, enzymes and adhesion molecules.

On the other hand, marine GAGs have different structures than those present in common mammal GAGs [3].

Green seaweeds from the order Ulvales, which comprise genera *Ulva* and *Enteromorpha* as major representatives, also produce sulfated polysaccharides rich in uronic acids. These polysaccharides have →4)-D-GlcA-β-(1→4)-L-Rha 3-sulfate-α-(1→ and →4)-L-IdoA-α-(1→4)-L-Rha 3-sulfate-α-(1→ as major disaccharide repeating units, however other sequences comprising 4-linked β-xylose units, sometimes 2-sulfated have also been reported [22]. In addition, some structural features of sulfated rhamnans with considerable amounts of uronic acids were isolated from green seaweeds of the genus *Monostroma* [23, 24] and from *Gayralia oxysperma* [25, 26].

These polysaccharides are heterogeneous with respect to chain length and sulfate content and must be purified to homogeneity before structural analysis is carried out.

Structural analysis of sulfated polysaccharides may be carried out by chemical methods: carbohydrate and sulfate content, monosaccharide composition methylation/ethylation and desulfation-methylation/ethylation, Smith degradation [27]. This chapter is mainly focused on the application of matrix-assisted laser-desorption ionization (MALDI) and electrospray ionization (ESI) mass spectrometry (MS), and nuclear magnetic resonance (NMR) spectroscopy in its wide variety of experimental methods.

MASS SPECTROMETRY SOFT METHODS FOR THE ANALYSIS OF SULFATED POLYSACCHARIDES

Mass spectrometry has become an important technique for carbohydrate analysis, mainly due to the development in the late 1980s of soft ionization methods which involve laser ionization of samples imbedded in a solid matrix, used in MALDI-MS and evaporation of solutions by electrospray, used in ESI-MS [28]. Both of these techniques are able to ionize large molecules such as proteins, nucleotides, synthetic polymers and polysaccharides. For polysaccharides MALDI- and ESI-MS provide structural information, which includes, besides molecular mass, sugar constituent, sequence, interresidue linkage position and substitution pattern [29].

However, application of mass spectrometry to sulfated polysaccharides has its limitations due to the labile nature of the sulfate group and the molecular weight; thus depolymerization methods in controlled conditions are often required before their analysis [30, 31].

1. Mass Spectrometry of Marine Sulfated Polysaccharides

Sulfated polysaccharides and oligosaccharides are difficult to study by MALDI-MS, as mentioned before, partly because of the labile nature of the sulfate group and desorption difficulties with increasing molecular weight. Thus, results can be erratic and several experimental factors should be taken into account, among others, selection of the matrix, the sample preparation method, and the type of analyzer used [32, 33]. However, MALDI-MS may be considered as a useful technique for analyzing low and medium molecular weight carbohydrates.

Fukuyama *et al.* (2002) [33] reported the analysis of several commercial neocarrabiose oligosaccharides by UV-MALDI-TOF-MS using 2,5-dihydroxy-benzoic acid (DHB) and *nor*-harmane as matrices. Experiments were carried out in the linear and reflectron modes and in both the positive- and negative-ion modes. In the positive- and negative-ion modes, with both matrices, peaks corresponding to $(M + Na)^+$ and $(M - Na)^-$ ions, respectively, were obtained, with only some signals due to glycosidic linkage cleavages. With DHB, abundant matrix signals were observed; on the other hand, *nor*-harmane afforded very few matrix signals in both ions modes, but more desulfation of the compounds took place. Post-source decay (PSD) fragmentation showed that most fragments derived from glycosidic linkage cleavages mainly of the *B*- and *Y*-type in the positive-ion mode and of the *B*-, *C*-, *Y*- and *Z*-type in the negative-ion mode (nomenclature according to ref [34]) (Fig. **2**). Electrospray-ionization time-of-flight mass spectrometry (ESI-TOF-MS) confirmed, with the observation of the $(M - Na)^-$ and the multiply charged anions, the identity and the purity of the samples.

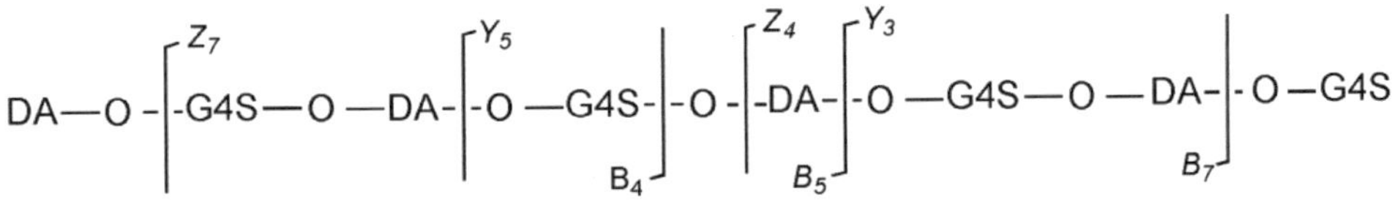

Fig. (2). PSD fragmentation pattern of neocarraoctaose $4^1,4^3,4^5,4^7$-tetrasulfate in the negative-ion mode using *nor*-harmane as matrix. Nomenclature according to Ref. [37].

The same sulfated oligosaccharides previously analyzed by UV-MALDI-TOF-MS were analyzed by positive-ion mode nano-ESI-MS [35]. CID analysis showed glycosidic cleavages of *B*- and *Y*-type as reported above. The labile sulfate substituents were retained enabling the unambiguous determination of the sequence of the sulfated oligosaccharides.

A partially cyclized mu/nu-carrageenan from *Gigartina skottsbergii* was submitted to autohydrolysis and the resulting mixture of oligosaccharides was fractionated. Structural analysis of the major fraction was carried out and a significant difference was found by ^{13}C NMR spectroscopy and UV-MALDI-TOF mass spectrometry: in the ^{13}C NMR spectrum, the reducing end-chain unit was the hydrated aldehyde of 3,6-anhydrogalactose, while the UV-MALDI-TOF mass spectrum showed an unimodal distribution of even and odd peaks, suggesting fragmentation of glycosidic linkages [36]. DHB and *nor*-harmane were used as matrices in the MALDI experiments.

Oligosaccharides of the carrabiose type were produced by partial hydrolysis of 3,6-anhydrogalactosidic linkages of kappa-carrageenan from *Kappahycus alvarezii*; the mixture of oligosaccharides was separated by size-exclusion and the oligosaccharides were characterized by one dimensional and 2 dimensional NMR spectroscopy and ESI-MS [38].

A set of sulfated red seaweed galactan derived oligosaccharides that included all galactosyl-sulfated isomers of the galactopyranosyl alditol disaccharides, plus a number of larger oligosaccharides were prepared. The oligosaccharides were analyzed by ESI-MS and by ESI-CID MS/MS. Spectra of G4S-DA-G4S-DA-G4S-DAOH, G4S-DA-G4S-DAOH, and G-DA2S-G-DAOH2S indicated that the location of the sulfate groups, which are close to the glycosidic bonds affect the rate of glycosidic cleavages. This effect depends of the monosaccharide unit and the specific sulfate attachment position. On the other hand, ESI-MS of the isomeric alditol disaccharides G4S-DAOH, G6S-LAOH, G2S-LAOH, and G3S-LAOH was achieved by the use of two strategies CID-MS/MS of the $(M - Na)^-$ and of the B_1 ion [39].

MALDI-TOF-MS of several oligosacharides and oligosaccharide alditols derived from agaroses, kappa- and iota-carrageenans using different matrices (DHB, *nor*-harmane, ferulic acid, and the liquid crystal matrices DHB - *n*-butylamine and ferulic acid - *n*-butylamine) were conducted [40]. These carbohydrates were selected as model compounds to study MALDI prompt and PSD fragmentation processes in both galactan groups. Particularly, sulfated oligosaccharide alditols showed in the negative-ion mode the molecular ion, $(M - Na)^-$, together with desulfation, while the sulfated oligosaccharides showed glycosidic *C*-cleavages and desulfation. Furthermore, monosulfated oligosaccharides and oligosaccharide alditols gave, in the negative-ion PSD experiments, cleavages of the glycosidic bonds similarly to what was observed for the sulfated neocarrabiose oligosaccharides [33]. On the other hand, for a disulfated oligosaccharide alditol mainly desulfation was observed. Best results were obtained by using DHB and/or *nor*-harmane as matrices.

The structural heterogeneity of carrageenan oligosaccharides released with enzymatic digestion from kappa/beta-carrageenan of the red seaweed *Tichocarpus crinitus* was investigated [41] The composition, sequence and sulfation pattern of the neocarrabiose oligosaccharides (degree of polymerization up to 20) was determined in the negative-ion mode by tandem MALDI-TOF-MS using arabinoosazone and DHB as matrices and ESI-MS; the addition of L-fucose to the arabinoosazone matrix reduced the in-source fragmentation. ESI-MS was applied to characterize structural features of a polysaccharide from tetrasporic *T. crinitus* after mild hydrolysis conditions [42]. Key fragments of the mixtures of oligosaccharides were subjected to tandem CID ESI-MS. In both publications, the authors suggested, as expected for both techniques, MALDI-MS and ESI-MS, more than one structural variant for some of the fragments.

A polysaccharide from sterile plants of *Anfeltiopsis flabelliformis* was depolymerized by mild acid hydrolysis to oligosaccharides, and the mixture was analyzed by positive- and negative-ion mode MALDI-TOF-MS, using DHB as matrix. The composition and sequence of the key fragments was elucidated by MALDI-MS/MS [43]. A similar procedure was used to study structural peculiarities of a low molecular weight sample obtained by mild acid hydrolysis of a kappa-carrageenan from *Chondrus armatus* [44].

The analysis of carrageenan oligosaccharides differing both in length and sulfation pattern was carried out by MALDI-TOF-MS using pyrenemethyl-guanidine as derivatizing agent and ionization efficiency enhancer and *p*-nitroaniline as matrix. In this way, they were able to observe a decasaccharide heptasulfated hybrid sequence kappa-kappa-mu-mu-kappa, being the highest *m/z* value 4330.2 in the positive-ion mode and 3781.6 in the negative-ion mode [45]. As reported before [36], the lowest *m/z* corresponded in both ionization modes to a monosulfated ion.

A purified food-grade kappa-carrageenan was subjected to degradation by four different methods: free radical depolymerization, mild acid hydrolysis, kappa-carrageenase digestion, and partial reductive hydrolysis. The structure of the oligosaccharides obtained by each procedure was accurately and comparatively elucidated by ESI-MS and CID MS/MS [46].

Negative-ion MALDI-TOF-MS was employed for the analysis of oligosaccharides obtained by partial acid hydrolysis of sulfated xyloarabinogalactan and positive-ion mode for neutral oligosaccharides derived from a neutral glucan from the green seaweed *Caulerpa racemosa.* In both experiments, DHB was the matrix [47].

UV-MALDI-TOF-MS was applied to native sulfated xylomannan and mannan fractions from *Nothogenia fastigiata* in order to determine their molecular weights and distribution profiles. The number-average molecular weight calculated from the spectra was similar to that determined by chemical end-group analysis for the lower molecular weight fractions (6000-10000 Da). For the other fractions, the number-average molecular weight was lower than that chemically determined; the increased difference was attributed to higher desorption difficulties and, consequently, mass-dependent discrimination. The best results were obtained using DHB as matrix, doped with 1-hydroxyisoquinoline, and with harmane and *nor*-harmane [48].

Recently, the use of MALDI-TOF-MS/MS and ESI-MS/MS was extended to study fucoidans extracted from brown seaweeds. Structural analysis of a fucoidan from *Fucus evanescens* was performed by MALDI-TOF and ESI-MS/MS [49]. The sample was deacetylated by treatment with aqueous ammonia and further depolymerized by a single-step solvolytic desulfation procedure in order to obtain oligosaccharides suitable for MS analysis. This desulfation procedure rendered mainly low-molecular-weight non-sulfated fragments, together with a minor percentage of sulfated oligosaccharides, along with the desulfated polysaccharide. The product of the desulfation treatment was extracted with ethanol, in order to obtain the oligosacharides to be analyzed by MALDI.TOF-MS and ESI-MS/MS. DHB was used as the MALDI matrix in the positive-ion mode. ESI-MS was recorded in the negative-ion mode. Precise information on the status of xylose, galactose and glucuronic acid was found: xylose was detected as $(Fuc_nXyl + Na)^+$ with $n = 2 - 6$ (signals could overlap with K^+ adducts of Fuc_nGal). ESI-MS/MS in the negative-ion mode indicated that xylose was sulfated on C-2 or C-4. Galactose was found to be sulfated on C-2 and probably C-4; galactose-containing oligosaccharides were also characterized: $(Gal_2SO_3)^-$, $(FucGalSO_3)^-$, and $(Fuc_2Gal_2SO_3)^-$. The most possible structures of glucuronic acid-containing oligosaccharides were Fuc-(1→4)-Fuc-(1→3)-GlcA and Fuc-(1→3)-GlcA.

Later, the fucoidan from the same seaweed was subjected to autohydrolysis to obtain low-molecular-weight fragments [50], and the mixture of oligosaccharides was analyzed by negative-ion mode MALDI-MS/MS using arabinooasazone as matrix.

Autohydrolysis was found to be a sulfate-retaining depolymerizing reaction, as the ratio of sulfate groups per monosaccharide remained almost the same as for the native fucoidan [51]: oligosaccharides with a degree of polymerization up to 6 and from 1 to 5 sulfate groups per molecule were produced. It was found that the sulfation pattern and the sequence of the mixture of oligosaccharides reflected the structure of the backbone elucidated by independent methods [51]. Moreover, D-galactose was shown to be multiply sulfated, and it was found to be part of the main chain. The following fragments were found: Gal 2-sulfate-(1→3)-Gal 2-

sulfate, Gal 2,4-disulfate-(1→4)-Fuc, and Fuc 2-sulfate-(1→4)-Gal 2-sulfate. The prevalence of oligosaccharides with even degree of polymerization in the reaction mixture after autohydrolysis, was attibuted by the authors to the fact that (1→3)-linkages hydrolyzed faster than (1→4)-linkages in the selected conditions.

The fucoidan of *Silvetia babingtonii* was also studied by the same procedure [52]. Similar structural features, but also some differences, were found by comparison with fucooligosaccharides from *F. evanescens*. Analysis of the fucoidans from the brown algae *Coccophera langsdorfii* and *Sacharina gurjanovae* was also reported [53, 54].

A highly sulfated fucan from the brown seaweed *Laminaria cichorioides* was also investigated [55]. Application of autohydrolysis for depolymerization of the fucan rendered multisulfated fucooligosaccharides with a degree of polymerization of 2-5, including mono- and disulfated fucose residues. The major feature of the tandem MALDI-MS, using arabinooasazone as matrix, and ESI-MS of the fucan fragments was the presence of ^{0,2}A-/^{0,2}X-ions from cross-ring cleavages characteristic of (1→4)-type linkages. Both techniques were carried out in the negative-ion mode.

The results obtained showed that the fucose residues of the fucan were predominantly 3-linked. Analysis of tandem mass spectra revealed that sulfate groups occupied the C-2 or C-2/C-4 positions. There were no traces of 3-sulfation when mono- and disulfated fucose residues were analyzed.

On the other hand, the same group of researchers used ESI-MS for the analysis of four fucoidan fractions extracted from brown alga *Costaria costata* at different life-stages [56]. They depolymerized the samples by autohydrolysis and analyzed by ESI-MS and ESI-MS/MS only single charged negative ions. It was confirmed that monosaccharide content and structure of fucoidans changed following its life stage. Generative alga in general produced a highly sulfated galactofucan having lower molecular weight along with a less sulfated mannoglucuronofucan with higher molecular weight, which was extensively synthesized by vegetative alga.

Recently, the structure of a galactofucan from the brown alga *Alaria augusta* was also reported [57]: the results obtained by negative-ion mode tandem MALDI-TOF-MS and ESI-MS confirmed the data obtained by methylation analysis and ^{13}C NMR of the main chain structure.

2. Mass Spectrometry of Sulfated Glycosaminoglycans

GAG oligosaccharides, obtained by multiple degradative steps, should be purified to homogeneity before carrying structural analysis. However, purification may be skipped over using on-line separations coupled with mass spectrometry [58 - 60]. As the scope of the chapter is the structural analysis sulfated polysaccharides, mass spectrometry analyses of hyaluronan will not be included.

Oligosaccharide profiling by mass spectrometry after enzymatic or chemical digestion determines the compositions, abundances and sequences of sulfated GAG oligosaccharides. However, the labile nature of the sulfate groups must be kept in mind.

Several groups have developed ionic liquid matrices, which appear to reduce the loss of sulfate from highly sulfated HS/heparin oligosaccharides. Other authors used UV-MALDI-TOF to determine masses of oligosaccharides obtained after nitrous acid depolymerization [61]. IR-MALDI, considered softer than UV-MALDI, has been employed in the analysis of HS and CS oligosaccharides.

The analysis of heparin-like oligosaccharides by negative ESI-MS with minimum loss of sulfate was reported [62]. Experiments were conducted using the optimized solvent conditions, maximizing the charge state of the sulfated oligosaccharides and minimizing the extent of sodium adduction. Low energy CID MS/MS were obtained for the highest charge state of five oligosaccharide structures and fragmentation processes were investigated.

LC (liquid chromatography)/MS removes the background from the biological matrix enabling the comparison among different samples. Several chromatography systems have been developed, being the most widely used reversed phase ion pairing for the separation of GAG oligosaccharides [60]. However, as highly sulfated compounds may be irreversibly bound to the stationary phase, hydrophilic chromatography is an alternative in this case [60, 63]. Amino and amide stationary phases were used.

It is complicated to sequence sulfated oligosaccharides, due to the tendency of the sulfate groups to be lost as SO_3. However, the extent of such losses is considerably lowered for deprotonated sulfate groups: pairing with a metal cation has a stabilizing effect.

Kailemia *et al.* (2012) [64] showed for the first time the complete structural characterization of the highly sulfated heparin based drug Arixtra (Fig. **3**). The use of NaOH as a component of electrospray (ESI) enabled the production of a precursor in which all ionizable protons were replaced by Na^+. This precursor was uniquely suitable for MS/MS analysis leading to the formation of abundant glycosidic and cross-ring fragment ions, which allowed full characterization of the heparin oligosaccharide using a single CID spectrum.

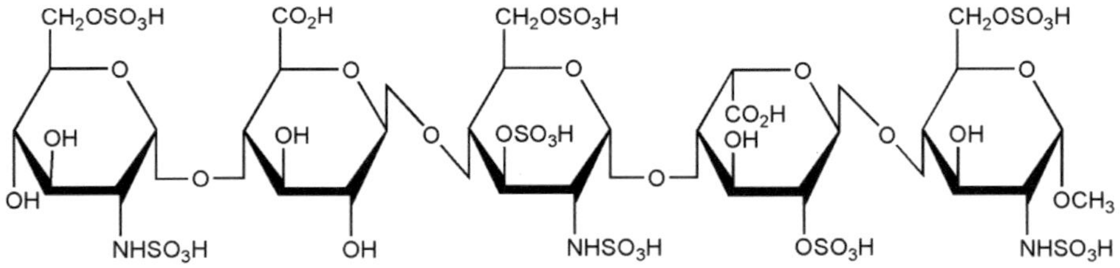

Fig. (3). Chemical structure of Arixtra.

A review on applications of mass spectrometry to structural analysis of marine oligosaccharides, which also covers the characterization of GAG oligosaccharides from marine animals, was published [29]. In addition, a review about methods for heparin characterization, very difficult to be undertaken due to its high degree of sequence microheterogeneity and size polydispersity, appeared recently, being MALDI-MS and ESI-MS important tools to be used [30].

NUCLEAR MAGNETIC RESONANCE SPECTROSCOPY FOR STRUCTURAL DETERMINATION OF SULFATED POLYSACCHARIDES

Nuclear magnetic resonance (NMR) spectroscopy is one of the most advanced and useful techniques for structural analysis of biomolecules, including sulfated polysaccharides, giving very important information about sequence, interresidue linkage position and substitution pattern.

A wide variety of experimental methods is available for polysaccharide and oligosaccharide NMR, which comprise uni- or multidimensional, as well as homo- and heteronuclear experiments. These virtually unlimited experiments enable assessment of a large number of physicochemical aspects from the samples, however, only the role of NMR chemical shifts in structural studies of primary structure of sulfated polysaccharides will be discussed here. An excellent review on this subject focused in GAGs is that of Pomin [2].

Chemical shift values, observed by lines in different positions in the NMR spectrum, are useful for the diagnostics of the local chemical properties that surround measurable atomic nucleus of a given molecule subjected to NMR analysis. Only specific types of nuclear isotopes can be screened, because of their quantum spin-numbers (one half) associated with technical properties of the NMR instruments. The magnetically sensitive nuclei of most of the glycans are ^{1}H, ^{13}C, and ^{15}N, the latter, for GAGs.

The resonances in the NMR spectrum change according to the amounts and positions of sulfates, as well as to non-sulfated or desulfated sites. The high electronic density of sulfate groups causes a deshielding effect of the neighboring atomic nuclei, such as ^{13}C, ^{15}N, ^{13}C-bound ^{1}H, or ^{15}N-bound ^{1}H. As a consequence, when these sulfated polysaccharides atomic nuclei are placed close to sulfate esters, they suffer an electronic deshielding process induced by sulfation. This ends up in displacing their respective NMR chemical shifts toward a spectral region of resonances of higher frequencies. In addition, usually, β-effects are observed, and this generally leads to an upfield resonance shift when close to the sulfate groups. This is seen particularly in ^{13}C-atoms of sulfated polysaccharides.

Other structural features of sulfated polysaccharides, such as α/β-anomericity, glycosylation sites, methylation, presence of pyruvic acid ketals, N-acetylation/--sulfation, C-5 epimerization of uronic acids, and adjacent residue type, also lead to either deshielding or shielding effects and thus, downfield or upfield chemical shifts displacements. However, here, focus on substitution by sulfate groups will be undertaken.

One dimensional ^{1}H NMR is the most used and quickest NMR method considering experimental time and sensitivity. It is also the most readily available NMR experimental type for collection of chemical shift-based data and the most reliable experiment for integration and determination of nuclear amounts.

One dimensional ^{1}H NMR spectrum of complex carbohydrates, such as sulfated polysaccharides or their derivatives gives rise to signals coming only from the non-exchangeable protons when the samples are dissolved in deuterium oxide (D_2O), the most common solvent used in NMR experiments to avoid residual signal from the water. The non-exchangeable ^{1}H-signals arise from hydrogens directly bound to carbons in these molecules. Hydrogens attached to highly electronegative atoms, such as oxygen and nitrogen, in hydroxyl and amide/amine groups exchange rapidly with the solvent. Consequently, the exchangeable protons in the ^{1}H-related NMR experiments performed under the regular experimental conditions, are undetected. This decreases signal density and resonance overlapping tendency, thus, making peak identification and structural determination easier.

The low signal dispersion is typical of complex carbohydrates, whose NMR resonances are essentially condensed in a very limited spectral range, usually from 6.0 to 3.0 ppm for ^{1}H-chemical shifts. Few exceptions arise from side chain chemical groups attached to the sugar rings, such as methyl protons of acetyl groups or pyruvic acid ketals, as well as those from deoxy-sugars, as rhamnose and fucose, which have ^{1}H-chemical shifts at ~1.4-2.0 ppm. The low chemical shift dispersion of the major carbon-bound ^{1}H resonances in carbohydrates makes difficult to achieve complete assignments by one dimensional ^{1}H NMR.

Table 2. ^{1}H NMR chemical shifts and coupling constants for methyl α-D-galactopyranoside and its monosulfated derivatives[a].

Compound	Chemical shifts (ppm)[b]							
	H1	H2	H3	H4	H5	H6	H6′	OCH_3
α-D-galactopyranoside	4.837	3.819	3.811	3.968	3.897	3.741	3.752	3.414
α-D-galactopyranoside 2-sulfate	5.131	4.443	3.940	4.048	3.923	3.749	3.763	3.428
α-D-galactopyranoside 3-sulfate	4.899	3.982	4.477	4.324	3.935	3.751	3.759	3.430
α-D-galactopyranoside 4-sulfate	4.867	3.855	3.941	4.713	4.030	3.810	3.755	3.425
α-D-galactopyranoside 6-sulfate	4.842	3.830	3.828	4.018	4.148	4.214	4.158	3.421
	Coupling constants (Hz)							
	$J_{1,2}$	$J_{2,3}$	$J_{3,4}$	$J_{4,5}$	$J_{5,6}$	$J_{5,6'}$	H6,6′	OCH_3
α-D-galactopyranoside	3.2	10.2	2.7	1.2	4.1	8.3	- 11.8	
α-D-galactopyranoside 2-sulfate	3.8	10.2	3.7	1.0	4.2	8.1	- 11.8	
α-D-galactopyranoside 3-sulfate	3.8	10.3	3.2	1.1	4.1	8.2	- 11.8	
α-D-galactopyranoside 4-sulfate	3.9	10.4	3.3	Nd	4.1	8.3	- 12.0	
α-D-galactopyranoside 6-sulfate	3.0	10.3	3.3	0.8	4.1	8.3	- 10.8	

[a]Data from Ruiz Contreras *et al.* 1988 [65]. [b]Relative to internal acetone at 2.225 ppm.

Comparison of the ^{1}H-NMR data (Table **2**) of the monosulfated derivatives of methyl α-D-galactopyranose with those of the non-sulfated compound demonstrated that the sulfate group deshields the geminal and vicinal protons. The secondary sulfates cause α-effects of 0.6-0.7 ppm, whereas the primary sulfate group shows 0.4-0.5 ppm for this effect. The β-effects are 0.1-0.4 ppm depending on the axial or equatorial orientation of the vicinal proton. The resonance for an equatorial proton next to an equatorial sulfate group is shifted more downfield as compared to other situations. The values of $J_{H,H}$ for all the galactosides are similar, indicating that sulfation does not cause significant distortion of the $^4C_1$ chair conformation or change in the rotamer populations around the C-5 – C-6 bond [65]. These results indicate that non-exchangeable protons attached to sulfated carbons are usually among the resolved signals in sulfated polysaccharides spectra.

Other well-resolved resonances exist along with those from the sulfated sites. They are the anomeric signals of both β-D-galactopyranose and β-D-glucopyranose units, and their uronic acid and 2-deoxy-2-amino-derivatives, as well as β- and α-D-mannopyranose, β-D-xylopyranose, between others (4.4-5.0 ppm). On the other hand, anomeric signals corresponding to α-D-galactopyranose, and α-D-glucopyranose, as well as those corresponding to α-L-fucopyranose, α-L-fucofuranose, β-L-arabinopyranose, α-L-rhamnopyranose, appear at δ 5.0-5.6 [2, 20, 26, 66].

Usually, chemical shifts in oligosaccharides are easier for assignments than in native polymers, whose NMR resonances usually show enhanced line broadening. This happens most likely because of the large rotational correlation times typically found at high-MW molecules. This gives longer relaxation times and thus broader lines in their NMR spectrum and the tendency to signal overlap [2]. Therefore, it is harder to carry out chemical shift and structural assignments using only one dimensional ^{1}H NMR spectra of polysaccharides. Moreover, the first assignments were based on the studies of low molecular weight structures [67]. However, in 1977, Welti [68] made the full assignment of the ^{1}H NMR spectra of agarose and kappa-carrageenan, and segments of iota-carrageenan and agarose sulfate. These results are in agreement with generalizations mentioned before concerning the geminal and vicinal proton deshielding effect of sulfate groups. Similar conclusions arise from the comparison of ^{1}H NMR assignment for mu- and nu-carrageenans [69, 70], heparin [71], and so on.

^{1}H-based NMR methods are fast and sensitive, since this nucleus has the highest natural abundance, together with the highest relative NMR receptivity. When one dimensional ^{1}H-signals are resolved, their integration values are very reliable in terms of the nuclear amount. This is particularly useful for quantification of different structural features of a polysaccharide sample, *i.e.,* the amount of kappa and iota structural units in a hybrid kappa-2 type carrageenan could easily be quantified from the area corresponding to the anomeric signals of the 3,6-anhydro-α-D-galactose and its 2-sulfated derivative. This is an important industrial parameter of the quality of this hydrocolloid [72].

The use of two dimensional ^{1}H experiments applied to structural determination of sulfated polysaccharides, mostly homonuclear correlation spectroscopy (COSY) and total correlation spectroscopy (TOCSY), generalized in the last two decades, greatly increased the provided structural information as consequence, on one hand of overcoming the signal overlapping, and on the other, correlating a given signal with those of the neighboring protons, although these techniques are usually combined with heteronuclear experiments [2, 73 - 76].

^{13}C NMR spectroscopy produced a tremendous step forward for structural studies of sulfated polysaccharides and it was developed particularly during the eighties, although the first studies were carried out in the previous decade [77, 78]. It is important to mention the enormous work carried out by A.I. Usov and other researchers at the Zelinsky Institute in Moscow, who applied NMR spectroscopy, and especially ^{13}C NMR spectroscopy combined with excellent chemistry for structural analysis of seaweed sulfated polysaccharides [10, 79 - 82].

The ^{13}C NMR data for sulfated methyl galactosides are summarised in Table **3**. Comparison with the non-sulfated methyl galactoside shows specific downfield shifts of 6-8 ppm for the signals of the carbon atoms bearing the sulfate group. A sulfate group at C-6 gives rise to the smallest downfield shift (6.33 ppm), whereas sulfate groups at C-3 or C-4 causes the largest shifts (8.28 and 8.24 ppm, respectively). An equatorial (C-3) or axial (C-4) orientation does not influence markedly the magnitude of the shift. The signals from the vicinal carbon atoms were shifted upfield by 1-2 ppm. An axial sulfate group (C-4) caused a smaller upfield shift (0.6-1.0 ppm) of the resonance of the vicinal carbon atoms than an equatorial sulfate group (1.8-2.2 ppm) [65].

Table 3. ^{13}C NMR chemical for methyl α-D-galactopyranoside and its monosulfated derivatives[a].

Compound	**Chemical shifts (ppm)[b]**						
	C1	**C2**	**C3**	**C4**	**C5**	**C6**	**OCH_3**
α-D-galactopyranoside	100.67	69.47	70.76	70.50	71.99	62.50	56.31
α-D-galactopyranoside, 2-sulfate	98.63	76.64	68.61	70.60	71.81	62.32	56.32
α-D-galactopyranoside, 3-sulfate	100.43	67.39	79.04	68.75	71.63	62.29	56.25
α-D-galactopyranoside, 4-sulfate	100.53	69.53	69.78	78.74	71.35	62.44	56.42
α-D-galactopyranoside, 6-sulfate	100.62	69.30	70.23	70.49	69.73	68.83	56.36

[a]Data from Ruiz Contreras *et al.* 1988 [65]. [b]Relative to internal acetone at 31.55 ppm.

Regular polysaccharides, like agarose or kappa- and iota-carrageenan, were shown to give ^{13}C-NMR spectra which consisted of 12 signals, corresponding to the 12 carbon atoms of the repeating disaccharide unit. The two most downfield resonances belonging, as expected, to anomeric carbons have an important peculiarity: whereas the C-l signal of 3-linked β-D-galactose residues are in approximately the same position in the spectra of agarose and carrageenans, the chemical shifts of the C-l signal of 3,6-anhydrogalactose residues in these two types of polysaccharides differ by up to 5 ppm. This difference is an example of a well-known influence of the absolute configuration of the monosaccharide residues on the glycosylation effect in disaccharide units. Hence, ^{13}C-NMR spectroscopy provides evidence on the absolute configuration of 3,6-anhydrogalactose in red seaweed galactans and allows to attribute an unknown polysaccharide to the agar or carrageenan group immediately after its isolation and without its destruction. Such an attribution may be also done when the 4-linked monosaccharide residues in the galactan molecule have no 3,6-anhydro residues [79 - 82]. Further work [83 - 92] resulted in reports of new spectra of characterized carrageenans, agarans, and/or oligosaccharides derived from them.

As indicated before, this approach was focused on the characterization of 'pure' samples resulting in spectra with 12 peaks. Methylation studies have shown that sulfated galactans also contain additional minor structural features, which may influence important properties. In order to identify these minor components, it is necessary to have the whole set of possible absorptions produced by all the diads potentially present. Another approach arises from calculation for the units not yet determined and an up-to-date compilation of the ^{13}C NMR chemical shift data produced. A computer program was developed as an aid to matching experimental data with the chemical shift database [93]. This system was very useful, however, in the case of lambda-carrageenan; the anomeric resonances were predicted to appear at 103.7 and 94.7 ppm, while the chemical shifts obtained for this polymer at 70°C were 103.9 and 92.0 ppm. This discrepancy in chemical shift values between the anomeric resonance of the 4-linked unit in the carrageenan molecule and the predicted one was suggested to be due to differences in the conformation. The lower than-expected chemical shift for this anomeric carbon suggested that the strong shielding effect of the 2-sulfate of the neighboring unit on the anomeric carbon of the α-galactose 2,6-disulfate unit of lambda-carrageenan was caused by a modification of the preferred torsional angles around the glycosidic bond. Other discrepancies appeared in nearly all the predicted values for the other chemical shifts for this structure, thus indicating conformational changes induced by sulfation [94].

On the other hand, it was found that signals from the anomeric carbon atoms vicinal to a sulfated C-2 are sometimes shifted downfield instead of upfield, as previously generalized. This was observed by comparison of the anomeric signals of the α-D-galactose unit of mu- and nu-carrageenans [94, 95]. Valuable structural information may be obtained not only from the resonance positions in the ^{13}C-NMR spectra, but also from the integrated intensities of signals. The intensity of every signal depends not only on the amount of the corresponding carbon atoms in the specimen, but also on their relaxation time and nuclear Overhauser enhancement. However, under the usual recording conditions, the influence of these additional factors is equalized considerably, so the proportion of integrated intensities of signals may be used as an analytical tool [80].

^{13}C NMR spectra of many disaccharide repeating units usually found in the red algal galactans were completely interpreted some time ago and published as tables, which may be used to identify the corresponding structural elements in new polysaccharides [27, 90, 91, 96 - 98].

Many other kinds of sulfated polysaccharides were studied in a similar way, giving rise to an enormous amount of information, very useful for the study of new products or for identification of the known structures isolated from new sources.

However, it should be kept in mind that ^{13}C NMR spectroscopy is not a very sensitive method and small amounts of structures in a complex sample can easily stay unnoticed. Moreover, ^{13}C NMR spectroscopic data should be complemented by chemical analysis of the sample, as well as other data, like linkage analysis and mass spectrometric data.

Sets of ^{13}C NMR data from model compounds and matrices were used to analyze the structures of red algal galactans with complex ^{13}C NMR spectra [99, 100]. Although it is an interesting approach, this method cannot replace a whole set of experimental data needed to determine the structure of these products.

In polymers, although qualitatively and quantitatively informative, the ^{1}H and ^{13}C resonances related to sulfation types, as well as to minor components, give just a rough snapshot of the overall structural composition. No information about sequencing can be raised therein. On the other hand, the combined ^{1}H- and ^{13}C-chemical shifts from cross-peaks of intermediate-sized sulfated oligosaccharides can be of great diagnostic potential for the study of sequences of structural units of the parent polysaccharide [2].

Nowadays, the NMR spectrum containing carbon directly observed is not so commonly used in structural analysis of sulfated polysaccharides. Besides the very low sensitivity, it only provides ^{13}C-chemical shifts since no internuclear correlations are involved. Assignments based only on this technique are difficult. Although one dimensional ^{13}C-signals are considerably narrower than those of one dimensional ^{1}H, two dimensional ^{13}C-based NMR heteronuclear spectra, like heteronuclear single quantum coherence (HSQC), heteronuclear multiple quantum coherence (HMQC), and heteronuclear multiple bound coherence (HMBC) are under more utility, because they provide much more reliable information in terms of chemical shift assignments. Correlation between ^{13}C- and ^{1}H-atoms and respective chemical shifts can be generated from these experiments. ^{13}C-signals are usually observed narrower than ^{1}H-signals because, in general, ^{13}C atoms in sugar rings are mostly secondary carbons attached to paramagnetic oxygens. The presence of the paramagnetic oxygens speeds up the spin-relaxation of neighboring atoms, including ^{13}C. Therefore, the longitudinal relaxation of ^{13}C-atoms in regular hexose rings is usually faster than those of ^{1}H. This gives rise to lower tendencies of resonance overlap in ^{13}C-NMR spectra of sulfated polysaccharides [2]. Nonetheless, it is from the $^{1}H/^{13}C$ NMR experiments that information regarding anomericity, residue type, linkage patterns, and sulfation sites of sulfated polysaccharide molecules is straightforwardly extracted, since the specific structure-related resonances occur in just certain regions of the two dimensional spectra. Besides, in the ^{1}H chemical shift correlation seen in these heteronuclear ^{13}C-based NMR experiments, the magnetization transfer back to ^{1}H during the INEPT step in those pulse sequences is a key procedure to enhance sensitivity. The sensitivity enhancement by INEPT in ^{13}C-based experiments is in the factor of 4 [2].

Numerous examples appeared in the literature where ^{1}H/^{13}C two dimensional NMR techniques were the key in the elucidation of the structure of different sulfated polysaccharides. However, when new complex structures are under investigation, these spectra should be carried out on the native polymer, as well as on modified products derived from it, like the desulfated counterpart and partially degraded products [7, 20, 26, 75, 76, 101 - 104].

NMR has proven to be an invaluable method for characterizing GAGs, as well as for the analysis of unfractionated heparin (UFH) and low molecular weight heparins (LMWH) pharamceuticals. The utility of NMR for GAG characterization was aptly demonstrated by the 2008 contamination of the pharmaceutical heparin supply chain with oversulfated chondroitin sulfate. This event resulted in numerous reported deaths and highlighted the need for improved quality control methods. The identity of the contaminant was elucidated using ^{1}H and ^{13}C NMR spectra, as well as the two-dimensional HSQC spectra, which provide useful information about heparin structural features and substitution patterns and can be used to identify and quantify modifications such as those found in LMWH [105].

Although HSQC is routinely applied for the characterization and quality assurance of pharmaceutical heparins, ^{15}N NMR has been underutilized for GAG characterization.

Recently, solution conditions have been reported that facilitate detection of the prominent heparin sulfamate groups and allow the determination of ^{1}H and ^{15}N chemical shifts using the [^{1}H, ^{15}N] HSQC experiment. The rate of chemical exchange of the sulfamate proton with water is minimized in the basic pH regime and is affected by the local negative charge density. The determination of the NH exchange rates for a set of structurally defined heparin-derived oligosaccharides led to the characterization of the first reported intramolecular hydrogen bond involving the sulfamate NH group adjacent to the 3-sulfate in the pentasaccharide Arixtra (Fig. **3**) [106].

However, until now, ^{15}N NMR is not so popular, and its use is restricted to a very small number of research groups with access to this kind of experiments.

CONFLICT OF INTEREST

The author (editor) declares no conflict of interest, financial or otherwise.

ACKNOWLEDGEMENTS

Both authors are Research Members of the National Research Council of Argentina (CONICET). This work was supported by grants of CONICET (11220110100208/2012-2014) and the University of Buenos Aires (20020130100216BA/2014-2017, and 20020130100576BA/2014-2017).

REFERENCES

[1] V.H. Pomin, and P.A. Mourão, "Structure, biology, evolution, and medical importance of sulfated fucans and galactans", *Glycobiology,* vol. 18, no. 12, pp. 1016-1027, 2008. [http://dx.doi.org/10.1093/glycob/cwn085] [PMID: 18796647]

[2] V.H. Pomin, "NMR chemical shifts in structural biology of glycosaminoglycans", *Anal. Chem.* , vol. 86, no. 1, pp. 65-94, 2014. [http://dx.doi.org/10.1021/ac401791h] [PMID: 23909370]

[3] V.H. Pomin, "Marine medicinal glycomics", *Front. Cell. Infect. Microbiol.* , vol. 4, p. 5, 2014. [http://dx.doi.org/10.3389/fcimb.2014.00005] [PMID: 24524028]

[4] C.A. Stortz, M.R. Cases, and A.S. Cerezo, "Red seaweed galactans: Methodology for the structural determination of corallinans. A different agaroid", In: *Techniques in Glycobiology.* Marcel Dekker: New York, 1997, pp. 567-593.

[5] R.S. Aquino, A.M. Landeira-Fernandez, A.P. Valente, L.R. Andrade, and P.A. Mourão, "Occurrence of sulfated galactans in marine angiosperms: evolutionary implications", *Glycobiology,* vol. 15, no. 1, pp. 11-20, 2005. [http://dx.doi.org/10.1093/glycob/cwh138] [PMID: 15317737]

[6] P.V. Fernández, M.P. Raffo, J. Alberghina, and M. Ciancia, "Polysaccharides from the green seaweed Codium decorticatum. Structure and cell wall distribution", *Carbohydr. Polym.* , vol. 117, pp. 836-844, 2015. [http://dx.doi.org/10.1016/j.carbpol.2014.10.039] [PMID: 25498707]

[7] P.X. Arata, I. Quintana, D.J. Canelón, B.E. Vera, R.S. Compagnone, and M. Ciancia, "Chemical structure and anticoagulant activity of highly pyruvylated sulfated galactans from tropical green seaweeds of the order Bryopsidales", *Carbohydr. Polym.* , vol. 122, pp. 376-386, 2015. [http://dx.doi.org/10.1016/j.carbpol.2014.10.030] [PMID: 25817682]

[8] P.A. Mourão, and A.S. Perlin, "Structural features of sulfated glycans from the tunic of Styela plicata (Chordata-Tunicata). A unique occurrence of L-galactose in sulfated polysaccharides", *Eur. J. Biochem.* , vol. 166, no. 2, pp. 431-436, 1987. [http://dx.doi.org/10.1111/j.1432-1033.1987.tb13534.x] [PMID: 3609020]

[9] A-P. Alves, B. Mulloy, J.A. Diniz, and P.A. Mourão, "Sulfated polysaccharides from the egg jelly layer are species-specific inducers of acrosomal reaction in sperms of sea urchins", *J. Biol. Chem.* , vol. 272, no. 11, pp. 6965-6971, 1997. [http://dx.doi.org/10.1074/jbc.272.11.6965] [PMID: 9054385]

[10] A.I. Usov, and M.I. Bilan, "Fucoidans-sulfated polysaccharides of brown algae", *Russ. Chem. Rev.* , vol. 78, pp. 785-799, 2009. [http://dx.doi.org/10.1070/RC2009v078n08ABEH004063]

[11] D.O. Croci, A. Cumashi, N.A. Ushakova, M.E. Preobrazhenskaya, A. Piccoli, L. Totani, N.E.

Ustyuzhanina, M.I. Bilan, A.I. Usov, A.A. Grachev, G.E. Morozevich, A.E. Berman, C.J. Sanderson, M. Kelly, P. Di Gregorio, C. Rossi, N. Tinari, S. Iacobelli, G.A. Rabinovich, and N.E. Nifantiev, Consorzio Interuniversitario Nazionale per la Bio-Oncologia (CINBO), Italy, "Fucans, but not fucomannoglucuronans, determine the biological activities of sulfated polysaccharides from Laminaria saccharina brown seaweed", *PLoS One,* vol. 6, no. 2, p. e17283, 2011. [http://dx.doi.org/10.1371/journal.pone.0017283] [PMID: 21387013]

[12] M. Rabanal, N.M. Ponce, D.A. Navarro, R.M. Gómez, and C.A. Stortz, "The system of fucoidans from the brown seaweed Dictyota dichotoma: chemical analysis and antiviral activity", *Carbohydr. Polym.* , vol. 101, pp. 804-811, 2014. [http://dx.doi.org/10.1016/j.carbpol.2013.10.019] [PMID: 24299842]

[13] V.H. Pomin, "Review: an overview about the structure-function relationship of marine sulfated homopolysaccharides with regular chemical structures", *Biopolymers,* vol. 91, no. 8, pp. 601-609, 2009. [http://dx.doi.org/10.1002/bip.21200] [PMID: 19353634]

[14] A.C. Vilela-Silva, A.P. Alves, A.P. Valente, V.D. Vacquier, and P.A. Mourão, "Structure of the sulfated α-L-fucan from the egg jelly coat of the sea urchin Strongylocentrotus franciscanus: patterns of preferential 2-O- and 4-O-sulfation determine sperm cell recognition", *Glycobiology,* vol. 9, no. 9, pp. 927-933, 1999. [http://dx.doi.org/10.1093/glycob/9.9.927] [PMID: 10460834]

[15] A.A. Kolender, C.A. Pujol, E.B. Damonte, M.C. Matulewicz, and A.S. Cerezo, "The system of sulfated α-(1>3)-linked D-mannans from the red seaweed Nothogenia fastigiata: structures, antiherpetic and anticoagulant properties", *Carbohydr. Res.* , vol. 304, no. 1, pp. 53-60, 1997. [http://dx.doi.org/10.1016/S0008-6215(97)00201-2] [PMID: 9403994]

[16] M. Pérez Recalde, M.D. Noseda, C.A. Pujol, M.J. Carlucci, and M.C. Matulewicz, "Sulfated mannans from the red seaweed Nemalion helminthoides of the South Atlantic", *Phytochemistry,* vol. 70, no. 8, pp. 1062-1068, 2009. [http://dx.doi.org/10.1016/j.phytochem.2009.05.001] [PMID: 19493552]

[17] P. Mandal, C.A. Pujol, M.J. Carlucci, K. Chattopadhyay, E.B. Damonte, and B. Ray, "Anti-herpetic activity of a sulfated xylomannan from Scinaia hatei", *Phytochemistry,* vol. 69, no. 11, pp. 2193-2199, 2008. [http://dx.doi.org/10.1016/j.phytochem.2008.05.004] [PMID: 18572208]

[18] M.A. Cardoso, M.D. Noseda, M.T. Fujii, R.G. Zibetti, and M.E. Duarte, "Sulfated xylomannans isolated from red seaweeds Chondrophycus papillosus and C. flagelliferus (Ceramiales) from Brazil", *Carbohydr. Res.* , vol. 342, no. 18, pp. 2766-2775, 2007. [http://dx.doi.org/10.1016/j.carres.2007.08.024] [PMID: 17889841]

[19] P.V. Fernández, J.M. Estevez, A.S. Cerezo, and M. Ciancia, "Sulfated β-D-mannan from the green seaweed Codium vermilara", *Carbohydr. Polym.* , vol. 87, pp. 916-919, 2012. [http://dx.doi.org/10.1016/j.carbpol.2011.06.063]

[20] P.V. Fernández, I. Quintana, A.S. Cerezo, J.J. Caramelo, L. Pol-Fachin, H. Verli, J.M. Estevez, and M. Ciancia, "Anticoagulant activity of a unique sulfated pyranosic (1->3)-β-L-arabinan through direct interaction with thrombin", *J. Biol. Chem.* , vol. 288, no. 1, pp. 223-233, 2013. [http://dx.doi.org/10.1074/jbc.M112.386441] [PMID: 23161548]

[21] N.S. Gandhi, and R.L. Mancera, "The structure of glycosaminoglycans and their interactions with proteins", *Chem. Biol. Drug Des.* , vol. 72, no. 6, pp. 455-482, 2008. [http://dx.doi.org/10.1111/j.1747-0285.2008.00741.x] [PMID: 19090915]

[22] M. Lahaye, and A. Robic, "Structure and functional properties of ulvan, a polysaccharide from green seaweeds", *Biomacromolecules,* vol. 8, no. 6, pp. 1765-1774, 2007. [http://dx.doi.org/10.1021/bm061185q] [PMID: 17458931]

[23] W. Mao, H. Li, Y. Li, H. Zhang, X. Qi, H. Sun, Y. Chen, and S. Guo, "Chemical characteristic and

anticoagulant activity of the sulfated polysaccharide isolated from Monostroma latissimum (Chlorophyta)", *Int. J. Biol. Macromol.* , vol. 44, no. 1, pp. 70-74, 2009. [http://dx.doi.org/10.1016/j.ijbiomac.2008.10.003] [PMID: 19007806]

[24] H. Li, W. Mao, Y. Chen, S. Ren, X. Qi, Y. Chen, C. Zhao, N. Li, C. Wang, C. Lin, M. Yan, and J. Shan, "Sequence analysis of the sulfated rhamno-oligosaccharides derived from a sulfated rhamnan", *Carbohydr. Polym.* , vol. 90, no. 3, pp. 1299-1304, 2012. [http://dx.doi.org/10.1016/j.carbpol.2012.06.076] [PMID: 22939344]

[25] J.E. Cassolato, M.D. Noseda, C.A. Pujol, F.M. Pellizzari, E.B. Damonte, and M.E. Duarte, "Chemical structure and antiviral activity of the sulfated heterorhamnan isolated from the green seaweed Gayralia oxysperma", *Carbohydr. Res.* , vol. 343, no. 18, pp. 3085-3095, 2008. [http://dx.doi.org/10.1016/j.carres.2008.09.014] [PMID: 18845298]

[26] J. Ropellato, M.M. Carvalho, L.G. Ferreira, M.D. Noseda, C.R. Zuconelli, A.G. Gonçalves, D.R. Ducatti, J.C. Kenski, P.L. Nasato, S.M. Winnischofer, and M.E. Duarte, "Sulfated heterorhamnans from the green seaweed Gayralia oxysperma: partial depolymerization, chemical structure and antitumor activity", *Carbohydr. Polym.* , vol. 117, pp. 476-485, 2015. [http://dx.doi.org/10.1016/j.carbpol.2014.09.089] [PMID: 25498661]

[27] A.A. Kolender, and M.C. Matulewicz, "Sulfated polysaccharides from the red seaweed Georgiella confluens", *Carbohydr. Res.* , vol. 337, no. 1, pp. 57-68, 2002. [http://dx.doi.org/10.1016/S0008-6215(01)00283-X] [PMID: 11755912]

[28] D.J. Harvey, "Mass spectrometry-Ionization methods overview", *Reference module in Chemistry, Molecular Sciences and Chemical Engineering.*

[29] Y. Lang, X. Zhao, L. Liu, and G. Yu, "Applications of mass spectrometry to structural analysis of marine oligosaccharides", *Mar. Drugs,* vol. 12, no. 7, pp. 4005-4030, 2014. [http://dx.doi.org/10.3390/md12074005] [PMID: 24983643]

[30] C.J. Jones, S. Beni, J.F. Limtiaco, D.J. Langeslay, and C.K. Larive, "Heparin characterization: challenges and solutions", *Annu. Rev. Anal. Chem. (Palo Alto, Calif.),* vol. 4, pp. 439-465, 2011. [http://dx.doi.org/10.1146/annurev-anchem-061010-113911] [PMID: 21469955]

[31] S.D. Anastyuk, N.M. Shevchenko, and V.I. Gorbach, *"Fucoidan analysis by tandem MALDI-TOF and ESI mass spectrometry", Natural Products from Marine Algae: Methods and Protocols.* Springer Science: New York, 2015, pp. 299-312. [http://dx.doi.org/10.1007/978-1-4939-2684-8_19]

[32] D.J. Harvey, "Matrix-assisted laser desorption/ionization mass spectrometry of carbohydrates", *Mass Spectrom. Rev.* , vol. 18, no. 6, pp. 349-450, 1999. [http://dx.doi.org/10.1002/(SICI)1098-2787(1999)18:6<349::AID-MAS1>3.0.CO;2-H] [PMID: 10639030]

[33] Y. Fukuyama, M. Ciancia, H. Nonami, A.S. Cerezo, R. Erra-Balsells, and M.C. Matulewicz, "Matrix-assisted ultraviolet laser-desorption ionization time-of-flight mass spectrometry of sulfated neocarrabiose oligosaccharides", *Carbohydr. Res.* , vol. 337, pp. 1553-1562, 2002. [http://dx.doi.org/10.1016/S0008-6215(02)00230-6] [PMID: 12350325]

[34] B. Domon, and C.E. Costello, "A systematic nomenclature for carbohydrate fragmentations in FAB-MS/MS spectra of glycoconjugates", *Glycoconj. J.* , vol. 5, pp. 397-409, 1988. [http://dx.doi.org/10.1007/BF01049915]

[35] J.T. Aguilan, F.M. Dayrit, J. Zhang, M.R. Niñonuevo, and C.B. Lebrilla, "Structural analysis of κ-carrageenan [corrected] sulfated oligosaccharides by positive mode nano-ESI-FTICR-MS and MS/MS by SORI-CID", *J. Am. Soc. Mass Spectrom.* , vol. 17, no. 1, pp. 96-103, 2006. [http://dx.doi.org/10.1016/j.jasms.2005.09.009] [PMID: 16352442]

[36] M. Ciancia, Y. Sato, H. Nonami, A.S. Cerezo, R. Erra-Balsells, and M.C. Matulewicz, "Autohydrolysis of a partially cyclized mu/nu-carrageenan and structural elucidation of the oligosaccharides by chemical analysis, NMR spectroscopy and UV-MALDI mass spectrometry",

ARKIVOC, vol. xii, pp. 319-331, 2005.

[37] S.H. Knutsen, D.E. Myslabodski, B. Larsen, and A.I. Usov, "A modified system of nomenclature for red algal galactans", *Bot. Mar.* , vol. 37, pp. 163-169, 1994. [http://dx.doi.org/10.1515/botm.1994.37.2.163]

[38] D.R. Ducatti, F.G. Colodi, A.G. Gonçalves, M.E. Duarte, and M.D. Noseda, "Production of agaro- and carra-oligosaccharides by partial acid hydrolysis", *Brazilian Journal of Pharmacognosy,* vol. 21, pp. 296-304, 2011.

[39] A.G. Gonçalves, D.R. Ducatti, T.B. Grindley, M.E. Duarte, and M.D. Noseda, "ESI-MS differential fragmentation of positional isomers of sulfated oligosaccharides derived from carrageenans and agarans", *J. Am. Soc. Mass Spectrom.* , vol. 21, no. 8, pp. 1404-1416, 2010. [http://dx.doi.org/10.1016/j.jasms.2010.03.045] [PMID: 20462767]

[40] M.K. Fatema, H. Nonami, D.R. Ducatti, A.G. Gonçalves, M.E. Duarte, M.D. Noseda, A.S. Cerezo, R. Erra-Balsells, and M.C. Matulewicz, "Matrix-assisted laser desorption/ionization time-of-flight (MALDI-TOF) mass spectrometry analysis of oligosaccharides and oligosaccharide alditols obtained by hydrolysis of agaroses and carrageenans, two important types of red seaweed polysaccharides", *Carbohydr. Res.* , vol. 345, no. 2, pp. 275-283, 2010. [http://dx.doi.org/10.1016/j.carres.2009.10.009] [PMID: 20005511]

[41] S.D. Anastyuk, A.O. Barabanova, G. Correc, E.L. Nazarenko, D.L. Davydova, W. Helbert, P.S. Dmitrenok, and I.M. Yermak, "Analysis of structural heterogeneity of κ/β-carrageenan oligosaccharides from Tichocarpus crinitus by negative ion ESI and tandem MALDI mass spectrometry", *Carbohydr. Polym.* , vol. 86, pp. 546-554, 2011. [http://dx.doi.org/10.1016/j.carbpol.2011.04.081]

[42] A.O. Byankina Barabanova, E.V. Sokolova, S.D. Anastyuk, V.V. Isakov, V.P. Glazunov, A.V. Volodko, I.M. Yakovleva, T.F. Soloveva, and I.M. Yermak, "Polysaccharide structure of tetrasporic red seaweed Tichocarpus crinitus", *Carbohydr. Polym.* , vol. 98, no. 1, pp. 26-35, 2013. [http://dx.doi.org/10.1016/j.carbpol.2013.04.063] [PMID: 23987313]

[43] A.O. Kravchenko, S.D. Anastyuk, V.V. Isakov, E.V. Sokolova, V.P. Glazunov, and I.M. Yermak, "Structural peculiarities of polysaccharide from sterile form of Far Eastern red alga Ahnfeltiopsis flabelliformis", *Carbohydr. Polym.* , vol. 111, pp. 1-9, 2014. [http://dx.doi.org/10.1016/j.carbpol.2014.04.022] [PMID: 25037322]

[44] A.A. Kalitnik, P.A. Marcov, S.D. Anastyuk, A.O. Barabanova, V.P. Glazunov, S.V. Popov, Y.S. Ovodov, and I.M. Yermak, "Gelling polysaccharide from Chondrus armatus and its oligosaccharides: the structural peculiarities and anti-inflammatory activity", *Carbohydr. Polym.* , vol. 115, pp. 768-775, 2015. [http://dx.doi.org/10.1016/j.carbpol.2014.04.070] [PMID: 25439960]

[45] K. Ohara, J-C. Jacquinet, D. Jouanneau, W. Helbert, M. Smietana, and J.J. Vasseur, "Matrix-assisted laser desorption/ionization mass spectrometric analysis of polysulfated-derived oligosaccharides using pyrenemethylguanidine", *J. Am. Soc. Mass Spectrom.* , vol. 20, no. 1, pp. 131-137, 2009. [http://dx.doi.org/10.1016/j.jasms.2008.09.004] [PMID: 18838278]

[46] Y. Sun, B. Yang, Y. Wu, Y. Liu, X. Gu, H. Zhang, C. Wang, H. Cao, L. Huang, and Z. Wang, "Structural characterization and antioxidant activities of κ-carrageenan oligosaccharides degraded by different methods", *Food Chem.* , vol. 178, pp. 311-318, 2015. [http://dx.doi.org/10.1016/j.foodchem.2015.01.105] [PMID: 25704717]

[47] K. Chattopadhyay, U. Adhikari, P. Lerouge, and B. Ray, "Polysaccharides from Caulerpa ramosa. Purification and structural features", *Carbohydr. Polym.* , vol. 68, pp. 407-415, 2007. [http://dx.doi.org/10.1016/j.carbpol.2006.12.010]

[48] R. Erra-Balsells, A.A. Kolender, M.C. Matulewicz, H. Nonami, and A.S. Cerezo, "Matrix-assisted ultraviolet laser-desorption ionization time-of-flight mass spectrometry of sulfated mannans from the red seaweed Nothogenia fastigiata", *Carbohydr. Res.* , vol. 329, no. 1, pp. 157-167, 2000.

[http://dx.doi.org/10.1016/S0008-6215(00)00174-9] [PMID: 11086695]

[49] S.D. Anastyuk, N.M. Shevchenko, E.L. Nazarenko, P.S. Dmitrenok, and T.N. Zvyagintseva, "Structural analysis of a fucoidan from the brown alga Fucus evanescens by MALDI-TOF and tandem ESI mass spectrometry", *Carbohydr. Res.*, vol. 344, no. 6, pp. 779-787, 2009. [http://dx.doi.org/10.1016/j.carres.2009.01.023] [PMID: 19230864]

[50] S.D. Anastyuk, N.M. Shevchenko, S.P. Ermakova, O.S. Vishchuk, E.L. Nazarenko, P.S. Dmitrenok, and T.N. Zvyagintseva, "Anticancer activity in vitro of a fucoidan from the brown alga Fucus evanescens and its low-molecular fragments, structurally characterized by tandem mass spectrometry", *Carbohydr. Polym.*, vol. 87, pp. 186-194, 2012. [http://dx.doi.org/10.1016/j.carbpol.2011.07.036]

[51] M.I. Bilan, A.A. Grachev, N.E. Ustuzhanina, A.S. Shashkov, N.E. Nifantiev, and A.I. Usov, "Structure of a fucoidan from the brown seaweed Fucus evanescens C.Ag", *Carbohydr. Res.*, vol. 337, no. 8, pp. 719-730, 2002. [http://dx.doi.org/10.1016/S0008-6215(02)00053-8] [PMID: 11950468]

[52] S.D. Anastyuk, N.M. Shevchenko, P.S. Dmitrenok, and T.N. Zvyagintseva, "Structural similarities of fucoidans from brown algae Silvetia babingtonii and Fucus evanescens, determined by tandem MALDI-TOF mass spectrometry", *Carbohydr. Res.*, vol. 358, pp. 78-81, 2012. [http://dx.doi.org/10.1016/j.carres.2012.06.015] [PMID: 22824505]

[53] S.D. Anastyuk, T.I. Imbs, P.S. Dmitrenok, and T.N. Zvyagintseva, "Rapid mass spectrometric analysis of a novel fucoidan extracted from the brown alga *Coccophora langsdorfii*", *The Scientific World Journal,* vol. 2014, 2014. article ID 972450, 9 pages

[54] N.M. Shevchenko, S.D. Anastyuk, R.V. Menshova, O.S. Vishchuk, V.I. Isakov, P.A. Zadorozhny, T.V. Sikorskaya, and T.N. Zvyagintseva, "Further studies on structure of fucoidan from brown alga Saccharina gurjanovae", *Carbohydr. Polym.*, vol. 121, pp. 207-216, 2015. [http://dx.doi.org/10.1016/j.carbpol.2014.12.042] [PMID: 25659691]

[55] S.D. Anastyuk, N.M. Shevchenko, E.L. Nazarenko, T.I. Imbs, V.I. Gorbach, P.S. Dmitrenok, and T.N. Zvyagintseva, "Structural analysis of a highly sulfated fucan from the brown alga Laminaria cichorioides by tandem MALDI and ESI mass spectrometry", *Carbohydr. Res.*, vol. 345, no. 15, pp. 2206-2212, 2010. [http://dx.doi.org/10.1016/j.carres.2010.07.043] [PMID: 20813351]

[56] S.D. Anastyuk, T.I. Imbs, N.M. Shevchenko, P.S. Dmitrenok, and T.N. Zvyagintseva, "ESIMS analysis of fucoidan preparations from Costaria costata, extracted from alga at different life-stages", *Carbohydr. Polym.*, vol. 90, no. 2, pp. 993-1002, 2012. [http://dx.doi.org/10.1016/j.carbpol.2012.06.033] [PMID: 22840031]

[57] R.V. Menshova, S.D. Anastyuk, S.P. Ermakova, N.M. Shevchenko, V.I. Isakov, and T.N. Zvyagintseva, "Structure and anticancer activity in vitro of sulfated galactofucan from brown alga Alaria angusta", *Carbohydr. Polym.*, vol. 132, pp. 118-125, 2015. [http://dx.doi.org/10.1016/j.carbpol.2015.06.020] [PMID: 26256332]

[58] J. Zaia, "Principles of mass spectrometry of glycosaminoglycans". J. Biomacromol", *Mass Spectrom. (Tokyo),* vol. 1, pp. 3-36, 2005.

[59] J. Zaia, "On-line separations combined with MS for analysis of glycosaminoglycans", *Mass Spectrom. Rev.*, vol. 28, no. 2, pp. 254-272, 2009. [http://dx.doi.org/10.1002/mas.20200] [PMID: 18956477]

[60] J. Zaia, "Glycosaminoglycan glycomics using mass spectrometry", *Mol. Cell. Proteomics,* vol. 12, no. 4, pp. 885-892, 2013. [http://dx.doi.org/10.1074/mcp.R112.026294] [PMID: 23325770]

[61] L. Bultel, M. Landoni, E. Grand, A.S. Couto, and J. Kovensky, "UV-MALDI-TOF mass spectrometry analysis of heparin oligosaccharides obtained by nitrous acid controlled degradation and high performance anion exchange chromatography", *J. Am. Soc. Mass Spectrom.*, vol. 21, no. 1, pp. 178-

190, 2010.
[http://dx.doi.org/10.1016/j.jasms.2009.09.025] [PMID: 19892568]

[62] E.F. Naggar, C.E. Costello, and J. Zaia, "Competing fragmentation processes in tandem mass spectra of heparin-like glycosaminoglycans", *J. Am. Soc. Mass Spectrom.* , vol. 15, no. 11, pp. 1534-1544, 2004.
[http://dx.doi.org/10.1016/j.jasms.2004.06.019] [PMID: 15519220]

[63] J. Zaia, "Mass spectrometry and glycomics", *OMICS,* vol. 14, no. 4, pp. 401-418, 2010.
[http://dx.doi.org/10.1089/omi.2009.0146] [PMID: 20443730]

[64] M.J. Kailemia, L. Li, M. Ly, R.J. Linhardt, and I.J. Amster, "Complete mass spectral characterization of a synthetic ultralow-molecular-weight heparin using collision-induced dissociation", *Anal. Chem.* , vol. 84, no. 13, pp. 5475-5478, 2012.
[http://dx.doi.org/10.1021/ac3015824] [PMID: 22715938]

[65] R. Ruiz-Contreras, J.P. Kamerling, J. Breg, and F.G. Vliegenthart, "1H- and 13C-n.m.r. spectroscopy of synthetic monosulphated methyl α-D-galactopyranosides", *Carbohydr. Res.* , vol. 179, pp. 411-418, 1988.
[http://dx.doi.org/10.1016/0008-6215(88)84136-3]

[66] M.I. Bilan, E.V. Vinogradova, E.A. Tsvetkova, A.A. Grachev, A.S. Shashkov, N.E. Nifantiev, and A.I. Usov, "A sulfated glucuronofucan containing both fucofuranose and fucopyranose residues from the brown alga Chordaria flagelliformis", *Carbohydr. Res.* , vol. 343, no. 15, pp. 2605-2612, 2008.
[http://dx.doi.org/10.1016/j.carres.2008.06.001] [PMID: 18619579]

[67] K. Izumi, "Structural analysis of agar-type polysaccharides by NMR spectroscopy", *Biochim. Biophys. Acta,* vol. 320, no. 2, pp. 311-317, 1973.
[http://dx.doi.org/10.1016/0304-4165(73)90311-5] [PMID: 4750747]

[68] D. Welti, "Carrageenans. Part 12. The 300-MHz proton magnetic resonance of methyl β-D-galactopyranoside, methyl 3,6-anhydro-α-D-galactopyranoside, agarose, kappa-carrageenan, and segments of *iota*-carrageenan and agarose sulphate", *J. Chem. Res.* , no. (S), pp. 312-313, 1977.

[69] D. Jouanneau, P. Boulenguer, J. Mazoyer, and W. Helbert, "Enzymatic degradation of hybrid ι-/--carrageenan by Alteromonas fortis ι-carrageenase", *Carbohydr. Res.* , vol. 345, no. 7, pp. 934-940, 2010.
[http://dx.doi.org/10.1016/j.carres.2010.02.014] [PMID: 20227066]

[70] D. Jouanneau, M. Guibet, P. Boulenguer, J. Mazoyer, M. Smietana, and W. Helbert, "New insights into the structure of hybrid κ-/μ-carrageenan and its alkaline conversion", *Food Hydrocoll.* , vol. 24, pp. 452-461, 2010.
[http://dx.doi.org/10.1016/j.foodhyd.2009.11.012]

[71] A. Horne, and P. Gettins, "1H-N.m.r. spectral assignments for two series of heparin-derived oligosaccharides", *Carbohydr. Res.* , vol. 225, no. 1, pp. 43-57, 1992.
[http://dx.doi.org/10.1016/0008-6215(92)80038-3] [PMID: 1633604]

[72] H.J. Bixler, K. Johndro, and R. Falshaw, "Kappa-2 carrageenan: Structure and performance of commercial extracts II. Performance of two simulated dairy applications", *Food Hydrocoll.* , vol. 15, pp. 619-630, 2001.
[http://dx.doi.org/10.1016/S0268-005X(01)00047-9]

[73] D.L. Rabenstein, "Heparin and heparan sulfate: structure and function", *Nat. Prod. Rep.* , vol. 19, no. 3, pp. 312-331, 2002.
[http://dx.doi.org/10.1039/b100916h] [PMID: 12137280]

[74] A. Préchoux, and W. Helbert, "Preparation and detailed NMR analyses of a series of oligo---carrageenans", *Carbohydr. Polym.* , vol. 101, pp. 864-870, 2014.
[http://dx.doi.org/10.1016/j.carbpol.2013.10.007] [PMID: 24299849]

[75] M.I. Bilan, E.V. Vinogradova, A.S. Shashkov, and A.I. Usov, "Structure of a highly pyruvylated

galactan sulfate from the Pacific green alga Codium yezoense (Bryopsidales, Chlorophyta)", *Carbohydr. Res.* , vol. 342, no. 3-4, pp. 586-596, 2007. [http://dx.doi.org/10.1016/j.carres.2006.11.008] [PMID: 17134684]

[76] M.I. Bilan, A.A. Grachev, A.S. Shashkov, M. Kelly, C.J. Sanderson, N.E. Nifantiev, and A.I. Usov, "Further studies on the composition and structure of a fucoidan preparation from the brown alga Saccharina latissima", *Carbohydr. Res.* , vol. 345, no. 14, pp. 2038-2047, 2010. [http://dx.doi.org/10.1016/j.carres.2010.07.009] [PMID: 20701899]

[77] S.V. Yarotsky, A.S. Shashkov, and A.I. Usov, "Analysis of 13C NMR spectra of some red seaweed galactans", *Bioorg. Khim.* , vol. 3, pp. 1135-1137, 1977. [in Russian].

[78] G.K. Hamer, S.S. Bhattacharjee, and W. Yaphe, "Analysis of the enzymic hydrolysis products of agarose by13C-n.m.r. spectroscopy", *Carbohydr. Res.* , vol. 54, pp. C7-C10, 1977. [http://dx.doi.org/10.1016/S0008-6215(00)80567-4]

[79] A.I. Usov, S.V. Yarotsky, and A.S. Shashkov, "13C-nmr spectroscopy of red algal galactans", *Biopolymers,* vol. 19, pp. 977-990, 1980. [http://dx.doi.org/10.1002/bip.1980.360190504]

[80] A.I. Usov, "Sulfated polysaccharides of the red seaweeds", *Food Hydrocoll.* , vol. 6, pp. 9-23, 1992. [http://dx.doi.org/10.1016/S0268-005X(09)80055-6]

[81] A.I. Usov, "Structural analysis of red seaweed galactans of agar and carrageenan groups", *Food Hydrocoll.* , vol. 12, pp. 301-308, 1998. [http://dx.doi.org/10.1016/S0268-005X(98)00018-6]

[82] A.I. Usov, "Polysaccharides of the red algae", *Adv. Carbohydr. Chem. Biochem.* , vol. 65, pp. 115-217, 2011. [http://dx.doi.org/10.1016/B978-0-12-385520-6.00004-2] [PMID: 21763512]

[83] C. Bellion, G. Brigand, J.C. Prome, D. Welti, and S. Bociek, "Identification et caractérisation des précurseurs biologiques des carraghénanes par spectroscopie de r.m.n.-13C", *Carbohydr. Res.* , vol. 119, pp. 31-48, 1983. [http://dx.doi.org/10.1016/0008-6215(83)84043-9]

[84] C. Rochas, M. Rinaudo, and M. Vincendon, "Structural and conformational investigation of carrageenans", *Biopolymers,* vol. 19, pp. 2165-2175, 1980. [http://dx.doi.org/10.1002/bip.1980.360191203]

[85] C. Rochas, M. Rinaudo, and M. Vincendon, "Spectroscopic characterization and conformation of oligo kappa carrageenans", *Int. J. Biol. Macromol.* , vol. 5, pp. 111-115, 1983. [http://dx.doi.org/10.1016/0141-8130(83)90021-1]

[86] L.M. Morrice, M.W. McLean, W.F. Long, and F.B. Williamson, "Porphyran primary structure. An investigation using β-agarase I from Pseudomonas atlantica and 13C-NMR spectroscopy", *Eur. J. Biochem.* , vol. 133, no. 3, pp. 673-684, 1983. [http://dx.doi.org/10.1111/j.1432-1033.1983.tb07516.x] [PMID: 6861749]

[87] A.I. Usov, "NMR spectroscopy of red seaweed polysaccharides: agars, carrageenans, and xylans", *Bot. Mar.* , vol. 27, pp. 189-202, 1984. [http://dx.doi.org/10.1515/botm.1984.27.5.189]

[88] C.W. Greer, and W. Yaphe, "Characterization of Hybrid (Beta-Kappa-Gamma) Carrageenan from Eucheuma gelatinae J. Agardh (Rhodophyta, Solieriaceae) Using Carrageenases, Infrared and 13C-Nuclear Magnetic Resonance Spectroscopy", *Bot. Mar.* , vol. 27, pp. 473-478, 1984.

[89] C.W. Greer, C. Rochas, and W. Yaphe, "lota-carrageenan Oligosaccharides as Model Compounds for Structural Analysis of lota-carrageenan by 13C-NMR Spectroscopy", *Bot. Mar.* , vol. 28, pp. 9-14, 1985. [http://dx.doi.org/10.1515/botm.1985.28.1.9]

[90] M. Lahaye, W. Yaphe, and C. Rochas, "13C-n.m.r.-spectral analysis of sulfated and desulfated

polysaccharides of the agar type", *Carbohydr. Res.* , vol. 143, pp. 240-245, 1985. [http://dx.doi.org/10.1016/S0008-6215(00)90713-4]

[91] M. Lahaye, and W. Yaphe, "Effects of seasons on the chemical structure and gel strength of Gracilaria pseudoverrucosa agar (Gracilariaceae, Rhodophyta)", *Carbohydr. Polym.* , vol. 8, pp. 285-301, 1988. [http://dx.doi.org/10.1016/0144-8617(88)90067-7]

[92] M. Lahaye, W. Yaphe, M.T. Phan Viet, and C. Rochas, "13C-n.m.r. spectroscopic investigation of methylated and charged agarose oligosaccharides and polysaccharides", *Carbohydr. Res.* , vol. 190, pp. 249-265, 1989. [http://dx.doi.org/10.1016/0008-6215(89)84129-1]

[93] C.A. Stortz, and A.S. Cerezo, "The 13C NMR spectroscopy of carrageenans: calculation of chemical shifts and computer-aided structural determination", *Carbohydr. Polym.* , vol. 18, pp. 237-242, 1992. [http://dx.doi.org/10.1016/0144-8617(92)90088-8]

[94] C.A. Stortz, B.E. Bacon, R. Cherniak, and A.S. Cerezo, "High-field NMR spectroscopy of cystocarpic and tetrasporic carrageenans from Iridaea undzdosa", *Carbohydr. Res.* , vol. 261, pp. 317-326, 1994. [http://dx.doi.org/10.1016/0008-6215(94)84027-X]

[95] M. Ciancia, M.C. Matulewicz, P. Finch, and A.S. Cerezo, "Determination of the structures of cystocarpic carrageenans from Gigartina skottsbergii by methylation analysis and NMR spectroscopy", *Carbohydr. Res.* , vol. 238, pp. 241-248, 1993. [http://dx.doi.org/10.1016/0008-6215(93)87016-L] [PMID: 8431935]

[96] R. Falshaw, and R.H. Furneaux, "Carrageenans from the tetrasporic stages of *Gigartina clavifera* and *Gigartina alveata* (Gigartinaceae, Rhodophyta)", *Carbohydr. Res.* , vol. 347, pp. 83-94, 1995.

[97] M.E. Duarte, M.D. Noseda, M.A. Cardoso, S. Tulio, and A.S. Cerezo, "The structure of a galactan sulfate from the red seaweed Bostrychia montagnei", *Carbohydr. Res.* , vol. 337, no. 12, pp. 1137-1144, 2002. [http://dx.doi.org/10.1016/S0008-6215(02)00095-2] [PMID: 12062529]

[98] H.J. Prado, M. Ciancia, and M.C. Matulewicz, "Agarans from the red seaweed Polysiphonia nigrescens (Rhodomelaceae, Ceramiales)", *Carbohydr. Res.* , vol. 343, no. 4, pp. 711-718, 2008. [http://dx.doi.org/10.1016/j.carres.2007.12.024] [PMID: 18262175]

[99] I.J. Miller, and J.W. Blunt, "New 13C NMR methods for determining the structure of algal polysaccharides. Part I. The effect of substitution on the chemical shifts of simple diad galactans", *Bot. Mar.* , vol. 43, pp. 239-250, 2000. a

[100] I.J. Miller, and J.W. Blunt, "New 13C NMR methods for determining the structure of algal polysaccharides. Part II. Galactans consisting of mixed diads", *Bot. Mar.* , vol. 43, pp. 251-261, 2000. b

[101] M.I. Bilan, A.S. Shashkov, and A.I. Usov, "Structure of a sulfated xylofucan from the brown alga Punctaria plantaginea", *Carbohydr. Res.* , vol. 393, pp. 1-8, 2014. [http://dx.doi.org/10.1016/j.carres.2014.04.022] [PMID: 24879011]

[102] L.G. Ferreira, M.D. Noseda, A.G. Gonçalves, D.R. Ducatti, M.T. Fujii, and M.E. Duarte, "Chemical structure of the complex pyruvylated and sulfated agaran from the red seaweed Palisada flagellifera (Ceramiales, Rhodophyta)", *Carbohydr. Res.* , vol. 347, no. 1, pp. 83-94, 2012. [http://dx.doi.org/10.1016/j.carres.2011.10.007] [PMID: 22055816]

[103] M. Perez Recalde, D.J. Canelón, R.S. Compagnone, M.C. Matulewicz, A.S. Cerezo, and M. Ciancia, "Carrageenan and agaran structures from the red seaweed Gymnogongrus tenuis", *Carbohydr. Polym.* , vol. 136, pp. 1370-1378, 2016. [http://dx.doi.org/10.1016/j.carbpol.2015.10.007] [PMID: 26572482]

[104] D.J. Canelón, M. Ciancia, A.L. Suárez, R.S. Compagnone, and M.C. Matulewicz, "Structure of highly substituted agarans from the red seaweeds Laurencia obtusa and Laurencia filiformis", *Carbohydr. Polym,* vol. 101, pp. 705-713, 2014.

[105] M. Guerrini, D. Beccati, Z. Shriver, A. Naggi, K. Viswanathan, A. Bisio, I. Capila, J.C. Lansing, S. Guglieri, B. Fraser, A. Al-Hakim, N.S. Gunay, Z. Zhang, L. Robinson, L. Buhse, M. Nasr, J. Woodcock, R. Langer, G. Venkataraman, R.J. Linhardt, B. Casu, G. Torri, and R. Sasisekharan, "Oversulfated chondroitin sulfate is a contaminant in heparin associated with adverse clinical events", *Nat. Biotechnol.*, vol. 26, no. 6, pp. 669-675, 2008. [http://dx.doi.org/10.1038/nbt1407] [PMID: 18437154]

[106] D.J. Langeslay, C.N. Beecher, A. Naggi, M. Guerrini, G. Torri, and C.K. Larive, "Characterizing the microstructure of heparin and heparan sulfate using N-sulfoglucosamine 1H and 15N NMR chemical shift analysis", *Anal. Chem.*, vol. 85, no. 2, pp. 1247-1255, 2013. [http://dx.doi.org/10.1021/ac3032788] [PMID: 23240897]

CHAPTER 12

Physicochemical, Antimicrobial and Mechanical Properties of Thermoplastic Materials Based on Biopolymers with Application in the Food Industry

Kelvia Álvarez[1,2], **Lucía Famá**[3] and **Tomy J. Gutiérrez**[1,2,4,*]

[1] *Department of Analytical Chemistry, Faculty of Pharmacy, Central University of Venezuela (UCV), PO Box 40109, Caracas 1040-A, Venezuela*

[2] *Institute of Food Science and Technology, Faculty of Sciences, Central University of Venezuela (UCV), PO Box 47097, Caracas 1041-A, Venezuela*

[3] *LPMC, IFIBA-CONICET, Dep. Physics, Faculty of Exact and Natural Sciences, University of Buenos Aires, University City (1428), Pab. 1, Buenos Aires, Argentina*

[4] *Thermoplastic Composite Materials (CoMP) Group, Institute of Research in Materials Science and Technology (INTEMA), Faculty of Engineering, University of Mar del Plata and National (UNdMP) and National Council of Scientific and Technical Research (CONICET), Colón 10850, Mar del Plata 7600, Buenos Aires, Argentina*

Abstract: Currently, there is great interest in the study of new ecological technologies searching a harmonious lifestyle with the environment. In this sense, numerous investigations on thermoplastic materials have been conducted as a significant alternative due to their promising applications. In this regard, thermoplastics materials from different biodegradable and edible polymer sources, which can be used in food packaging, films or coatings, have been evaluated as an alternative to replace synthetic materials to contribute to environmental pollution. The aim of this chapter was to study the physicochemical, mechanical and structural properties of biopolymer films for use in the food industry. In this chapter, different components used for the production of biodegradable and edible coatings: biopolymers, plasticizers, antimicrobials and antioxidants are reviewed.

Keywords: Biopolymers, Edible films, Food industry, Physicochemical and mechanical properties, Thermoplastic material.

INTRODUCTION

The current lifestyle has led the consumers change their eating habits. For this reason, the consumption of processed foods has increased noticeably in the last

* **Corresponding author Tomy J. Gutiérrez:** Institute of Food Science and Technology, Faculty of Sciences, Central University of Venezuela, PO Box 47097, Caracas 1041-A, Venezuela; Tel: +58 212 7534403; Fax: +58 212 7533971; E-mails: tomy.gutierrez@ciens.ucv.ve; tomy_gutierrez@yahoo.es

Martin Masuelli & Denis Renard (Eds.)

decades, and this has also increased the consumption and disposal of food packaging. This is how the need to reduce the environmental impact caused by these materials is highlighted. Therefore, the use of biopolymers-based thermoplastic materials have been proposed as an alternative to the use of synthetic materials made from the petroleum, since these are ecological and edible materials. Furthermore, these thermoplastic materials have improved food safety and quality, because they reduce the permeability to gases (O_2, CO_2, H_2O and ethylene), affecting the food spoilage. In addition, these biomaterials are an excellent vehicle for the incorporation of antimicrobial agents and natural antioxidants that allow to ensure the food safety. In the same way, the additives incorporated can improve the firmness, brightness and color of food, and even in processed foods.

In this chapter, we focus on the analysis of the different biopolymers used in the production of edible films such as the thermoplastic materials, as well as the physicochemical, mechanical, antimicrobial and structural properties of these edible plastics will be analyzed.

1- Biodegradable and Edible Films Based on Biopolymers

Most materials used for food packaging are petroleum-based polymers, due to their wide availability, low cost, great tensile strength and strain at break, as well as the oxygen barrier properties and heat sealability. However, these materials are not biodegradable, causing serious environmental problems. As a result, the demands have been directed towards the use of biodegradable materials obtained from agricultural products, such as the wastes from the food industry and from natural resources of low cost. Among the possible biopolymers used for the manufacture of films; carbohydrates, starches, pectic materials, vegetable and animal proteins are the most commonly used [1].

A film consists of a preformed thin layer of protective material formed for different components which vary depending on the functionality of the film. In several cases, it is applied on food products to reduce their exposure from the environment, as well as to establish barriers to moisture and oxygen, and solutes movement. They may also be fulfill the function of transport of several food additives, such as antimicrobials, antioxidants, antisoftening, micronutrients and bioactive components, which offer properties that can ensure or improve the food quality, safety and functionality [2 - 7].

Edible and biodegradable films have also been used to reduce the deleterious effects produced by processing conditions and food storage. This because the semipermeable barrier of the films can extend food lifetime, due to the reduction of moisture loss, solute migration, gases exchange, respiration rates and oxidative reactions [8].

2- Film Components

The basic formulation of edible films includes a forming agent, which allows to obtain a three dimensional network or a matrix capable of self-supported, that at the same time preferably should be translucent and homogeneous. The components used can be classified into three categories: hydrocolloids (polysaccharides and proteins), lipids (fatty acids and waxes) and composites (hydrocolloids and lipids mixtures or combinations of components of the same group) [9, 10]. In this chapter the characteristics and properties of thermoplastic materials based on biopolymers will emphasized.

2.1- Polysaccharides

Polysaccharides have been the focus of research in recent decades, in order to be used as base material for the formation of biodegradable and edible films. Beyond their importance in contributing to the conservation of the environment and the possibility of extending the food shelf life, a remarkable reason of the use of polysaccharides is their low cost compared to the materials obtained from conventional sources as petroleum.

Polysaccharides are composed of monosaccharides linked by glycosidic bonds [11]. According to the monomers present in the molecules, they are classified as: *Homopolysaccharide* in the case of only one type of monosaccharide, and *Heteropolysaccharides* when the molecules are composed of more than one type of monosaccharides. On the other hand, there is another classification depending on their structure: linear structure (agar, methylcellulose, hydroxypropyl methyl cellulose, microcrystalline cellulose, inulin, carrageenan, pectin, gellan gum, carboxymethylcellulose, chitosan), branched structure (arabic gum, ghatti), or any combination, *e.g.* starches [11].

Polysaccharides most commonly used for the preparation of edible films or coatings are starch, alginate, pectin, algae extracts, gums, chitosan and different types of cellulose [12]. An important advantage of the systems formed by this type of component is their hydrophilic nature which provides a water vapor barrier, thus allowing the coatings avoid the moisture loss from the foods [8, 13].

The characteristics of the most important polysaccharides used for the preparation of thermoplastic materials to be used in the food industry are described below:

2.1.1- Starch

Currently, the use of starches is the focus of study in the food industry, because is of easy access, since it can be obtained from various renewable sources, such as cereal grains (corn, wheat, rice), leguminous seeds (lentils), tubers (potato and cassava), and it represents one of the most important raw materials for industrial use [14 - 21]. Furthermore, its low cost compared to other alterative biomaterials, starch makes a very tempting material for potential use in the food industry.

Starch is a semicrystalline polymer, which generally is deposited in plant tissues as insoluble granules, with particle size between 2 and 100 μm, and have elliptical, spherical, angular, simple or compound, and irregular shapes [22]. It is composed by anhydroglucose units linked together primarily through glycosidic bonds, and are composed mainly of two polymers of D-glucose: amylose (linear bonds) and amylopectin (branched bonds) [23].

In the starch granules, both amylose and amylopectin molecules are arranged radially, forming a series of concentric layers. In these layers, there are amorphous and crystalline regions. Crystalline regions are the chains associated with helix formations, such as short chains of amylopectin organized in clusters and linear amylose, while the areas of branch points are of amorphous character [24].

Amylose consists of repeating anhydroglucose cells. It is essentially a linear molecule of glucose units with α-1,4 bonds, which have an average molecular weight of 60,000 Dalton and a degree of polymerization between 100 and 10,000 units. Each molecule has a reducing end (C=O) and a non-reducing end (OH). Due to the large amount of hydroxyl groups contained therein, this imparts hydrophilic properties to the polymer, resulting in a high affinity for moisture. The linear nature, mobility and the presence of large amount of OH groups along the polymer chains, makes that amylose molecules tend to be oriented parallel to each other and closer together to allow the formation of hydrogen bonding between the adjacent chains. Currently, the three-dimensional (3D) conformation for amylose is the more favorable and accepted structure, which consist in a double helix that is able to be packaged in regular arrangements giving a characteristic X-ray diffraction pattern, which may be related to water content of these materials [25]. Likewise, this macromolecule melts between 100 °C-120 °C [22].

Amylopectin has a highly branched structure, which contains short cells of 10-60 units of anhydrous glucose linked by α-1,4 bonds. Furthermore, it has side chains with 15-45 units of glucose, which has as average an 5% of α-1,6 bonds at points of branch [26]. The molecular weight of amylopectin is about 1,000 times greater than amylose, and ranging from 1 x 10^7 to 5 x 10^8 g/mol [27]. Its branched nature and larger size reduce their mobility in solution and eliminate the possibility of significant levels of interchain hydrogen bonds. The amylopectin chains are arranged radially within the granule with its non-reducing terminal ends oriented toward the surface, and these are arranged alternating crystalline regions with amorphous regions (as a double helix) [28].

According to amylose content, the starches can be classified into different groups: waxy starches, containing very small quantities of amylose (*e.g.* amaranth), about 7%; normal starches, between 17 and 24%, and high amylose starches with 70% or more [29, 30].

The starch characteristics, as well as their functional properties of the materials containing, largely depend on the way in which amylose and amylopectin are associated and distributed [28]. The amylose content, length and location of the branches of amylopectin are determined for their properties, such as water absorption, gelation, pasta conformation, retrogradation and susceptibility to enzymatic attack [31, 32].

In addition, the starches contain minor components, such as protein (0.05 to 0.5 wt.%), lipids (0.1 to 1.7 wt.%), inorganic substances (0.1 to 0.3 wt.% ash) and non-starch polysaccharides [24], which can interact with many additives or food components [33, 34]. This makes them widely used in the food industry, because they offer protection of products against shock and vibration that occur during transport. In addition, they can provide antioxidant and/or antimicrobial on the foods.

For example, the application of biodegradable films based on cassava starch with added natural antimicrobial (cinnamon and clove) were investigated by Kechichian *et al.* [35] on the slices of bread. The authors performing microbial counts on the slices of bread covered with these films, obtaining that the mold and yeast counts in the slices of bread covered with films containing cinnamon or clove powder and stored at room temperature for 7 days were equal or slightly higher than in control, concluding that the expected antimicrobial effect was not achieved under the experimental conditions used.

The authors also explained that possibly an increase in water activity (a_w) of the films occurred during storage of the coated bread slices and being this the possible reason, whereby, the microbial counts were similar or slightly higher in slices covers compared to the control slices.

Garcia *et al.* [36] showed that pet food packaged by films based on cassava starch preserved the fresh and natural appearance of the food for a longer time than the control, retarding moisture migration, transport of gases and reducing the exchange of odors.

2.1.2- Cellulose and Derivatives

Cellulose is composed of the units of D-glucose linked by β(1, 4) glycosidic bonds [37], while its derivatives are polysaccharides formed by linear chains of β(1, 4) glucoside units with substitutions methyl hydroxypropyl or carboxyl groups. Only four forms derived from cellulose are used for preparing coatings and edible films: hydroxypropyl cellulose (E463; HPC), hydroxypropyl methylcellulose (E464; HPMC), carboxymethyl cellulose (E466; CMC) or methylcellulose (E461; MC). Films from cellulose derivatives are characterized by a lower water vapor barrier, due to the hydrophilic nature of the polysaccharides, and poor mechanical properties. In order to improve moisture barrier, the hydrophobic components such as fatty acids, it can be incorporated, thus generating composite films [38].

Cellulose and its derivatives have been used to develop edible films for different types of food. In this regard, Franklin *et al.* [39] determined the effectiveness of cellulose/hydroxypropyl methylcellulose based films with the addition of 1, 25, 75 or 100 mg/mL of nisin to control the population of *L. monocytogens* inoculated on the surface of hot dogs sausages, vacuum packed. The authors reported that the films based on cellulose containing 75 and 100 mg/mL of nisin had a significant decrease ($p \leq 0.05$) in the population of *L. monocytogenes* for 60 days of storage, which was greater than 2 log CFU/g. Santiago-Silva *et al.* [40] evaluated the effectiveness of cellulose films with 25% and 50% of pediocin on sliced ham inoculated with *L. innocua* and *Salmonella sp.* They found that films were more effective against the first one, resulting in a reduction of 2 log in CFU/g of *L. innocua* population on the cellulose films with 50% of pediocin, after 15 days of storage at 12 °C, while the films with 25% and 50% of pediocin caused a reduction of 0.5 log CFU/g in the population of *Salmonella sp.* after 12 days of storage at 12 °C, compared to the control.

2.1.3- Alginate

Alginates are polysaccharides extracted from the brown algae of *Phaephyceae* class [10], which have very good film-forming properties. Their great potential to form films is due to their properties as hydrocolloids, which impart thickening, stabilizing, film formers, gel-producing and stabilizing emulsions properties. Thus, divalent cations (calcium, magnesium, manganese, iron or aluminum) are commonly used as gelling agents in the formation of alginate films. Among the desirable properties attributed to the application of alginate films are; moisture retention, improvements in the texture properties, juiciness, color and food smell. Edible films made from alginates are strong and resistant, but exhibit poor resistance to water vapor permeability, due to their hydrophilic nature [38].

The effectiveness for the application of films from alginate to be used as a food packaging has been demonstrated by several investigators. Cutter and Siragusa [41], Natrajan and Sheldon [42], and Millette *et al.* [43], reduced the *Brochothrix thermosphacta* (> 3.0 log CFU/g), *Salmonella enterica* ser. *Typhimurium* (> 4.0 log CFU/g) and *Staphylococcus aureus* (> 2.5 log CFU/g) populations in the ground beef, chicken skin and meat fillets, respectively, using calcium alginate films with nisin (from 5-100,000 mg/mL) during their storage under refrigeration. Similarly, Marcos *et al.* [44] reported a bacteriostatic effect against *L. monocytogenes* inoculated in the sliced ham during storage for 60 days at 1 °C, using alginate based films with enterocins A and B (2,000 AU/cm). Oussalah *et al.* [45] evaluated the effectiveness of alginate-based films containing essential oil of oregano, cinnamon or savory to the 1% (w/v), on inoculated populations of *S. enterica* ser. *Typhimurium* or *E. coli* O157:H7 in meat slices stored for 5 days at 4 °C. The authors reported that films containing oregano essential oil or cinnamon essential oil were effective against *S. enterica* ser. *Typhimurium* (> 1 log cycle); while materials with oregano essential oil were more effective (> 2 log cycles) against *E. coli* O157:H7. In the same way, Oussalah *et al.* [46] studied the effect of the application of alginate based films containing 1% (w/v) essential oils of oregano, cinnamon or savory, on ham and bologna slices, inoculated with *S. enterica* ser. *Typhimurium* or *L. monocytogenes*. The authors concluded that films containing cinnamon oil were more effective, achieving reductions in both populations of microorganisms to more than 2 log CFU/g.

2.1.4- Chitosan

Chitosan is a copolymer consisting of units of β-(1-4)-2-acetamido-D-glucose and β-(1-4)-2-amino-D-glucose, and it is obtained from deacetylation of chitin [47], likewise, is the second most abundant polysaccharide in nature and is found in the exoskeleton of crustaceans, fungal cell walls and other biological materials [37, 48]. Chitosan is non-toxic, biodegradable and biocompatible [49]. Some authors suggest that it is a good alternative as a base material for the production of antimicrobial films, due to their good film forming properties, its ability to retain water and inhibit various bacterial enzyme systems [50, 51]. Elsabee and Abdou [47] indicated that the chitosan based films offer a selective permeability to gases (CO_2 and O_2) and have a significant mechanical properties. Its use in some food is often limited, due to their high water vapor permeability, which is considered a disadvantage; nevertheless, this limitation has been resolved by the addition of other hydrophobic compounds (neutral lipids, fatty acids, *etc.*).

Several studies have demonstrated the effectiveness of edible films prepared from chitosan to improve the conservation of a variety of foods. Ouattara *et al.* [52] evaluated the effectiveness of edible chitosan films with acetic or propionic acid, with or without the addition of lauric acid or cinnamaldehyde to preserve bologna, ham and pastrami, vacuum packed and stored under refrigeration. The effectiveness of the films to inhibit microbial growth was studied against native lactic acid bacteria, *Enterobacteriaceae* and against *Lactobacillus sakei liquefaciens* or *Serratia* inoculated onto the surface of the products. The authors concluded that the growth of lactic acid bacteria was not affected. Nevertheless, the growth of the *Enterobacteriaceae* and *Serratia liquefaciens* was delayed or completely inhibited after the application of the film. Siripatrawan and Noipha [53] were able to inhibit completely microbial growth in pork sausages with chitosan films using green tea extracts, extending the useful life of the product refrigerated at 4 °C. Duan *et al.* [54] succeeded in reducing populations of *L. monocytogenes*, *E. coli*, and *Pseudomonas fluorescens* on the mozzarella cheese surface using films based on chitosan with lysozyme during storage at 10 °C.

2.2- Proteins

The investigations of protein-based films have presented better physicochemical and mechanical properties compared to those prepared from polysaccharides and fat, because the proteins have a unique structure, which confers a wider range of functional properties. Although, it is known that the use of proteins is limited in developing films, because the proteins are denatured at high temperatures (> 50 °C). However, this phenomenon has been overcome due to changes in the properties of proteins by chemical and enzymatic methods, in combination with hydrophobic materials or some polymers, or by using a physical method [55].

Among the proteins widely used for film formation is the casein, this because the caseinates present favorable characteristics for its use on foods that need to be visually appealing. Caseinates are good film formers emulsified by their amphiphilic nature; their disordered structure and its ability to form hydrogen bonds, as well as have antimicrobial properties. Kristo *et al.* [56] characterized sodium caseinate films using as carriers of preservatives (sodium lactate, potassium sorbate, and nisin). The authors demonstrated antimicrobial action, especially when potassium sorbate was used, this modified barrier properties, water vapor and thermo mechanical behavior of the films. Moreover, Mendes de Souza *et al.* [57] studied the behavior of lysozyme incorporated to a modified film with sodium caseinate (using chemical or biochemical crosslinking agents). The results showed that the active caseinate films modified by pH change and glyoxal, efficiently delayed the release of lysozyme, being a promising way to apply antimicrobial agents during food storage.

Other type of proteins used in the elaboration of edible films are the whey proteins, this because they have excellent barriers to O_2. Further, their mechanical properties can be improved by adding a plasticizer such as glycerol.

It have also been made zein-based films, although it is worth remembering that zein is a prolamin and is the main storage protein of corn, and it is characterized by be a relatively hydrophobic thermoplastic material and, therefore, form strong films, gloss, resistant to microbial attack, insoluble in water as well as antioxidant properties and adhesion capacity. An example of this is the work of Cheng *et al.* [58], who worked with zein and chitosan as base material, supplemented with phenolic compounds (ferulic acid or gallic acid) and dicarboxylic acid (adipic acid or succinic acid), obtaining as results antioxidant activity, demonstrated by DPPH and free radical scavenger tests, as well as antimicrobial activity against *Stapylococcus aureus* and *E. coli.*

2.3- Plasticizers

Plasticizers are typically used in the preparation of biodegradable films by their potential to reduce the brittle character of many films based on polysaccharides, since they can increase the flexibility and shear strength of these materials, this because they reduce effectively the intra and intermolecular interactions between polymeric chains.

The plasticizers most frequently used in films applicable in the food industry are:

- Mono-, di-, and oligosaccharides: mainly glucose, syrups and honey.
- Polyols: sorbitol, glycerol, polyethylene glycols, *etc.*
- Lipids and derivatives: fatty acids, monoglycerides, surfactants and ester derivatives.

The basic requirement is that the plasticizer must be compatible with the polysaccharide matrix. This implies a similarity in the intermolecular forces between the two components. In general, plasticizers are of low molecular weight compared to the polymers commonly used for these materials [59, 60]. Those, having features such as smaller, high polarity, and polar molecules, with distance between the polar groups within the molecule, generally impart greater plasticizing effects on a polymer system, increasing the free volume and reducing the intermolecular attractions between the adjacent polymer chains by reducing hydrogen bonding between the polymeric chains [61].

Table 1. Antimicrobial agents commonly used in films and coatings based on biopolymer.

Classification of antimicrobial agent	Antimicrobial agent	Biopolymeric matrix	Chemical compounds present that exerts the antimicrobial effect	Inhibited microorganism	Effect on the films	References
Plant origin	Oregano essential oil (*Oreganum heracleoticum L.*) and thyme essential oil (*Thymus vulgaris L.*)	Soy protein	Carvacrol and thymol	*E. coli* *E. coli* O157:H7 *S. aureus*	---------	Emiroğlu *et al.* [71]
	Mexican lemon essential oil (*Citrus aurantifolia swingle)*	Pectin	Limonene, α-pinene, β-pinene, myrcene and α-terpineol.	*Escherichia coli* O157:H7 *Salmonella Typhimurium* *Bacillus cereus* *Staphylococcus aureus* *Listeria monocytogenes*	---------	Aldana *et al.* [72]
	Thyme essential oil (*Thymus vulgaris L.*) (TMEO) and (*Thymus piperella*) (TPEO)	Chitosan	(TMEO): Camphor, 1,8-cineole y mircene. (TPEO): carvacrol, para-*cimene*, γ-terpinene and α-terpineol.	*Serratia marcenscens* *Listeria innocua*	---------	Ruiz-Navaja *et al.* [73]

(Table 1) contd.....

Classification of antimicrobial agent	Antimicrobial agent	Biopolymeric matrix	Chemical compounds present that exerts the antimicrobial effect	Inhibited microorganism	Effect on the films	References
Animal origin	Lactoperoxidase system	Defatted soy flour	Gluconolactone	*Salmonella*	Color, tensile strength, and moisture modify the mechanical behavior. Increases in color and elongation, while decrease elasticity as result of an increased the protein concentration in the film. The sensory properties of the product containing the films were not significantly affected.	Lee and Min [74],
Animal origin	Ovotransferrin, EDTA, potassium sorbate	κ-Carrageenan	Ovotransferrin contains mannose and glucosamine. It has ability to sequester iron cation, giving antimicrobial capacity.	*E. coli* *S. typhimurium* *S. aureus* *C. albicans*	Total aerobic count decreased by using ovotransferrin and EDTA.	Seol *et al.* [75]

(Table 1) contd.....

Classification of antimicrobial agent	Antimicrobial agent	Biopolymeric matrix	Chemical compounds present that exerts the antimicrobial effect	Inhibited microorganism	Effect on the films	References
Microbial origin	Natamycin	Cassava starch	Macrolide. Antimicrobial destruction *via* oxidative mechanisms.	*Saccharomyces cerevisiae*	Natamycin presence tends to decrease the strength at break and Young's modulus and to increase the tension at break. The color also is affected by the presence of this antimycotic. It recommend to use lower concentration of natamycin to avoid these effects in the films.	Resa *et al.* [76]
	Nisin	Hydroxypropylmethylcellulose	Nisin, natural antimicrobial produced by fermentation of layers of *Lactococcus lactis subsp. lactis*	*Listeria* *Enterococcus* *Staphylococcus* *Bacillus spp*	Increases in thickness due to the formation of salt crystals, resulting in an inhomogeneous surface, reducing the transparency of the films. The WVP showed not significantly different, but if a slight upward trend.	Imran *et al.* [77]
Bacteriocins	Enterocina synthesized by *E. faecium* SM21	Gelatin	Dioleilfosfoglicerol (DOPG) Dioleilfosfocolina (DOPC)	*L. monocytogenes* *S. aureus* *B. cereus*	The bacteriocins incorporation did not affect the functional properties of the films, since the mechanical and barrier properties were not altered.	Ibarguren *et al.* [78]

(Table 1) contd.....

Classification of antimicrobial agent	Antimicrobial agent	Biopolymeric matrix	Chemical compounds present that exerts the antimicrobial effect	Inhibited microorganism	Effect on the films	References
Others	Phenolic compound and dicarboxylic acid	Zein / chitosan	Ferulic acid or gallic acid, adipic acid or succinic acid	*S. aureus* *E. coli.*	Films with phenolic components and gallic acid resulted be more permeable due to the hydrophilic films. The color was affected by darkening caused by the phenolic compounds, as well as reduced the gloss due to the incorporation of adipic acid. Greater tensile strength and strain at break with the addition of succinic, and maximum elongation for the films containing ferulic acid.	Cheng *et al.* [58]
Traditional	Organic acids	Protein isolate: Argentine Anchovy (*E. anchoita*)	Sorbic acid (SA) Benzoic acid (BA)	*E. coli* O157:H7 *L. monocytogenes*	Increase in the concentrations of SA or BA led to thicker and opaque films, increments in breaking elongation, and decrease in resistance, as well as increase in water vapor permeability, due to the rough surface of the films caused by the addition of benzoic acid.	da Rocha *et al.* [79]

The most commonly used plasticizers in food coatings are polyols: propylene glycol [62], glycerol [61], sorbitol [63], polyethylene glycol [64], oligosaccharides (sucrose) [65], and water. Likewise, the concentration of plasticizer, affects the moisture content of the films and greatly influences on the material properties [66].

2.4- Antimicrobials

Today, the incorporation of antimicrobials in thermoplastic materials is very common, because they act primarily by inhibiting and/or reducing the growth of microorganisms. Antimicrobials are chemical components that are naturally or intentionally added to food, food packaging, surfaces in contact with food, or food processing environments to inhibit the growth of microorganisms [67]. Moreover, the focus of action of antimicrobials are the pathogens microorganisms, which produce toxins or cause infections, and deteriorative microorganisms, which are metabolic, which impart disagreeable odors and flavors, as well as generate problems of texture and color of the products [68].

There are some considerations for the selection of antimicrobials for use in thermoplastic materials. To do this, it must take into account the effectiveness of antimicrobial against the microorganism target, the food components and interactions that occur between biopolymer-antimicrobial. These interactions can modify the antimicrobial activity and the characteristics of the film, being the key factors in the development of antimicrobial films and coatings [69]. Further, the diffusion of the antimicrobial film depends on the size, shape (linear, branched or cyclic), and the polarity of the diffusing molecules, as well as, on the chemical structure of the film and the degree of molecular crosslinking [70]. The different interactions between the polymer matrix and the antimicrobial agent can lead to the inhibition against pathogen microorganisms. According to Davidson *et al.* [67], the ideal antimicrobial should be: (1) effective at low concentrations in their natural form, (2) economic according to its use, (3) does not provoke sensory changes in the product, (4) inhibits a wide range of pathogens and spoilage organisms, and (5) not is toxic.

Antimicrobials can be classified in two groups: traditional and naturals. Traditional antimicrobials group are considered those chemical substances included within most of the regulations, and in the natural case are the substances obtained or derived from materials or biological processes, whose safety is attributed to when ingested are degraded by the body.

Table **1** shows recent studies on edible biopolymeric films, where antimicrobial agents are incorporated taking into consideration the specific active component, agent content, microorganisms that were inhibited, and the physical and chemical effects caused in the evaluated films.

2.5- Antioxidants

Antioxidants are substances able to delay or prevent the development of rancidity in food, as well as any deterioration of their sensory attributes related to the taste and smell (*flavor*). They can inhibit or retard the oxidation of food in the following ways: inactivating free radicals, in whose case are named "primary antioxidants", *e.g.* vitamin E (α-tocopherol) or either by bonding or capture metal ions, converting hydroperoxides to not radical species, and inactivating the action of oxygen or absorbing UV radiation, in which case are named "secondary antioxidants". Secondary antioxidants are usually effective only when a smaller secondary component is present as occurs in the case of citric acid, which is a sequestering agent that only can act when the metal ions are present, or in the case of ascorbic acid, which is effective in the presence of tocopherols or other primary antioxidants [80].

Several investigations have demonstrated that edible films have excellent capacity for the transport of antioxidant compounds, showing that the lipid oxidation of food can be delayed or inhibited by the incorporation of antioxidants in the film-forming solutions. Wu *et al.* [81, 82] reported that the addition of tocopherols on the composite films based on pre-cooked starch/alginate used to cover meat burgers caused the inhibition of the lipid oxidation of the product for several days. Gomez-Stake *et al.* [83] studied the effect of the incorporation of oregano or rosemary extracts in gelatin-based films on the lipid oxidation of frozen smoked sardines, noting that the application of the films caused a reduction in the lipid oxidation of sardines, which was demonstrated by analyzing peroxide and thiobarbituric acid reactive substances (TBARS).

The possibility to use antioxidants as additives in these kinds of edible biopolymers is influenced by the availability of water, which in turn is affected by the product moisture and moisture from the environment. This mechanism itself can have positive effects on the preservation of quality, due to the limited availability of oxygen in the coated product. In some cases, addition of antioxidants may involve greater protection, due to the improvements in the oxygen barrier properties of the films [84, 85]. However, reduction in the molecular mobility of antioxidants can occur. Therefore, the antioxidant effect is due to the oxygen-barrier effect, as concluded by Atarés *et al.* [86].

3- Processing Techniques

Not only the materials used for forming biopolymer-based thermoplastic films are a key factor determining the physicochemical properties of the resulting material, but the processing conditions of these materials also have an important effect [87]. The main techniques that have been tested successfully by years for the production of edible films are coating, casting, extrusion and mixtures.

The film-forming process by the mixture is one of the major techniques used for the preparation of composites based on biopolymers and additives, which involves in the dispersion of the polymer, generally in some diluent (solvent). In many cases, it includes a step involving heating of the mixture, pH modification, or incorporation of additives; and a second step, which produces solidification of the material through cooling and/or drying. The speed and temperature at which these stages are carried out have much influence on the physicochemical properties of the resulting films. During drying, the solvent is evaporated leading to a decrease in the solubility of the polymer until the polymer chains are aligned for the film forming. In general, fast drying of biopolymers-based solutions restricts the mobility of the polymer chains, which produces a limitation in the development of intermolecular associations in the film structure [88]. A fast drying can cause defects such as cracks [89]. In the mixing process, the choice of substrate is also crucial to obtain films that can easily be handled, after that, the solvent is evaporated.

The method of film-forming by extrusion was an originally developed to produce synthetic materials or nanocomposites, but nowadays it is also widely used in the plastics industry for biopolymer based materials. The extrusion process depends on the glass and melting temperatures of the polymer and the characteristics of the additive, which are produced while heating the systems, under the condition of adding little amount of water.

Compared with the method for mixing, the extrusion technique is attractive for industrial processes, since it does not require the addition of solvent, which then must be evaporated, thus providing less time for obtaining the material. For example, to form multilayer films, the extrusion technique can be very useful. However, the differences in the characteristics of each component of these films can lead to numerous defects during formation of the material. These defects are particularly detrimental to the mechanical and barrier properties. Furthermore, if a significant number of composite materials based on biopolymers have been successfully prepared with the mixing method, this technique is limited to biopolymers that can be dissolved in suitable solvents. In addition, another disadvantage of the mixing method with respect to the extrusion technique is that in some cases, the use of a high-power ultrasound is required to incorporate an additive.

4- Properties Evaluated

The main characteristics that are evaluated in the edible and biopolymeric films are: optical properties, water solubility, water sorption-desorption, thickness, microstructure, crystallinity, compatibility of the components in composite materials, thermal behavior, gas barrier properties (CO_2/O_2/ethylene), mechanical behavior and antimicrobial properties. Some of the most important results of the mechanical, antimicrobial, structural and physicochemical properties of these biomaterials are explained below:

4.1- Physicochemical Properties

There are several parameters evaluated in the biopolymer films in order to establish their physicochemical behavior. Some of the tests usually carried out for this purpose are; water activity, water solubility, water vapor permeability, thickness, density, among others. Thickness is an important parameter of the edible films made from biopolymers, since this parameter directly affects its other properties, as well as the properties of the foods that are covered with these biomaterials. For example, the thickness of the films based on biopolymer affects the permeability of gases (CO_2/O_2/ethylene), while this influences the effectiveness of protection of the foods [90]. Likewise, thickness depends on the biopolymer used as base, and on the additives incorporated in the films [91].

The optical properties are related to the characteristics of the material surface. In the case of edible films, it should pay close attention to these properties that do not affect any crucial aspect of the foods, because it can impact the consumer's acceptability for the product [92]. The surface and microstructure of the films play an important role in their optical properties, *e.g.* a smoother surface has higher brightness [93].

Properties of biopolymers depend on its structure. Mostly biopolymers have amorphous structure (vitreous type), nevertheless, some linear and cross-linked biopolymers can also form ordered zones, *i.e.* they partially have a crystalline structure. The increase of the crystalline phase mainly occurs due to the high interaction between the biopolymer chains, which limits the flexibility of these materials. Likewise, an attractive foodstuff for the consumers requires a uniform surface without defects in the films. Generally, the control of the surface of the films is difficult because when functional ingredients (antimicrobials, antioxidants, coloring agents, emulsifiers, flavor, nutraceutical) are incorporated for other purposes, they often lead to the loss of the homogeneity in the film.

On the other hand, the moisture content and water solubility are some of the main problems of the biopolymer-based films. The water solubility of the films can provide a perception on the behavior of the biopolymeric films in an aqueous medium, and this is a measure of its resistance to water, determining the biodegradability of the films when used as the packaging material [94]. In fact, the solubility is a desired property in many cases, *e.g.* for the encapsulation of food additives. Normally, the effects of the incorporation of additives in the solubility of the films depend on the type of ingredient, its concentration and hydrophilicity.

When an edible film is applied to a food, the concentration of gases, water, aroma and oils on the surface of the food can change during storage, this is usually due to the metabolism of foods, microbial respiration, solubility and permeability of the material. Therefore, the efficiency of these films largely depends on their barrier properties, which in turn depend on the chemical composition and the structure of the film-forming biopolymers, as well as of the product characteristics and storage conditions. The water vapor permeability through an edible film is the sum of three factors: a) sorption (investigated by isotherms), b) diffusion of water into the biopolymer and c) water desorption. The value of water vapor permeability of a material is essential when it observes the capacity of the material to retard the moisture loss from the product to the environment or vice versa, since this characteristic affects the product quality [81, 82]. It is expected that the hydrophilic compounds increase water solubility and water vapor permeability of the films, while the hydrophobic compounds decrease these parameters [95].

The oxygen permeability (OP) of the materials used for food packaging is of great importance for their conservation, since oxygen is a key factor in various deterioration reactions [96]. In general, the films based on a biopolymeric matrix offer a good barrier to oxygen, since the hydrogen bonding network of its structure is perfectly packaged [97 - 99].

Some recently published studies on the influence of additives, such as; antimicrobials, antioxidants or oils on chitosan-based films showed that thickness and color had the following trend: an increase in the thickness is related to a decrease in the transparency of films, leading to their opacity due to loss of matrix homogeneity [70]. These trends modify the structure of the material, which generally passes from a smooth and homogeneous structure to a porous and non-homogeneous structure. Nevertheless, the barrier properties, moisture content, solubility and mechanical behavior showed insignificant trends. In these cases, both the polysaccharides, used as the base and the additive, were fundamental in observing this behavior [70]. For example, hydrophilicity of the additives are essential for moisture, water solubility and water vapor permeability (WVP), and as a consequence, there characteristics modify the mechanical performance of biodegradable films. Likewise, a hydrophilic additive incorporated into a filmic material tends to cause a decrease in the water vapor barrier properties, moisture content and water solubility of the material, resulting in higher flexibility and less strength at break [70].

In this sense, Rubilar *et al.* [100] indicated that the addition of carvacrol in chitosan based films caused a reduction in the WVP, due to the hydrophobic nature of the additive. In contrast, the incorporation of grape seed extract (GSE) allowed to observe an opposite behavior in this property, probably due to the hydrophilic nature of the additive [101]. Furthermore, the GSE originated less crystalline films, resulting in an increased of WVP. According to Sperling [88], and Miller and Krochta [102], a polymer with high crystallinity is generally less permeable to gases, because of its ordered structure. In both the systems, the oxygen permeability (OP) was greater with the addition of additives, probably due to the microstructural changes, which led to obtain more porous films. As the gas permeability depends on the crystallinity of the films, an increase of this parameter led to a decrease in the OP [98, 102 - 104]. Similar results were reported by Altiok *et al.* [105] in materials with high content of thyme oil, and by Bonilla *et al.* [106] after the addition of essential oils of thyme and basil. This behavior was previously explained by Byun *et al.* [107] with the addition of α-tocopherol, which indicated that this type of additive might improve the WVP by the hydrophobic character. However, in a similar study, Martins *et al.* [99] found an increase in the WVP, this despite hydrophobic character of the additive, thus, attributing this finding to the decrease in the crystallinity of the films. Pranoto *et al.* [108], showed that when garlic oil was added to the alginate-based edible films, the WVP did not experience significant changes. However, the addition of potassium sorbate (KS) caused a tendency to increase the WVP. The authors indicated that, probably, the antimicrobial agent contributed to the loss of the compact structure of the material, thus increasing the possibility of the transfer of the water vapor through the film. With both the additives, the appearance of the material tended to take a yellow color and was more opaque, but showed no significant changes in the color difference (ΔE). With the addition of resveratrol, WVP and OP were not affected despite the modification of the structure of the film, which was caused by the formation of a more disordered network, due to the additive [109]. Likewise, the green tea extract (GTE) [110] and tea tree essential oil (TTO) [111], caused decrease in the WVP, possibly due to the hydrogen bonding interactions between the chitosan network and polyphenolic compounds of the additives, thus limiting the availability of hydrophilic groups to form hydrogen bonding with water, leading to a decreased affinity of the chitosan film by water [112]. With the addition of GTE, the films were of yellow and red color. The incorporation of TTO contributed to the surface roughness of the film, which reduced its brightness. The decrease in the moisture content, water solubility and

WVP with the addition of oils was shown by Ojagh *et al.* [113] in the films based on chitosan with the essential oil of cinnamon, and by Georgantelis *et al.* [114] with the addition of olive oil.

Similar behaviors were reported by Ghasemlou *et al.* [115] in starch-based films containing two essential oils: *Zataria multiflora Boiss* (ZEO) or *Mentha pulegium* (MEO), and were explained according to Sánchez-González *et al.* [111] and Shen *et al.* [112], due to the hydrogen bonding interactions between the starch network and these polyphenolic compounds. The addition of ZEO or MEO showed no significant difference in the OP, with the exception of the addition of the highest concentration of both oils (3% v/v). This could be due to the swelling of the polymer in the presence of the volatile oils in high concentrations [116]. Han *et al.* [117] reported similar results in starch-beeswax films, suggesting that the oxygen can penetrate between the particles of the wax, thus facilitating the way for the case of higher concentrations of additive. Furthermore, the increase in the amount of the wax decreased the surface energy of the starch films. This, together with the increased of surface hydrophobicity of the material, accelerated the oxygen absorption. In the starch films with ZEO, a smoother surface was observed, nevertheless, holes were also found; and the addition of high concentrations of MEO showed a coarse microstructure. Generally, ZEO seems to be better distributed in the polymeric matrix compared to MEO. Nevertheless, Souza *et al.* [118] reported increases in both the barrier properties (WVP and OP) in the starch films with the incorporation of cinnamon essential oil, which has a continuous, regular, smooth and uniform structure, after the addition of this oil. Similar results in the WVP were presented by Attarian *et al.* [119] in the films with honey, cinnamon powder, clove powder, orange essential oil, pepper and coffee powder. This typical behavior in the WVP was possibly due to the fact that many of these additives were powders and were not fully incorporated in the film, which caused a discontinuity. Also Kechichian *et al.* [35] developed films with the mixtures of clove powder, red pepper powder, honey, propolis extract, coffee powder and orange essential oil as additives, showing a similar behavior in this property.

On the other hand, potassium sorbate (KS) has been used in great amount of works due to its antimicrobial effect. For example, Shen *et al.* [112] observed increases both in the WVP and in the OP of starch-KS films. This behavior was also reported by Famá *et al.* [120]. Therefore, one of the factors affecting this parameter is the polymer structure [102], since a polymer with low contribution to the crystalline phase can increase the OP, due to the addition of KS. Hence, the addition of KS weakened the intermolecular forces between the adjacent starch chains, thus facilitating the chain mobility which increased the free volume between the polymer molecules, causing increase in the OP [121].

The effect of KS was also studied by Sayanjali *et al.* [122] in the edible films based on carboxymethyl cellulose (CMC). The presence of the antimicrobial caused greater flexibility, thus suggesting the weakening of the polymer-polymer interactions, leading to an increase in the WVP, indicating that this additive acted as a plasticizer [120]. In addition, Gemili *et al.* [123] reported that films with L-ascorbic acid and L-tyrosine acid showed higher porosity, including holes. However, high levels of KS showed a decrease in the WVP, due to the clogging of the pores of the matrix [122]. A similar case was reported by Sanchez-Gonzalez *et al.* [124], where the films based on methyl cellulose (MC) containing *Lactobacillus plantarum* showed significant changes in the WVP. This fact was attributed to the presence of these bacteria that induced discontinuities in the matrix, which could promote the transfer of water molecules. However, films containing this additive did not suffer significant changes in the brightness and transparency. Pastor *et al.* [109] with the addition of resveratrol to a MC matrix observed that the WVP and OP were modified only with high concentrations of the additive, leading to structural changes in the semi-crystalline arrangement of the polysaccharide, and the appearance of resveratrol crystals, resulting in being more opaque and less bright.

Discontinuities in the microstructure of the films based on hydroxypropyl methyl cellulose (HPMC) in the presence of various saturated and unsaturated fatty acids were observed by Jimenez *et al.* [125]. The films containing lauric acid (LA), meristic acid (MA) and palmitic acid (PA) presented more than one layer in its structure, corresponding to the lipids crystallized as successive layers [126], which limited the WVP and led to reductions in the opacity and brightness.

Moreover, Imran *et al.* [77] observed increase in the thickness values with the addition of nisin (N), due to the formation of salt crystals, thus producing an inhomogeneous surface, which reduced the transparency of the films. The WVP showed no significant differences, but a slight upward trend. According to Khwaldia *et al.* [127] restructuring the hydroxypropyl methyl cellulose (HPMC) was the possible factor of this behavior. When Akhtar *et al.* [128] incorporated betacyanins, they showed a more porous surface, due to their inhomogeneous distribution in the termoplastic material. This additive by its hydrophilic nature led to increase moisture absorption and WVP (due to its phenolic components), and decrease the OP.

In the alginate films, incorporating oregano essential oil (OEO) [129], ginger extracts [130], and garlic oil as a natural antibacterial agent [108], led to an increase in the thickness of the films, leading to color changes as a result of the decrease in transparency and increase in the tendency to become yellow. The studies on WVP showed different behaviors in each case. The incorporation of

OEO led to a tendency to decrease the WVP, possibly due to the microstructure produced by the internal crosslinking caused by the additive, which led to a more compact structure with the oil content. Therefore, it is generally accepted that the addition of hydrophobic lipids to hydrophilic polymer films improves the water vapor properties. For example, previous studies on chitosan films with essential oils have shown a similar trend in the WVP [131, 132]. However, other studies on the incorporation of essential oils and natural extracts showed no improvements in this property [86, 108]. Such is the case of ginger extracts, which did not significantly affected the WVP [130]; and similar results were also reported in the edible films based on apple puree, and alginate with essential oils and plant extracts [133].

A poor performance was found in the films with garlic oil [108], probably due to the hydrophilic characteristic of the additive. These authors also studied the materials of chitosan-garlic oil, determining that the additive affected the film's appearance: and made it more yellow and less transparent.

Increase in the moisture content, water solubility and WVP were also expressed by Bierhalz *et al.* [134] in the case of adding natamycin in the alginate films. These results were associated with the formation of a more flexible packaging and due to an increase in the free volume of the polymer structure [108]. The incorporation of natamycin showed a nonuniform distribution of the antimicrobial agent on the surface of the film, which resulted in the appearance of crystals, possibly due to the low solubility of natamycin in the film-forming solution. The presence of crystals of the antimicrobial agent was also observed in the cellulose films containing natamycin and nisin [135], and in the methyl cellulose films (MC) with natamycin [136]. Changes in the surface contributed to increase the opacity. Du *et al.* [137] showed that cinnamon and clove essential oils bud in the edible films based on apple puree and pectin, tended to be yellow, which increased the ΔE. Regarding to the WVP, different authors have explained that films having this matrix were not very good against moisture, but the addition of lipids could potentially improve this property [2]. Contrary to what was observed in the WVP of these systems in general, the authors showed that these materials are excellent for oxygen barriers.

4.2- Mechanical Properties

In order to develop a material with specific applications, it is essential that its physicochemical properties must be compatible with its use. In the case of the food industry, food packaging materials must provide compatibility for the products intended to be coated, since their influence on the product's performance and acceptance thereof by consumers is fundamental. For these reasons, the material must have a certain level of resistant to ensure the integrity of food products during handling and storage.

The study of the mechanical properties of the materials with perspectives to be used in this type of industry, is a topic of great importance and has been extensively studied in recent decades [138 - 140].

In general, the mechanical properties of the materials are determined by the components used in their manufacture, as well as by the conditions of its production process [87]. In the systems based on biopolymers with additives, intra- and intermolecular interactions determine their mechanical response [141]. Different biopolymer sources and various additives used for film-forming are very relevant factors, since they bring important consequences, as they can significantly alter their mechanical properties [142, 143].

The knowledge of the mechanical behavior of a material implies at least know the dependence of the elastic modulus, loss tangent, stress and strain at break with the different variables involved in their behavior.

The most frequently test used to evaluate the effort required to break a material is the stress-strain assay. From these curves, the following parameters: Young´s modulus (E), tensile strength (TS) strain at break (EB) and toughness (T), can be obtained. These parameters are useful for describing the mechanical properties of a film, since they serve to evaluate the stretch ability, flexibility/hardness, storage and/or energy loss. Furthermore, an adequate strength can ensure the integrity of the material [144].

Table **2** shows the parameters of uniaxial tensile of thermoplastic materials made from different polysaccharides and various types of additives.

Table 2. Some values of the mechanical parameters of thermoplastic materials based on biopolymers: Young´s modulus (E), tensile strength (TS) and strain at break (EB).

Base	Antimicrobial / antioxidant	E (MPa)	TS (MPa)	EB (%)	References
Alginate	Garlic oil	-	38.7 - 66.1	2.7 - 4.8	Pranoto *et al.* [108]
	Oregano oil	-	31.1 - 71	2.2 - 3.7	Benavides *et al.* [129]
	Ginseng extract	70.6 – 74.3	10.8 - 11.3	21.8 - 25.4	Norajit *et al.* [130]
	White Ginseng	63.6	8.1	24.2	
	Red ginseng	64.9	13.8	28	
	Natamycin	-	107 - 122.5	6.6 - 6.8	Bierhalz *et al.* [134]

(Table 2) contd.....

Base	Antimicrobial / antioxidant	E (MPa)	TS (MPa)	EB (%)	References
Chitosan	α-tocopherol	-	15 - 40	20 - 55	Martins *et al.* [99]
	Carvacol-grape seed extract	-	22.6 - 48.9	10.9 - 28	Rubilar *et al.* [100]
	Garlic oil	-	29 - 37	2.5 - 3.5	Pranoto *et al.* [108]
	Potassium sorbate		13.5 - 37	3.1 - 10	
	Nisin		13.6 - 37	3.5 - 30.7	
	Resveratrol	1550 - 2739	43 - 94	5,1 - 14	Pastor *et al.* [109]
	Green tea extract	-	25 - 28.4	54.6 - 60.7	Siripatrawan and Harte [110]
	Gallic acid	-	9.2 - 23.8	11 - 33.2	Sun *et al.* [145]
Starch	Combination: cinnamon, cloves, pepper, honey, propolis extract, coffee, orange essential oil	-	1 - 4.7	41 - 164.7	Kechichian *et al.* [35]
	Potassium sorbate	-	7 - 34	0.6 - 2.5	Shen *et al.* [112]
	Zataria multiflora Boiss	-	8 - 15	50 - 170	Ghasemlou *et al.* [115]
	Mentha pulegium	-	2 - 10	40 - 100	
	Cinnamon essential oil	-	1 - 3.8	128.8 - 264	Souza *et al.* [118]
	Potassium sorbate	1.3 - 5.2	0.3 - 0.7	30 - >80	Famá *et al.* [120]
	Nisin	-	9.6 - 11.4	12 - 28	Avedaño [146]
Cellulose	Nisin	421 - 2331	16 - 63	13 - 50	Imran *et al.* [77]
	Resveratrol	1604 - 1903	49 - 66	4 - 15	Pastor *et al.* [109]
	Potassium sorbate	-	8.8 - 17.6	6.6 - 28.8	Sayanjali *et al.* [122]
	L-ascorbic	165 - 2527	6.2 - 86.3	2.6 - 6.4	Gemili *et al.* [123]
	Lactobacillus plantarum	381 - 535	26.7 - 31.6	29 - 34	Sánchez-González *et al.* [124]
Cellulose	L-tyrosine acid	274 - 2716	2.5 - 73	0.9 - 3.6	Sánchez-González *et al.* [124]
	Lauric acid	13.8	17	1.5	Jiménez *et al.* [125]
	Meristic acid	17.9	27	2	Jiménez *et al.* [125]
	Palmitic acid	19.3	33	3	Jiménez *et al.* [125]
	Stearic acid	18.7	34	4	Jiménez *et al.* [125]
	Lauric acid	13.8	17	1,5	Jiménez *et al.* [125]
	Betacyanins	1102 - 2492	39.9 - 64.5	4.3 - 8.7	Akhtar *et al.* [128]
Pectin	Natamycin	-	48.5 - 65.5	4.7 – 4.8	Bierhalz *et al.* [134]

Among the most commonly polysaccharides studied as a material base for the formation of biodegradable and edible films are alginate, starch, chitosan, pectin and different types of cellulose.

In particular, alginate based films generally have a well-known mechanical response when some types of antioxidants, antimicrobial agents or essential oils are incorporated. The research papers show increase in the strain at the break and reduction of tensile strength with the addition of antioxidants in the biopolymers-based films [108, 129, 130, 134]. Specifically, this behavior was observed by Benavides *et al.* [129] in the films with oregano essential oil and by Bierhalz *et al.* [134], with the addition of natamycin. Similarly, this occurred in the work performed by Norajit *et al.* [130] when the ginger extracts were incorporated. In addition, Pranoto *et al.* [108] appreciated this fact with the addition of garlic oil as a natural antibacterial agent.

Instead, the behavior of the mechanical properties of films based on chitosan, turned out to be different when various additives were used. For example, Martins *et al.* [99] found that the addition of α-tocopherol significantly decreased the values of tensile strength and strain at break. This behavior was also demonstrated by Pastor *et al.* [109] in the systems with different concentrations of resveratrol. In addition, they observed a decrease in the Young´s modulus. It is worth noting that the trend was more evident when the concentration of resveratrol was greater. Similar results have been reported by Rubilar *et al.* [100] in the systems containing different proportions of carvacrol and grape seed extract, and some results have also been reported in the films with added oleic acid [147]. In these researches, the authors concluded that the reduction of these quasi-static mechanical parameters can be explained, due to the addition of a hydrophobic agent in the chitosan films, thus stimulating the development of discontinuous structures, which generate the least mobility of the biopolymer chains, *i.e.* causing less flexibility and tensile strength in the material [148, 149].

Many studies indicate that the addition of oils or different bioactive agents in the polysaccharide-based films, leads to decrease the TS and EB values. Nevertheless, some studies have shown that it is not always feasible to observe this behavior, because there are many the factors involved in the development of these systems. For example, Siripatrawan and Harte [110], reported atypical results in the quasi-static properties of the films with green tea extract (GTE). Likewise, the concentrations below 5% of GTE showed no significant differences, while a greater content of GTE, increased both TS and EB. The improvements in the films containing high concentrations of GTE were attributed to the interactions between the chitosan matrices with the polyphenolic compounds from green tea extract. The increase in the tensile properties of the chitosan-olive oil films were also reported by Georgantelis *et al.* [114], Zivanovic and Chi [131], and Ponce *et al.* [150]. This behavior was attributed to the developed interactions between the lipid phase and carbohydrates. Therefore, a strong interaction between the polymer and the additive might have caused the crosslinking effect, and decreased the free volume and molecular mobility of the polymer [151]. Ojagh *et al.* [113] confirmed the trend of increase in the TS when the effect of the addition of the cinnamon essential oil was studied. However, a decrease in the EB was also obtained. According to the authors, this was due to the formation of a more compact structure as a result of the addition of the cinnamon essential oil. Likewise, in the work carried out by Pranoto *et al.* [108] on chitosan-based films, with the addition of garlic oil, potassium sorbate (KS) and nisin (N) were reported decreases in the TS, especially with the addition of KS. Likewise, lower TS values were related to higher EB values, with the exception of the chitosan-garlic oil films, which were similar to the matrix without the addition of garlic essential oil.

On the other hand, Sun *et al.* [152] observed an increase in the TS for low concentrations of gallic acid, and a decrease in the TS for higher contents of 1.5%. The increase in the TS at low concentrations was attributed to the intermolecular hydrogen bonding interactions between the NH_3- groups of chitosan and OH-groups of gallic acid [152], and at higher concentrations, the decrease in the TS was possibly due to the rupture of the internal structure of the films, caused by the excess of additive [134]. This phenomenon was previously reported by other researchers in similar systems such as olive oil-chitosan and oleic acid-chitosan [147, 151]. In the case of materials with higher concentrations of gallic acid, they explained the decrease the TS as was explained by Bierhalz *et al.* [134], by the possible rupture of the internal structure of the films, due to the excess of the additive. Likewise, in the study of Sun *et al.* [145], a decrease in the EB values was observed, indicating that the incorporation of gallic acid in the chitosan films had given rise to a strong interaction between the additive and the matrix, thus restricting the movement of the matrix chains.

Additionally, the starches are polysaccharides frequently chosen for film formation with application in the food industry. This type of biopolymeric films have reported very different results in the tensile properties. Reductions both in the TS and the EB with increased additive concentration was observed in films with cinnamon essential oil [118]. This was because the oil reduced the intermolecular interactions between the polymer chains in the films based on starch containing different combinations of antimicrobial agents such as honey, cinnamon powder, clove powder, propolis, orange essential oil, pepper powder, coffee powder and propolis extract [35, 112, 119]. Kechichian *et al.* [35] attributed these results to the presence of various powder ingredients, which lead to weaker interactions between the polymer chains, thus creating possible breakpoints resulting in nonhomogeneous matrix. They also explained that the previous treatment of the additive was very important before being used.

Shen *et al.* [112] using potassium sorbate and starch, found decrease in both the TS and EB, while Famá *et al.* [120] obtained films with lower E and TS values, but with a significant increase in the strain at break, concluding that the presence of the antimicrobial agent increased the flexibility of the polymer.

Likewise, increased EB and decreased TS were found by Ghasemlou *et al.* [115], in the starch films with *Zataria multiflora Boiss* (ZEO) or *Mentha pulegium* (MEO), demonstrating a similar behavior to the plasticizing effect of the essential oils. The conclusion was that the effect of the structures created between the lipids and polymer reduces the cohesion of the forces of the starch network, leading to decrease the resistance [125].

In the case of the addition of nisin (N) in the starch films, Avedaño [146] found no significant differences in the straight force, with the exception of the system with higher nisin content, which showed a very low TS value. In all cases, reductions in the EB were shown, demonstrating that the materials with N had little flexibility and were highly fragile.

Moreover, it is known that the films formed by cellulose have advantages in terms of the mechanical strength compared to other biodegradable materials based on polysaccharides. This is because the addition of any ingredient could harm the physical properties of these materials. In general, it has been observed that the addition of antioxidants, antimicrobial or essential oils decreases the three parameters studied based on the tensile properties: TS, E and EB. In particular, Sanchez-Gonzalez *et al.* [124], obtained a decrease in the E and TS, without noticeable changes in the EB in the films based on methyl cellulose (MC) with the addition of *L. plantarum*. Pastor *et al.* [109] found decrease in both the TS and EB and an increase in the elastic modulus of films based on MC containing resveratrol. Taking this into account, the importance of the choice of base polysaccharide is evident, beyond the kind of additive. Sayanjali *et al.* [122] also observed an increase in the E of the films based on carboxymethylcellulose (CMC) and potassium sorbate (KS), concluding that this antimicrobial caused a reduction in the mobility of the polymer chains [70]. Nevertheless, greater amounts of KS weakened the structure of the polymer, leading to lower elastic modulus. This behavior would be consistent with previous studies on the films based on chitosan with the addition of KS and other preservatives [108].

Likewise, the addition of L-ascorbic in the films of cellulose acetate (CA) caused no significant changes in the TS, while the use of L-tyrosine produced a reduction in this parameter. The elastic modulus was also affected by the addition of both antioxidants, with random changes in its behavior [123].

Among the investigations realized on the mechanical properties of the film based on hydroxypropyl methyl cellulose (HPMC) is the work done by Akhtar *et al.* [128], which reported a decrease in both the TS and EB with the addition of betacyanins. The same behavior was observed by Jiménez *et al.* [125] with the addition of various saturated and unsaturated fatty acids.

Imran *et al.* [77] also observed a reduction in the TS of the HPMC-nisin films, but with a tendency to increase the EB. The authors attributed this behavior to milk proteins and carbohydrates present in the preparation, which could interact with the HPMC, thus modifying its mechanical behavior.

Moreover, Sánchez-González *et al.* [149] after incorporating tea tree essential oil (TTO) obtained a reduction in the TS and E, unchanged from the EB. Other authors explicated this behavior in the chitosan-based films with added essential oils [108, 111, 131], showing that this effect was caused by structural discontinuities due to the additive. These authors also observed a reduction in the E, a slight increase in the TS, and no significant changes were observed in the EB of the HPMC-*L. plantarum* films.

Finally, the studies on pectin-based films are less common in the literature [130]. Bierhalz *et al.* [134] incorporated natamycin as an active agent in the pectin-based films, showing that the additive caused a significant reduction in the TS without affecting the EB. Similarly, this mechanical behavior was attributed to the structural changes caused by the excess of antimicrobial agent [153]. Moreover, Du *et al.* [137] studied the effects of the cinnamon and clove essential oils on the edible films based on methoxy pectin finding and suggested that the incorporation of the oils led to reductions in both the TS and E; being this consistent with the results reported by Bierhalz *et al.* [134].

4.3- Structural Properties

The structural properties, chemical composition and interactions between the constituents of food, limit the manufacture of handicrafts and industrial products, even affecting its quality and safety. Typically, it is believed that a structural analysis is the observation of scanning electron microscopy or transmission images. However, techniques such as confocal laser scanning microscopy, atomic force microscopy, optical microscopy, X-ray diffraction, differential scanning calorimetry, thermogravimetric analysis, surface tension and dynamic-mechanical tests, allow obtaining more information on the possible application of a biopolymeric materials. Likewise, a proper understanding of the relationship between structure and function helps us to design new structures in the future to manufacture high quality, healthy and tasty food products.

Moreover, the components used in the manufacture of edible films can influence on the structural properties. Likewise, these properties somehow explain the interactions that occur between their constituents. Falguera *et al.* [154] studied the combination of various polysaccharides, proteins and lipids, with the purpose of evaluating these edible and biodegradable films, in order to determine the properties of each compound and the synergy between them. They analyzed that the polysaccharides and protein are macromolecules suitable for the formation of thermoplastic materials, since they possess good mechanical and structural properties, though such biomatrices do not have barriers against moisture transfer. Escamilla-García *et al.* [155] evaluated the interactions between zein and chitosan using Raman spectroscopy and X-ray diffraction, observing an increase in the band intensity justified by the amide groups of glutamine of zein and the hydroxyl groups of chitosan by the formation of a complex through hydrogen bonding. Additionally, both compounds interact provoking the disappearance of cysteine and lysine occurs, which are the components of the chitosan and the zenia, respectively. Otherwise, this was observed in the films made from quinoa protein-chitosan-sunflower oil, where the hydrophobic interactions generated the formation of micropores and microfractures detected by SEM [156].

Other authors as Mali *et al.* [157] conducted studies on biopolymerc films obtained from yam starch, which were observed by the SEM, suggesting that the incorporation of glycerol as the plasticizing agent allowed to obtain homogeneous structures. Meanwhile, Gutiérrez *et al.* [141] evaluated two starches (cassava and cush-cush yam) with different amylose contents for the edible film-forming plasticized with glycerol, determining that a more open and compact structure was associated with the increase in hydrogen bonding interaction between the amylose and glycerol, which allowed obtaining films with higher elongation.

Finally, in Table **3** some structural features observed in the edible films made from biopolymers are indicated.

CONCLUDING REMARKS

Films made from biopolymers have proven to be a good alternative to improve the quality and safety of different foods, because they offer protection against the environment, as they have barrier properties against the transfer of gases, such as $CO_2/O_2/H_2O$/ethylene. In addition, they have mechanical and antimicrobial properties, which limit the growth of various microorganisms, nevertheless, these properties can be affected or vary depending on the compounds used for formulation. Furthermore, their high flexibility properties without impairing the tensile strength make these films highly promising for use in food packaging.

Therefore, it is recommended that when the ingredients (antioxidant, anti-microbial, micronutrients, *etc.*) to be incorporated into the formulation of the film are selected, its composition and the characteristics of the type of food in which it will be applied must be taken into account to ensure a successful outcome, along with the purpose or function of the film and the influence that these ingredients can make on the properties of the films.

Table 3. Structural characteristics observed in films derivate from biopolymers.

Matrix	Techniques used	Characteristics observed in the films	References
Cush-cush yam and cassava starches	Optical microscopy (LM) and Scanning electron microscopy (SEM).	The films based on "waxy" starch showed a poor gelatinization compared to the cassava starch, which had a higher amylose content. A more open, smooth and compact structure was observed in films based on cassava starch, this possibly due to the greater glycerol-amylose interactions.	Gutiérrez *et al.* [141]
Yam starch	Polarized light micro-scopy (PLM) and Scanning electron microscopy (SEM).	Employing polarized light microscopy found that the starch is completely gelatinized before the film-forming. Moreover, SEM images revealed that the films with starch-glycerol showed smooth surfaces without pores or cracks, and a compact structure.	Mali *et al.* [157]
Potato starch	Confocal laser scanning microscopy (CLSM) and Transmission electron microscopy (TEM).	They observed that films with a high amylose content were denser, stable and with a strong molecular orientation. In the amylopectin-based films the TEM images were not obtained due to inlays that they did not allow carry out this analysis.	Rindlav-Westling *et al.* [158]
Corn starch	Attenuated total reflec-tance Fourier transform infrared spectroscopy (ATR/FTIR), X-ray photoelectron spectro-scopy (XPS), X-ray diffraction (XRD), calorímetro diferencial de barrido (DSC), and Scanning electron microscopy (SEM).	They observed the formation of a compact structure, due to the greater hydrogen bonding interactions between the starch and the SiO_2 nanoparticles.	Xiong *et al.* [159]

(Table 3) contd.....

Matrix	Techniques used	Characteristics observed in the films	References
Protein-chitosan	Attenuated total reflectance Fourier transform infrared spectroscopy (ATR/FTIR), X-ray diffraction (XRD) and scanning electron microscopy (SEM).	The addition of sunflower oil allowed to observe a poor permeability to water vapor, thus reducing the tensile strength, due to the formation of micropores and microfractures detected by SEM.	Valenzuela *et al.* [156]
Hydroxypropyl methylcellulose (HPMC)	Light microscopy (LM), Atomic force microscopy (AFM) and Scanning electron microscopy (SEM).	They evaluated the behavior of the films by incorporating mixtures of surfactants. They obtained complex structures, and the same were repeated in different scales. Likewise, they noted the formation of lamellar structures, showing lipids finely distributed within the polymeric matrix.	Jiménez *et al.* [125] and Villalobos *et al.* [160]
β-casein	Scanning electron microscopy (SEM) and Attenuated total reflectance Fourier transform infrared spectroscopy (ATR/FTIR).	Using SEM was observed the cryo-fracture zone of the films, appreciating a smooth and compact structure. The exposed surface to drying was porous and granulated, this due to water evaporation and migration of glycerol. Employing FTIR were not observed structural alterations.	Mauer *et al.* [161]

CONFLICT OF INTEREST

The author (editor) declares no conflict of interest, financial or otherwise.

ACKNOWLEDGEMENTS

The authors would like to thank the Consejo Nacional de Investigaciones Científicas y Técnicas (CONICET) (Postdoctoral fellowship internal PDTS-Resolution 2417), Universidad de Buenos Aires (UBACyT) and Dr. Mirian Carmona-Rodríguez.

REFERENCES

[1] V. Alves, N. Costa, L. Hilliou, F. Larotonda, M. Gonçalves, A. Sereno, and I. Coelhoso, "Design of biodegradable composite films for food packaging", *Desalination,* vol. 199, no. 1-3, pp. 331-333, 2006.
[http://dx.doi.org/10.1016/j.desal.2006.03.078]

[2] T.H. McHugh, and E. Senesi, "Apple wraps: a novel method to improve the quality and extend the shelf life of fresh-cut apples", *J. Food Sci.,* vol. 65, no. 3, pp. 480-485, 2000.
[http://dx.doi.org/10.1111/j.1365-2621.2000.tb16032.x]

[3] O. Martín-Belloso, M.A. Rojas-Graü, and R. Soliva-Fortuny, "Delivery of flavor and active ingredients using edible films and coatings", In: *Edible films and coating for food applications,* M.E.

Embuscado, K.C. Huber, Eds., Springer Science + Business Media, LLC, EEUU: New York, 2009, pp. 295-313.
[http://dx.doi.org/10.1007/978-0-387-92824-1_10]

[4] Y. Zhao, "Edible Coatings for enhancing quality and health benefits of berry fruits", In: *Flavor and Health Benefits of Small Fruits,* M.C. Qian, A.M. Rimando, Eds., ACS Symposium Series, EEUU: Washington DC, USA, 2010.
[http://dx.doi.org/10.1021/bk-2010-1035.ch018]

[5] G.V. Barbosa-Cánovas, "Películas y recubrimientos comestibles en contexto con otras tecnologías novedosas de conservación y procesado de alimentos", In: *En: Películas y recubrimientos comestibles. Propiedades y aplicaciones en alimentos, G.I. Olivas-Orozco, G.A. González-Aguilar, O. Martín-Belloso.*R. Soliva-Fortuny, Clave Editorial: D.F., México, 2012, pp. 13-20.

[6] A. Chiralt, M.J. Fabra, and L. Sánchez-González, "Propiedades de las películas comestibles", In: *En: Películas y recubrimientos comestibles. Propiedades y aplicaciones en alimentos, G.I., Olivas-Orozco, G.A. González-Aguilar.*O. Martín-Belloso, and R. Soliva-Fortuny, Clave Editorial: D.F., México, 2012, pp. 21-57.

[7] R.M. Raybaudi-Massilia, and J. Mosqueda-Melgar, "Polysaccharides as carriers and protectors of additives and bioactive compounds in foods", In: *The Complex World of Polysaccharides,* D.N. Karunaratn, Ed., InTech, Rijeka: Croacia, 2012, pp. 429-454.

[8] A. Gennadios, M.A. Hanna, and L.B. Kurth, ""Application of edible coatings on meats, poultry and seafoods: A Review", LWT -", *Food Sci. Technol. (Campinas.),* vol. 30, no. 4, pp. 337-350, 1997.

[9] I.G. Donhowe, and O.R. Fennema, "The effects of plasticizers on crystallinity, permeability, and mechanical properties of methylcellulose films", *J. Food Process. Preserv.,* vol. 17, no. 4, pp. 247-257, 1993.
[http://dx.doi.org/10.1111/j.1745-4549.1993.tb00729.x]

[10] D.S. Cha, and M.S. Chinnan, "Biopolymer-based antimicrobial packaging: a review", *Crit. Rev. Food Sci. Nutr.,* vol. 44, no. 4, pp. 223-237, 2004.
[http://dx.doi.org/10.1080/10408690490464276] [PMID: 15462127]

[11] M. Izydorczyk, "Understanding the chemistry of food carbohydrates", In: *Food carbohydrates: chemistry, physical properties, and applications, S.W. Cui.*CRC Press, 2005, p. 53.
[http://dx.doi.org/10.1201/9780203485286.ch1]

[12] J.M. Krochta, and C. De Mulder-Johnston, "Edible and biodegradable polymer films: challenges and opportunities", *Food Technol.-Chicago,,* vol. 51, no. 2, pp. 61-74, 1997.

[13] J.J. Kester, and O.R. Fennema, "Edible films and coatings: a review", *Food Technol.-Chicago,* vol. 40, no. 12, pp. 47-59, 1986.

[14] R.L. Shogren, J.W. Lawton, W.M. Doane, and K.F. Tiefenbacher, "Structure and morphology of baked starch foams", *Polymer (Guildf.),* vol. 39, no. 25, pp. 6649-6655, 1998.
[http://dx.doi.org/10.1016/S0032-3861(97)10303-2]

[15] A. Sorrentino, G. Gorrasi, and V. Vittoria, "Potential perspectives of bionanocomposites for food packaging applications", *Trends Food Sci. Technol.,* vol. 18, no. 2, pp. 84-95, 2007.
[http://dx.doi.org/10.1016/j.tifs.2006.09.004]

[16] R.A. Talja, H. Helén, Y.H. Roos, and K. Jouppila, "Effect of various polyols and polyol contents on physical and mechanical properties of potato starch-based films", *Carbohydr. Polym.,* vol. 67, no. 3, pp. 288-297, 2007.
[http://dx.doi.org/10.1016/j.carbpol.2006.05.019]

[17] M.A. Bertuzzi, M. Armada, and J.C. Gottifredi, "Physiochemical characterization of starch based films", *J. Food Eng.,* vol. 82, no. 1, pp. 17-25, 2007.
[http://dx.doi.org/10.1016/j.jfoodeng.2006.12.016]

[18] D. Lourdin, G.D. Valle, and P. Colonna, "Influence of amylose content on starch films and foams",

Carbohydr. Polym., vol. 27, no. 4, pp. 261-270, 1995.
[http://dx.doi.org/10.1016/0144-8617(95)00071-2]

[19] C.A. Romero-Bastida, L.A. Bello-Pérez, M.A. García, M.N. Martino, J. Solorza-Feria, and N.E. Zarintzky, "Physicochemical and microstructural characterization of films prepared cellulosic by thermal and cold gelatinization from non-conventional sources of starches", *Carbohydr. Polym.*, vol. 60, no. 2, pp. 235-244, 2005.
[http://dx.doi.org/10.1016/j.carbpol.2005.01.004]

[20] Y. Lu, L. Tighzert, F. Berzin, and S. Rondot, "Innovative plasticized starch films modified with waterborne polyurethane from renewable sources", *Carbohydr. Polym.*, vol. 61, no. 2, pp. 174-182, 2005.
[http://dx.doi.org/10.1016/j.carbpol.2005.04.013]

[21] R.N. Tharanathan, "Biodegradable films and composite coatings: Past, present and future", *Trends Food Sci. Technol.*, vol. 14, no. 3, pp. 71-78, 2003.
[http://dx.doi.org/10.1016/S0924-2244(02)00280-7]

[22] J.M. Swinkels, "Industrial starch chemistry", *AVEBE brochure Foxhol The Netherland,* pp. 8-24, 1987.

[23] J.H. Li, T. Vasanthan, R. Hoover, and B.G. Rossnagel, "Starch from hull-less barley: IV. Morphological and structural changes in waxy, normal and high-amylose starch granules during heating", *Food Res. Int.*, vol. 37, no. 5, pp. 417-428, 2004.
[http://dx.doi.org/10.1016/j.foodres.2003.09.016]

[24] Q. Liu, "Understanding starches and their role in foods", In: *Food Carbohydrates: Chemistry, Physical Properties and Applications.*CRC Press Boca Raton FL: USA, 2005.
[http://dx.doi.org/10.1201/9780203485286.ch7]

[25] H.F. Zobel, "Starch granule structure", In: *Developments in carbohydrate chemistry, R.J. Alexander, and H.F. Zobel.*The American Association of Cereal Chemists St Paul Minnesota: USA, 1994, pp. 1-36.

[26] M.J. van der Maarel, B. van der Veen, J.C. Uitdehaag, H. Leemhuis, and L. Dijkhuizen, "Properties and applications of starch-converting enzymes of the alpha-amylase family", *J. Biotechnol.*, vol. 94, no. 2, pp. 137-155, 2002.
[http://dx.doi.org/10.1016/S0168-1656(01)00407-2] [PMID: 11796168]

[27] S. You, S.G. Stevenson, M.S. Izydorczyk, and K.R. Preston, "Separation and characterization of barley starch polymers by a flow field-flow fractionation technique in combination with multiangle light scattering and differential refractive index detection", *Cereal Chem.*, vol. 79, no. 5, pp. 624-630, 2002.
[http://dx.doi.org/10.1094/CCHEM.2002.79.5.624]

[28] E. Zavareze, and A. Guerra Días, "Impact of heat-moisture treatment and annealing in starches: A review", *Carbohydr. Polym.*, vol. 83, no. 2, pp. 317-328, 2011.
[http://dx.doi.org/10.1016/j.carbpol.2010.08.064]

[29] M.R. Calixto Cotos, and I.A. Salas, "Aislamiento y caracterización parcial del almidón nativo de Amarantus caudatus Linneo", *Revista de la Sociedad Química del Perú,* vol. 70, no. 1, pp. 48-54, 2004.

[30] L.A. Bello-Pérez, M.G. Méndez-Montealvo, and E. Agama-Acevedo, *"Carbohidratos en alimentos regionales Iberoamericanos", En: Almidón: definición, estructura y propiedades.*Editorial Universidad de San Pablo: Brasil, 2006, pp. 20-30.

[31] J.L. Jane, "Structure of starch granules", *J. Appl. Glycosci.*, vol. 54, no. 1, pp. 31-36, 2007.
[http://dx.doi.org/10.5458/jag.54.31]

[32] L. Copeland, J. Blazek, H. Salman, and M.C. Tang, "Form and functionality of starch", *Food Hydrocoll.*, vol. 23, no. 6, pp. 1527-1531, 2009.

[http://dx.doi.org/10.1016/j.foodhyd.2008.09.016]

[33] G.A. Baker, and P.J. Crosbie, "Consumer preferences for food safety attributes: A market segment approach", *Agribusiness,* vol. 10, no. 4, pp. 319-324, 1994. [http://dx.doi.org/10.1002/1520-6297(199407/08)10:4<319::AID-AGR2720100405>3.0.CO;2-J]

[34] M.A. Garcia, M.N. Martino, and N.E. Zaritzky, "Plasticized starch-based coatings to improve strawberry (Fragaria× ananassa) quality and stability", *J. Agric. Food Chem.,* vol. 46, no. 9, pp. 3758-3767, 1998. [http://dx.doi.org/10.1021/jf980014c]

[35] V. Kechichian, C. Ditchfield, P. Veiga-Santos, and C.C. Tadini, "“Natural antimicrobial ingredients incorporated in biodegradable films based on cassava starch”, LWT -", *Food Sci. Technol. (Campinas.),* vol. 43, no. 7, pp. 1088-1094, 2010.

[36] N. García, L. Famá, L. Ribba, M. Aranguren, A. Dufresne, and S. Goyanes, *Recubrimientos Biodegradables Renovables.*INNOVAR-Concurso Nacional de Innovaciones. Ministerio de Ciencia, Tecnología e Innovación Productiva: Ministerio de la Nación, Argentina, 2010.

[37] T. Bourtoom, "Edible films and coatings: characteristics and properties", *International Food Research Journal,* vol. 15, no. 3, pp. 237-248, 2008.

[38] A. Dhanapal, P. Sasikala, R. Lavanya, V. Kavitha, G. Yazhini, and M. Shakila Banu, "Edible films from polysaccharides", *Food Science and Quality Management,* vol. 3, pp. 9-18, 2012.

[39] N.B. Franklin, K.D. Cooksey, and K.J. Getty, "Inhibition of Listeria monocytogenes on the surface of individually packaged hot dogs with a packaging film coating containing nisin", *J. Food Prot.,* vol. 67, no. 3, pp. 480-485, 2004. [http://dx.doi.org/10.4315/0362-028X-67.3.480] [PMID: 15035361]

[40] P. Santiago-Silva, N.F. Soares, J.E. Nóbrega, M.A. Júnior, K.B. Barbosa, A.C. Volp, E.R. Zerdas, and N.J. Würlitzer, "Antimicrobial efficiency of film incorporated with pediocin (ALTA® 2351) on preservation of sliced ham", *Food Contr.,* vol. 20, no. 1, pp. 85-89, 2009. [http://dx.doi.org/10.1016/j.foodcont.2008.02.006]

[41] C.N. Cutter, and G.R. Siragusa, "Growth of Brochothrix thermosphacta in ground beef following treatments with nisin in calcium alginate gels", *Food Microbiol.,* vol. 14, no. 5, pp. 425-430, 1997. [http://dx.doi.org/10.1006/fmic.1997.0114]

[42] N. Natrajan, and B.W. Sheldon, "Inhibition of Salmonella on poultry skin using protein- and polysaccharide-based films containing a nisin formulation", *J. Food Prot.,* vol. 63, no. 9, pp. 1268-1272, 2000. [http://dx.doi.org/10.4315/0362-028X-63.9.1268] [PMID: 10983804]

[43] M. Millette, C. Le Tien, W. Smoragiewicz, and M. Lacroix, "Inhibition of Staphylococcus aureus on beef by nisin-containing modified alginate films and beads", *Food Contr.,* vol. 18, no. 7, pp. 878-884, 2007. [http://dx.doi.org/10.1016/j.foodcont.2006.05.003]

[44] B. Marcos, T. Aymerich, J.M. Monfort, and M. Garriga, "High-pressure processing and antimicrobial biodegradable packaging to control Listeria monocytogenes during storage of cooked ham", *Food Microbiol.,* vol. 25, no. 1, pp. 177-182, 2008. [http://dx.doi.org/10.1016/j.fm.2007.05.002] [PMID: 17993392]

[45] M. Oussalah, S. Caillet, S. Salmiéri, L. Saucier, and M. Lacroix, "Antimicrobial effects of alginate-based film containing essential oils for the preservation of whole beef muscle", *J. Food Prot.,* vol. 69, no. 10, pp. 2364-2369, 2006. [http://dx.doi.org/10.4315/0362-028X-69.10.2364] [PMID: 17066914]

[46] M. Oussalah, S. Caillet, S. Salmiéri, L. Saucier, and M. Lacroix, "Antimicrobial effects of alginate-based films containing essential oils on Listeria monocytogenes and Salmonella typhimurium present in bologna and ham", *J. Food Prot.,* vol. 70, no. 4, pp. 901-908, 2007.

[http://dx.doi.org/10.4315/0362-028X-70.4.901] [PMID: 17477259]

[47] M.Z. Elsabee, and E.S. Abdou, "Chitosan based edible films and coatings: a review", *Mater Sci Eng C Mater Biol Appl,* vol. 33, no. 4, pp. 1819-1841, 2013. [http://dx.doi.org/10.1016/j.msec.2013.01.010] [PMID: 23498203]

[48] P.K. Dutta, S. Tripathi, G.K. Mehrotra, and J. Dutta, "Perspectives for chitosan based antimicrobial films in food applications", *Food Chem.,* vol. 114, no. 4, pp. 1173-1182, 2009. [http://dx.doi.org/10.1016/j.foodchem.2008.11.047]

[49] M. Ye, H. Neetoo, and H. Chen, "Control of Listeria monocytogenes on ham steaks by antimicrobials incorporated into chitosan-coated plastic films", *Food Microbiol.,* vol. 25, no. 2, pp. 260-268, 2008. [http://dx.doi.org/10.1016/j.fm.2007.10.014] [PMID: 18206768]

[50] D.H. Young, H. Köhle, and H. Kauss, "Effect of chitosan on membrane permeability of suspension-cultured Glycine max and Phaseolus vulgaris cells", *Plant Physiol.,* vol. 70, no. 5, pp. 1449-1454, 1982. [http://dx.doi.org/10.1104/pp.70.5.1449] [PMID: 16662696]

[51] P. Darmadji, and M. Izumimoto, "Effect of chitosan in meat preservation", *Meat Sci.,* vol. 38, no. 2, pp. 243-254, 1994. [http://dx.doi.org/10.1016/0309-1740(94)90114-7] [PMID: 22059662]

[52] B. Ouattar, R.E. Simard, G. Piett, A. Bégin, and R.A. Holley, "Inhibition of surface spoilage bacteria in processed meats by application of antimicrobial films prepared with chitosan", *Int. J. Food Microbiol.,* vol. 62, no. 1-2, pp. 139-148, 2000. [http://dx.doi.org/10.1016/S0168-1605(00)00407-4] [PMID: 11139014]

[53] U. Siripatrawan, and S. Noipha, "Active film from chitosan incorporating green tea extract for shelf life extension of pork sausages", *Food Hydrocoll.,* vol. 27, no. 1, pp. 102-108, 2012. [http://dx.doi.org/10.1016/j.foodhyd.2011.08.011]

[54] J. Duan, S.I. Park, M.A. Daeschel, and Y. Zhao, "Antimicrobial chitosan-lysozyme (CL) films and coatings for enhancing microbial safety of mozzarella cheese", *J. Food Sci.,* vol. 72, no. 9, pp. M355-M362, 2007. [http://dx.doi.org/10.1111/j.1750-3841.2007.00556.x] [PMID: 18034728]

[55] T. Bourtoom, "Edible protein films: properties enhancement", *International Food Research Journal,* vol. 16, no. 1, pp. 1-9, 2009.

[56] E. Kristo, K.P. Koutsoumanis, and C.G. Biliaderis, "Thermal, mechanical and water vapor barrier properties of sodium caseinate films containing antimicrobials and their inhibitory action on Listeria monocytogenes", *Food Hydrocoll.,* vol. 22, no. 3, pp. 373-386, 2008. [http://dx.doi.org/10.1016/j.foodhyd.2006.12.003]

[57] P. Mendes de Souza, A. Fernández, G. López-Carballo, R. Gavara, and P. Hernández-Muñoz, "Modified sodium caseinate films as releasing carriers of lysozyme", *Food Hydrocoll.,* vol. 24, no. 4, pp. 300-306, 2010. [http://dx.doi.org/10.1016/j.foodhyd.2009.10.005]

[58] S.Y. Cheng, B.J. Wang, and Y.M. Weng, "“Antioxidant and antimicrobial edible zein/chitosan composite films fabricated by incorporation of phenolic compounds and dicarboxylic acids”, LWT -", *Food Sci. Technol. (Campinas.),* vol. 63, no. 1, pp. 115-121, 2015.

[59] H.E. Bair, "Thermal Analysis of Additives", In: *Mechanical Properties of Solid Polymers,* E.A. Turi, Ed., Academic Press New York, EEUU: New York, 1981.

[60] I.M. Ward, "Mechanical Properties of Solid Polymers", In: *Thermal Characterization of polymeric materials*2nd Wiley and Sons, EEUU, 1985, p. 475.

[61] P. Myllarinen, R. Partanen, J. Sepalla, and P. Forsell, "Effect of glicerol on behavior of amylose and amylopectin films", *Carbohydr. Polym.,* vol. 50, no. 4, pp. 355-361, 2002. [http://dx.doi.org/10.1016/S0144-8617(02)00042-5]

[62] J.H. Jagannath, C. Nanjappa, D.K. Das Gupta, and A.S. Bawa, "Studies on the stability of an edible film and its use for the preservation of carrot (Daucus carota)", *Int. J. Food Sci. Technol.,* vol. 41, no. 5, pp. 498-506, 2006.
[http://dx.doi.org/10.1111/j.1365-2621.2005.01038.x]

[63] M.A. Cerqueira, A.M. Lima, B.W. Souza, J.A. Teixeira, R.A. Moreira, and A.A. Vicente, "Functional polysaccharides as edible coatings for cheese", *J. Agric. Food Chem.,* vol. 57, no. 4, pp. 1456-1462, 2009.
[http://dx.doi.org/10.1021/jf802726d] [PMID: 19170503]

[64] T. Bourtoom, M.S. Chinnan, P. Jantawat, and R. Sanguandeekul, "Effect of plasticizer type and concentration on the properties of edible films from water soluble fish proteins", *Food Sci. Technol. Int.,* vol. 12, pp. 119-126, 2006.
[http://dx.doi.org/10.1177/1082013206063980]

[65] P. Veiga-Santos, L.M. Oliveira, M.P. Cereda, and A.P. Scamparini, "Sucrose and inverted sugar as plasticizer. Effect on cassava starch-gelatin film mechanical properties, hydrophilicity and water activity", *Food Chem.,* vol. 103, no. 2, pp. 255-262, 2007.
[http://dx.doi.org/10.1016/j.foodchem.2006.07.048]

[66] J.M. Krochta, *Protein-based Films and Coatings.,* A. Gennadios, B. Raton, Eds., CRC Press New York: USA, 2002, pp. 1-41.

[67] P.M. Davidson, F.J. Critzer, and T.M. Taylor, "Naturally occurring antimicrobials for minimally processed foods", *Annu. Rev. Food Sci. Technol.,* vol. 4, pp. 163-190, 2013.
[http://dx.doi.org/10.1146/annurev-food-030212-182535] [PMID: 23244398]

[68] P.M. Davidson, and T.M. Taylor, "Chemical Preservatives and Natural Antimicrobial Compounds", In: *Food Microbiology: Fundamentals and Frontiers.,* M.P. Doyle, L.R. Beuchat, Eds., 3rd ed ASM Press: Washington, D.C., 2007, pp. 713-746.
[http://dx.doi.org/10.1128/9781555815912.ch33]

[69] C.A. Campos, L.N. Gerschenson, and S.K. Flores, "Development of edible films and coatings with antimicrobial activity", *Food Bioprocess Tech.,* vol. 4, no. 6, pp. 849-875, 2011.
[http://dx.doi.org/10.1007/s11947-010-0434-1]

[70] A. Cagri, Z. Ustunol, and E.T. Ryser, "Antimicrobial, mechanical, and moisture barrier properties of low pH whey protein-based edible films containing paminobenzoic or sorbic acids", *J. Food Sci.,* vol. 66, no. 6, pp. 865-870, 2001.
[http://dx.doi.org/10.1111/j.1365-2621.2001.tb15188.x]

[71] Z.K. Emiroğlu, G.P. Yemiş, B.K. Coşkun, and K. Candoğan, "Antimicrobial activity of soy edible films incorporated with thyme and oregano essential oils on fresh ground beef patties", *Meat Sci.,* vol. 86, no. 2, pp. 283-288, 2010.
[http://dx.doi.org/10.1016/j.meatsci.2010.04.016] [PMID: 20580990]

[72] D.S. Aldana, S. Andrade-Ochoa, C.N. Aguilar, J.C. Contreras-Esquivel, and G.V. Nevárez-Moorillón, "Antibacterial activity of pectic-based edible films incorporated with Mexican lime essential oil", *Food Contr.,* vol. 50, pp. 907-912, 2015.
[http://dx.doi.org/10.1016/j.foodcont.2014.10.044]

[73] Y. Ruiz-Navajas, M. Viuda-Martos, E. Sendra, J.A. Pérez-Álvarez, and J. Fernández-López, "In vitro antibacterial and antioxidant properties of chitosan edible films incorporated with Thymus moroderi or Thymus piperella essential oils", *Food Contr.,* vol. 30, no. 2, pp. 386-392, 2013.
[http://dx.doi.org/10.1016/j.foodcont.2012.07.052]

[74] H. Lee, and S.C. Min, "“Antimicrobial edible defatted soybean meal-based films incorporating the lactoperoxidase system”, LWT -", *Food Sci. Technol. (Campinas.),* vol. 54, no. 1, pp. 42-50, 2013.

[75] K.H. Seol, D.G. Lim, A. Jang, C. Jo, and M. Lee, "Antimicrobial effect of κ-carrageenan-based edible

film containing ovotransferrin in fresh chicken breast stored at 5°C", *Meat Sci.,* vol. 83, no. 3, pp. 479-483, 2009.
[http://dx.doi.org/10.1016/j.meatsci.2009.06.029] [PMID: 20416673]

[76] C.P. Resa, L.N. Gerschenson, and R.J. Jagus, "Effect of natamycin on physical properties of starch edible films and their effect on Saccharomyces cerevisiae activity", *Food Bioprocess Tech.,* vol. 6, no. 11, pp. 3124-3133, 2013.
[http://dx.doi.org/10.1007/s11947-012-0960-0]

[77] M. Imran, S. El-Fahmy, A-M. Revol-Junelles, and S. Desobry, "Cellulose derivative based active coatings: Effects of nisin and plasticizer on physicochemical and antimicrobial properties of hydroxypropyl methylcellulose films", *Carbohydr. Polym.,* vol. 81, no. 2, pp. 219-225, 2010.
[http://dx.doi.org/10.1016/j.carbpol.2010.02.021]

[78] C. Ibarguren, G. Céliz, A.S. Díaz, M.A. Bertuzzi, M. Daz, and M.C. Audisio, "Gelatine based films added with bacteriocins and a flavonoid ester active against food-borne pathogens", *Innov. Food Sci. Emerg.,* vol. 28, pp. 66-72, 2015.
[http://dx.doi.org/10.1016/j.ifset.2015.01.007]

[79] M. da Rocha, M.R. Loiko, E.C. Tondo, and C. Prentice, "Physical, mechanical and antimicrobial properties of Argentine anchovy (Engraulis anchoita) protein films incorporated with organic acids", *Food Hydrocoll.,* vol. 37, pp. 213-220, 2014.
[http://dx.doi.org/10.1016/j.foodhyd.2013.10.017]

[80] M.H. Gordon, "The development of oxidative rancidity in food", In: *Antioxidants in food, Practical applications, J. Pokorny.,* N. Yanishlieva, M. Gordon, Eds., Woodhead Publishing Limited Ediciones Cambridge: England, 2001, pp. 7-21.
[http://dx.doi.org/10.1201/9781439823057.pt1]

[81] Y. Wu, J-W. Rhim, C.L. Weller, F. Hamouz, S.L. Cuppett, and M. Schnepf, "Moisture loss and lipid oxidation for precooked beef patties stored in edible coatings and films", *J. Food Sci.,* vol. 65, no. 2, pp. 300-304, 2001.
[http://dx.doi.org/10.1111/j.1365-2621.2000.tb15997.x]

[82] Y. Wu, C.L. Weller, F. Hamouz, S. Cuppett, and M. Schnepf, "Moisture loss and lipid oxidation for precooked ground-beef patties packaged in edible starch-alginate-based composite films", *J. Food Sci.,* vol. 66, no. 3, pp. 486-493, 2001.
[http://dx.doi.org/10.1111/j.1365-2621.2001.tb16137.x]

[83] J. Gómez-Estaca, P. Montero, B. Giménez, and M.C. Gómez-Guillén, "Effect of functional edible films and high pressure processing on microbial and oxidative spoilage in cold-smoked sardine (Sardina pilchardus)", *Food Chem.,* vol. 105, no. 2, pp. 511-520, 2007.
[http://dx.doi.org/10.1016/j.foodchem.2007.04.006] [PMID: 26059129]

[84] E. Ayranci, and S. Tunc, "A method for the measurement of the oxygen permeability and the development of edible films to reduce the rate of oxidative reactions in fresh foods", *Food Chem.,* vol. 80, no. 3, pp. 423-431, 2003.
[http://dx.doi.org/10.1016/S0308-8146(02)00485-5]

[85] E. Ayranci, and S. Tunc, "The effect of edible coatings on water and vitamin C loss of apricots (Armeniaca vulgaris Lam.) and green peppers (Capsicum annuum L.)", *Food Chem.,* vol. 87, no. 3, pp. 339-342, 2004.
[http://dx.doi.org/10.1016/j.foodchem.2003.12.003]

[86] L. Atarés, C. De Jesús, P. Talens, and A. Chiralt, "Characterization of SPI-based edible films incorporated with cinnamon or ginger essential oils", *J. Food Eng.,* vol. 99, no. 3, pp. 384-391, 2010.
[http://dx.doi.org/10.1016/j.jfoodeng.2010.03.004]

[87] L. Bokobza, "Multiwall carbon nanotube elastomeric composites: a review", *Polymer (Guildf.),* vol. 48, no. 17, pp. 4907-4920, 2007.

[http://dx.doi.org/10.1016/j.polymer.2007.06.046]

[88] L.H. Sperling, *Introduction to Physical Polymer Science.*2nd ed John Wiley & Sons Inc.: New York, USA, 2006, pp. 557-612.

[89] S. Obara, and J.W. McGinity, "Influence of processing variables on the properties of free films prepared from aqoues polymeric dispersions by a spray technique", *Int. J. Pharm.,* vol. 37, no. 1-2, pp. 849-853, 1995.

[90] L. Cisneros-Zevallos, and J.M. Krochta, "Dependence of coating thickness on viscosity of coating solution applied to fruits and vegetables by dipping method", *J. Food Sci.,* vol. 68, no. 2, pp. 503-510, 2003.
[http://dx.doi.org/10.1111/j.1365-2621.2003.tb05702.x]

[91] S. Mali, M.V. Grossmann, M.A. Garcia, M.M. Martino, and N.E. Zaritzky, "Barrier, mechanical and optical properties of plasticized yam starch films", *Carbohydr. Polym.,* vol. 56, no. 2, pp. 129-135, 2004.
[http://dx.doi.org/10.1016/j.carbpol.2004.01.004]

[92] L.A. Kunte, A. Gennadios, S.L. Cuppett, M.A. Hanna, and C.L. Weller, "Cast films from soy protein isolates and fractions", *Cereal Chem.,* vol. 74, no. 2, pp. 115-118, 1997.
[http://dx.doi.org/10.1094/CCHEM.1997.74.2.115]

[93] G. Ward, and A. Nussinovitch, "“Peel gloss as a potential indicator of banana ripening”, LWT -", *Food Sci. Technol. (Campinas.),* vol. 29, no. 3, pp. 289-294, 1996.

[94] R. Gnanasambadam, N.S. Hettiarachchy, and M. Coleman, "Mechanical and barrier properties of rice bran films", *J. Food Sci.,* vol. 62, no. 2, pp. 395-398, 1997.
[http://dx.doi.org/10.1111/j.1365-2621.1997.tb04009.x]

[95] G. Kavoosi, S.M. Dadfar, and A.M. Purfard, "Mechanical, physical, antioxidant, and antimicrobial properties of gelatin films incorporated with thymol for potential use as nano wound dressing", *J. Food Sci.,* vol. 78, no. 2, pp. E244-E250, 2013.
[http://dx.doi.org/10.1111/1750-3841.12015] [PMID: 23317304]

[96] R. Sothornvit, and N. Pitak, "Oxygen permeability and mechanical properties of banana films", *Food Res. Int.,* vol. 40, no. 3, pp. 365-370, 2007.
[http://dx.doi.org/10.1016/j.foodres.2006.10.010]

[97] M.A. Cerqueira, A.M. Lima, J.A. Teixeira, R.A. Moreira, and A.A. Vicente, "Suitability of novel galactomannans as edible coatings for tropical fruits", *J. Food Eng.,* vol. 94, no. 3-4, pp. 372-378, 2009.
[http://dx.doi.org/10.1016/j.jfoodeng.2009.04.003]

[98] A.I. Bourbon, A.C. Pinheiro, M.A. Cerqueira, C.M. Rocha, M.C. Avides, M.A. Quintas, and A.A. Vicente, "physicochemical characterization of chitosan-based edible films incorporating bioactive compounds of different molecular weight", *J. Food Eng.,* vol. 106, no. 2, pp. 111-118, 2011.
[http://dx.doi.org/10.1016/j.jfoodeng.2011.03.024]

[99] J.T. Martins, M.A. Cerqueira, and A.A. Vicente, "Influence of α-tocopherol on physicochemical properties of chitosan-based films", *Food Hydrocoll.,* vol. 27, no. 1, pp. 220-227, 2012.
[http://dx.doi.org/10.1016/j.foodhyd.2011.06.011]

[100] J.F. Rubilar, R.M. Cruz, H.D. Silva, A.A. Vicente, I. Khmelinskii, and M.C. Vieira, "Physico-mechanical properties of chitosan films with carvacrol and grape seed extract", *J. Food Eng.,* vol. 115, no. 4, pp. 466-474, 2013.
[http://dx.doi.org/10.1016/j.jfoodeng.2012.07.009]

[101] M.B. Vásconez, S.K. Flores, C.A. Campos, J. Alvarado, and L.N. Gerschenson, "Antimicrobial activity and physical properties of chitosan-tapioca starch based edible films and coatings", *Food Res. Int.,* vol. 42, no. 7, pp. 762-769, 2009.
[http://dx.doi.org/10.1016/j.foodres.2009.02.026]

[102] K.S. Miller, and J.M. Krochta, "Oxygen and aroma barrier properties of edible films: a review", *Trends Food Sci. Technol.,* vol. 8, no. 7, pp. 228-237, 1997. [http://dx.doi.org/10.1016/S0924-2244(97)01051-0]

[103] B.W. Souza, M.A. Cerqueira, A. Casariego, A.M. Lima, J.A. Teixeira, and A.A. Vicente, "Effect of moderate electric fields in the permeation properties of chitosan coatings", *Food Hydrocoll.,* vol. 23, no. 8, pp. 2110-2115, 2009. [http://dx.doi.org/10.1016/j.foodhyd.2009.03.021]

[104] B.W. Souza, M.A. Cerqueira, J.T. Martins, A. Casariego, J.A. Teixeira, and A.A. Vicente, "Influence of electric fields on the structure of chitosan edible coatings", *Food Hydrocoll.,* vol. 24, no. 4, pp. 330-335, 2010. [http://dx.doi.org/10.1016/j.foodhyd.2009.10.011]

[105] D. Altiok, E. Altiok, and F. Tihminlioglu, "Physical, antibacterial and antioxidant properties of chitosan films incorporated with thyme oil for potential wound healing applications", *J. Mater. Sci. Mater. Med.,* vol. 21, no. 7, pp. 2227-2236, 2010. [http://dx.doi.org/10.1007/s10856-010-4065-x] [PMID: 20372985]

[106] J. Bonilla, L. Atarés, M. Vargas, and A. Chiralt, "Effect of essential oils and homogenization conditions on properties of chitosan based films", *Food Hydrocoll.,* vol. 26, no. 1, pp. 9-16, 2012. [http://dx.doi.org/10.1016/j.foodhyd.2011.03.015]

[107] Y. Byun, Y.T. Kim, and S. Whiteside, "Characterization of an antioxidant polylactic acid (PLA) film prepared with α-tocopherol, BHT and polyethylene glycol using film cast extruder", *J. Food Eng.,* vol. 100, no. 2, pp. 239-244, 2010. [http://dx.doi.org/10.1016/j.jfoodeng.2010.04.005]

[108] Y. Pranoto, V.M. Salokhe, and S.K. Rakshit, "Physical and antibacterial properties of alginate-based edible film incorporated with garlic oil", *Food Res. Int.,* vol. 38, no. 3, pp. 267-272, 2005. [http://dx.doi.org/10.1016/j.foodres.2004.04.009]

[109] C. Pastor, L. Sánchez-González, A. Chiralt, M. Cháfer, and C. González-Martínez, "Physical and antioxidant properties of chitosan and methylcellulose based films containing resveratrol", *Food Hydrocoll.,* vol. 30, no. 1, pp. 272-280, 2013. [http://dx.doi.org/10.1016/j.foodhyd.2012.05.026]

[110] U. Siripatrawan, and B.R. Harte, "Physical properties and antioxidant activity of an active film from chitosan incorporated with green tea extract", *Food Hydrocoll.,* vol. 24, no. 8, pp. 770-775, 2010. [http://dx.doi.org/10.1016/j.foodhyd.2010.04.003]

[111] L. Sánchez-González, C. González-Martínez, A. Chiralt, and M. Cháfer, "Physical and antimicrobial properties of chitosan–tea tree essential oil composite films", *J. Food Eng.,* vol. 98, no. 4, pp. 443-452, 2010. [http://dx.doi.org/10.1016/j.jfoodeng.2010.01.026]

[112] X.L. Shen, J.M. Wu, Y. Chen, and G. Zhao, "Antimicrobial and physical properties of sweet potato starch films incorporated with potassium sorbate or chitosan", *Food Hydrocoll.,* vol. 24, no. 4, pp. 285-290, 2010. [http://dx.doi.org/10.1016/j.foodhyd.2009.10.003]

[113] S.M. Ojagh, M. Rezaei, S.H. Razavi, and S.M. Hosseini, "Development and Evaluation of a novel biodegradable film made from chitosan and cinnamon essential oil with low affinity toward water", *Food Chem.,* vol. 102, no. 1, pp. 193-198, 2010. [http://dx.doi.org/10.1016/j.foodchem.2009.10.006]

[114] D. Georgantelis, I. Ambrosiadis, P. Katikou, G. Blekas, and S.A. Georgakis, "Effect of rosemary extract, chitosan and α-tocopherol on microbiological parameters and lipid oxidation of fresh pork sausages stored at 4°C", *Meat Sci.,* vol. 76, no. 1, pp. 172-181, 2007. [http://dx.doi.org/10.1016/j.meatsci.2006.10.026] [PMID: 22064204]

[115] M. Ghasemlou, N. Aliheidari, R. Fahmi, S. Shojaee-Aliabadi, B. Keshavarz, M.J. Cran, and R. Khaksar, "Physical, mechanical and barrier properties of corn starch films incorporated with plant essential oils", *Carbohydr. Polym.,* vol. 98, no. 1, pp. 1117-1126, 2013. [http://dx.doi.org/10.1016/j.carbpol.2013.07.026] [PMID: 23987453]

[116] R. Avenainen, "Packaging-flavour interactions", In: *Novel food Packaging Techniques,* R. Avenainen, Boca Raton, Eds., Woodhead Publishing: FL, USA, 2003, pp. 144-171.

[117] J.H. Han, G.H. Seo, I.M. Park, G.N. Kim, and D.S. Lee, "Physical and mechanical properties of pea starch edible films containing beeswax emulsions", *J. Food Sci.,* vol. 70, no. 6, pp. 290-296, 2006. [http://dx.doi.org/10.1111/j.1750-3841.2006.00088.x]

[118] A.C. Souza, G.E. Goto, J.A. Mainardi, A.C. Coelho, and C.C. Tadini, "“Cassava starch composite films incorporated with cinnamon essential oil: Antimicrobial activity, microstructure, mechanical and barrier properties”, LWT -", *Food Sci. Technol. (Campinas.),* vol. 54, no. 2, pp. 346-352, 2013.

[119] A.C. Attarian, V. Kechichian, V. Veiga-Santos, C. Ditchfield, and C.C. Tadini, "Effect of Antimicrobial Edible Additives on Cassava Starch Biobased Films Characterization", In: *2006 CIGR Section VI International Symposium on Futer of Food Engineering Warsaw*, Poland, 2006.

[120] L. Famá, A.M. Rojas, S. Goyanes, and L. Gerschenson, "“Mechanical properties of tapioca - starch edible films containing sorbates”, LWT -", *Food Sci. Technol. (Campinas.),* vol. 38, no. 6, pp. 631-639, 2005.

[121] I. Arvanitoyannis, C.G. Biliaderis, H. Ogawa, and N. Kawasaki, "Biodegradable films made from low-density polyethylene (LDPE), rice starch and potato starch for food packaging applications: Part 1", *Carbohydr. Polym.,* vol. 36, no. 2-3, pp. 89-104, 1998. [http://dx.doi.org/10.1016/S0144-8617(98)00016-2]

[122] S. Sayanjali, B. Ghanbarzadeh, and S. Ghiassifar, "“Evaluation of antimicrobial and physical properties of edible film based on carboxymethyl cellulose containing potassium sorbate on some mycotoxigenic Aspergillus species in fresh pistachios”, LWT -", *Food Sci. Technol. (Campinas.),* vol. 44, no. 4, pp. 1133-1138, 2011.

[123] S. Gemili, A. Yemenicioğlu, and S.A. Altınkaya, "Development of antioxidant food packaging materials with controlled release properties", *J. Food Eng.,* vol. 96, no. 3, pp. 325-332, 2010. [http://dx.doi.org/10.1016/j.jfoodeng.2009.08.020]

[124] L. Sánchez-González, J. Quintero Saavedra, and A. Chiralt, "Physical properties and antilisterial activity of bioactive edible films containing Lactobacillus plantarum", *Food Hydrocoll.,* vol. 33, no. 1, pp. 92-98, 2013. [http://dx.doi.org/10.1016/j.foodhyd.2013.02.011]

[125] A. Jiménez, M.J. Fabra, P. Talens, and A. Chiralt, "Effect of lipid self-association on the microstructure and physical properties of hydroxypropyl-methylcellulose edible films containing fatty acids", *Carbohydr. Polym.,* vol. 82, no. 3, pp. 585-593, 2010. [http://dx.doi.org/10.1016/j.carbpol.2010.05.014]

[126] K. Larsson, and P. Dejmek, "Crystal and liquid crystal structures of lipids", In: *Food emulsions,* K. Larsson, S.E. Friberg, Eds., New York/Basel: Maecer Dekker, Inc.: NY, USA, 1990.

[127] K. Khwaldia, M. Linder, S. Banon, and S. Desobry, "Effects of mica, carnauba wax, glycerol, and sodium caseinate concentrations on water vapor barrier and mechanical properties of coated paper", *J. Food Sci.,* vol. 70, no. 3, pp. 192-197, 2005. [http://dx.doi.org/10.1111/j.1365-2621.2005.tb07135.x]

[128] M-J. Akhtar, M. Jacquot, M. Jamshidian, M. Imran, E. Arab-Tehrany, and S. Desobry, "Fabrication and physicochemical characterization of HPMC films with commercial plant extract: Influence of light and film composition", *Food Hydrocoll.,* vol. 31, no. 2, pp. 420-427, 2013. [http://dx.doi.org/10.1016/j.foodhyd.2012.10.008]

[129] S.R. Benavides, R. Villalobos-Carvajal, and J.E. Reyes, "Physical, mechanical and antibacterial

properties of alginate film: Effect of the crosslinking degree and oregano essential oil concentration", *J. Food Eng.*, vol. 110, no. 2, pp. 232-239, 2012. [http://dx.doi.org/10.1016/j.jfoodeng.2011.05.023]

[130] K. Norajit, M.K. Kim, and G.H. Ryu, "Comparative studies on the characterization and antioxidant properties of biodegradable alginate films containing ginseng extract", *J. Food Eng.*, vol. 98, no. 3, pp. 377-384, 2010. [http://dx.doi.org/10.1016/j.jfoodeng.2010.01.015]

[131] S. Zivanovic, and A.F. Chi, "Draughon, "Antimicrobial activity of chitosan films enriched with essential oils", *J. Food Sci.*, vol. 70, no. 1, pp. 45-51, 2005. [http://dx.doi.org/10.1111/j.1365-2621.2005.tb09045.x]

[132] M.H. Hosseini, S.H. Razavi, and M.A. Mousavi, "Antimicrobial, physical and mechanical properties of chitosan-based films incorporated with thyme, clove and cinnamon essential oils", *J. Food Process. Preserv.*, vol. 33, no. 6, pp. 727-743, 2009. [http://dx.doi.org/10.1111/j.1745-4549.2008.00307.x]

[133] M.A. Rojas-Graü, R.J. Avena-Bustillos, C. Olsen, M. Friedman, O.R. Henika, O. Martín-Belloso, Z. Pan, and T.H. McHugh, "Effects of plant essential oils and oils compounds on mechanical, barrier and antimicrobial properties of alginate-apple puree edible films", *J. Food Eng.*, vol. 81, no. 3, pp. 634-641, 2007. [http://dx.doi.org/10.1016/j.jfoodeng.2007.01.007]

[134] A.C. Bierhalz, M. Altenhofen da Silva, and T. Guenter Kieckbusch, "Natamycin release from alginate/pectin films for food packaging applications", *J. Food Eng.*, vol. 110, no. 1, pp. 18-25, 2012. [http://dx.doi.org/10.1016/j.jfoodeng.2011.12.016]

[135] A.C. Pires, N.F. Soares, N.J. Andrade, L.H. Silva, G.P. Camilloto, and P.C. Bernardes, "Development and evaluation of active packaging for sliced mozzarella preservation", *Packag. Technol. Sci.*, vol. 21, no. 7, pp. 375-383, 2008. [http://dx.doi.org/10.1002/pts.815]

[136] H. Türe, E. Erog˘lu, B. Özen, and F. Soyer, "Physical properties of biopolymers containing natamycin and rosemary extract", *Int. J. Food Sci. Technol.*, vol. 44, no. 2, pp. 402-408, 2009. [http://dx.doi.org/10.1111/j.1365-2621.2008.01785.x]

[137] W.X. Du, C.W. Olsen, R.J. Avena-Bustillos, T.H. McHugh, C.E. Levin, and M. Friedman, "Effects of allspice, cinnamon, and clove bud essential oils in edible apple films on physical properties and antimicrobial activities", *J. Food Sci.*, vol. 74, no. 7, pp. M372-M378, 2009. [http://dx.doi.org/10.1111/j.1750-3841.2009.01282.x] [PMID: 19895483]

[138] L. Averous, and N. Boquillon, "Biocomposites based on plasticized starch: thermal and mechanical behaviours", *Carbohydr. Polym.*, vol. 56, no. 2, pp. 111-122, 2004. [http://dx.doi.org/10.1016/j.carbpol.2003.11.015]

[139] J.P. Kerry, M.N. OGrady, and S.A. Hogan, "Past, current and potential utilisation of active and intelligent packaging systems for meat and muscle-based products: A review", *Meat Sci.*, vol. 74, no. 1, pp. 113-130, 2006. [http://dx.doi.org/10.1016/j.meatsci.2006.04.024] [PMID: 22062721]

[140] K.C. Huber, and M.E. Embuscado, *Edible films and coatings for food applications.*Springer: Nueva York, NY, USA, 2009, p. 403. [http://dx.doi.org/10.1007/978-0-387-92824-1]

[141] T.J. Gutiérrez, M.S. Tapia, E. Pérez, and L. Famá, "Structural and mechanical properties of edible films made from native and modified cush-cush yam and cassava starch", *Food Hydrocoll.*, vol. 45, pp. 211-217, 2015. [http://dx.doi.org/10.1016/j.foodhyd.2014.11.017]

[142] L. Famá, C. Bernal, and S. Goyanes, "Application of carbon nanotubes (CNTs) in the development of nanocomposites based on biodegradable polymers", In: *Carbon nanotubes: synthesis and*

*properties.*Nova Science Publishers Inc.: USA, 2014, pp. 157-184.

[143] L. Famá, S. Goyanes, V. Pettarin, and C. Bernal, "Mechanical behavior of starch/carbon nanotubes composites", In: *Handbook of Polymer nanocomposites. Processing, Performance and Application: Volume B: Carbon Nanotube Based Polymer Composites,* K.K. Kar, S.K. Rana, J.K. Pandey, Eds., Springer-Verlag GmbH: USA, 2014.

[144] M.M. Murillo-Martínez, R. Pedroza-Islas, C. Lobato-Calleros, A. Martinez-Ferez, and E.J. Vernon-Carter, "Designing W-1/O/W-2 double emulsions stabilized by protein-polysaccharide complexes for producing edible films: rheological, mechanical and water vapour properties", *Food Hydrocoll.,* vol. 25, no. 4, pp. 577-585, 2011. [http://dx.doi.org/10.1016/j.foodhyd.2010.06.015]

[145] X. Sun, Z. Wang, H. Kadouh, and K. Zhou, "The antimicrobial, mechanical, physical and structural properties of chitosan-gallic acid films", *LWT-Food Sci. Technol. (Campinas.),* vol. 57, no. 1, pp. 83-89, 2014.

[146] G.L. Avedaño Cetina, "Diseño y evaluación de las propiedades mecánicas y de barrera de un biopolímero obtenido a partir de almidón de papa para ser empleado en empaques para alimentos", In: *Universidad Nacional abierta y a distancia UNAD Escuela de Ciencias Básicas e Ingeniería de Alimentos.*DUITAMA, 2009.

[147] M. Vargas, A. Albors, A. Chiralt, and C. González-Martínez, "Characterization of chitosan-oleic acid composite films", *Food Hydrocoll.,* vol. 23, no. 2, pp. 536-547, 2009. [http://dx.doi.org/10.1016/j.foodhyd.2008.02.009]

[148] C. Péroval, F. Debeaufort, D. Despré, and A. Voilley, "Edible arabinoxylan-based films. 1. Effects of lipid type on water vapor permeability, film structure, and other physical characteristics", *J. Agric. Food Chem.,* vol. 50, no. 14, pp. 3977-3983, 2002. [http://dx.doi.org/10.1021/jf0116449] [PMID: 12083869]

[149] L. Sánchez-González, M. Vargas, C. González-Martínez, A. Chiralt, and M. Cháfer, "Characterization of edible films based on hydroxypropylmethylcellulose and tea tree essential oil", *Food Hydrocoll.,* vol. 23, no. 8, pp. 2102-2109, 2009. [http://dx.doi.org/10.1016/j.foodhyd.2009.05.006]

[150] A.G. Ponce, S.I. Roura, C.E. del Valle, and M.R. Moreira, "Antimicrobial and antioxidant activities of edible coatings enriched with natural plant extracts: In vitro and in vivo studies", *Postharvest Biol. Technol.,* vol. 49, no. 2, pp. 294-300, 2008. [http://dx.doi.org/10.1016/j.postharvbio.2008.02.013]

[151] M. Pereda, G. Amica, and N.E. Marcovich, "Development and characterization of edible chitosan/olive oil emulsion films", *Carbohydr. Polym.,* vol. 87, no. 2, pp. 1318-1325, 2012. [http://dx.doi.org/10.1016/j.carbpol.2011.09.019]

[152] Y. Sun, Y. Liu, Y.Z. Li, M.Z. Lv, P.W. Li, H.L. Xu, and L. Wang, "Preparation and characterization of novel curdlan/chitosan blending membranes for antibacterial applications", *Carbohydr. Polym.,* vol. 84, no. 3, pp. 952-959, 2011. [http://dx.doi.org/10.1016/j.carbpol.2010.12.055]

[153] J.H. Han, "Antimicrobial food packaging", *Food Technol.-Chicago,* vol. 54, no. 3, pp. 56-65, 2000.

[154] V. Falguera, J.P. Quintero, A. Jiménez, J.A. Muñoz, and A. Ibarz, "Edible films and coatings: structures, active functions and trends in their use", *Trends Food Sci. Technol.,* vol. 22, no. 6, pp. 292-303, 2011. [http://dx.doi.org/10.1016/j.tifs.2011.02.004]

[155] M. Escamilla-García, G. Calderón-Domínguez, J.J. Chanona-Pérez, R.R. Farrera-Rebollo, J.A. Andraca-Adame, I. Arzate-Vázquez, J.V. Mendez-Mendez, and L.A. Moreno-Ruiz, "Physical and structural characterisation of zein and chitosan edible films using nanotechnology tools", *Int. J. Biol. Macromol.,* vol. 61, pp. 196-203, 2013. [http://dx.doi.org/10.1016/j.ijbiomac.2013.06.051] [PMID: 23831381]

[156] C. Valenzuela, L. Abugoch, and C. Tapia, "Quinoa protein-chitosan-sunflower oil edible film: Mechanical, barrier and structural properties", *LWT - Food Sci. Technol. (Campinas.),* vol. 50, no. 2, pp. 531-537, 2013.

[157] S. Mali, M.V. Grossmann, M.A. Garcia, M.N. Martino, and N.E. Zaritzky, "Microstructural characterization of yam starch films", *Carbohydr. Polym.,* vol. 50, no. 4, pp. 379-386, 2002. [http://dx.doi.org/10.1016/S0144-8617(02)00058-9]

[158] A.S. Rindlav-Westling, M. Stading, A-M. Hermansson, and P. Gatenholm, "Structure, mechanical and barrier properties of amylose and amylopectin films", *Carbohydr. Polym.,* vol. 36, no. 2-3, pp. 217-224, 1998. [http://dx.doi.org/10.1016/S0144-8617(98)00025-3]

[159] H. Xiong, S. Tang, H. Tang, and P. Zou, "The structure and properties of a starch-based biodegradable film", *Carbohydr. Polym.,* vol. 71, no. 2, pp. 263-268, 2008. [http://dx.doi.org/10.1016/j.carbpol.2007.05.035]

[160] R. Villalobos, J. Chanona, P. Hernández, G. Gutiérrez, and A. Chiralt, "Gloss and transparency of hydroxypropyl methylcellulose films containing surfactants as affected by their microstructure", *Food Hydrocoll.,* vol. 19, no. 1, pp. 53-61, 2005. [http://dx.doi.org/10.1016/j.foodhyd.2004.04.014]

[161] L.J. Mauer, D.E. Smith, and T.P. Labuza, "Water vapor permeability, mechanical, and structural properties of edible β-casein films", *Int. Dairy J.,* vol. 10, no. 5-6, pp. 353-358, 2000. [http://dx.doi.org/10.1016/S0958-6946(00)00061-3]

CHAPTER 13

Influence of Nanostructures in the Physicochemical Properties of Polysaccharide Based Biocomposites: Characterization and Applications

Joaquín A. González, María Emilia Villanueva and **Guillermo J. Copello***

Universidad de Buenos Aires (UBA), Facultad de Farmacia y Bioquímica, Cátedra de Química Analítica Instrumental, IQUIMEFA (UBA-CONICET), Junín 956, C1113AAD Buenos Aires, Argentina

Abstract: The design and engineering of innovative bionanocomposites for a wide variety of applications should be based on the understanding of the relationship between their nanoscale structure and their chemical and mechanical properties. The understanding of the chemistry and structure at the nanometric level imply an essential knowledge for future developments of novel materials based on polysaccharides and nanostructured materials toward several applications (for instance bioremediation, medical applications, biotechnology, food industry, *etc.*). The study of the physico-chemical characteristics of a material would imply not only its basic description but could be significantly relevant in the optimization of its synthesis toward the obtaining of tailored properties or the control of the material behavior in a particular application. Herein, we present the introduction to the most used techniques in the characterization of bionanocomposites, such as electron microscopies, FTIR, Raman spectroscopy, NMR, Rheology behavior, DSC, TGA, SAXS and SANS. The basics of these techniques and key approaches in the characterization of bionanocomposites are presented for the understanding of the relevance of the characterization step in a material evaluation.

Keywords: Biocomposite, DSC, FTIR, Nanostructure, NMR, Polysaccharide, Raman, Rheology, SAXS, TGA.

INTRODUCTION

In the past decade, the advances in the nanotechnology field increased. It has been pointed out that this rise was driven by the spread of instrumental techniques capable of analyzing the structure of a material in the nanometric scale, regarding

* **Corresponding author Guillermo J. Copello:** Universidad de Buenos Aires (UBA), Facultad de Farmacia y Bioquímica, Cátedra de Química Analítica Instrumental, IQUIMEFA (UBA-CONICET), Junín 956, C1113AAD Buenos Aires, Argentina; Tel: +54-011-5287-4336; E-mail: gcopello@ffyb.uba.ar

Martin Masuelli & Denis Renard (Eds.)

both its physical and chemical properties. Tailored nanocomposite materials can be synthesized if the control of the heterogeneity of the structure on the nanometer scale is possible. From the molecular scale morphology plays a key role in achieving macroscopic molecular and supramolecular assemblies. The synthetic goal is to prepare multiphase micro- or nanoheterogeneous systems with desired and tunable properties. With this aim, a detailed multidisciplinary characterization is mandatory in order to establish the correlations among the composition, mechanical, viscoelastic, optical, and other macroscopic properties [1].

Regarding a material, that comprises a polymeric matrix and a nanostructure, including either nanoparticles or nanodomains (*i.e.* nanopores or nanometric structural levels of organization), the characterization often involves the separate study of the polymer and the nanostructure. With advances in the bionano-composites technology, it became more relevant to study the influence of nanostuctures in the physicochemical properties of the materials. When it comes to a biopolymer, the typical procedures of characterization are Nuclear Magnetic Resonance (NMR), Infrared (FTIR) and Raman spectroscopy for functional groups and carbon backbone analysis, size exclusion chromatography and light scattering methods for molecular weight determination, or rheological analysis for the study of their mechanical behavior, among others. Regarding nanoparticles, similar techniques for the study of their chemistry are often used, such as FTIR or Raman spectroscopy, also specific techniques focused on their nanosize are also employed. These techniques involve light scattering techniques, such as Static and Dynamic Light Scattering (SLS and DLS respectively), which are useful for size, distribution and polydispersity analysis. In addition, microscopy techniques, such as Transmission Electron Microscopy (TEM), Secondary Electron Microscopy (SEM) and Atomic Force Microscopy (AFM), are used for obtaining internal or topographical images, and Electrophoretic Light Scattering, for zeta potential measurements, among others.

Several of these techniques are used both for biopolymer and nanoparticle analysis and could bring information not only from each separate component of the bionanocomposite but could help to understand the interaction among them. In this chapter, we will try to cover most of these techniques from a point of view focused on obtaining information about the interaction of the biopolymer, mainly a polysaccharide, and the nanostructure. Herein we will present this subject in relation to Electron Microscopies, FTIR, Raman spectroscopy, NMR, Rheological methods, DSC, TGA, SAXS and SANS. The basics of these techniques and key approaches in the characterization of bionanocomposites are presented for the understanding of the potential of the characterization in a material assessment.

Electron Microscopies

With respect to physicochemical characterization, the analysis of a material or its components does not need to follow a mandatory order of techniques. Nevertheless, the techniques most likely to be performed in the first place are the Electron Microscopies (EM). The EMs bring information about morphology, topography, internal structure and other features that can be "observed" directly from an image. Therefore, they are not powerful techniques by themselves but they are powerful for the routine analysis and also when they are complemented with other techniques.

The significance of the EM in bionanocomposites science and technology arises principally from two causes. First, by using EM, information of the morphology of polymer matrix, the nanostructure and the contact interaction between them can be simultaneously assessed with micrometer to nanometer resolution. Second, electron microscopy allows studying the response of the structural details towards applied load, chemical and temperature exposure, time degradation, *etc.*, enabling the design of tailored materials. In addition, EM is the only technique, which provides direct evidence of intercalation and exfoliation of the nanostructure in the polymer matrix allowing straightforward quantification of morphological features of the polymer nanocomposites [2].

The main difference among EMs is the position of the detector which would lead to different sources of the image and, therefore, different information. Secondary Electron Microscopy (SEM) uses the sample holder as an anode towards which the electrons are accelerated. When the electron beam impacts the sample, an inelastic interaction can lead to the generation of secondary electrons that are expelled from the sample. In their path, they can reach a detector placed above the sample in a particular angle. Since the secondary electrons are only generated in outer layers of the sample, this image reflects the topography of the sample. If a new detector is placed near the lens, close to the electron beam path, a Backscattered Electron (BSE) image could be obtained. In this case, the detected electrons are the same electrons from the electron beam after elastically (or almost elastically) interacting with the sample nuclei and being backscattered to the detector. Since the probability of backscattering is higher with higher atomic numbers, the image presents a map of atomic number contrast. Thus, when inorganic nanoparticles became the part of a composite together with a biopolymer, they could be clearly detected as brighter islands surrounded by a darker polymer matrix [2].

From another point of view, a definitive elemental mapping could be obtained if an energy dispersive detector is placed in the sample chamber. From this technique, called Energy Dispersive Spectroscopy (EDS), two modes of information can be obtained. The first, an energy dispersive spectrum from which the quantities of each element could be integrated. For bionanocomposites samples, special care has to be taken when it comes to quantify the organic moieties. EDS spectrum is generated by X-rays produced by the sample after the electron beam impact. The energy of these X-rays is related to the element from which it was generated and its quantity is related to the element abundance. However, the efficiency of X-ray generation of light elements, such as C, N or H, is low and very susceptible to experimental conditions. Thus, quantification of samples containing a high organic content could be misleading. Similar information can be obtained if a Wavelength Dispersive detector (WDS) is attached to the SEM. The second mode of information is the elemental mapping. An EDS image can be obtained together with the SEM image by selecting the element of interest. In example, if Ag nanoparticles are included within a starch matrix, the EDS mapping image can show the localization of Ag nanoparticles as a brighter or colored contrast. This image could be obtained simultaneously with every element that is present in the sample in sufficient quantity to be detected. Due to the slow response of WDS detector, elemental mapping is an exclusive feature of EDS.

Transmission Electron Microscopy (TEM) uses a different instrumental disposition that places the sample in the middle of the electron beam path toward the anode, where the detector is placed. The image in TEM is generated by the transmitted electrons, which are elastically or inelastically scattered. If the electron is transmitted unaltered through the sample, the image will be bright, but if there is an object in the path, the electron will be deviated and the image will appear as a dark spot in the location of the object. Since the image is generated by electrons that go through all the sample thickness, the generated image contains internal structure information. Therefore, TEM samples a bigger volume than SEM and is more accurate when the evaluation of nanoparticles dispersion in a material is desired. On the other hand, the sample preparation in TEM is more laborious than in SEM.

In TEM, a typical hydrogel or soft polymer requires to be embedded in a polymeric resin, stained with heavy metals for atomic number contrast and cuts in ten to hundreds of nanometer slices, whereas SEM just requires gold coating for insulating samples and in some cases freeze-drying and/or fixation. Regarding resolutions, TEM is ahead of SEM since HR-TEM (High Resolution TEM) presents below Angstrom resolution whereas SEM is roughly at the low nanometer level [3]. In addition, if the module is attached to the TEM instrument, EDS could be performed on the samples.

Nuclear Magnetic Resonance spectroscopy

A very powerful method for polysaccharides analysis is Nuclear Magnetic Resonance (NMR) spectroscopy [4, 5]. It is a non-destructive method providing a lot of information in a reasonable amount of time [6]. Its major drawbacks are the low sensitivity (when compared to *i.e.* Mass spectrometry) and high cost of the equipment.

NMR spectroscopy theory is based on the principle of nuclear spin, an intrinsic angular momentum that every magnetic nucleus possesses. When a molecule is placed in a magnetic field, this momentum is aligned in either the same direction as or the opposite direction to the field, resulting in the expression of two defined energy states. The energy gap between these states, ΔE, depends on the applied magnetic field and the chemical environment of the nucleus in a molecule. The applied magnetic field shows the main influence in the ΔE, setting the frequency of resonance of a particular nucleus family (*i.e.* ^{1}H or ^{13}C, considering their gyromagnetic ratio). Meanwhile chemical environment influences the ΔE in a lower extent but enough to let nuclei involved in different chemical groups to be differentiated (*i.e.* a H from a -OH from a H-CH_2-), by an effect known as the chemical shift. As the resonance frequency of a certain nucleus is dependent on the magnetic field strength, a recalculation to the chemical shift (in parts per million, ppm) is performed [7, 8].

Most elements have at least one naturally occurring NMR active isotope; unfortunately, it is not always the most common one. The natural abundance of ^{1}H is close to 100% and it is therefore easily detected by NMR spectroscopy. The detectable ^{13}C and ^{15}N isotopes are less abundant, only 1.1% and 0.4%, respectively, leading to longer experiment times and the need of higher sample amounts. These are particularly relevant when it comes to biopolymer analysis. Although ^{1}H-NMR spectra are obtained in a shorter time and with high sensitivity and resolution. The relevance of the ^{13}C spectra leads the researchers to perform longer determinations that require higher amount of samples and therefore a high associated cost.

Magnetic nuclei are affected by the surroundings and especially by the closeness of other magnetic nuclei resulting in spin-spin (or *J)* couplings. This feature leads to split resonances that can reveal in example how many neighboring protons a certain proton has, and whether another NMR active nucleus, like ^{19}F, is attached nearby. For some nuclei, such as ^{13}C, the coupling pattern can complicate the spectrum and therefore decoupling is often applied.

Solid-state NMR spectroscopy (SS-NMR) deals with samples that can not be solubilized in liquid media, having extremely long motional correlation times in solution or need to be analyzed in the solid state. It is an effective technique that is widely used to characterize macromolecular systems, especially amorphous and mixed systems that cannot be characterized by scattering methods. The principles of SS-NMR are analogous to liquid-state NMR but the instrumentation requires some adaptation. A nuclear spin experiences a great number of interactions. The main three are dipolar and quadrupolar interactions and chemical shift anisotropy. In liquids, most of these interactions will average out because of the rapid time-averaged molecular motion (Brownian motion). In a solid state, the resultant of these interaction leads to broad and unresolved lines. Hopefully, these interactions depend on the orientation of the sample and their expression can be minimized by making the sample spin in the magic angle. As a resultant, the interactions are averaged. By spinning the sample (usually at a frequency of 1 to 100 kHz) at the magic angle θm (~54.74°) with respect to the direction of the magnetic field, the peaks appear narrow and with higher resolution.

Due to the presence of angular-dependent anisotropic interactions, the spectral resolution of solid-state NMR spectra is several orders of magnitude lower than that of high-resolution NMR in liquids. Again, this is critical in the analysis of ^{13}C. However, the pulsed techniques improve ^{13}C signal by the cross-polarization sequence in which polarization from abundant nuclei like ^{1}H can be transferred to low abundant nuclei, such as ^{13}C, in order to enhance signal to noise ratio and reduce delay time between successive pulses.

The introduction of two-dimensional and higher-dimensional NMR techniques offer fundamental advantages [9]. First, as in liquids, the introduction of a new frequency dimension provides a means to increase the spectral resolution. Even more important, multidimensional spectroscopy also provides routes to new information, unavailable from 1D spectra even in the limit of high resolution. Experiments can be designed to evidence the correlation between different spin interactions providing structural information, or to relate various states taken up by the molecular unit during different times by exchange and in this way dynamic processes in real time can be examined. Progress in multidimensional SS-NMR has been hampered by experimental and conceptual difficulties, which have now been overcome.

Analysis of Polysaccharide Nanocomposites

Besides the well-known use of NMR in elucidating polysaccharides structure (identities of the monosaccharide units, their relative amounts, how many sugars are present in the repeating unit, their connectivities, chain branching, *etc.*), this technique may provide other relevant information about polysaccharide composites such us the interaction between the polysaccharide and other components of a material or the chain mobility in the composite.

Interaction Between Nanocomposite Components

NMR can be useful in determining molecular interaction between the components within the composite. The methodology is simple: individual ^{13}C-NMR spectra are obtained from the composite and its different components. If new signals appear in the composite spectra, it can be concluded that the components interact forming new chemical bonds abundantly enough to be detected. When nanocomposites are being studied, the interaction between the nanomaterial and the polymeric structure can be described in this way.

Liu *et al.* developed a glycerol/starch/Na^+-montmorillonite starch based nano-composite and studied the interaction between glycerol and starch and their mobility by proton NMR spectroscopy [10]. The assay was performed by drying the samples and dissolving them in DMSO. The ^{1}H-NMR spectra of extruded pure starch and the composites with different amount of glycerol were obtained at different temperatures. The results showed the formation of a new peak in the glycerol-starch composites, this peak was present because of the –OH group exchanges/interactions between high amylose chain and glycerol. The interaction between starch and glycerol was temperature dependent; and it could be hypothesized that the interaction between starch/glycerol can be grouped into two main types, stable interactions and thermal dependent interactions. The peak intensity of NMR spectrum can be directly related to the mobility of corresponding component. In this case, the authors concluded that the mobility of glycerol molecules was lower than that of starch chains due to Na^+-montmorillonite network influence. In other words, glycerol molecules were strongly 'locked' within the montmorillonite network. Glycerol mobility increased within its concentration.

González *et al.* prepared chitin hybrid materials reinforced with graphene oxide nanosheets [11]. These nanocomposites were characterized with solid-state ^{13}C-CP-MAS NMR (Cross-Polarization-Magic Angle Spinning Nuclear Magnetic Resonance) in order to elucidate if the formation of new molecular bonds took place. Chitin gels, Chitin–graphene oxide nanosheets nanocomposite and commercial chitin powder were analyzed. When comparing the obtained spectra no additional signals or peak shifts could be observed. On the other hand, the relative intensity of the signals corresponding to C3 and C5, varied among the spectra. The variation of these signals was attributed to the possibility of the formation of different intra and interchain hydrogen bonds, which are related to the three polymorphic crystalline structures of chitin, α, β and γ. The authors found a correlation between the loss of the α type of chain arrangement and the increase of the nanosheet quantity in the material.

Leroux *et al.* incorporated alginate between the sheets of a double layered hydroxide *via* direct coprecipitation [12]. Solid-state ^{13}C-CP-MAS NMR experiments were performed in the alginate and the nanocomposite. Even in a qualitative analysis, from the comparison of both spectra, it could be surmised that the polymer conformation was not modified once immobilized between the inorganic layers. In addition to this, the similarity between both spectra suggested that the interaction between the organic guest and the inorganic framework was weak [13].

Molecular Mobility of the Biopolymer Chains

The spin-spin relaxation time (T_2) is particularly sensitive to changes in molecular mobility. The T_2 value of a material in the solid state differs in several orders of magnitude from that in the liquid state. This characteristic can be used for differentiate polysaccharide molecules in the more mobile liquid state from those in a more "immobile" solid-like state [14].

The nature of molecular mobility in a polymer system can be probed easily *via* a T_2 NMR experiment. Spin-spin relaxation, or T_2, probes how fast a nucleus loses transverse magnetization. Signal intensity decays as a function of local inhomogeneities in the magnetic field mainly due to perturbation by nuclei through space or dipolar interactions. This signal attenuation or the characteristic relaxation rate for this process is called the spin-spin relaxation or T_2 time. The T_2 time can be used to describe molecular motion and is particularly sensitive to chain dynamics in polymer systems above their glass transition temperature (T_g). Generally, the faster the decay rate the more rigid, or fewer degrees of freedom, the chemical group associated with that decay has [15].

Terenzi *et al.* investigated polymer mobility in nanocomposites made from cellulose nanofibers in a xyloglucan biopolymer matrix at different hydration states [16]. The nanocomposites were subjected to detailed ^2H and ^{13}C-NMR relaxation studies. Since the ^2H NMR signal arises from heavy water and the ^{13}C signal from the polysaccharides, molecular water and polymer dynamics was investigated separately. In the neat components, ^2H T_2 data were consistent with water clustering at the cellulose nanofibers surface. A description of water interaction with the nanoscale phases was obtained. At low hydration, water molecules at the cellulose nanofibers/xyloglucan interface exhibit higher water mobility than in neat cellulose nanofibers or xyloglucan, due to locally high water concentration. At the same time, cellulose nanofibers-associated interface segments of xyloglucan show slower NMR dynamics than that in neat xyloglucan.

Goetz *et al.* developed nanocomposites by cross-linking cellulose nanowhiskers with poly(methyl vinyl ether-co-maleic acid) and polyethylene glycol [15]. ^{13}C-CP-MAS NMR measurements and ^{1}H T_2 NMR experiments were carried out. NMR studies showed that cross-linking occurs between the matrix and cellulose nanowhiskers *via* an esterification reaction. Proton NMR T_2 relaxation experiments provided information on the mobility of the polymer chains within the matrix, which can be related to the structure of the cross-linked nanocomposite. ^{1}H T_2 relaxation can be used on this cellulose nanowhiskers nanocomposite system to infer information on the length of chains between junctions of cross-linking points, cross-linking density, and sterical constrains from surrounding chains.

Particle Thickness

In 2006, Schmidt-Rohr *et al.* described a new NMR approach characterizing the thickness of phosphate, silicate, carbonate and other nanoparticles in organic-inorganic nanocomposites [17]. The particle thickness was probed using the strongly distance-dependent dipolar couplings between the abundant protons in the organic phase and a particular nucleus, such as ^{31}P, ^{29}Si, ^{13}C, ^{27}Al, ^{23}Na, in the inorganic phase. This approach requires pulse sequences with heteronuclear dephasing only for the polymer or surface protons that experience strong homonuclear interactions, but not by dispersed -OH or water protons in the inorganic phase, which have long spin-spin relaxation times, T_2. Although this methodology has not been found in the literature applied to polysaccharide nanocomposites, it successfully describes organic-inorganic systems and, therefore, be applied in these bionanocomposites. For example, Hu *et al.* successfully characterized the interfacial species and revealed enhanced interactions between mineral and organic matrix, concomitant with the size effects in a hydroxyapatite nanocrystals in synthetic nanocomposites, fabricated with self-assembling block copolymer templates [18]. The surface-to-volume ratios determined by NMR spectroscopy and long-range $^{31}P\{^{1}H\}$ dipolar dephasing showed that increments in citrate concentrations influences the thicknesses of the hydroxyapatite nanocrystals.

Infrared Spectroscopy

The Infrared Spectroscopy (IR) is one of the most widespread techniques used in the characterization of inorganic and organic substances. The infrared spectrum includes the wavenumbers ranging from 12,800 to 10 cm^{-1}, however in a practical vision it can be divided into near, middle and far infrared radiation. Nowadays the mayor part of the qualitative and quantitative applications of IR are performed in a portion of the middle infrared region, between 400-4000 cm^{-1}. The development of more practical and versatile instruments has made IR spectroscopy a technique that can be found in most of the research laboratories. It is an ideal technique for routine analysis due to the facility and quickness of the experiments and the relative low cost of the assays and equipment. Moreover, the infrared spectrum can be obtained from any kind of sample in terms of its physicochemical state: gas, liquid or solid [19].

The Fourier Transform Infrared Spectroscopy (FTIR) instrument consists in a spectrophotometer, which measures the amount of infrared radiation that passes through the sample and reaches the detector. The low energy of the irradiation, in comparison with other radiations such as X-rays or UV, is able to be absorbed by the molecular bondings of the substances. Nevertheless, this absorption process has one selection rule: the molecular bonding must have a dipolar moment different from zero ($\mu \neq 0$). The dipole moment of such a molecule changes as the bond expands and contracts during vibration. On the other hand, a homonuclear diatomic molecule (for example O_2, N_2, *etc.*) is not capable to absorb the infrared radiation because its dipole moment remains zero no matter how the vibration affects the bonding. The interactions of infrared radiation with matter may be understood in terms of changes in molecular dipoles associated with vibrations and rotations. An understanding of molecular symmetry and group theory is important when initially assigning infrared bands. By the selection rules that are associated with symmetries, we can explain whether the transition is forbidden or not and we can predict and interpret the bands present in the IR spectrum. The detailed explanation of this theory exceeds the scope of this work and it can be found in a more specific literature [20, 21].

The high selectivity of IR arises in the complex pattern of the spectra plotted as absorbance (or percentage of transmittance) *versus* frequency (cm^{-1}). The peaks and valleys and the relative intensity of transmittance among these peaks are the features used to identify the presence of an organic molecule. In addition, the wavenumber range between 500 and 1500 cm^{-1} is known as the fingerprint region where the pattern is almost unique for any given molecule [19].

For the obtaining of an IR spectra, the solid samples usually require a pretreatment where they have to be grounded and diluted with an infrared-nonabsorbent powder (such as highly pure NaCl or KBr). Then the mixture is compacted into the shape of a pill and put into the spectrophotometer. Attenuated Total Reflectance (ATR) is a specific sampling method used for the obtaining of IR spectra without any special preparation even for liquid samples. The material is settled over a crystal, which has a great infrared reflectance (for example, Ge, diamond or ZnSe); then the beam light passes through the crystal and penetrates a few microns of the sample surface several times depending on the incidence angle. The resultant radiation is detected and analyzed as a typical FTIR spectrum. The advantages of ATR over FTIR rely on the simplicity of the sampling, the possibility of determining an IR spectrum of a sample with high infrared absorptivity and it is ideal for the study of coated materials. Nevertheless, since the penetration of the beam in the sample depends on the wavelength and the refraction index, many of the samples present low sensitivity compared with the ones prepared by the KBr method. Also, the IR range of sampling is limited by the crystal used. In example, Ge and ZnSe crystals go up to 800 cm^{-1} and 650 cm^{-1} respectively, whereas the KBr pellet goes up to 400 cm^{-1}.

Regarding to nanomaterials with a polysaccharide matrix, most of the related articles include an IR experiment looking for changes in the polysaccharide original spectrum due to the addition of nanoparticles, cross-linking reactions, adsorption processes or chemical treatments, among others [22 - 25]. Since the origin of each signal arises from different vibrations of the functional groups (*i.e.* -OH, $-CH_2-$, -CHO, -COOH, $-NH_2$, *etc.*), the localization and the description of the changes that those signals may suffer would describe the chemical and interaction processes in which they are involved. Those changes can be expressed as shifts, decrease or increase in the relative intensity and appearance or disappearance of one or more peaks. Several phenomena may be attributed to changes in the IR spectra; they depend on the kind of physical or chemical events that the sample was exposed to. For example, Zhu *et al.* synthesized a novel magnetic Fe_2O_3/chitosan adsorbent composite where the polysaccharide matrix was cross-linked by chemical reaction with glutaraldehyde [26]. The FTIR results showed the disappearance of the peak at 1597 cm^{-1} and the formation of a broad doublet peak at 1655 - 1560 cm^{-1} related to the consumption of $-NH_2$ chitosan groups to form Schiff base. Moreover, the doublet at 1655 - 1560 cm^{-1}, attributable to the -OH stretching from chitosan, suffered a broadening and a shift toward lower wavenumber values. From this appreciation, the authors suggested a non-covalent interaction between the Fe_2O_3 nanoparticles and the carbohydrate hydroxyl groups. This work exhibits how FTIR means a powerful tool for the characterization of composites in terms of mechanism of interaction. Nevertheless, special care has to be taken in processing and interpreting the spectra since in some cases peak shifts and relative intensity variation could arise form baseline deviation and sample preparation.

In another work performed by Asadinezhad *et al.,* medical-grade polyvinyl chloride (PVC) samples were covalently coated, previous activation with acrylic acid, with chitosan and chitosan/pectin [27]. All the materials were characterized and proved for prevention of bacterial adhesion. Among the techniques applied, ATR-FTIR was used to characterize the surface chemistry of the pristine PVC, the grafted PVC with acrylic acid and the final coated PVC samples. This protocol allowed the authors to control each step of the synthesis and the changes on the material surface. In the first steps, the co-polymerization of acrylic acid onto the PVC surface gave FTIR signals within the 1100 - 1300 cm^{-1} range related to C–O stretching vibrations in carboxylic compounds. The following spectra contained characteristic nitrogen groups bands from chitosan amine groups confirming the presence of the coating chains.

Raman Spectroscopy

The Raman Spectroscopy is a high-resolution photonic technique that provides valuable chemical and structural information from any given organic and inorganic compound. The Raman analysis is based on the scattered light by the sample after its irradiation with a monochromatic laser (for example, near UV or near IR can be used as light sources). The most common light sources are lasers produced by argon ion or krypton ion [28, 29]. Once the light beam impinges on a molecule different electronic transitions and scattering processes it might occur:

- The photons can be absorbed by the molecule and it increases its kinetic energy;
- The sample, originally in the fundamental rotational, vibrational and electronic state, is excited by the incident light beam and reaches a virtual excited state. Then it relaxes and emits a scattered photon. Those photons may maintain their original frequency (energy). This elastic process is known as Rayleigh scattering;
- The scattered photons might have a different energy in comparison with the incident photons. These inelastic processes are named as Raman Stokes Scattering when the frequency of the resultant photons is lower than initial, and Raman Anti-Stokes Scattering if the scattered photons have higher frequency;
- Some materials can fluoresce.

Since all those scattering processes take place at the same time, optic filters block the main band corresponding to the Rayleigh scattering. When a sample is able to fluoresce, the laser beam energy can be attenuated to diminish fluorescence, but in most cases the use of higher wavelength lasers need to be used to avoid fluorescence. Consequently, the Raman spectrum consists in the plot of scattering bands with different intensities *vs* the emitted radiation frequency expressed as wavenumber (cm^{-1}) (or Raman shift) [19]. Due to the Maxwell-Boltzmann ground energy distribution, 99% of the molecules are located in the lowest energy level at room temperature. In addition, the Anti-Stokes phenomenon occurs when the molecule is in an energetic basal level higher than the fundamental one (1% of the molecules). Therefore, the probability of the Stokes effect is 100 times higher than Anti-Stokes. That condition justifies the use of the spectrum region that only show the Raman Stokes scattering. In some exceptional cases, the sample is able to fluoresce and that emission can mask the Stokes signals making almost impossible to get the Raman spectrum. In such case, the use of the Anti-Stokes part of the spectrum is justified on spite of the lower sensitivity.

This technique can be considered as complementary to IR because the necessary conditions for a compound to present Raman scattering are completely different to the IR selection rules. Unlike IR, the interaction between molecules and light in Raman spectroscopy is due to the dipole rotational and vibrational energy levels of the ground electronic state while in IR the absorptions only happens when there is a dipole moment different from zero between two atoms. These aspects explain why IR is very powerful characterizing the functional groups present in the external part of the molecules. Meanwhile, Raman exhibits better the crystalline orientation of the molecular backbone where the atom composition is quite homogeneous.

Although Raman Spectroscopy is not as widely implemented in the study of polysaccharide-based nanomaterials as IR is, the potential applications of this technique make it a promising tool for the structural characterization of novel composites. Most of the scientific works apply Raman for the identification of the components since their vibrational signals are specific to all the chemical bondings and the symmetry of each molecule. Therefore, the obtained spectra, especially between 500 and 2000 cm^{-1}, represent the fingerprints useful for the identification of the materials components. Another application include the measurement of nanostructures such as nanowires, nanowhiskers, nanoplatelets, *etc*. In the case of composites with polysaccharide chains with different polymorphic orientations, Raman Spectroscopy may be useful to determine which of those structures is the predominant because each conformation has its own rotational and vibrational behavior.

Yuen *et al.* used this technique for the study of four types of non-starch polysaccharides including cellulose, guar gum, locust bean gum and xanthan gum and their modifications after carboxymethylation [30]. Raman and FTIR spectra of the native and modified polysaccharides were acquired and analyzed. Using the area of some specific bands of the FTIR and Raman spectra the authors could correlate the degree of carboxymethylation of each polysaccharide with those spectroscopic areas. When compared to traditional wct chemistry methods (colorimetric), the spectroscopic methods developed in this study are relatively simple, fast, without the use of toxic chemicals, requiring smaller sample size, and are ubiquitous for different types of modifications. Furthermore, Raman spectroscopy has the added advantage of being non-destructive, if the beam energy is low, and not requiring sample preparation.

The use of Raman spectroscopy has been lately boosted by the rising of novel graphene derivatives nanomaterials (such as graphene and graphene oxide platelets, carbon nanotubes, fullerenes, among others). The introduction of these derivatives within polysaccharide matrices has generated a wide offer of nanomaterials with several application fields (for example wound healing materials, biosensors, adsorbent matrices, materials for drug delivery, *etc.*). The work of Peregrino *et al.* consisted in the development of a biofilm nanocomposite composed by acetylated starch and graphene oxide nanosheets [31]. The Raman results showed a change in the graphene oxide signals; the *D* intensity signal (I_D, 1340 cm^{-1}) slightly decrease respect to the *G* signal (I_G, 1600 cm^{-1}) when the nanosheets were added to the starch matrix. The authors could not formulate an explanation about those results. However, in a subsequent study, performed by Kumar *et al.,* where a bionanocomposite composed of chitosan/graphene oxide was developed, the authors suggested that the addition of graphene oxide to polysaccharide matrices produces a change in the Raman spectrum of the graphene oxide, specially an increment of the I_D/I_G, which responds to a partial reduction of the exfoliated nanosheets across the polysaccharide matrix [32]. Moreover, that ratio increment would indicate the size reduction of the in-plane sp^2 domains together with further unorganization probably related to the reduction process [33].

Small Angle Scattering

Small-angle scattering techniques (SAS) have become important tools when it comes to the characterization of biopolymers and nanosized or nanostructured materials. SAS techniques are scattering methods where collimated radiation is scattered by structures that are larger than the incident radiation wavelength. The radiation interacts with the sample and receives thereby a momentum transfer $\hbar q$. By this process, radiation suffers an energy change and a direction change. The energy change is detected by variations in the intensity, I (q) (a.u.) at different scattering vectors, q, (Å^{-1}), which are function of the angle given by the direction change, following: q=4πsenθ/λ, where 2θ = is the scattering angle and λ is the radiation wavelength. As a result the scattering profile is expressed in $I(q)$ *vs* q plot [34].

SAS differ from other scattering techniques in the angle of detection of the scattered radiation, which is typically between 0-2° from the incident beam. SAS is used more often in biopolymers research, such as polysaccharides, than techniques like wide angle X-ray scattering (WAXS, angle ~2-90°) because scattering at WAXS working angles is given more efficiently by crystalline or semicrystalline objects formed by periodically ordered structures with dimensions of less than 1 nm. Biopolymers are typically amorphous and usually the information that can be obtained from a WAXS experiment is limited to composites that contain inorganic nanostructures, such as nanoclays, or crystalline polymers [35].

SAS techniques can give information about the particle size and shape, aggregation behavior, phase changes as well as information on overall organization. They could also bring spatial orientation and time resolved information. Small angle X-ray scattering (SAXS) and small angle neutron scattering (SANS) are two SAS techniques, each with their own advantages and disadvantages. SAXS has the advantage of enabling time resolved measurements since scattering is determined by the interactions between the incident X-ray beam and the electrons of a material; on the other hand, neutron scattering relies on interactions with the nucleus and as such requires long data collection times [36].

SAXS analysis can bring a variety of features with respect to the structural information of a material that modern instruments and Synchrotron facilities have extended to a range between 1 to 1000 nm (such as its objects size, the presence of periodic structures, volume fraction, and inhomogeneity). The sample can be tested without any treatment and still give information from structural features of dense phases. In the case of polymer nanocomposites, one of the biggest advantages of a typical SAXS instrument or synchrotron radiation is that they probe materials with nanometer resolution (1-100 nm) and at the same time, it uses large sample size, covering large number of the nanostructures. Also, it is sensitive to the electronic density contrast between the nanodomains and the matrix, in which such domains are dispersed [37, 38]. Therefore, SAXS information is given by the bulk material where large number of objects and organized structures are measured. In contrast to microscopy techniques, where a limited number of fields can be analyzed (each one at a time). Thus, biased data collection is minimized in SAXS providing a more accurate description of the sample.

Samples in SAXS can be dried powders, liquid suspensions, solids in the form of bulk materials or films, and even gums or highly hydrated polymeric gels. Any sample that can be placed in a sample holder in a way that it can present sufficient scattering without blocking the beam can be analyzed by SAXS. A special care has to be taken when it comes to organic polymeric samples, such as polysaccharides, proteins and DNA, and this is the avoidance of radiation damage. This appears when the sample is subject to high energy of radiation for a given time period that will depend on the sample. Radiation damage is frequently detected by collecting several spectra of the same sample that show enough difference, between the first and the last, that one can infer that radiation have altered the samples affecting its chemical structure. Therefore, the maximum number of exposures that a sample should be subject to is the maximum number in which radiation damage is not detected.

The analysis of SAXS profiles can be done from several approaches, either by a direct way using Scherrer's equation for determining some object size or periodicity distance (in example lamellar distance), by observing Guinier or Kratky plots, Porod's law or by the numerous mathematical models that are used to fit data in order to interpret the patterns from form or structure factors (in example Solid spheres, rods, core shell particles, Beaucage unified fit, *etc.*) [34, 39, 40]. The analysis of all these features can be done from a single profile since usually smaller structures, such as nanoparticles, influence the scattering profiles in a higher q than larger structures, such as mass fractals, and vice-versa. Therefore, if the assay is done within a wide q range, the scattering driven by small structures will not interfere with the scattering of larger structures, or at least it will be subtle.

In order to show the potential of SAXS techniques in the bionanocomposites analysis herein we present some applications present in literature. Bell *et al.* have used SAXS for a kinetic study of the nucleation and growth of the Ag nanoparticles in a maltodextrin matrix [41]. This analysis was done by fitting SAXS data to a "scattering polydisperse spheres" model. The analysis showed that the growth mechanism was consistent with an initial nucleation step, and a slow growth of the initial nuclei from 0.35 to 0.9 nm diameter at 135 min. After 135 min, rapid growth ensues resulting in a monodisperse size of near 10 nm.

In a different approach, Chen *et al.* study the internal structure variations with respect to media pH of heparin loaded chitosan nanoparticles by SAXS [42]. The scattering curves were characterized by a monotonically decayed pattern without discernible peaks, and the intensity at low q ($q < 0.37$ nm) was found to decrease with increasing pH. In a two-phase system scatter, intensity is proportional to the square of the electron density contrast of the systems. Therefore, the negative slope of the scattering profile at low q ranges was attributed to the decrease of the electron density contrast between the heparin/chitosan domains and the chitosan matrix with increasing pH. This was consistent with TEM images where the dissagregation of the nanoparticles could be observed and with the increasing release of heparin from test nanoparticles at higher pH.

González *et al.* used the SAXS technique to obtain structural information in the nanometric scale from chitin-graphene oxide (Chi:nGO) and chitin-reduced graphene oxide (Chi:rGO) gels [33]. In this work, the SAXS profile allowed the authors to identify two structural levels in the chitin gel network, as can be seen in Fig. (**1**). By analyzing the power-law behavior in the low q range ($q < 0.05$ Å^{-1})

and the short knee-like middle regime with lower slope (α), fitted by the Beaucage unified equation they compared the changes in the two structural levels of the chitin network. From the low q range, they find that the addition of nGO or rGO did not alter the 2.5 power-law slope, indicative of fractal objects. Nevertheless, this slope was different from the chitin powder slope that was the typical for surface fractals (3.5) (inset Fig. **1a**). On the other hand, the modeling by the Beaucage equation showed that the R_g of the short knee-like middle regime did not vary upon the addition of the nanosheets. Nevertheless, the intensity of this regime showed to decrease with the increase of nGO and rGO content in the gel. This was explained proposing a filler behavior by the nanosheets that do not induce new conformation of chitin chains but interferes with the organization of this structural level of the polysaccharide.

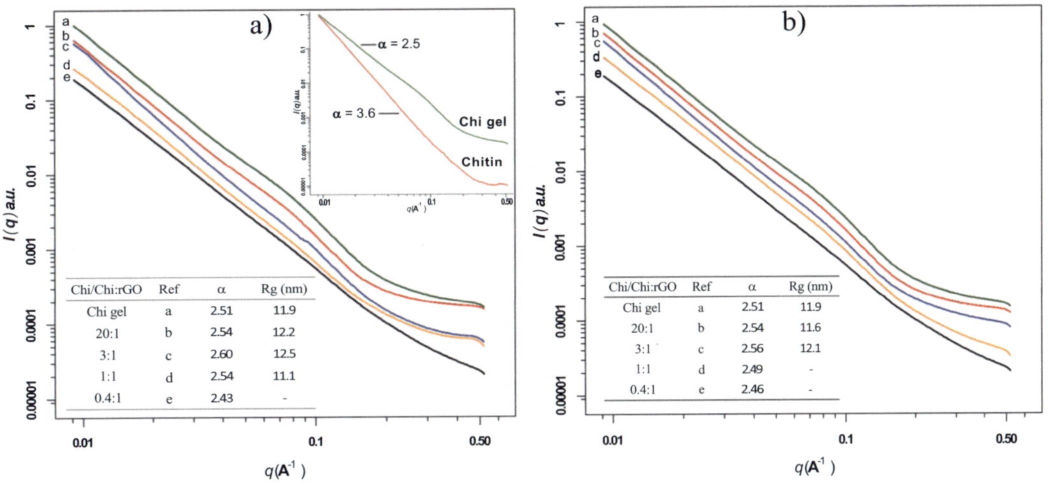

Fig. (1). SAXS profiles of Chi:nGO (**a**) and Chi:rGO (**b**) hybrids. The insets in both figures show the R_g and α of each hybrid gel. Reproduced from Ref [33]. with permission from The Royal Society of Chemistry.

SAXS technique could also bring information about structures orientation along an axis. This information is achieved by observing the 2D scattering patterns obtained ideally by a round detector. The SAXS 2D scattering pattern for a non-oriented structure is the image of concentric isotropic circles in relation to the scattering vector. If structural orientation is present, the 2D image becomes elliptic leading to an anisotropic pattern. Nakagawa *et al.* have analyzed SAXS 2D patterns during the production of casein-pectin nanoparticles by freezing [43]. The major interest in this study was to induce protein self-aggregation or complexation in the cryoconcentrated phase during freezing. The 2D profiles suggested that the freezing solutions contained anisotropic structures along a horizontal plane, which is suggestive of the presence of ice microstructures. The authors concluded that the freezing process not only leads to casein-pectin nanoparticles but also is driving the orientation of the nanoparticle internal structure.

SAXS is usually compared to SANS mostly in its experimental and instrumental advantages and disadvantages than in the information that both could bring because it is essentially very similar. The first difference arises due to the high flux of the X-ray beam from the synchrotron and the modern 2D detectors that allow the capture of a wide range of q values, data acquisition of a SAXS profile is usually done within seconds. This allows kinetic structural studies, such as biomaterial assembly, material response to media stimuli (pH, temperature, *etc.*). SANS has two main limitations regarding this task. First, high-flux neutrons are required, and second, the covering of a wide range of q values can only be achieved by sequentially changing the detector-to-sample distance, which is a time consuming procedure. Thus, in general, kinetic studies are much less feasible with SANS, especially if a nuclear reactor is used as a source of cold neutrons and the sample contains low scatters, a common feature of most biomaterials, such as hydrogels, which usually contain less than 1% w/w of material in water [35, 39].

SAXS also advantages SANS in the need of smaller sample sizes and easier or null sample preparation, for example since due to high neutron capture cross-section of H atom, SANS samples must be prepared in deuterated water. Nevertheless, in some cases this could be an advantage of SANS since in SAXS the X-ray beam interacts with electrons while in the case of SANS, the neutron beam interacts with the nuclei within the bulk of the sample. Therefore, a material with a high electron density such as an inorganic nanoparticle scatters X-rays strongly while biopolymers are low scatters. In the case of SANS, the amount of scattering can simply be enhanced by using contrast matching agents, in the case of aqueous solutions, by using mixtures of H_2O and D_2O [44].

Additional advantage of SAXS is its higher monochromaticity (both from a synchrotron and X-ray tube), while SANS data usually have certain contribution from neutron wavelength smearing. However, SANS is much less harmful for most biomaterials, while doing SAXS, the user should always be aware of possible radiation damage [35].

Differential Scanning Calorimetry

The Differential Scanning Calorimetry (DSC) represents one of the most important techniques used toward the thermal transition's determination of organic and inorganic materials. Thus, when a bionanocomposite is being developed, DSC could be of great help in determining how the proportion of the components in the material is affecting its overall thermal behavior. This changes could be desirable or not. In addition, when the nanostuctured moiety is in low proportion important changes in the thermal behavior of a biopolymer can be observed. Therefore, DSC is a key technique when it comes to evaluate the compromise situation between choosing components proportion and loss or gain of a particular thermal behavior.

The instrument consists in a calorimeter, which measures the differential heat flow involved in the heating (or cooling) of a sample and an inert reference (usually an empty sample holder) at a constant temperature rate. The results are commonly presented as heat flow normalized by the sample mass (W/g) against temperature (°C). The information obtained from the DSC curve refers to the enthalpy (ΔH) related to internal processes that the sample suffers as the temperature increases or decreases. Those internal processes, *i.e.* physicochemical processes, can be classified as glass transitions, crystallization and melting processes [45]. The temperature where the glass transition occurs is known as T_g and is the temperature where an amorphous solid, such as a polysaccharide, increases its freedom of motion and decreases its viscosity. The T_g is usually detected as a rise in the baseline of the thermogram curve due to a change in the sample heat capacity without a defined phase transition. It is important to notice that the glass transition is reversible and the substance can completely recover its initial molecular conformation if it is cooled. At higher temperatures, the mobility of the molecules may be enough to generate a characteristic arrangement of a crystalline form. This is appreciated as an exothermic peak in the DSC plot where the temperature at the maximum heat flow corresponds to the crystallization temperature (T_c). Finally, if the sample is heated beyond this point it may reach the melting temperature (T_m), which is visible as an endothermic peak in the curve.

In the polysaccharide based nanocomposites field, DSC means a grateful tool for the physicochemical characterization of the materials in terms of description of the components. Some of those determinations allow the study of the mechanisms of interaction among the components, impurities detection, level of reinforcement, degree of hydration, thermal degradations (under air or N_2 atmosphere depending on the kind of reaction, if it needs O_2 or not to occur), among other applications.

A good example of DSC as a tool for the characterization of a nanocomposite is the work carried out by Cao *et al.* They developed a hybrid material composed by plasticized starch mixed with increasing amounts of oxidized multiwalled carbon nanotubes (MWCNTs) [46]. The resulting materials were characterized through DSC and the authors suggested that addition of MWCNTs affects the T_g and the ΔH associated to the melting process (ΔH_m). The higher values of T_g at increasing MWCNTs proportions might be attributed to the occurrence of intermolecular interactions between starch and stiff MWCNTs, which reduced the flexibility of molecular chains of starch in contact with the MWCNTs surface. Moreover, as the additive percentage increases the ΔH_m is lower, therefore the recrystallization of the starch needs more energy due to the higher hydrogen bonds established between MWCNTs and the starch chains.

In other study, the work done by González *et al.* shows how DSC can be used to characterize a chitin based hybrid composite in terms of the degree of hydration and decomposition depending on the amount of graphene oxide nanosheets (nGO) acting as reinforcing nanofillers [11]. The authors found that the water evaporation endothermic peak shift towards higher temperatures while nGO proportion is higher suggesting that the interaction water-solid is superior in those materials with more nGO (Fig. **2**). Finally, around 200 °C it can be appreciated the decomposition of the oxidized groups of nGO and appears a sharp exothermic event at 340 °C owing to the deacetylation of chitin [47]. This last reaction occurred only in the chitin commercial powder sample but not in the chitin and hybrid gels. The authors suggested that this decomposition probably did not occur as a sharp and defined event due to the structural arrangement of the chitin chains interacting with the nanofiller in the nanocomposite. Thus, the chitin chains in this material would acquire a different structure in comparison with the commercial chitin. Another effect observed by the authors is that this new molecular environment favors a broadening of the deacetylation enthalpy, which was proposed to be related to the presence of oxidized groups in the graphene oxide nanosheets. When the experiment was carried out in a protective N_2 atmosphere, the exothermic deacetylation peak disappeared allowing the T_m to be determined (Fig. **2b**). Again, a T_m change for the bionanocomposite was evidenced with respect to the chitin powder. This event was explained by the nGO dependent chitin structural arrangement.

Thermogravimetric Analysis

Thermogravimetric Analysis (TGA) is included in the group of Thermal Analysis techniques as well as DSC. In TGA, the changes in the weight of a sample are monitored while it is subjected with a specific temperature-heating program (for example, it can be linear, gradually isothermic, completely isothermic, *etc.*). The experiments are carried out with the control of the sample atmosphere. The physical and/or chemical processes that the sample could suffer depend on the type of gases surrounding it. Since the most common reaction that may happen in TGA is the oxidation, the use of a clean oxygen atmosphere (or N_2, He, Ar, *etc.*) is a good way to slow down the sample oxidation.

TGA measurements provide valuable information that can be used to select materials end-use applications, predict product performance and improve product quality. The technique is particularly useful for the following types of measurements: thermal stabilities, moisture and volatiles content, oxidative stabilities, decomposition kinetics, effects of reactive atmospheres on materials and filler contents. Most of the assays mentioned above are plotted as percentage of weight *versus* temperature (°C) [48].

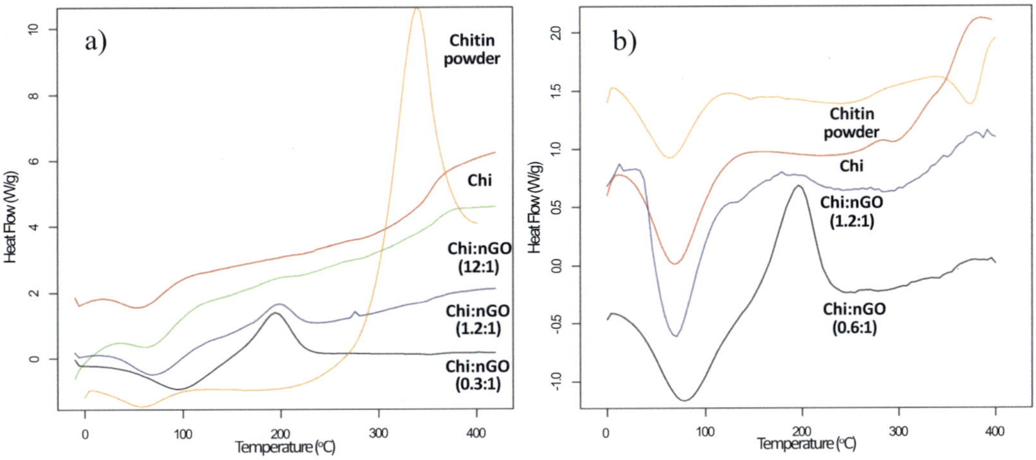

Fig. (2). DSC thermograms of Chi gels and Chi:nGO hybrids under air **a**) and N_2**b**) atmospheres. Values between brackets correspond to chitin to nGO mass relations. Adapted from Ref [11]. with permission from The Royal Society of Chemistry.

In their work, Islam *et al.* used TGA to prove if the addition of montmorillonite clay nanoparticles within an electrospun pollulan/poly vinyl alcohol (PULL/PVA) matrix can improve the thermal stability of the nanocomposite [49]. Effectively, the resulting curves showed an increment in the decomposition temperature when the percentage of nanoclay was higher. The authors suggested that this result might be attributed to the higher chain compactness of the polymer blend due to the interaction between the PULL/PVA blend and clay.

One important advantage of the TGA instrument is the possibility of coupling it with a Fourier Transformation Infrared Spectrometer (FT-IR) or Mass Spectrometer (MS) gas detector. Therefore, the gases produced by the oxidative decomposition of the sample can be qualitative and quantitatively detected at a certain time/temperature. This feature of TGA lets us know which compounds are detached from the sample during its heating process and how much of those compounds were originally present in the sample. Moreover, the temperature of detachment can be related with mechanism of interaction among the components. In example, if a polysaccharide based nanocomposite releases CO_2 molecules that originally belong to the polysaccharide matrix as carboxyl groups, changes in the oxidative decomposition temperature may be attributed to interactions between the carboxylates groups and the nanoobject. In the TGA-MS experiment it will be resulted in a peak of the mass spectrum at m/z = 44 at the same temperature of decarboxylation. In a real case, Unsoy *et al.* used a TGA instrument coupled to a FT-IR detector in order to quantify the amount of chitosan bounded to coated Magnetic Nanoparticles [50]. Briefly, the heated sample suffered a weight loss around 700 °C due to the complete oxidative degradation of the organic chitosan layer which volatilized as CO_2 and is detected and quantified at 2000-2400 cm^{-1} in the FT-IR detector.

Rheological Methods

In the field of polysaccharide-based nanocomposites the rheological behavior is often studied since the addition of nanostructures within the polymeric chains frequently induces the reinforcement of the nanocomposite in relation with the pristine polysaccharide by acting as fillers. In order to study the viscoelastic properties of the system, rheological measurements may involve the application of large forces or shear stress to a material that can cause permanent structural damage or shear thinning. The general objective of these techniques is the determination of the rheological behavior of a given fluid under a given flow condition through the experimental measurement of macroscopic quantities using specific instruments [51]. Although all rheological methods are used to predict the mechanical behavior of a material, each method brings different information.

Uniaxial Compression and Tension

The uniaxial compression and tension test provides a simple and effective way to characterize a material's response to loading. By subjecting a sample to a controlled compressive or tensile displacement along a single axis, the change in dimensions and resulting load can be measured to calculate a stress-strain profile. From the obtained curve, elastic and plastic material properties can then be determined. The main mechanical parameters that can be derived from a uniaxial compression or tension test include Hencky's strain (ε_h) and Young's modulus (E). Whereas dynamic testing can be performed on very soft gels, uniaxial compression tests usually require relatively firm gels. This technique has been applied in polysaccharide nanocomposites by several authors [52 - 54].

Tensile testing of a nanocomposite film can be done according to ASTM D882-02 or ASTM D638-03. The latter method can be used to test materials of thickness up to 14 mm. The first method is preferred to do tensile testing of materials in the form of thin sheets of less than 1 mm in thickness. During tensile testing, a rectangular specimen is placed in the grips of movable and stationary fixtures in a testing machine capable of pull the sample away from the stationary fixture at a constant velocity. The sample is pulled apart until it breaks. The applied load (force) and the resulting elongation of the specimen are thus measured [55, 56]. These techniques have been used to characterize polysaccharide nanocomposites. For example, Villanueva *et al.* characterized the tensile strength of a fabric containing chitin nanowhiskers- coated with tetraethoxysilane [57]. The maximum tensile strength and elongation at break were recorded, concluding that the mechanical properties did not change after the coating application.

Texture Profile Analysis

In an instrumental texture profile analysis (TPA), a sample of specific dimensions is compressed uniaxially; the compressive force is then removed and the sample is re-compressed. During the test, compressive force is recorded as a function of the amount of compression (distance). Thus, two force *vs* distance plots or TPA curves would be obtained [58]. Several instrumental texture profile parameters may be derived from the TPA curves (Fig. **3**): the maximum force (*H*), which occurs at the end of the first compression, equates to hardness; the force of the first maximum (*F*) is called fracturability; the work done to compress the sample on the first and second cycle is given as the area under the respective curves (A_1 and A_2), and the ratio A_2/A_1 is related to cohesiveness (*C*); the distance *S* is called springiness; and the negative area 3 is the adhesiveness or stickiness. Textural characteristics such as gumminess (hardness x cohesiveness) and chewiness (hardness x cohesiveness x springiness) are derived functions [59].

Pasting Behavior

In order to evaluate the pasting behavior of a material, changes in viscosity during programmed heating and cooling can be studied using a variety of instruments such as the Brabender amylograph, which is a rotational viscometer. It has a variable-speed drive system ranging from approximately 20 to 300 rpm at the rotating cup. Temperature can be controlled by a head transfer probe, connected to an external water-circulating thermostat. Several standard and special rotors of different geometries are available; once the rotor is dipped into the fluid and rotated at a given speed, the corresponding torque is recorded by means of interchangeable torsion-spring cartridges inserted in the measuring head. Each torsion spring is characterized by a spring constant, whose value can be found on a table provided with the device. These values represent the spring full deflection, and the total deflection scale is currently divided into 1000 units, known as the Brabender units. The experimental deflection of the spring over a range of shear rate is recorded on a moving strip chart, and the results obtained can be expressed directly in Brabender units of stress at the given rpm value [51]. Considerable quantities of samples are required.

These techniques have been used to characterize different nanocomposites, mainly those based on starch. For example, Yu *et al.* prepared a nanocomposite using ZnO nanoparticles stabilized by carboxymethylcellulose sodium (CMC) as the filler in a glycerol plasticized-pea starch matrix [60]. Pasting curves of pea starch suspension and ZnO nanoparticles-CMC-glycerol plasticized starch were measured during a heating-cooling cycle to evaluate the effect of ZnO–CMC nanoparticles. The presence of ZnO-CMC nanoparticles improved the pasting viscosity. In other work, Chang *et al.* studied the pasting behavior of a starch-chitin nanoparticles (CNP) nanocomposite using a rapid viscosity analyzer [61]. The pasting profiles of the starch slurries with different CNP concentrations were similar. When granular starch began to plasticize and dissolve in water, the starch slurry turned into a starch solution, and the viscosity increased. When chitin nanoporticles were added, the peak viscosity of the starch slurry decreased and shifted to a lower temperature. This meant that the addition of chitin nanoparticles is propitious to the plasticization of starch and to a decrease in the interaction between starch molecules. The pasting viscosity increased at higher amounts of chitin nanoparticles.

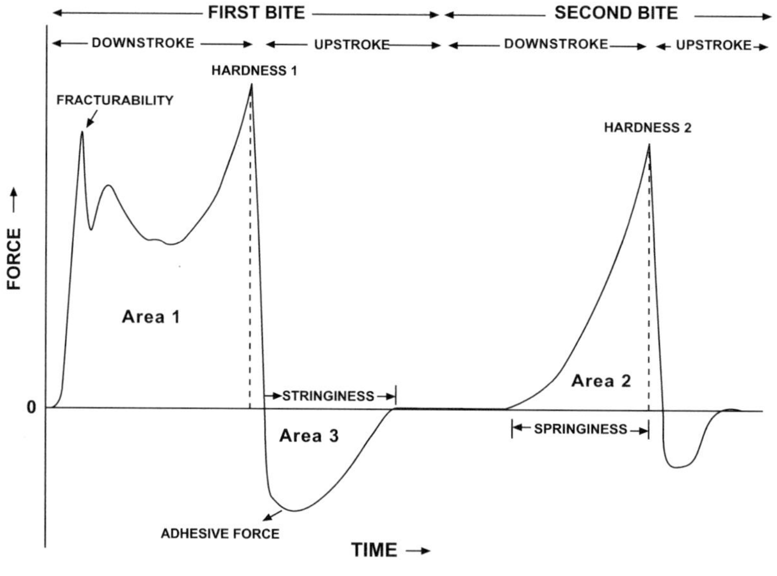

Fig. (3). A generalized TPA curve (From [59] Copyright© 2000 by John Wiley Sons, Inc. Reprinted by permission of John Wiley & Sons, Inc.).

Penetrometer

Another technique used to characterize mechanical properties is the penetrometer, which is a device to test the firmness of the material [62]. A needle, cone or sphere penetrates the sample with a given force for a predetermined time, and the depth of penetration is measured. The main difficulties with this technique are that the area of contact between penetrometer and sample does not remain constant during the test and that the sample is displaced in a direction opposite to that in which the penetrometer moves [63].

Dynamic Oscillatory Rheometry

The viscoelastic properties of materials are determined by transient methods. The transient methods include stress relaxation (application of constant and instantaneous strain and measuring decaying stress with respect to time), such as in dynamic oscillatory rheometry, and creep (application of constant and instantaneous stress and measuring increasing strain with time), such as in the static methods which are described below [64].

In dynamic oscillatory rheometry the dynamic test is performed applying a small sinusoidal strain (or stress) and measuring the resulting stress (or strain). The sample is continuously stressed, with a strain that is never large enough to destroy the structure.

For a dynamic sinusoidal experiment, two moduli can be defined: (i) stress in-phase/strain or storage modulus (G') and (ii) stress out-of-phase/strain or loss modulus (G''). Storage modulus (G') is a measure of the energy stored in the material and recovered from it per cycle. On a molecular basis, the magnitude of G' is dependent upon what rearrangements can take place within the period of oscillation, and is taken as an indication of the solid or elastic character of the material. Another parameter which is often useful in indicating the physical behavior of a system is the loss tangent ($\tan \delta$) (damping phase). It is the ratio of the energy lost to the energy stored for each cycle of the deformation, since $\tan \delta = G''/G'$. It is a useful indicator of the relative contributions of the viscous (G'') and elastic (G') components to the viscoelastic properties of a material [14]. Therefore, if $\tan \delta$ is high, the viscous component of the material will be predominant, whereas if it is lower the material will have a more solid-elastic behavior. The viscoelastic behavior of a material can also be characterized by dynamic mechanical thermal analysis. This is done by measuring the response of a material to an oscillating force as a function of temperature [65].

Mabrouk *et al.* prepared a nanocomposite based on poly(styrene-co-hexylacrylate) copolymer and cellulose whiskers as the nanosize filler by *in situ* miniemulsion polymerization and their melt rheological behaviors were investigated under dynamic shear conditions [66]. The effects of γ-methacryloxypropyl triethoxysilane (MPS) content along with the whisker loading were explored. In the absence of whiskers, a transition from a liquid- to a solid-like behavior was observed when the polymer was synthesized in the presence of MPS. When cellulose nanofiller was added, the storage modulus, G', and the dynamic viscosities, η, of the nanocomposites increased monotonically with whisker content and the resulting materials displayed a solid-like behavior. Above 2 wt % loading, a percolated interconnected whisker–whisker network is built up, producing a jump in the storage modulus and strong shear-thinning behavior of the viscosity. However, as nanocomposites were prepared in the presence of 3% of MPS, no enhancement nor in the storage modulus nor in the viscosity was observed up to 5 wt % of whisker loading. Such a phenomenon was ascribed to inhibition of build-up of the whisker network. The non-linear viscoelastic behavior of the nanocomposites was also investigated and analyzed in terms of the breakdown of different networks, namely the filler-filler and the polymer-filler networks.

Static Experiments

Static experiments can be used to obtain information about the mechanical characteristics. One of the static experiments is a creep test, which is used to investigate the viscoelastic structure of materials over medium and long time scales. In a creep experiment, a constant stress is applied to the sample and the resultant displacement or deformation is measured against time. If required, the stress can be removed, and the relaxation of the sample measured. The results are generally expressed in terms of the creep compliance against time [14]. Creep and creep recovery data provide considerable information about material properties, especially if the deformations during the test remain sufficiently small so that the behavior is linear respecting the material's viscoelasticity. Such data normally show clearly, whether the material is non-crosslinked and thus whether viscous flow occurs under stress [67].

The second type of static experiment to measure viscoelasticity is stress relaxation, which is usually conducted at relatively low deformation. Here, the polymer is deformed (usually by compression) to a given value of the strain and the stress necessary to maintain that strain is measured as a function of time. As the sample relaxes (*i.e.* as the chains change their conformations, disentangle or slide over one another, and so on), the stress decreases. Stress-relaxation experiments are somewhat easier to perform than creep experiments [14].

CONCLUDING REMARKS

The characterization techniques result in vitals tools to understand the physicochemical characteristics of a bionanocomposite, which would not only help to predict its behavior but are incredibly relevant in the design of a tailored material. The characterization techniques above presented give a glance from a very broad spectrum of properties. From electron microscopies, morphology can be analyzed. Oughting to the dissimilar fundamentals of the spectroscopic techniques, a wide range of information can be obtained, from functional groups and chemical structure information from FTIR, Raman spectroscopy and NMR techniques, to information about the particle size and shape, aggregation behavior, *etc.*, from SAS techniques. Thermal techniques, such as DSC and TGA, give an insight about thermal stability, phase transition and reaction temperatures. Finally, the rheological methods provide information related to the solid-liquid behavior of materials and, therefore, bring information about their mechanical features.

CONFLICT OF INTEREST

The author (editor) declares no conflict of interest, financial or otherwise.

ACKNOWLEDGEMENTS

J.A.G. is grateful for his doctoral fellowship granted by Universidad de Buenos Aires. M.E.V. is grateful for her doctoral fellowship granted by Consejo Nacional de Investigaciones Científicas y Técnicas. The authors are grateful to the Brazilian Synchrotron Light Laboratory (LNLS) for SAXS facilities. This work was supported with grants from Universidad de Buenos Aires (UBACYT 20020130100780BA) and Consejo Nacional de Investigaciones Científicas y Técnicas (PIP 11220120100657CO).

REFERENCES

[1] J. Brus, D. Hlavatá, and A. Strachota, "Self-Organization, Structure, Dynamic Properties, and Surface Morphology of Silica/Epoxy Films As Seen by Solid-State NMR, SAXS, and AFM", *Macromolecules,* vol. 37, no. 4, pp. 1346-1357, 2004. [http://dx.doi.org/10.1021/ma035608h]

[2] R. Adhikari, and G.H. Michler, "Polymer Nanocomposites Characterization by Microscopy", *Polym. Rev. (Phila. Pa.),* vol. 49, no. 3, pp. 141-180, 2009. [http://dx.doi.org/10.1080/15583720903048094]

[3] J.P. Patterson, M.P. Robin, C. Chassenieux, O. Colombani, and R.K. OReilly, "The analysis of solution self-assembled polymeric nanomaterials", *Chem. Soc. Rev.,* vol. 43, no. 8, pp. 2412-2425, 2014. [http://dx.doi.org/10.1039/c3cs60454c] [PMID: 24519401]

[4] T.J. Bootten, P.J. Harris, L.D. Melton, and R.H. Newman, "Solid-state 13C-NMR spectroscopy shows that the xyloglucans in the primary cell walls of mung bean (Vigna radiata L.) occur in different domains: a new model for xyloglucan-cellulose interactions in the cell wall", *J. Exp. Bot.,* vol. 55, no. 397, pp. 571-583, 2004. [http://dx.doi.org/10.1093/jxb/erh065] [PMID: 14966211]

[5] A. Domján, J. Bajdik, and K. Pintye-Hódi, "Understanding of the Plasticizing Effects of Glycerol and PEG 400 on Chitosan Films Using Solid-State NMR Spectroscopy", *Macromolecules,* vol. 42, no. 13, pp. 4667-4673, 2009. [http://dx.doi.org/10.1021/ma8021234]

[6] K.A. Verwaest, T.N. Vu, K. Laukens, L.E. Clemens, H.P. Nguyen, B. Van Gasse, J.C. Martins, A. Van Der Linden, and R. Dommisse, "(1)H NMR based metabolomics of CSF and blood serum: a metabolic profile for a transgenic rat model of Huntington disease", *Biochim. Biophys. Acta,* vol. 1812,

no. 11, pp. 1371-1379, 2011.
[http://dx.doi.org/10.1016/j.bbadis.2011.08.001] [PMID: 21867751]

[7] J. Cavanagh, W.J. Fairbrother, A.G. Palmer III, M. Rance, and N.J. Skelton, *Protein NMR Spectroscopy: Principles and Practice.*Academic Press: San Diego, USA, 1995.

[8] R.A. de Graaf, *In Vivo NMR Spectroscopy: Principles and Techniques.*John Wiley & Sons: Witlshire, Great Britain, 2007.
[http://dx.doi.org/10.1002/9780470512968]

[9] K. Schmidt-Rohr, and H.W. Spiess, "CHAPTER ONE - Introduction", In: *Multidimensional Solid-State NMR and Polymers.,* K. Schmidt-Rohr, H.W. Spiess, Eds., Academic Press: San Diego, USA, 1994, pp. 1-12.
[http://dx.doi.org/10.1016/B978-0-08-092562-2.50006-0]

[10] H. Liu, D. Chaudhary, S. Yusa, and M.O. Tadé, "Glycerol/starch/Na+-montmorillonite nanocomposites: A XRD, FTIR, DSC and 1H NMR study", *Carbohydr. Polym.,* vol. 83, no. 4, pp. 1591-1597, 2011.
[http://dx.doi.org/10.1016/j.carbpol.2010.10.018]

[11] J.A. González, M.F. Mazzobre, M.E. Villanueva, L.E. Díaz, and G.J. Copello, "Chitin hybrid materials reinforced with graphene oxide nanosheets: chemical and mechanical characterisation", *RSC Advances,* vol. 4, no. 32, pp. 16480-16488, 2014.
[http://dx.doi.org/10.1039/c3ra47986b]

[12] F. Leroux, J. Gachon, and J-P. Besse, "Biopolymer immobilization during the crystalline growth of layered double hydroxide", *J. Solid State Chem.,* vol. 177, no. 1, pp. 245-250, 2004.
[http://dx.doi.org/10.1016/j.jssc.2003.08.013]

[13] E. Mostafa Moujahid, J-P. Besse, and F. Leroux, "Synthesis and characterization of a polystyrene sulfonate layered double hydroxide nanocomposite. In-situ polymerization vs. polymer incorporation", *J. Mater. Chem.,* vol. 12, no. 11, pp. 3324-3330, 2002.
[http://dx.doi.org/10.1039/B205837P]

[14] A.A. Karim, M. Norziah, and C. Seow, "Methods for the study of starch retrogradation", *Food Chem.,* vol. 71, no. 1, pp. 9-36, 2000.
[http://dx.doi.org/10.1016/S0308-8146(00)00130-8]

[15] L. Goetz, M. Foston, A.P. Mathew, K. Oksman, and A.J. Ragauskas, "Poly(methyl vinyl ether-c--maleic acid)-polyethylene glycol nanocomposites cross-linked in situ with cellulose nanowhiskers", *Biomacromolecules,* vol. 11, no. 10, pp. 2660-2666, 2010.
[http://dx.doi.org/10.1021/bm1006695] [PMID: 20857999]

[16] C. Terenzi, K. Prakobna, L.A. Berglund, and I. Furó, "Nanostructural effects on polymer and water dynamics in cellulose biocomposites: (2)h and (13)c NMR relaxometry", *Biomacromolecules,* vol. 16, no. 5, pp. 1506-1515, 2015.
[http://dx.doi.org/10.1021/acs.biomac.5b00330] [PMID: 25853702]

[17] K. Schmidt-Rohr, A. Rawal, and X-W. Fang, "A new NMR method for determining the particle thickness in nanocomposites, using T2,H-selective X1H recoupling", *J. Chem. Phys.,* vol. 126, no. 5, p. 054701, 2007.
[http://dx.doi.org/10.1063/1.2429069] [PMID: 17302492]

[18] Y-Y. Hu, X.P. Liu, X. Ma, A. Rawal, T. Prozorov, M. Akinc, S.K. Mallapragada, and K. Schmidt-Rohr, "Biomimetic Self-Assembling Copolymer–Hydroxyapatite Nanocomposites with the Nanocrystal Size Controlled by Citrate", *Chem. Mater.,* vol. 23, no. 9, pp. 2481-2490, 2011.
[http://dx.doi.org/10.1021/cm200355n]

[19] D.A. Skoog, F.J. Holler, and S.R. Crouch, *Principles of Instrumental Analysis.*6th ed Thomson Brooks/Cole: D.F., México, 2007.

[20] G. Raj, G. R. A. B. V. Jain, A. Bhagi, and V. Jain, *Group theory and Symmetry in Chemistry.,* Krishna

Prakashan, Ed., .

[21] P. Atkins, J. de Paula, and R. Friedman, *Physical Chemistry: Quanta, Matter, and Change.*OUP: Oxford, England, 2013.

[22] N. Lin, J. Huang, and A. Dufresne, "Polysaccharide Nanocrystals-Based Materials for Advanced Applications", *Polysacch.-Based Nanocrystals Chem. Appl,* pp. 219-254, 2014.

[23] Y. Zheng, J. Monty, and R.J. Linhardt, "Polysaccharide-based nanocomposites and their applications", *Carbohydr. Res.,* vol. 405, pp. 23-32, 2015.
[http://dx.doi.org/10.1016/j.carres.2014.07.016] [PMID: 25498200]

[24] G.J. Copello, M.E. Villanueva, J.A. González, S. López Egües, and L.E. Díaz, "TEOS as an improved alternative for chitosan beads cross-linking: A comparative adsorption study", *J. Appl. Polym. Sci.,* vol. 131, no. 21, 2014.
[http://dx.doi.org/10.1002/app.41005]

[25] J.A. González, M.E. Villanueva, L.L. Piehl, and G.J. Copello, "Development of a chitin/graphene oxide hybrid composite for the removal of pollutant dyes: Adsorption and desorption study", *Chem. Eng. J.,* vol. 280, pp. 41-48, 2015.
[http://dx.doi.org/10.1016/j.cej.2015.05.112]

[26] H-Y. Zhu, R. Jiang, L. Xiao, and W. Li, "A novel magnetically separable γ-Fe2O3/crosslinked chitosan adsorbent: preparation, characterization and adsorption application for removal of hazardous azo dye", *J. Hazard. Mater.,* vol. 179, no. 1-3, pp. 251-257, 2010.
[http://dx.doi.org/10.1016/j.jhazmat.2010.02.087] [PMID: 20334972]

[27] A. Asadinezhad, I. Novák, M. Lehocký, F. Bílek, A. Vesel, I. Junkar, P. Sáha, and A. Popelka, "Polysaccharides coatings on medical-grade PVC: a probe into surface characteristics and the extent of bacterial adhesion", *Molecules,* vol. 15, no. 2, pp. 1007-1027, 2010.
[http://dx.doi.org/10.3390/molecules15021007] [PMID: 20335959]

[28] H.J. Bowley, D.L. Gerrard, J.D. Louden, G. Turrell, D.J. Gardiner, and P.R. Graves, *Practical raman spectroscopy.*Springer Science & Business Media: Berlin, Germany, 1989.

[29] A. Kudelski, "Analytical applications of Raman spectroscopy", *Talanta,* vol. 76, no. 1, pp. 1-8, 2008.
[http://dx.doi.org/10.1016/j.talanta.2008.02.042] [PMID: 18585231]

[30] S-N. Yuen, S-M. Choi, D.L. Phillips, and C-Y. Ma, "Raman and FTIR spectroscopic study of carboxymethylated non-starch polysaccharides", *Food Chem.,* vol. 114, no. 3, pp. 1091-1098, 2009.
[http://dx.doi.org/10.1016/j.foodchem.2008.10.053]

[31] P.P. Peregrino, M.J. Sales, M.F. da Silva, M.A. Soler, L.F. da Silva, S.G. Moreira, and L.G. Paterno, "Thermal and electrical properties of starch-graphene oxide nanocomposites improved by photochemical treatment", *Carbohydr. Polym.,* vol. 106, no. 1, pp. 305-311, 2014.
[http://dx.doi.org/10.1016/j.carbpol.2014.02.008] [PMID: 24721083]

[32] S. Kumar, and J. Koh, "Physiochemical and optical properties of chitosan based graphene oxide bionanocomposite", *Int. J. Biol. Macromol.,* vol. 70, pp. 559-564, 2014.
[http://dx.doi.org/10.1016/j.ijbiomac.2014.07.019] [PMID: 25077836]

[33] J.A. González, M.E. Villanueva, M.L. Peralta Ramos, C.J. Perez, L.L. Piehl, and G.J. Copello, "Chitin based hybrid composites reinforced with graphene derivatives: a nanoscale study", *RSC Advances,* vol. 5, no. 78, pp. 63813-63820, 2015.
[http://dx.doi.org/10.1039/C5RA13563J]

[34] O. Glatter, O. Kratky, Ed., *Small-Angle X-ray Scattering.*Academic Press: New York, USA, 1982.

[35] L. L. Hyland, M. B. Taraban, and Y. B. Yu, "Using Small-Angle Scattering Techniques to Understand Mechanical Properties of Biopolymer-Based Biomaterials", *Soft Matter,* vol. 9, no. 43, p. 10.1039/C3SM51209F, 2013.
[http://dx.doi.org/10.1039/c3sm51209f]

[36] Dmitri I Svergun, and Michel H J Koch, "Small-angle scattering studies of biological macromolecules in solution", *Rep. Prog. Phys.*, vol. 66, no. 10, p. 1735, 2003. [http://dx.doi.org/10.1088/0034-4885/66/10/R05]

[37] N.F. Magalhães, K. Dahmouche, G.K. Lopes, and C.T. Andrade, "Using an organically-modified montmorillonite to compatibilize a biodegradable blend", *Appl. Clay Sci.*, vol. 72, no. 0, pp. 1-8, 2013. [http://dx.doi.org/10.1016/j.clay.2012.12.008]

[38] L.N. Carli, O. Bianchi, G. Machado, J.S. Crespo, and R.S. Mauler, "Morphological and structural characterization of PHBV/organoclay nanocomposites by small angle X-ray scattering", *Mater Sci Eng C Mater Biol Appl*, vol. 33, no. 2, pp. 932-937, 2013. [http://dx.doi.org/10.1016/j.msec.2012.11.023] [PMID: 25427508]

[39] L.A. Feigin, D.I. Svergun, and G.W. Taylor, *Structure analysis by small-angle X-ray and neutron scattering.*Plenum Press: New York, USA, 1987. [http://dx.doi.org/10.1007/978-1-4757-6624-0]

[40] G. Beaucage, "Approximations Leading to a Unified Exponential/Power-Law Approach to Small-Angle Scattering", *J. Appl. Cryst.*, vol. 28, no. 6, pp. 717-728, 1995. [http://dx.doi.org/10.1107/S0021889895005292]

[41] N.S. Bell, D.R. Dunphy, T.N. Lambert, P. Lu, and T.J. Boyle, "In situ characterization of silver nanoparticle synthesis in maltodextrin supramolecular structures", *Colloids Surf. B Biointerfaces,* vol. 134, pp. 98-104, 2015. [http://dx.doi.org/10.1016/j.colsurfb.2015.06.030] [PMID: 26162978]

[42] M-C. Chen, H-S. Wong, K-J. Lin, H-L. Chen, S-P. Wey, K. Sonaje, Y-H. Lin, C-Y. Chu, and H-W. Sung, "The characteristics, biodistribution and bioavailability of a chitosan-based nanoparticulate system for the oral delivery of heparin", *Biomaterials,* vol. 30, no. 34, pp. 6629-6637, 2009. [http://dx.doi.org/10.1016/j.biomaterials.2009.08.030] [PMID: 19767097]

[43] K. Nakagawa, and M. Kagemoto, "Characterization of casein-based nanoparticles formed upon freezing by in situ SAXS measurement", *Colloids Surf. B Biointerfaces,* vol. 103, pp. 366-374, 2013. [http://dx.doi.org/10.1016/j.colsurfb.2012.10.052] [PMID: 23261557]

[44] K.R. Juggernauth, *"Structure-Property Relationships in Clay/Polymer Dispersions: Characterizing Bulk and Nanoscale Changes"*, Ph.D. Dissertation, University of Michigan, Michigan USA, 2013.

[45] J.M. Sturtevant, "Biochemical Applications of Differential Scanning Calorimetry", *Annu. Rev. Phys. Chem.,* vol. 38, no. 1, pp. 463-488, 1987. [http://dx.doi.org/10.1146/annurev.pc.38.100187.002335]

[46] X. Cao, Y. Chen, P.R. Chang, and M.A. Huneault, "Preparation and properties of plasticized starch/multiwalled carbon nanotubes composites", *J. Appl. Polym. Sci.,* vol. 106, no. 2, pp. 1431-1437, 2007. [http://dx.doi.org/10.1002/app.26799]

[47] L.S. Guinesi, and É.T. Cavalheiro, "The use of DSC curves to determine the acetylation degree of chitin/chitosan samples", *Thermochim. Acta,* vol. 444, no. 2, pp. 128-133, 2006. [http://dx.doi.org/10.1016/j.tca.2006.03.003]

[48] A.W. Coats, and J.P. Redfern, "Thermogravimetric analysis. A review", *Analyst (Lond.),* vol. 88, no. 1053, pp. 906-924, 1963. [http://dx.doi.org/10.1039/an9638800906]

[49] M.S. Islam, M.S. Rahaman, and J.H. Yeum, "Electrospun novel super-absorbent based on polysaccharide-polyvinyl alcohol-montmorillonite clay nanocomposites", *Carbohydr. Polym.,* vol. 115, pp. 69-77, 2015. [http://dx.doi.org/10.1016/j.carbpol.2014.08.086] [PMID: 25439870]

[50] G. Unsoy, S. Yalcin, R. Khodadust, G. Gunduz, and U. Gunduz, "Synthesis optimization and characterization of chitosan-coated iron oxide nanoparticles produced for biomedical applications", *J.*

Nanopart. Res., vol. 14, no. 11, pp. 1-13, 2012. [http://dx.doi.org/10.1007/s11051-012-0964-8] [PMID: 22448125]

[51] Lapsin and Pricl, *Rheology of industrial polysaccharides.*Springer Science & Business Media: Dordrecht, Germany, 1995.

[52] N.L. Yusof, L.Y. Lim, and E. Khor, "Flexible chitin films: structural studies", *Carbohydr. Res.*, vol. 339, no. 16, pp. 2701-2711, 2004. [http://dx.doi.org/10.1016/j.carres.2004.09.008] [PMID: 15519329]

[53] S. Chillo, S. Flores, M. Mastromatteo, A. Conte, L. Gerschenson, and M.A. Del Nobile, "Influence of glycerol and chitosan on tapioca starch-based edible film properties", *J. Food Eng.*, vol. 88, no. 2, pp. 159-168, 2008. [http://dx.doi.org/10.1016/j.jfoodeng.2008.02.002]

[54] S.E. Whitney, M.J. Gidley, and S.J. McQueen-Mason, "Probing expansin action using cellulose/hemicellulose composites", *Plant J.*, vol. 22, no. 4, pp. 327-334, 2000. [http://dx.doi.org/10.1046/j.1365-313x.2000.00742.x] [PMID: 10849349]

[55] ASTM D882-02, Standard Test Method for Tensile Properties of Thin Plastic Sheeting*Available at: https://www.astm.org/DATABASE.CART/HISTORICAL/D882-02.htm.*

[56] ASTM D638-03, Standard Test Method for Tensile Properties of Plastics*Available at: https://www.astm.org/DATABASE.CART/HISTORICAL/D638-03.htm.*

[57] M.E. Villanueva, A. Salinas, L.E. Díaz, and G.J. Copello, "Chitin nanowhiskers as an alternative antimicrobial controlled release carrier", *New J. Chem.*, 2014.

[58] W.M. Breene, "Application of texture profile analysis to instrumental food texture evaluation*", *J. Texture Stud.*, vol. 6, no. 1, pp. 53-82, 1975. [http://dx.doi.org/10.1111/j.1745-4603.1975.tb01118.x]

[59] M. Pons, and S.M. Fiszman, "Instrumental texture profile analysis with particular reference to gelled systems", *J. Texture Stud.*, vol. 27, no. 6, pp. 597-624, 1996. [http://dx.doi.org/10.1111/j.1745-4603.1996.tb00996.x]

[60] J. Yu, J. Yang, B. Liu, and X. Ma, "Preparation and characterization of glycerol plasticized-pea starch/ZnO-carboxymethylcellulose sodium nanocomposites", *Bioresour. Technol.*, vol. 100, no. 11, pp. 2832-2841, 2009. [http://dx.doi.org/10.1016/j.biortech.2008.12.045] [PMID: 19217775]

[61] P.R. Chang, R. Jian, J. Yu, and X. Ma, "Starch-based composites reinforced with novel chitin nanoparticles", *Carbohydr. Polym.*, vol. 80, no. 2, pp. 420-425, 2010. [http://dx.doi.org/10.1016/j.carbpol.2009.11.041]

[62] Q. Hu, Y. Fang, Y. Yang, N. Ma, and L. Zhao, "Effect of nanocomposite-based packaging on postharvest quality of ethylene-treated kiwifruit (Actinidia deliciosa) during cold storage", *Food Res. Int.*, vol. 44, no. 6, pp. 1589-1596, 2011. [http://dx.doi.org/10.1016/j.foodres.2011.04.018]

[63] P. Sherman, "Techniques for assessing the rheological properties of toiletry and cosmetic products", *J. Soc. Cosmet. Chem.*, vol. 17, no. 8, pp. 439-465, •••.

[64] S. Gunasekaran, and M.M. Ak, "Dynamic oscillatory shear testing of foods— selected applications", *Trends Food Sci. Technol.*, vol. 11, no. 3, pp. 115-127, 2000. [http://dx.doi.org/10.1016/S0924-2244(00)00058-3]

[65] P. Kumar, K.P. Sandeep, S. Alavi, and V.D. Truong, "A review of experimental and modeling techniques to determine properties of biopolymer-based nanocomposites", *J. Food Sci.*, vol. 76, no. 1, pp. E2-E14, 2011. [http://dx.doi.org/10.1111/j.1750-3841.2010.01919.x] [PMID: 21535671]

[66] A.B. Mabrouk, A. Magnin, M.N. Belgacem, and S. Boufi, "Melt rheology of nanocomposites based on acrylic copolymer and cellulose whiskers", *Compos. Sci. Technol.,* vol. 71, no. 6, pp. 818-827, 2011. [http://dx.doi.org/10.1016/j.compscitech.2011.01.012]

[67] T.L. Smith, and N.W. Tschoegl, "Rheological properties of wheat flour doughs", *Rheol. Acta,* vol. 9, no. 3, pp. 339-344, 1970. [http://dx.doi.org/10.1007/BF01975400]

CHAPTER 14

Intrinsic Viscosity of Strong Linear Polyelectrolytes in Solutions of Low Ionic Strength and Its Interpretation

Georges M. Pavlov[1,2,*] and **Alexander S. Gubarev**[2]

[1] *Institute of Macromolecular Compounds, Russian Academy of Science, Russian Federation*

[2] *St. Petersburg State University, Saint Petersburg, Russian Federation*

Abstract: Evaluation of reliable experimental quantities characterizing the isolated strong polyelectrolyte chain in solution at low ionic strengths is a challenge for researchers working in polymer science and biophysics. A simple method for estimating the intrinsic viscosity as initial slopes of the $\ln\eta_r$ and η_{sp} *vs. c* dependencies is discussed. Using several examples, we have demonstrated the adequacy of this estimated intrinsic viscosity. The results determined in salt free water solutions for homologous series of sodium poly(styrene-4-sulfonate) in 30-fold range of molar mass, as well as for the polyions with the persistence length of the corresponding bare chains differing 30 times are discussed. The "apparent intrinsic viscosities" of poly(styrene-4-sulfonate) samples at non-zero polymer concentration in salt-free solutions are compared with the directly measured intrinsic viscosities at different ionic strengths.

Keywords: Dilute solutions of charged polymers, Extremely low ionic strength, Intrinsic viscosity, Molecular hydrodynamic methods, Polyelectrolytes.

INTRODUCTION

Polyelectrolytes are polymers bearing various amounts of charges on their chains. Polyelectrolytes represent a broad class of polymers differing in charge density and topology of chains; they exhibit very different physical properties [1, 2] and they are the objects of permanent study due to their important role in living nature and in various technological processes [3 - 8]. The studies of conformational properties of linear polyelectrolyte are also of fundamental interest. Despite that, there is huge and continuing interest in polyelectrolytes as they remain one of the least understood states of soft matter. For linear highly charged polyelectrolyte,

* **Corresponding author Georges M. Pavlov:** Institute of Macromolecular Compounds, RAS; St. Petersburg, St. Petersburg State University, Russian Federation; Tel: 07(812)3286869; E-mail: g.m.pavlov@spbu.ru

Martin Masuelli & Denis Renard (Eds.)

the question of the conformational status of charged chain at extremely low ionic strengths including the salt free solutions remains unclear till today, both theoretically and experimentally.

So far, according to the observation of Stevens and Kremer, polyelectrolytes are a typical example of a situation in which the experiments are carried out in semidilute regime, while the theories are best designed to dilute regime [5]. This utterance is pertinent until now. At the same time, it should be noted the disparity between the number of theoretical and experimental studies of highly charged macromolecules in solution, with the obvious predominance of the number of theoretical studies. This situation may be explained as follows. Due to the large size of charged macromolecules in dilute solutions at low ionic strengths, it is difficult to obtain characteristics that can be regarded as those of isolated macromolecules; experimental studies should be carried out at extremely low concentrations of polyions. Often, in these concentration ranges, the instruments used for measuring the corresponding experimental responses are not sensitive. Getting the adequate experimental characteristics of isolated polyions still remains a huge challenge for researchers till today.

One of the ways to characterize the isolated macromolecules of polyions in solution of extremely low ionic strength is the study of their viscous flow. The procedure for determination of intrinsic viscosity of polyelectrolytes in extremely low ionic strengths was worked up and reported earlier [9, 10].

This contribution describes: a) history and basic principles of the intrinsic viscosity in polymer science, b) the proposed method of estimating this characteristic for any kind of polymers, c) its application for the study of charged linear polymers, d) the ways of the interpretation of obtained results and comparison with the values obtained by conventional methods.

Hydrodynamic Methods and Essential Hydrodynamic Characteristics

The methods of molecular hydrodynamics (viscometry, velocity sedimentation, translational diffusion), as well as osmometry were actually the first experimental methods which enabled us to understand the nature of the polymeric material and realize that the basic structural unit of a polymer substance is a macromolecule [11 - 13].

Recently, valuable information about the conformations of the macromolecules were obtained also by scattering techniques, both X-ray and neutron scattering [14 - 18]. Nowadays the transmission and scanning electron as well as atomic force microscopy (TEM, SEM, and AFM, respectively) have become well-known, due to their unique potential to visualize individual macromolecules and to distinguish between their topologies and conformations [19 - 22]. However, the study of macromolecules in solution, particularly of their homologous series, by the methods of molecular hydrodynamics, *especially the intrinsic viscosity measurements*, remains the indispensable base for understanding the conformational organization of the various types of macromolecules in solution. These methods have demand in molecular biophysics, in soft matter science, nano-, polymer- and colloid sciences [23 - 29].

The main methods of molecular hydrodynamics are analytical ultracentrifugation (velocity and equilibrium sedimentations), translational diffusion and viscous flow studies in dilute polymer solutions. Size exclusion chromatography (SEC) can be also included in this group of methods. These methods allow researchers to obtain a certain set of hydrodynamic characteristics, such as velocity sedimentation coefficient (s), translational diffusion coefficient (D), intrinsic viscosity ($[\eta]$) and some others. These characteristics are related to the molar mass and size of macromolecules in various ways. The molar mass can be determined within the framework of the methods of molecular hydrodynamics from the combination of the velocity sedimentation and translational diffusion data analysis, or from experiments on sedimentation equilibrium. The molar mass distribution of samples is determined usually from the SEC and/or from the velocity sedimentation analysis. Thus, the combination of hydrodynamic methods allows almost complete characterization of macromolecules, and determination of their gross conformation. The above techniques are applicable for studying neutral polymers or polyelectrolytes with the charge effects being virtually suppressed by the addition of low molecular mass salt. The principal content of this chapter is devoted to the discussion of characteristics of polyions in solutions with very low ionic strength (when polyelectrolyte effects are clearly observed). Molar mass of the charged macromolecules was determined by sedimentation-diffusion analysis of polyelectrolyte solutions with virtually vanishing polyelectrolyte effects. Needless to say that the knowledge of the molar mass of charged macromolecules is necessary for the interpretation of the results obtained in solutions of extremely low ionic strength. In this chapter, the paramount importance is paid to intrinsic viscosity as an important hydrodynamic characteristic.

The study of the viscosity of cellulose nitrate solutions by H. Staudinger led to the understanding of the structure of the polymer material and resulted in introduction of the term “macromolecule”. Thus, the important result of Staudinger’s work was the following statement [30]:

$$\lim_{(c\to 0)}(\eta_{\mathrm{sp}}/c) \equiv [\eta] \sim M, \tag{1}$$

where $\eta_{sp} = \eta_r - 1$ is the *specific viscosity* of the solution and $\eta_r = (\eta/\eta_o)$ its *relative viscosity*, η_o and η are the dynamic viscosities of solvent and solution, correspondingly; c is the polymer concentration in solution, and M is molar mass of dispersed polymer. It follows from Equation (1) that $[\eta]$ is a measure of molar mass of dispersed species. The statistical analysis shows that the order of the so called "viscosity average molar mass" (M_η) is dependent on the scaling index value b_η (see below). However, for all possible macromolecular conformations, M_η value is close to the weight average molar mass value M_w, at least when $M_w/M_n \leq 2$, where M_n is the number average molar mass [12].

By lucky chance, the object selected by Staudinger for the viscous flow investigation (highly substituted cellulose nitrate) matched the hypothesis of virtually direct proportion between M and $[\eta]$, as was shown later in numerous publications, for instance in [31 - 33].

Thus, fairly simple way to assess the molar mass of the polymer molecules was proposed. This achievement launches the development of analysis of the viscous flow of dilute polymer solutions and elaboration of various methods for evaluating the intrinsic viscosity of polymers. It should be noted that this trend continues to grow up today [34 - 36].

Dynamic viscosity of a fluid (one component and/or multicomponent) is a measure of its resistance to flow. Different layers of the laminar flowing liquid are moving at different velocities. As a result, velocity gradient or shear rate appears in moving solvent/liquid. Macromolecule immersed in such moving solvent will be involved in rotational motion; additional energy dissipation will arise due to friction between macromolecules and solvent molecules, and the dynamic viscosity of the solution will increase in comparison with the solvent.

The difference between dynamic viscosity of solution and solvent $(\eta-\eta_0)$ characterizes increase in dynamic viscosity of the solution due to the presence of macromolecules. This difference normalized to the dynamic viscosity of the solvent $(\eta-\eta_0)/\eta_0 \equiv \eta_{sp}$ is a dimensionless quantity called *specific viscosity* (η_{sp}). The ratio η_{sp}/c between specific viscosity and polymer concentration c expressed in g/cm^3 has the inverse dimension (cm^3/g) and is called *reduced viscosity*. The reduced viscosity extrapolated to zero concentration is termed "intrinsic viscosity of the polymer" ($[\eta]$).

The Huggins equation (Fig. **1a** and **1c**) is the most known relationship used to determine the intrinsic viscosity [37]. This equation can be derived from a series expansion in the concentration of the dynamic viscosity:

$$\eta = \eta_0(1 + k_\eta c + k'_\eta c^2 + \ldots), \tag{2}$$

$$\eta_{sp}/c = (\eta_r - 1)/c = k_\eta + k'_\eta c + \ldots = [\eta] + k_H[\eta]^2 c + \ldots, \tag{3}$$

where $k_H = k_\eta'/[\eta]^2$ is a dimensionless value called the Huggins parameter.

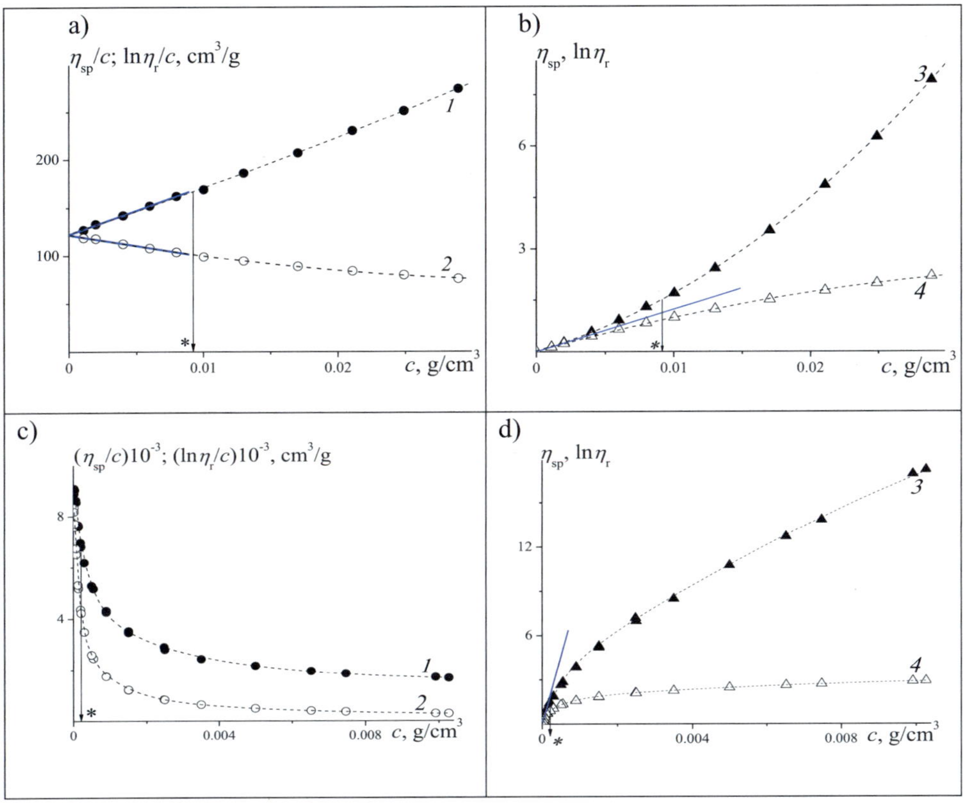

Fig. (1). Different representation of the dilute solutions viscosity data of uncharged (a, b) and charged (c, d) polymers in water salt free solutions. The polymers are neutral linear polyvinylpyrrolidone (PVP, M = 400 000 g/mol) (a, b) [39] and strong linear polyelectrolyte PSSNa-2 (M = 850 000 g/mol) (c, d) [9, 40]. Dependencies of the reduced specific viscosity (1) and inherent viscosity (2) at left side (a, c) and specific viscosity (3) and logarithm of relative viscosity (4) at right side (b, d) in function of polymer concentration are shown. The obvious difference of behavior in standard Huggins and Kraemer plots (a, c) are observed between neutral (a) and charged macromolecules (c) in pure water without adding low molecular salt. Dependencies of the specific viscosity (3) and logarithm of relative viscosity (4) for neutral and charged systems vary only by different level and sign of their curvatures (the right side of the Figure). The initial slopes of these dependencies (at $c \to 0$) are shown as solid lines and represent too the values of intrinsic viscosity (b and d). For uncharged PVP $[\eta]$ value acquired at (b) coincide with the value obtained by classical Huggins and Kraemer plots (a). The value of $[\eta]$ is determined as the initial slope for polyelectrolyte in H_2O solutions without low molecular salt, is shown in (d), which is also averaged from initial slopes of $\ln\eta_r$ (c) and η_{sp} (c). The arrows indicate the solution concentration satisfying η_r = 2.5. The sample abbreviations correspond to those indicated in Tables **2-4**.

It should be noted that Kraemer postulated the following expression somewhat earlier [38]:

$$[\eta] = \lim_{(c\to 0)}\left(\frac{\ln \eta_r}{c}\right), \tag{4}$$

where $\ln\eta_r/c$ is the *inherent viscosity* of polymer.

It is easy to show that both relations (1) and (4) lead to equivalent results when $c\to 0$.

The expression $\ln\eta_r = \ln(1 + \eta_{sp})$ can be decomposed into a series of η_{sp}, when $\eta_{sp} \leq 1$, *i.e.* $\eta_r \leq 2$, as:

$$\ln \eta_r = \ln(1+\eta_{sp}) = \eta_{sp} - \eta_{sp}^2/2 + \eta_{sp}^3/3 - \ \dots . \tag{5}$$

Substituting the value of η_{sp} from the Huggins equation (3) into equation (5) and retaining only the quadratic terms of concentration, we obtain the following equation which is known as *the Kraemer equation* (Fig. **1a** and **c**):

$$\ln \eta_r / c = [\eta] + (k_H - 0.5)[\eta]^2 c + \dots , \tag{6}$$

The parameter (k_H - 0.5) is called the *Kraemer parameter* k_K and is uniquely linked to the Huggins parameter in the following way: $k_H = k_K + 0.5$. Consequently, both of these relationships (Eqs. 3 and 6) are mathematically equivalent when $c\to 0$. Therefore, it is reasonable to consider Huggins and Kraemer plots together, like *one twin plot*:

$$[\eta] = \lim_{(c\to 0)}(\eta_{sp}/c) = \lim_{(c\to 0)}(\ln \eta_r/c), \tag{7}$$

This definition of the $[\eta]$ value (Eq. **7**) means that the initial slopes of the $\eta_{sp}(c)$ and $\ln\eta_r(c)$ dependences are the desired estimations of intrinsic viscosity (Fig. **1b** and **d**). It is noteworthy that since $k_K < k_H$, the value of $\ln\eta_r/c$ is less dependent on concentration than the η_{sp}/c value; moreover, the strict relation $k_H = k_K + 0.5$ often fails for the neutral polymer systems.

Viscosity Measurement Equipment

There are a number of classical instruments for viscosity measurements, *e.g.* the capillary Ostwald and/or Ubbelohde viscometers, the falling ball or Hoeppler viscometers [12, 13, 34, 35, 41]. The Zimm-Crothers viscometers give the possibility to study viscous flow at very low shear rate [42]. Recently, automatic viscometers based on the principle of hydrostatic bridge have been developed; they allow on-line viscosity measurements in extremely dilute solutions emerging from the SEC columns [36, 43 - 45]. Regardless of the used method, the temperature during the measurements must be strictly constant, and continuous monitoring/check of the flow time of the solvent must be carried out.

Degree of Dilution

Determining the properties of individual macromolecules requires experiments in dilute solutions and in a wide interval of solute concentrations, sufficient for reliable extrapolation to $c \to 0$. The solutions under study should belong to the zone of dilute solutions. This means that the coils in the solution should be remote each other to the distance greater than their own size. The degree of dilution is usually characterized by dimensionless Debye parameter $c[\eta]$ (c is the polymer concentration expressed in g/cm^3, $[\eta]$ is the intrinsic viscosity expressed in cm^3/g). The product of the intrinsic viscosity and the concentration of the solution represents a good approximation of the volume fraction ϕ of the polymer substance in solution, with $\phi = nv_1/V = mv_1 \mathrm{N_A}/VM = c \times 0.36 < h^2 >^{3/2} \times \mathrm{N_A}/M = (0.36\mathrm{N_A}/\Phi)c[\eta] \approx c[\eta]$, where $v_1 = 0.36 < h^2 >^{3/2}$ is the volume occupied by macromolecular coil in solution, m is the overall mass of the polymer dispersed in overall volume of solution V, $< h^2 >$ is the mean square end-to-end distance of the polymer chain, Φ is the hydrodynamic Flory parameter [46]. The condition $c[\eta] = 1$ corresponds to the situation when each coil may contact with some others, and the coils start to overlap. The dilute solution regime corresponds to the inequality: $c[\eta] \ll 1$. Consequently, the value of concentration by itself does not characterize the degree of dilution of the solution. For example, the polymer solution with a concentration of ≈ 0.4 mg/ml $= 0.4 \times 10^{-3}$ g/cm^3 may not be diluted if the intrinsic viscosity $[\eta]$ is higher than 2000 cm^3/g (for instance, this is the case of the high molar mass polystyrene [47]).

Molecular Meaning of the Intrinsic Viscosity

In the previous sections, the methods of measurement and calculation of the intrinsic viscosity were discussed. It is worthy to notice that the dimension of this value is cm^3/g. The physical meaning of this quantity has been revealed by Einstein who calculated the intrinsic viscosity value for the suspension of hard spheres [11 - 13]. Einstein has shown that the increase in viscosity of suspension is related to the volume fraction of spheres in a solution. This result for the suspension of spheres can be presented as:

$$[\eta] = 2.5\phi = 2.5\mathrm{N_A}(v/M), \tag{8}$$

where ϕ is the volume fraction, v is the volume and M is the molar mass of a sphere.

This relationship was extended for the suspension of spheroids with asymmetry parameter p as,

$$[\eta] = \nu_S(p)\bar{\upsilon}, \qquad (9)$$

where $\bar{\upsilon}$ is the partial specific volume and $\nu_S(p)$ is the dimensionless Simha parameter [12, 13]. For chain molecules, this type of expression was generalized by Flory-Fox in the following form [46]:

$$[\eta] = \Phi < h^2 >^{3/2}/M. \qquad (10)$$

Hence, it becomes clear that the product ($[\eta]M$) characterizes hydrodynamic volume occupied by a macromolecule/particle in solution.

Kuhn-Mark-Houwink-Sakurada Relationships

H. Staudinger in his Nobel lecture [30] noted that W. Kuhn has pointed out more general expression between $[\eta]$ and M than a linear relationship, namely:

Table 1. The limits of the scaling index b_η, values in different range of molar masses and range of the $[\eta]$ values for different types of gross polymer conformations*.

Gross conformations	Types of polymer systems	Range of b_η	Range of Molar Masses	Range of $[\eta]$ cm³/g	Ref.
Globular/ Rigid sphere	globular proteins, hyperbranched macromolecules	≈0	*all*	*4-10*	[11, 13, 48, 49]
Thick macromolecules	brush-like linear polymer	$0.5 \geq b_\eta > 0$	*relatively short macromolecules*	*5-15*	[50, 51]
Θ-condition	flexible polymers	0.50	*all*	*10-100*	[52 - 55]
Rigid macromolecules without intrachain volume effect	rigid macromolecules	0.50	*very high*	*1 500-12 000*	[56 - 58]
Thermo-dynamically good solvent, non draining coil	flexible polymers	$0.85 \geq b_\eta > 0.5$	*very high*	*20-3300*	[55, 59 - 61]
Rigid macromolecules, crankshaft conformation	Cellulose, cellulose derivatives, aromatic polyamides	$1.0 \geq b_\eta > 0.5$	*moderate*	*20-800*	[31, 41, 62]

(Table 1) contd.....

Gross conformations	Types of polymer systems	Range of b_η	Range of Molar Masses	Range of $[\eta]$ cm^3/g	Ref.
Extra-rigid	Helical structures such as DNA, polysaccharides such as schizophyllan, xanthan	$1.8 \geq b_\eta > 1$	*range of relatively short macromolecules*	*40-100*	[56 - 58, 63]

* - the profusion of original references may be found in Polymer Handbook [33].

$$[\eta] \sim M^{b_\eta}, \tag{11}$$

where b_η is the exponent characterizing a polymer-solvent system [33]. Since then, a large number of polymer-solvent systems have been studied, the value of K_η and b_η were determined and given in various monographs and handbooks [11 - 13, 33].

The values of b_η are shown in Table **1** for various polymer homologous series.

In the modern connotations, this relationship is known as Kuhn-Mark-Houwink-Sakurada relationship or scaling relation. It should be noted that such relations hold for other hydrodynamic characteristics, namely, the translational diffusion coefficient D_0 and velocity sedimentation coefficient s_0:

$$\begin{aligned} [\eta] &= K_\eta M^{b_\eta}, \\ s_0 &= K_s M^{b_s} \\ D_0 &= K_D M^{b_D} \end{aligned} \tag{12}$$

Importantly to emphasize that makes sense to establish and consider this kind of correlations (Eq. 12) only for the polymer homologous series.

Polyelectrolytes

Polyelectrolytes are polymers whose repeating units bear electric charges; they represent a broad class of polymers differing in charge density and topology and exhibit very different physical properties [1, 2]. The study of conformational properties of linear polyelectrolyte is also of fundamental interest. For linear highly charged polyelectrolyte, the question of the conformational status of charged chain at extremely low ionic strengths or in salt free solutions still remains unclear. The presence of charges on the chain leads to intra- and inter-macromolecular interactions which are stronger and spread over much longer distances than those (interactions) in the uncharged macromolecules. In macromolecule carrying charges, strong electrostatic intrachain interactions appear both at short and long distances. First interpretation of the large size of the polyelectrolyte chains was grounded solely on the concept of the excluded volume effects, in other words, on the long-range electrostatic interactions only [64 - 66]. This is reasonable for the chains with low linear charge density, when the two successive neighboring charges are separated along the contour length by several statistical segments of bare (non charged) chain. The measure of the equilibrium rigidity of linear macromolecules is the Kuhn segment length A or the persistence length $a = A/2$. The length of the statistical segment is the inner scale of the macromolecule dimension.

Two additional scale lengths are introduced to characterize the electrostatic interactions in polyelectrolyte solutions: the Bjerrum length λ_B and the Debye-Hueckel screening length λ_{DH} [3 - 5]. The Debye-Hueckel screening length (λ_{DH}) is defined as a spread distance of influence of an electric field of a separate charge placed in the medium containing other charges: $\lambda_{DH} = (8\pi\lambda_B\mu)^{-1/2}$, where $\lambda_B = e^2/(4\pi\varepsilon_0\varepsilon k_B T)$ is the Bjerrum length characterizing the screening action of the solvent, e is the elementary charge, ε_0 is the electrical constant, ε is the dielectric permittivity of the medium, k_B is the Boltzmann constant, $\mu = (1/2)\sum n_i Z_i^2$ is the ionic strength of solution, n_i is the number of i-th ions per unit volume, and Z_i is the charge of the i-th ion in e units. In the case of one-one valent salt, $\mu = C \times 10^{-3} \times N_A$ [cm^{-3}], where C is the molar concentration of salt expressed in [mol × L^{-1}].

When the charges are located on average within the same statistical segment of the bare chain (statistical segment length of the skeleton without any charges), then the short-range interactions should be taken into account. The concept of the electrostatic contribution to the persistence length value was first developed by Odijk and also by Skolnick-Fixman [67 - 69]. The modeling of the electrostatic interactions in this concept is based on the Debye-Hueckel approximation and allows us to define an electrostatic persistence length of highly charged chain (proportional to the square of the Debye-Hueckel screening length): $a_{el} \sim \lambda^2_{DH}$. Later Barrat and Joanny and Dobrynin [70 - 73] introduced alternative models in which $a_{el} \sim \lambda_{DH}$. For the polyelectrolyte chains, the total persistence length is a sum of bare and electrostatic terms: $a_{tot} = a_{bare} + a_{el}$. The question of the contribution of the short-range electrostatic interaction into size and conformation of linear polyelectrolyte chains is under discussion and is still being improved [73 - 77]. One of the main question of the polyelectrolyte science is the separation of the contributions of the short- and long-range interactions into the size of the charged chains in function of the charge density and the ionic strength of the solution [68, 69, 78, 79].

Strategy of Polyelectrolyte Research

In the research of the conformational state of linear macromolecules (charged or neutral), the most fruitful approach is the study of homologous series of samples/fractions by different methods in a wide range of molar masses. Thus, the research strategy should generally include the following steps: 1) the synthesis of a number of samples of the same chemical structure and with the same linear charge density, 2) determination of the molar masses of samples/fractions, under conditions of substantially suppressed polyelectrolyte effects, 3) study of samples/fractions at low ionic strength, when the nature of the polyelectrolyte chains is fully manifested, 4) comparing the obtained results with molar masses/contour lengths of the charged chains to obtain the conformational parameters of charged chains at different ionic strengths.

Intrinsic Viscosity of Polyelectrolytes – Informative Feature of Isolated Charged Macromolecule

The main problem in the experimental studies of polyelectrolytes is to obtain the appropriate quantities characterizing the isolated charged macromolecules in solution with minimal ionic strength (salt free solution). At low ionic strength of solution and at low polyelectrolyte concentrations (c), the values of η_{sp}/c and $\ln\eta_r/c$ increase with decreasing c (instead of decreasing as it is usually the case with the uncharged linear polymers, compare Fig. (**1a** and **c**). This is caused by the polyelectrolyte effect. With dilution, the volume in which the counterions are spread increases, thus reducing the screening of the charges fixed on the chain and increasing their mutual repulsion. As a result, the size of the polyion increases. Thus, with decreasing ionic strength of the solution μ, the chain size increases by virtue of electrostatic repulsion between the monomeric units [6, 11].

Sometimes with further decreasing concentration, a maximum in the concentration dependences of η_{sp}/c and $\ln\eta_r/c$ is observed. The nature of this maximum was discussed in [80 - 82].

Fuoss suggested the empirical relationship which linearized the reciprocal $(\eta_{sp}/c)^{-1}$ *vs.* $c^{1/2}$ dependence to some extent [11, 83]:

$$c/\eta_{sp} = [\eta]^{-1} + (B[\eta])c^{1/2}, \quad (13)$$

where B is the slope coefficient.

In the $c/\eta_{sp}(c^{1/2})$ coordinates, the intercept is considered as a reciprocal intrinsic viscosity of a polyelectrolyte. The Fuoss plot is quite popular in the interpretation of viscometric data on polyelectrolytes; however, it should be noted that it does not always give linear dependences.

The *isoionic dilution method* was suggested as a more grounded procedure for the determination of $[\eta]$ of polyelectrolytes [11, 84, 85]. In this case, the initial salt-free solution is characterized by the ionic strength controlled by the polyelectrolyte concentration and by the degree of dissociation of ionic groups. The polyelectrolyte solution is diluted with a solution of a low molar mass salt whose ionic strength is equal to that of the initial salt free solution of the polyelectrolyte. In practice, the concentration of a low molar mass salt is selected so that the dependence of η_{sp}/c (or $\ln\eta_r/c$) with c is linear. Under- or over-salting of the solvent distorts the linearity of the indicated dependences. The resulting dependences are well reproduced by second-order polynomials with opposite signs of the coefficients at c^2. Therefore, the goal of optimization is to obtain a dependence satisfying the following condition: $\partial^2(\eta_{sp}/c)/\partial c^2 = 0$. In this case, the resulting dependence of η_{sp}/c (or $\ln\eta_r/c$) with c will be linear, allowing determination of $[\eta]$ at a fixed ionic strength.

In the works [86 - 88] the rheological dependence

$$\eta = \eta_0 \exp([\eta]c), \qquad \textbf{(14)}$$

was analyzed. This relationship may be reduced to the following equation:

$$[\eta]^* \equiv \partial \ln \eta_r / \partial c, \qquad \textbf{(15)}$$

In [9], the initial slope $\partial\ln\eta_r/\partial c$ at $c\to 0$ is viewed as the value of the intrinsic viscosity of a charged polymer. Note that the $\ln\eta_r$ dependencies have a monotonous character for both neutral and charged polymers, differing only in the degree of curvature (compare Fig. **1b** and **d**). In addition, it is important to establish that the peculiar behavior of $\ln\eta_r/c$ and/or η_{sp}/c *vs.* c plots is transformed to monotonic mode in $\ln\eta_r$ and/or η_{sp} *vs.* c coordinates. However, it should be noted that the peculiar behavior of η_{sp}/c and/or $\ln\eta_r/c$ *vs.* c plots is a hallmark of the polyelectrolyte nature of linear macromolecules; it is a very important and simple test allowing to detect that the linear macromolecules carry electrical charges.

It is clear that at $c\to 0$, the value of $[\eta]^*$ corresponds to the initial slope of the $\ln\eta_r$ with c plot coinciding with $[\eta]$ obtained from the Kraemer equation (Eq. 4) for any polymers (charged or neutral). It also turns out that the shape of the $\eta_{sp}(c)$ plot for linear polyelectrolyte is similar to that of the $\ln\eta_r(c)$ dependence; namely, it is curved upward (compare again Fig. **1b** and **d**). The comparison of the initial slopes of the dependences $\eta_{sp}(c)$ and $\ln\eta_r(c)$ at $c\to 0$ shows that they are close (Table **2**). Thus, in further discussion the averaged values of $\langle[\eta]\rangle = ((\partial\ln\eta_r/\partial c)_0 + (\partial\eta_{sp}/\partial c)_0)/2$ are used.

Table 2. The intrinsic viscosity values $[\eta]$ (cm^3/g) determined as initial slopes in dependences $\eta_{sp}(c)$ ($[\eta]_1$) and $\ln\eta_r(c)$ ($[\eta]_2$) at $c\to 0$. The suggested approach was tested on different samples in a wide range of molecular weights*.

Sample	$M10^{-3}$, g/mol	$[\eta]_1$ cm^3/g	$\Delta[\eta]_1$	$[\eta]_2$ cm^3/g	$\Delta[\eta]_2$	$\langle[\eta]\rangle$, cm^3/g	$\langle\Delta[\eta]\rangle$
PSSNa-1	1500	15400	300	15400	300	**15400**	300
PSSNa-2	847	9300	100	8900	60	**9100**	100
PSSNa-3	607	8660	60	8200	90	**8450**	80
PSSNa-4	605	5380	70	5100	30	**5260**	50
PSSNa-5	448	6230	20	5900	50	**6070**	40
PSSNa-6	375	5000	200	800	200	**4900**	200
PSSNa-7	124	740	40	660	40	**700**	40
PSSNa-8	84	280	10	260	10	**270**	10
PSSNa-9	53	149	7	135	7	**142**	7
PSSNa-10	51	182	5	167	6	**175**	6
PASA-*para*	240	549	7	533	8	**542**	8
PASA-*meta*	250	5220	20	4880	40	**5000**	200
PectinateNa**	46	1900	100	1800	100	**1900**	100
PVP***	400	123.0	0.7	122.7	0.5	**122.8**	0.6

*– the experimental data were taken in analysis in concentration range where $\eta_r < 1.5$ and the origin of coordinates (the point [0, 0]) has been included in analysis where it has been considered a reasonable continuation of experimental dependences;
** – the analyzed viscosity data for sodium salts of pectin is taken from [89];
*** – polyvinylpyrrolidone is used as an uncharged linear polymer example for test of the determination of intrinsic viscosity [9, 39].

Note that the method proposed in [9] was successfully applied to the viscosity data obtained for different linear polyelectrolytes [90 - 94]. This method may be called the "initial slopes method" to determine intrinsic viscosity of any kind of polymers.

It is important that comparative analysis between the value $[\eta]^* = \partial\ln\eta_r/\partial c$ and the intrinsic viscosity value obtained by the isoionic dilution shows that these two methods give virtually identical results (Fig. **7** in Ref. [9]). Consequently, in addition to the isoionic dilution method, the "initial slopes method" or "Staudinger-Kraemer method" can be recommended for determination of the intrinsic viscosity of polyelectrolyte in salt-free solutions or at low ionic strengths.

Characteristics of Isolated Charged Macromolecule

As was stated above, the most informative research works in the field of linear polymers are the studies of homologous series in a fairly wide range of molar masses by different methods. Main molecular characteristic, *viz.* the molar mass, may be evaluated from the study of the polyelectrolyte solutions with a sufficiently high ionic strength, when electrostatic interactions are mostly suppressed.

The criterion of sufficient screening of electrostatic interactions is the linear dependences of the Huggins and Kraemer plots satisfying the equations (3) and (6) in this case (see **1a**). The capacities of the method of the intrinsic viscosity estimation are illustrated by the study of the homologous series of sodium poly(styrene-4-sulfonate) (PSSNa). Table **3** summarizes the results of hydrodynamic study of PSSNa samples in 30-fold range of molar masses ($50<M\times10^{-3}<1500$, g/mol) and at different ionic strengths.

Table 3. Intrinsic viscosity, translational diffusion and velocity sedimentation coefficients and molecular masses of sodium poly(styrene-4-sulfonate) samples, determined in aqueous solutions.

Sample	$D_0\,10^7$, cm^2/s	$s_0\,10^{13}$, s	$M_0\,10^{-3}$, g/mol	$[\eta]$, cm^3/g		L/A	
	0.2M NaCl			H_2O*	0.2M NaCl	H_2O	0.2M NaCl
PSSNa-1	0.70	14.0	**1 500**	15400	385	2.8	900
PSSNa-2	0.91	10.8	**847**	9120	172	1.6	510
PSSNa-3	1.07	9.1	**607**	8450	170	1.1	360
PSSNa-4	1.10	9.3	**605**	5260	117	1.1	360
PSSNa-5	1.23	7.7	**448**	6065	117	0.8	270
PSSNa-6	1.44	7.5	**375**	4890	109	0.7	220
PSSNa-7	2.84	4.9	**124**	700	44	0.2	74
PSSNa-8	3.26	3.9	**84**	270	30	0.2	50
PSSNa-9	4.76	3.5	**53**	140	22	0.1	32
PSSNa-10	4.2	2.9	**51**	175	22	0.1	31

* - the values of $[\eta]$ are averaged of initial slopes of plots $\eta_{sp}(c)$ and $\ln\eta_r(c)$.

One of the first suitable steps in the hydrodynamic data interpretation, particularly the interpretation of the intrinsic viscosity data, is the use of the scale of the size of linear macromolecules. This scale is put up in the $[\eta]M_L$ *vs.* M/M_L coordinates as a result of the analysis of a rich library of experimental hydrodynamic data available in the literature [95, 96] and in references therein.

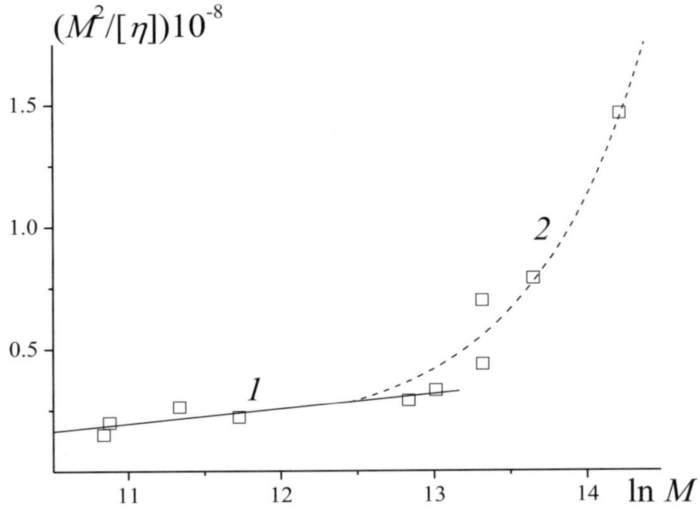

Fig. (2). Dependence of $M^2/[\eta]$ with ln$M$ for a homologous series of sodium poly(styrene-4-sulfonate) in H_2O solutions (Table **3**, columns 4 and 5). The linear part (1) of the plot corresponds to the behavior of a slightly bended rod. It is treated by (Eq. 16). According to this equation, the initial slope B is $45M_L^3/(2\pi N_A)$ and the intercept A is $-[45M_L^3/(2\pi N_A))]\times[\ln(M_L d)+ 0.697]$. The initial slope allows the evaluation of the values of mass per unit length M_L of linear chain. The intercept, in principle, allows the estimation of the value of the polymer chain diameter d. For the systems considered, the following estimations were obtained in the low molar masses region (1): $B = (6.1 \pm 1.6)\ 10^6$ and $A = -(4.8 \pm 1.9)10^7$ with the linear regression coefficient r = 0.891. The non-linear region (2) corresponds to the beginning of the coil formation of PSSNa chains.

The ascertainment of this scale and its use does not require any model assumptions; only the knowledge of the value of mass per unit length (M_L) is required. (Note again that the intrinsic viscosity is the most sensitive characteristic to changes in size of linear chains and thus to their conformational state.) This analysis of data presented in Table **3** shows that PSSNa macromolecules dissolved in the salt-free water occupy in solution a volume similar to that occupied by the extra rigid linear chains in solution [10]. Quantitatively, this volume corresponds to the length of the statistical segment (equal to 650 nm), and the corresponding relative contour lengths L/A are given in column 9 of Table **3**. Since the values of L/A are very small, the model and the theory of weakly bending rod can be used, and we can estimate an important structural parameter (the mass per unit length of PSSNa chains) [97, 98]:

$$M^2/[\eta] = [45M_L^3/(2\pi N_A)] \times [\ln M - \ln(M_L d) - 0.697]. \quad \textbf{(16)}$$

Fig. (**2**) shows the corresponding plot for a homologous series of sodium poly(styrene-4-sulfonate) in H_2O solutions; the initial slope (1) determines the value of $M_L = (8.0 \pm 0.7) \times 10^9$ g/(cm mol), which is in good congruity with the theoretical calculations of M_L^{theor} found in the range of $7.3 < M_L^{theor} \times 10^9$g/(cm mol) < 8.2 [10]. The hydrodynamic diameter of a polymer chain is evaluated with high error as $d = (2 \pm 1)$ nm, but nevertheless its average value is on a reasonable level. This is a very rare case observed for the extra-rigid linear macromolecules where the linear part of the curve obtained is so extended. In turn, this result (M_L value) indicates the adequacy of the estimates of the intrinsic viscosity values by the "initial slopes method" at $c \rightarrow 0$.

Behavior of Strong Polyelectrolytes with Different Structural Rigidity of Chains

Study of linear polyelectrolytes with different structural rigidity of the bare chains represents the tangible interest. For this purpose, we selected the aromatic sulfated polyamide with *para-* and *meta-* inclusion of the benzene ring into the main chain (PASA-*para* and PASA-*meta*) [99 - 101] as well as poly(styrene-4-sulfonate) (PSSNa). The chemical structures of the repeating units for these polymers are shown below:

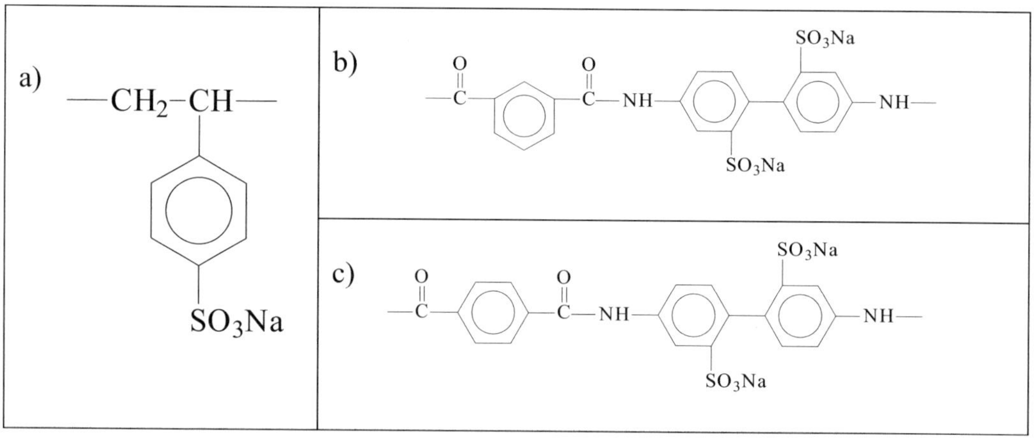

Scheme 1. The repeating units of polymers under study: (a) sodium poly(styrene-4-sulfonate); sulfated aromatic polyamides with *meta-* (b) and *para-* (c) inclusion of benzene ring into the main chain.

The structural components of statistical segment length for aromatic sulfated polyamides were calculated by methods of statistical mechanics (see Table **4**) [41, 102, 103]. The equilibrium rigidity of non-charged aliphatic chains (PSSNa at 0.2M NaCl solution) is 2.3 nm [12, 40]. Hence, the structure component of equilibrium rigidity in the compared polyelectrolytes changes by more than one order of magnitude. It should be noted that for all of these structures, the average distance between adjacent charges along the chain is about ≈ 0.7 nm, *i.e.*, it corresponds to the Bjerrum length. This observation also means that the studied polyelectrolytes can be classified as highly charged polyions.

The behavior of these chains in the salt-free solutions is compared in Fig. (**3**) which shows the dependences of lnη_r *vs.* *c*; the following qualitative conclusions may be drawn from this plot. At comparable contour length of chains, the most flexible chain (PSSNa) will be expanded to the greater extent. The initial slope of the curve lnη_r *vs.* *c*, *i.e.*, the value of intrinsic viscosity, is the highest in salt free solution for the PSSNa. The lowest value of the intrinsic viscosity in this series was observed for the chains with the highest structural rigidity (PASA-*para*). Indeed, the structurally rigid chain is already bulky, *i.e.* such the chain occupies a large volume even in the absence of charges. Charge effects for such kind of chains do not lead to significant changes in the volume occupied by the macromolecule in solution, *i.e.* in the intrinsic viscosity value. For PASA-*para* sample, [η] increases by only ≈15% in going from salt (0.2 M) to salt free solution. PASA-*meta* structure takes the intermediate position in this sample set. These data indicate the level of volume change of polyelectrolyte molecules, depending on the stiffness of the bare chains (structural rigidity) (Table **4**, columns 7 and 10).

Table 4. Molecular and hydrodynamic parameters of different (structural rigidity) linear polymers acquired in 0.2M NaCl and pure H_2O.

Sample	$[\eta]_{0.2M}$, cm^3/g	$M10^{-3}$, g/mol	A_{bare}, nm	L, nm	L/A	$v_{0.2M}$* 10^{-3} nm^3	λ, nm	$[\eta]H_2O$, cm^3/g	v H_2O* 10^{-3} nm^3	α_η	α_η^{Pst}
PASA-*para*	473	240	70	770	10	140	1.65	540	160	1.05	-
PASA-*meta*	209	250	9	710	85	65	1.50	5600	1 750	3.0	-
PSSNa-2	172	850	2.3	1000	450	180	0.252	9100	9 700	3.8	1.4
PSSNa-9/10	22	51	2.3	620	30	1.4	0.252	160	10	1.9	1.1

* - The volumes $v_{0.2M}$ and vH_2O occupied by a macromolecule were calculated in coil approximation using corresponding [η] values according to the following relation: $v \times 10^{-3} = (0.36[\eta]M/\Phi)$ nm^3; ** - swelling coefficient $\alpha_\eta = ([\eta]H_2O/[\eta]_{0.2M})^{1/3}$; *** - α_η^{Pst} were calculated in Ref. [55].

Changes in the volume occupied by polyions in solution of the extremely low ionic strength may be caused by the volume interaction as well as by increase in the equilibrium chain rigidity. The level of equilibrium rigidity of strongly charged polyelectrolyte was discussed qualitatively and quantitatively in a number of publications [9, 10, 40, 104, 105]. It was shown that, in principle, large size of strongly charged polyions molecules in salt free aqueous solution may be interpreted by increasing the total equilibrium chain rigidity (caused by the large contribution of short-range electrostatic interactions). The results presented in Table **4** show that this change is also dependent on the stiffness of the bare chain.

Note that the electrostatic interactions are not only strong, but they also spread over long distances, thus being long-range interactions. The problem is to discriminate the short- and long-range electrostatic intrachain interactions. This challenge remains unsolved, both theoretically and experimentally.

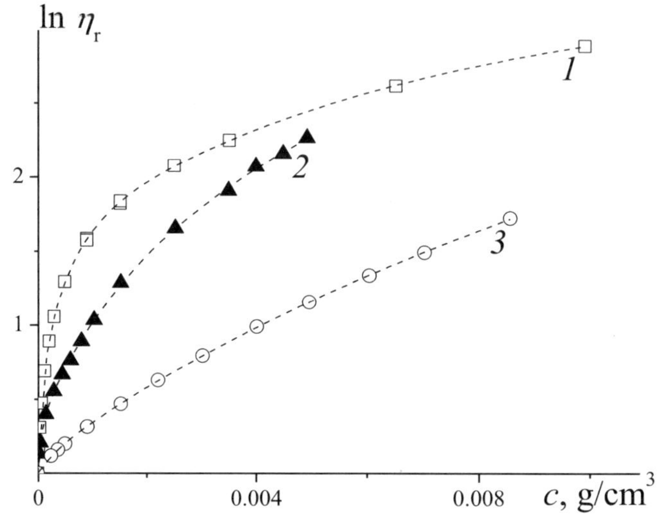

Fig. (3). Logarithm of relative viscosity lnη_r of polyelectrolytes solutions in pure water without adding low molecular salts. The polyelectrolytes chains have the significantly different structural (bare) rigidities: (*1*) – PSSNa-2, (*2*) - PASA-*meta*, (*3*) - PASA-*para*.

Now let us assume that the decisive influence on the size of highly charged chains is exerted by the long-range electrostatic effects (following W. Kuhn [64] and O. Ptitsyn [65]). We will also suggest, as a first approximation, that the size of chains at 0.2M NaCl solution does not differ significantly from the size of the unperturbed chain. On the basis of these assumptions, the lower limit of the swelling coefficient α_η of the polyions in salt-free aqueous solutions is evaluated and presented in the last column of the Table **4**. For comparison, the last column in Table **4** shows values of α_η^{Pst} corresponding to uncharged polystyrene chains during the transition from θ-conditions in cyclohexane to toluene (which is a thermodynamically good solvent for polystyrene). The values of α_η corresponding to virtually similar degree of polymerization are compared. From this comparison, we can get an insight into the limits of varying swelling coefficient of strong polyelectrolyte chains. The last row of Table **4** contains data on samples of PSSNa-9/10, whose contour length L is comparable to that of PASA-*para* sample, however the swelling coefficient α_η is still two times higher than the one for PASA-*para*. This fact proves the great influence of bare chain structural rigidity on polyelectrolyte potential to increase its volume in extremely low ionic strengths.

"Apparent Intrinsic Viscosity" [η]* at Nonzero Polymer Concentration

As follows from the literature data and from the data presented in the above sections, the most significant changes in the size of polyelectrolyte chains with increasing ionic strength of the solution can be expected for the polyelectrolyte with flexible bare chain [4, 10, 80]. In this paragraph the values of "apparent intrinsic viscosity" [η]* will be compared with [η] values obtained by direct measurement of the intrinsic viscosity in solutions of different ionic strengths. The comparison was made for two samples of PSSNa with different molar masses: 1 500 000 and 50 000 g/mol and shown in Fig. (**4**). Thus, the points (1 and 3) correspond to the values obtained at infinite dilution at a given ionic strength, reached by the use of NaCl in separately accomplished experiments: $\partial \ln\eta_r/\partial c = [\eta]_0^{\mu}$ at $c \to 0$.

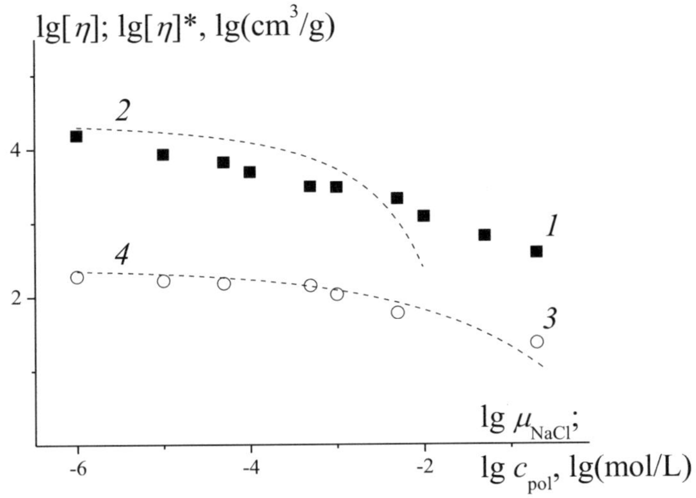

Fig. (4). Comparison of the values $[\eta] = (\partial\ln\eta_r/\partial c)_0$ at $c \to 0$ (1 and 3) obtained from independently prepared solutions in different ionic strength solvents, dashed lines 2 and 4 describe the dependences of $[\eta]^*_i = (\partial\ln\eta_r/\partial c)_{ci}$ calculated as the tangent at each point of the initial dependences $\ln\eta_r(c)$ obtained by measuring PSSNa solutions in pure water. 1 and 2 correspond to sample with M = 1 500 000 and 3 and 4 – M = 50 000 g/mol.

The dashed lines (2 and 4) correspond to the values $(\partial\ln\eta_r/c)_{ci} = [\eta]^*_i$ derived from the $\ln\eta_r(c)$ curves obtained from the viscometric study of the samples in salt free solution by serial dilution of stock solution with distilled water. To combine datasets (1, 3) and (2, 4) on one x-scale representing the ionic strength of solution: for (1, 3) the logarithm of NaCl concentration has been considered as x-values, and for (2, 4) the ionic strength of polymer solution in salt free water was determined as a molar polyelectrolyte concentration.

Fig. (**4**) reveals palatable and amazing correlation between $(\partial \ln\eta_r/c)_0$ at $c\rightarrow 0$ (intrinsic viscosity) characterizing the individual charged macromolecule and $(\partial \ln\eta_r/c)_{ci} \equiv [\eta]^*_i$ ("apparent" intrinsic viscosity) obtained at the different concentration of PSSNa in salt free solution, especially for low molar mass sample. At a finite polymer concentration, $[\eta]^*$ characterizes the specific volume occupied by the probe macromolecule placed into a solution with a uniform distribution of all other macromolecules. This approach can be considered as an experimental realization of the mean field concept [106 - 108]. The value of $[\eta]^*$ changes with varying polyelectrolyte concentration. The increase in polyelectrolyte concentration in aqueous salt free solutions leads at least to the appearance of two effects. First, the change in the ionic strength of the solution may be reached at low concentration by increasing the polyion concentration, and at higher concentration by approaching the different charged coils to the distance comparable to the Bjerrum length. At higher concentration of polyion it may be accompanied by the further counterions condensation [109]. Second, a certain contraction of molecular coils due to the electrostatic repulsion between neighboring charged macromolecules will occur. It is interesting to note that when a certain concentration of polyelectrolyte is achieved, the value of $[\eta]^*$ begins to decrease considerably. This occurs at a concentration that approximately corresponds to the condition $c[\eta]^*\approx 0.5$; this decrease is most clearly observed for high molar mass sample.

The slope method based on (Eq. 15) allows to estimate the value with the dimension of cm^3/g which could be called "apparent intrinsic viscosity". We are not sure that this value is the only and unique characteristic of one isolated macromolecule surrounded by many homologues. However, it is associated with the effects of compaction of coil of charged flexible-chain polymers in going to moderate polymer concentration. Once again, the situation is the opposite of that observed for uncharged chains (compare Fig. **1b** and **d**). For uncharged chains, the value of $(\ln\eta_r/c)_{c\neq 0}$ increases with the concentration; this fact is associated with the emergence of network structure in the solution. This observation/phenomenon should be further investigated both experimentally and theoretically.

CONCLUDING REMARKS

The intrinsic viscosity is a cornerstone value of molecular physics of polymers. This value is the most frequently used in characterization of linear polymers, especially synthetic. The use of classical Huggins (η_{sp}/c *vs.* c) and Kraemer ($\ln\eta_r/c$ *vs.* c) plots is a common method to evaluate the appropriate value of intrinsic viscosity. However, these plots become non-linear in the case of charged polymers and do not allow determining the $[\eta]$ value. Thus, these plots may serve as a test for the presence of charges in linear chains. However, "primary dependences", namely, (η_{sp} *vs.* c) and ($\ln\eta_r$ *vs.* c) are similar in shape both for the neutral and charged polymers. The initial slope of "primary plots" at $c\rightarrow 0$ is regarded as an estimate of the $[\eta]$ value, characterizing the individual macromolecules of any type. The problem of the adequacy of the obtained intrinsic viscosity values in salt free polyelectrolyte solutions was subjected to scrutiny. Firstly, the estimates obtained by initial slop method were compared with the results obtained by the isoionic dilution method. This comparison demonstrated satisfactory agreement between the values obtained by two different methods. In addition, the interpretation of the values of intrinsic viscosity obtained by the initial slope method for a wide homologous series of sodium poly(styrene-4-sulfonate) leads to an adequate assessment of the mass per unit length of the polymer chain and a reasonable estimate for the statistical segment length. It is also a confirmation of the adequacy of the method proposed for intrinsic viscosity estimation. Significant effect of the structural rigidity of the linear polymer chain (bare chain rigidity) on the intrinsic viscosity was also demonstrated in salt free solutions. Furthermore, so-called the apparent viscosity values ($\ln\eta_r/c$ at $c_i\neq 0$) are matched with the intrinsic viscosities at μ_i obtained by isoionic dilution method. Thus, the initial slop method can be used for systematic monitoring of the intrinsic viscosity values of the charged polymer chains.

CONFLICT OF INTEREST

The author (editor) declares no conflict of interest, financial or otherwise.

ACKNOWLEDGEMENTS

The authors are grateful to E. F. Panarin, I. I. Gavrilova and Yu. A. Fedotov for kindly provided samples.

REFERENCES

[1] M. Schmidt, *Polyelectrolytes with Defined Molecular Architecture I.*Springer, 2004, p. 267. [http://dx.doi.org/10.1007/b10951]

[2] M. Schmidt, *Polyelectrolytes with Defined Molecular Architecture II.*Springer, 2010, p. 244.

[3] M. Hara, *Polyelectrolytes: Science and Technology.*Marcel Dekker, Inc.: New York, 1993, p. 416.

[4] H. Dautzenberg, W. Jaeger, J. Kötz, B. Philipp, C. Seidel, and D. Stscherbina, *Polyelectrolytes: Formation, Characterization, and Application.*Hanser-Gardner Publications: Munich, 1994, p. 343.

[5] M.J. Stevens, and K. Kremer, "The nature of flexible linear polyelectrolytes in salt free solution: A molecular dynamics study", *J. Chem. Phys.,* vol. 103, no. 4, pp. 1669-1690, 1995. [http://dx.doi.org/10.1063/1.470698]

[6] A.V. Dobrynin, and M. Rubinstein, "Theory of polyelectrolytes in solutions and at surfaces", *Prog. Polym. Sci.,* vol. 30, no. 11, pp. 1049-1118, 2005.

[http://dx.doi.org/10.1016/j.progpolymsci.2005.07.006]

[7] M. Rubinstein, and G.A. Papoian, "Polyelectrolytes in biology and soft matter", *Soft Matter,* vol. 8, no. 36, pp. 9265-9267, 2012. [http://dx.doi.org/10.1039/c2sm90104h]

[8] H. Eisenberg, *Biological macromolecules and polyelectrolytes in solution.*Clarendon Press: UC, 1976, p. 272.

[9] G.M. Pavlov, A.S. Gubarev, I.I. Zaitseva, and M.A. Sibileva, "Determination of intrinsic viscosity of polyelectrolytes in salt-free solutions", *Russ. J. Appl. Chem.,* vol. 79, no. 9, pp. 1407-1412, 2006. [http://dx.doi.org/10.1134/S1070427206090035]

[10] G.M. Pavlov, O.V. Okatova, A.S. Gubarev, I.I. Gavrilova, and E.F. Panarin, "Strong Linear Polyelectrolytes in Solutions of Extreme Concentrations of One–One Valent Salt. Hydrodynamic Study", *Macromol,* vol. 47, pp. 2748-2758, 2014.

[11] C. Tanford, *Physical Chemistry of Macromolecules.*Wiley: New York, 1961, p. 710.

[12] V.N. Tsvetkov, V.E. Eskin, and S.Y. Frenkel, *Structure of Macromolecules in Solution.*Nat. Lend. Libr. Sci. Techn: Boston, 1971, p. 762.

[13] C.R. Cantor, and P.R. Schimmel, *Biophysical Chemistry.*W. H. Freeman & Company: San Francisco, 1980.

[14] M. Wintermantel, M. Gerle, K. Fischer, M. Schmidt, I. Wataoka, H. Urakawa, K. Kajiwara, and Y. Tsukahara, "Molecular Bottlebrushes", *Macromol,* vol. 29, no. 3, pp. 978-983, 1996.

[15] S. Lecommandoux, F. Chécot, R. Borsali, M. Schappacher, A. Deffieux, A. Brûlet, and J.P. Cotton, "Effect of Dense Grafting on the Backbone Conformation of Bottlebrush Polymers: Determination of the Persistence Length in Solution", *Macromol.,* vol. 35, no. 23, pp. 8878-8881, 2002. [http://dx.doi.org/10.1021/ma0203344]

[16] L. Feuz, P. Strunz, T. Geue, M. Textor, and O. Borisov, "Conformation of poly(L-lysine)-graft-poly(ethylene glycol) molecular brushes in aqueous solution studied by small-angle neutron scattering", *Eur Phys J E Soft Matter,* vol. 23, no. 3, pp. 237-245, 2007. [http://dx.doi.org/10.1140/epje/i2007-10180-9] [PMID: 17619817]

[17] M. Kikuchi, L.T. Lien, A. Narumi, Y. Jinbo, Y. Izumi, K. Nagai, and S. Kawaguchi, "Conformational Properties of Cylindrical Rod Brushes Consisting of a Polystyrene Main Chain and Poly(n-hexyl isocyanate) Side Chains", *Macromol,* vol. 41, no. 17, pp. 6564-6572, 2008.

[18] G. Cheng, Y.B. Melnichenko, G.D. Wignall, F. Hua, K. Hong, and J.W. Mays, "Small Angle Neutron Scattering Study of Conformation of Oligo(ethylene glycol)-Grafted Polystyrene in Dilute Solutions: Effect of the Backbone Length", *Macromol.,* vol. 41, no. 24, pp. 9831-9836, 2008. [http://dx.doi.org/10.1021/ma801370q]

[19] S.S. Sheiko, and M. Möller, "Visualization of macromoleculesa first step to manipulation and controlled response", *Chem. Rev.,* vol. 101, no. 12, pp. 4099-4124, 2001. [http://dx.doi.org/10.1021/cr990129v] [PMID: 11740928]

[20] T. Stephan, S. Muth, and M. Schmidt, "Shape Changes of Statistical Copolymacromonomers: From Wormlike Cylinders to Horseshoe- and Meanderlike Structures", *Macromol.,* vol. 35, no. 27, pp. 9857-9860, 2002. [http://dx.doi.org/10.1021/ma025711r]

[21] I.I. Potemkin, A.R. Khokhlov, S. Prokhorova, S.S. Sheiko, M. Möller, K.L. Beers, and K. Matyjaszewski, "Spontaneous Curvature of Comblike Polymers at a Flat Interface", *Macromol.,* vol. 37, no. 10, pp. 3918-3923, 2004. [http://dx.doi.org/10.1021/ma021519d]

[22] D. Lanson, M. Schappacher, A. Deffieux, and R. Borsali, "Synthesis of Comblike Poly(styrene--isoprene) Block Copolymers and Their Properties in Good and Selective Solvents", *Macromol.,* vol.

39, no. 20, pp. 7107-7114, 2006.
[http://dx.doi.org/10.1021/ma061169l]

[23] H.G. Elias, *Macromolecules: Structure and properties.*vol. Vol. 1. Plenum Press, 1977, p. 1131. [http://dx.doi.org/10.1007/978-1-4615-7364-7]

[24] F.W. Billmeyer, *Textbook of polymer science.*3rd ed Wiley Interscience: New York, 1984, p. 578.

[25] P. Munk, and T.M. Aminabhavi, *Introduction to Macromolecular Science.*2nd ed Wiley: New York, 2002.

[26] S.E. Harding, "The intrinsic viscosity of biological macromolecules. Progress in measurement, interpretation and application to structure in dilute solution", *Prog. Biophys. Mol. Biol.,* vol. 68, no. 2-3, pp. 207-262, 1997. [http://dx.doi.org/10.1016/S0079-6107(97)00027-8] [PMID: 9652172]

[27] K.E. van Holde, V. Johnson, and P. Ho, *Principles of Physical Biochemistry.*2nd ed Prentice Hall: Upper Saddle River, 1998.

[28] R.H. Hunter, *Foundations of Colloid Science.*Oxford University Press: Oxford, 2001.

[29] M. Bohdanecky, and J. Kovar, *Viscosity of Polymer Solutions.*Elsevier Scientific Publishing Company: Amsterdam, Oxford, New York, 1982, p. 285.

[30] H. Staudinger, *Nobel Lecture: Macromolecular Chemistry; Chemistry 1942-1962.*Elsevier: Amsterdam, 1953.

[31] M.L. Hunt, S. Newman, H.A. Scheraga, and P.J. Flory, "Dimensions and Hydrodynamic Properties of Cellulose Trinitrate Molecules in Dilute Solution", *J. Phys. Chem.,* vol. 60, no. 9, pp. 1278-1290, 1956. [http://dx.doi.org/10.1021/j150543a031]

[32] G.M. Pavlov, A.N. Kozlov, G.N. Martchenko, and V.N. Tsvetkov, "Sedimentation and viscosity of highly substituted cellulose nitrate", *Polym. Sci. B,* vol. 24, pp. 284-288, 1982.

[33] J. Brandrup, E.H. Immergut, and E.A. Grulke, *>Polymer Handbook, 2 Volumes Set*Wiley, 2003.

[34] *http://www.malvern.com/.*

[35] *http://www.anton-paar.com/.*

[36] J. Waters, and D. Leiske, "Characterization of hyaluronic acid with on-line differential viscometry, multiangle light scattering, and differential refractometry", *LC GC N. Am.,* vol. 23, no. 3, p. 302, 2005.

[37] M.L. Huggins, "The Viscosity of Dilute Solutions of Long-Chain Molecules. IV. Dependence on Concentration", *J. Am. Chem. Soc.,* vol. 64, no. 11, pp. 2716-2718, 1942. [http://dx.doi.org/10.1021/ja01263a056]

[38] E.O. Kraemer, "Molecular Weights of Celluloses and Cellulose Derivates", *Ind. Eng. Chem.,* vol. 30, no. 10, pp. 1200-1203, 1938. [http://dx.doi.org/10.1021/ie50346a023]

[39] G.M. Pavlov, E.F. Panarin, E.V. Korneeva, C.V. Kurochkin, V.E. Baikov, and V.N. Ushakova, "Hydrodynamic properties of poly(1-vinyl-2-pyrrolidone) molecules in dilute solution", *Makromol. Chem.,* vol. 191, no. 12, pp. 2889-2899, 1990. [http://dx.doi.org/10.1002/macp.1990.021911205]

[40] G.M. Pavlov, A.S. Gubarev, I.I. Gavrilova, and E.F. Panarin, "Conformations of Sodium Poly(styrene-4-sulfonate) Macromolecules in Solutions with Different Ionic Strengths", *Polym. Sci. A,* vol. 53, no. 11, pp. 1003-1011, 2011. [http://dx.doi.org/10.1134/S0965545X11110095]

[41] V.N. Tsvetkov, *Rigid-chain polymers. Consult. Bureau.*Plenum: London, 1989, p. 490.

[42] B.H. Zimm, and D.M. Crothers, "Simplified Rotating Cylinder Viscometer for DNA", *Proc. Natl.*

Acad. Sci. USA, vol. 48, no. 6, pp. 905-911, 1962.
[http://dx.doi.org/10.1073/pnas.48.6.905] [PMID: 16590954]

[43] M.A. Haney, "A differential viscometer 1", *Am. Lab.,* vol. 17, no. 3, pp. 41-56, 1985.

[44] M.A. Haney, "A differential viscometer 2", *Am. Lab.,* vol. 17, no. 4, pp. 116-126, 1985.

[45] M.A. Haney, "The differential viscometer 2. Online viscosity detector for size-exclusion chromatography", *J. Appl. Polym. Sci.,* vol. 30, no. 7, pp. 3037-3049, 1985.
[http://dx.doi.org/10.1002/app.1985.070300725]

[46] P.J. Flory, and T.G. Fox, "Treatment of Intrinsic Viscosities", *J. Am. Chem. Soc.,* vol. 73, no. 5, pp. 1904-1908, 1951.
[http://dx.doi.org/10.1021/ja01149a002]

[47] G. Meyerhoff, and B. Appelt, "A Low Shear Viscometer with Automated Recording and Application to High Molecular Weight Polystyrene Solutions", *Macromol,* vol. 12, no. 5, pp. 968-971, 1979.
[http://dx.doi.org/10.1021/ma60071a037]

[48] G.M. Pavlov, K. Knop, O.V. Okatova, and U.S. Schubert, "Star-Brush-Shaped Macromolecules: Peculiar Properties in Dilute Solution", *Macromol.,* vol. 46, no. 21, pp. 8671-8679, 2013.
[http://dx.doi.org/10.1021/ma400160f]

[49] I. Perevyazko, J. Seiwert, M. Schömer, H. Frey, U.S. Schubert, and G.M. Pavlov, "Hyperbranched Poly(ethylene glycol) Copolymers: Absolute Values of the Molar Mass, Properties in Dilute Solution, and Hydrodynamic Homology", *Macromol.,* vol. 48, no. 16, pp. 5887-5898, 2015.
[http://dx.doi.org/10.1021/acs.macromol.5b01020]

[50] S.Y. Magarik, G.M. Pavlov, and G.A. Fomin, "Hydrodynamic and Optical Properties of Homologous Series of Styrene-Methyl Methacrylate Graft Copolymers", *Macromol.,* vol. 11, no. 2, pp. 294-300, 1978.
[http://dx.doi.org/10.1021/ma60062a003]

[51] G.M. Pavlov, A.M. Breul, M.D. Hager, and U.S. Schubert, "Hydrodynamic and Molecular Study of Poly{4-[4-(hexyloxy)phenyl]ethynylphenyl methacrylate} in Dilute Solutions and Conformational Peculiarities of Brush-Like Macromolecules", *Macromol. Chem. Phys.,* vol. 213, no. 9, pp. 904-916, 2012.
[http://dx.doi.org/10.1002/macp.201100653]

[52] F. Abe, Y. Einaga, T. Yoshizaki, and H. Yamakawa, "Excluded-volume effects on the mean-square radius of gyration of oligo- and polystyrenes in dilute solutions", *Macromol.,* vol. 26, no. 8, pp. 1884-1890, 1993.
[http://dx.doi.org/10.1021/ma00060a014]

[53] F. Abe, K. Horita, Y. Einaga, and H. Yamakawa, "Excluded-volume effects on the mean-square radius of gyration and intrinsic viscosity of oligo- and poly(methyl methacrylates)", *Macromol.,* vol. 27, no. 3, pp. 725-732, 1994.
[http://dx.doi.org/10.1021/ma00081a016]

[54] F. Peeters, and H. Smits, "De invloed van beperkte wederzijdse doordringing van 'po'-klumens of hun sedimentatiesnelheid in verdunde oplossingen", *Bull. Soc. Chim. Belg.,* vol. 90, no. 2, pp. 111-122, 1981.

[55] H.L. Wagner, "The Mark–Houwink–Sakurada Equation for the Viscosity of Atactic Polystyrene", *J. Phys. Chem. Ref. Data,* vol. 14, no. 4, pp. 1101-1106, 1985.
[http://dx.doi.org/10.1063/1.555740]

[56] T. Yanaki, T. Norisuye, and H. Fujita, "Triple Helix of Schizophyllum commune Polysaccharide in Dilute Solution. 3. Hydrodynamic Properties in Water", *Macromol.,* vol. 13, no. 6, pp. 1462-1466, 1980.
[http://dx.doi.org/10.1021/ma60078a019]

[57] T. Sato, T. Norisuye, and H. Fujita, "Double-stranded helix of xanthan: dimensional and

hydrodynamic properties in 0.1 M aqueous sodium chloride", *Macromol.,* vol. 17, no. 12, pp. 2696-2700, 1984.
[http://dx.doi.org/10.1021/ma00142a043]

[58] T. Norisuye, "Semiflexible polymers in dilute solution", *Prog. Polym. Sci.,* vol. 18, no. 3, pp. 543-584, 1993.
[http://dx.doi.org/10.1016/0079-6700(93)90017-7]

[59] V.H. Lütje, and G. Meyerhoff, "Untersuchungen über das Reibungsverhalten von Polymethacrylsäuremethylestern in verschiedenen Lösungsmitteln", *Makromol. Chem.,* vol. 68, no. 1, pp. 180-193, 1963.
[http://dx.doi.org/10.1002/macp.1963.020680113]

[60] M. Tricot, J.P. Bleus, J-P. Riga, and V. Desreux, "Solution properties and unperturbed dimensions of poly(5-p-methyl methacrylate) and poly(2-tert-butylphenyl methacrylate)", *Makromol. Chem.,* vol. 175, no. 3, pp. 913-921, 1974.
[http://dx.doi.org/10.1002/macp.1974.021750318]

[61] T. Arai, F. Abe, T. Yoshizaki, Y. Einaga, and H. Yamakawa, "Hydrodynamic-Radius Expansion Factor of Polystyrene in Cyclohexane near the THETA Temperature", *Macromol.,* vol. 28, no. 16, pp. 5458-5464, 1995.
[http://dx.doi.org/10.1021/ma00120a009]

[62] G.M. Pavlov, "Investigation of cellulose and lignins by molecular hydrodynamic methods", In: *Cellulose and cellulose derivatives: Physico-chemical aspects and industrial applications.,* Y.E. Kennedy, G.O. Phillips, P.W. Williams, L. Piculell, Eds., Woodhead Pub. Ltd.: Cambridge, England, 1995, pp. 541-546.
[http://dx.doi.org/10.1533/9781845698539.6.541]

[63] D.M. Crothers, and B.H. Zimm, "Viscosity and sedimentation of the DNA from bacteriophages T2 and T7 and the relation to molecular weight", *J. Mol. Biol.,* vol. 12, no. 3, pp. 525-536, 1965.
[http://dx.doi.org/10.1016/S0022-2836(65)80310-2] [PMID: 5858429]

[64] W. Kuhn, O. Kunzle, and A. Katchalsky, "Verhalten polyvalenter fadenmolekelionen in losung", *Helv. Chim. Acta,* vol. 31, no. 7, pp. 1994-2037, 1948.
[http://dx.doi.org/10.1002/hlca.19480310716] [PMID: 18115082]

[65] O.B. Ptitsyn, "Theory of polyelectrolyte solutions. III. The influence of the non-uniform distribution of charges along the chain on the size and shape of macromolecules", *Vysokomol. Soedin.,* vol. 3, no. 9, pp. 1401-1405, 1961.

[66] N. Borochov, and H. Eisenberg, "Stiff (DNA) and Flexible (NaPSS) Polyelectrolyte Chain Expansion at Very Low Salt Concentration", *Macromol.,* vol. 27, no. 6, pp. 1440-1445, 1994.
[http://dx.doi.org/10.1021/ma00084a025]

[67] T. Odijk, "Polyelectrolytes near the rod limit", *J. Polym. Sci., B, Polym. Phys.,* vol. 15, no. 3, pp. 477-483, 1977.
[http://dx.doi.org/10.1002/pol.1977.180150307]

[68] T. Odijk, and A.C. Houwaart, "Theory of excluded-volume effect of a polyelectrolyte in a 1-1 electrolyte solution", *J. Polym. Sci., B, Polym. Phys.,* vol. 16, no. 4, pp. 627-639, 1978.
[http://dx.doi.org/10.1002/pol.1978.180160405]

[69] J. Skolnick, and M. Fixman, "Electrostatic persistence length of a wormlike polyelectrolyte", *Macromol.,* vol. 10, no. 5, pp. 944-948, 1977.
[http://dx.doi.org/10.1021/ma60059a011]

[70] J-L. Barrat, and J-F. Joanny, "Persistence Length of Polyelectrolyte Chains", *Europhys. Lett.,* vol. 24, no. 5, pp. 333-338, 1993.
[http://dx.doi.org/10.1209/0295-5075/24/5/003]

[71] A.V. Dobrynin, "Electrostatic Persistence Length of Semiflexible and Flexible Polyelectrolytes",

Macromol., vol. 38, no. 22, pp. 9304-9314, 2005.
[http://dx.doi.org/10.1021/ma051353r]

[72] J.L. Barrat, and J.F. Joanny, "Theory of polyelectrolyte solutions", *Adv. Chem. Phys.,* vol. 64, no. 94, pp. 1-66, 1996.
[http://dx.doi.org/10.1002/9780470141533.ch1]

[73] A.S. Gubarev, J.M. Carrillo, and A.V. Dobrynin, "Scale-Dependent Electrostatic Stiffening in Biopolymers", *Macromol.,* vol. 42, no. 15, pp. 5851-5860, 2009.
[http://dx.doi.org/10.1021/ma9008143]

[74] G.S. Manning, "The persistence length of DNA is reached from the persistence length of its null isomer through an internal electrostatic stretching force", *Biophys. J.,* vol. 91, no. 10, pp. 3607-3616, 2006.
[http://dx.doi.org/10.1529/biophysj.106.089029] [PMID: 16935960]

[75] J-M. Carrillo, and A.V. Dobrynin, "Polyelectrolytes in Salt Solutions: Molecular Dynamics Simulations", *Macromol.,* vol. 44, no. 14, pp. 5798-5816, 2011.
[http://dx.doi.org/10.1021/ma2007943]

[76] M. Fixman, "Electrostatic persistence length", *J. Phys. Chem. B,* vol. 114, no. 9, pp. 3185-3196, 2010.
[http://dx.doi.org/10.1021/jp1005742] [PMID: 20148527]

[77] M. Muthukumar, "Counterion adsorption theory of dilute polyelectrolyte solutions: apparent molecular weight, second virial coefficient, and intermolecular structure factor", *J. Chem. Phys.,* vol. 137, no. 3, p. 034902, 2012.
[http://dx.doi.org/10.1063/1.4736545] [PMID: 22830728]

[78] M. Fixman, and J. Skolnick, "Polyelectrolyte Excluded Volume Paradox", *Macromol.,* vol. 11, no. 5, pp. 863-867, 1978.
[http://dx.doi.org/10.1021/ma60065a004]

[79] M. Beer, M. Schmidt, and M. Muthukumar, "The Electrostatic Expansion of Linear Polyelectrolytes: Effects of Gegenions, Co-ions, and Hydrophobicity", *Macromol.,* vol. 30, no. 26, pp. 8375-8385, 1997.
[http://dx.doi.org/10.1021/ma9709821]

[80] H. Eisenberg, and J. Pouyet, "Viscosities of dilute aqueous solutions of a partially quaternized poly--vinylpyridine at low gradients of flow", *J. Polym. Sci., Polym. Phys. Ed.,* vol. 13, no. 68, pp. 85-91, 1954.

[81] K. Nishida, K. Kaji, T. Kanaya, and N. Fanjat, "Determination of intrinsic viscosity of polyelectrolyte solutions", *Polym.,* vol. 43, no. 4, pp. 1295-1300, 2002.
[http://dx.doi.org/10.1016/S0032-3861(01)00682-6]

[82] K. Nishida, K. Kaji, and T. Kanaya, "Theoretical calculation of reduced viscosity of polyelectrolyte solutions", *Polym.,* vol. 42, no. 21, pp. 8657-8662, 2001.
[http://dx.doi.org/10.1016/S0032-3861(01)00368-8]

[83] R.M. Fuoss, "Viscosity function for polyelectrolytes", *J. Polym. Sci., Polym. Phys. Ed.,* vol. 3, no. 4, pp. 603-604, 1948.

[84] D.T. Pals, and J.J. Hermans, "New method for deriving the intrinsic viscosity of polyelectrolytes", *J. Polym. Sci., Polym. Phys. Ed.,* vol. 5, no. 6, pp. 733-734, 1950.

[85] I. Sakurada, and N. Ise, "Uber die viskositat von polyelektrolytlosungen", *Makromol. Chem.,* vol. 40, pp. 126-139, 1960.
[http://dx.doi.org/10.1002/macp.1960.020400112]

[86] V.G. Baranov, Y.V. Brestkin, S.A. Agranova, and V.N. Pinkevich, "“Behavior of polystyrene macromolecules in stiffened good solvent”. Vysokomol. Soedin", *Ser. B,* vol. 28, no. 11, pp. 841-843, 1986.

[87] V.G. Baranov, S.Y. Frenkel, S.A. Agranova, Y.V. Brestkin, V.N. Pinkevich, and B.M. Shabsels, "“Concentrational dependence of viscosity of helix polypeptide solutions”. Vysokomol. Soedin", *Ser. B,* vol. 29, no. 10, pp. 745-748, 1987.

[88] V.I. Irzhak, *Fizicheskie svoistva vyazkouprugikh polimerov (Physical Properties of Viscoelastic Polymers), Sverdlovsk: Ural Sci.*Cent. RAS USSR, 1981, pp. 76-81.

[89] D.T. Pals, and J.J. Hermans, "Sodium salts of pectin and of carboxy methyl cellulose in aqueous sodium chloride. II Osmotic pressures", *Recl. Trav. Chim. Pays Bas,* vol. 71, no. 5, pp. 458-467, 1952. [http://dx.doi.org/10.1002/recl.19520710505]

[90] B.A. Wolf, "Polyelectrolytes Revisited: Reliable Determination of Intrinsic Viscosities", *Macromol. Rapid Commun.,* vol. 28, no. 2, pp. 164-170, 2007. [http://dx.doi.org/10.1002/marc.200600650]

[91] G.M. Pavlov, I.I. Zaitseva, A.S. Gubarev, E.V. Korneeva, I.I. Gavrilova, and E.F. Panarin, "Conformation of sodium poly(4-styrenesulfonate) macromolecules in aqueous solutions", *Dokl. Chem.,* vol. 419, pp. 111-112, 2008. [http://dx.doi.org/10.1134/S0012500808040083]

[92] J. Eckelt, A. Knopf, and B.A. Wolf, "Polyelectrolytes: Intrinsic Viscosities in the Absence and in the Presence of Salt", *Macromol.,* vol. 41, no. 3, pp. 912-918, 2008. [http://dx.doi.org/10.1021/ma702054f]

[93] A. Eich, and B.A. Wolf, "Intrinsic viscosities of polyelectrolytes: determination and modeling of the effects of extra salt", *ChemPhysChem,* vol. 12, no. 15, pp. 2786-2790, 2011. [http://dx.doi.org/10.1002/cphc.201100439] [PMID: 21919181]

[94] G.M. Pavlov, O.V. Okatova, N.N. Ulyanova, I.I. Gavrilova, and E.F. Panarin, "Characteristic features of the behavior of charged hydrophilic and hydrophobic macromolecules in solutions of different ionic strength", *Dokl. Chem.,* vol. 448, pp. 16-18, 2013. [http://dx.doi.org/10.1134/S0012500813010059]

[95] G.M. Pavlov, "Size and average density spectra of macromolecules obtained from hydrodynamic data", *Eur Phys J E Soft Matter,* vol. 22, no. 2, pp. 171-180, 2007. [http://dx.doi.org/10.1140/epje/e2007-00025-x] [PMID: 17377754]

[96] G.M. Pavlov, "Hydrodynamic of Macromolecules: Conformation Zoning of General Macromolecules", In: *Encyclopedia of Biophysics.,* G.C. Roberts, Ed., Springer Berlin Heidelberg: Berlin, 2013, pp. 1014-1024. [http://dx.doi.org/10.1007/978-3-642-16712-6_295]

[97] H. Yamakawa, and M. Fujii, "Intrinsic viscosity of wormlike chains. Determination of the shift factor", *Macromolecules,* vol. 7, no. 1, pp. 128-135, 1974. [http://dx.doi.org/10.1021/ma60037a024] [PMID: 4837983]

[98] S. Broersma, "Translational Diffusion Constant of a Random Coil", *J. Chem. Phys.,* vol. 51, no. 1, pp. 233-238, 1969. [http://dx.doi.org/10.1063/1.1671712]

[99] G.M. Pavlov, E.V. Korneeva, Y.A. Fedotov, G.E. Polushina, and I.E. Andreeva, "Molecular characteristics of sulfonated aromatic polyamide based on isophthalic acid", *Russ. J. Appl. Chem.,* vol. 69, no. 5, pp. 734-737, 1996.

[100] G.M. Pavlov, E.V. Korneeva, and Y.A. Fedotov, "Draining and volume effects in the chains of semi-rigid sulfated aromatic polyamide", *Polym. Sci. A,* vol. 39, pp. 1979-1985, 1997.

[101] G.M. Pavlov, A.S. Gubarev, I.I. Zaitseva, and Y.A. Fedotov, "Spontaneous birefringence in films of some phenyl-containing polymers", *Polym. Sci. B,* vol. 49, no. 7-8, pp. 191-195, 2007.

[102] T.M. Birshtein, "Flexibility of polymer chains containing plane cyclic moieties", *Polym. Sci. A,* vol.

19, no. 1, p. 54, 1977.

[103] H. Benoit, "Calcul de l'écart quadratique moyen entre les extrémités de diverses chaînes moléculaires de type usuel", *J. Polym. Sci., Polym. Phys. Ed.,* vol. 3, no. 3, pp. 376-388, 1948.

[104] G.M. Pavlov, O.V. Okatova, I.I. Gavrilova, N.N. Ul'yanova, and E.F. Panarin, "Sizes and conformations of hydrophilic and hydrophobic polyelectrolytes in solutions of various ionic strengths", *Polym. Sci. A,* vol. 55, no. 12, pp. 699-705, 2013. [http://dx.doi.org/10.1134/S0965545X13120079]

[105] G.M. Pavlov, O.V. Okatova, N.N. Ul'yanova, I.I. Gavrilova, and E.F. Panarin, "Size of linear polyelectrolytes with different charge density in salt-free aqueous solutions", *Dokl. Chem.,* vol. 454, pp. 13-16, 2014. [http://dx.doi.org/10.1134/S0012500814010042]

[106] S.A. Baeurle, and E.A. Nogovitsin, "Challenging scaling laws of flexible polyelectrolyte solutions with effective renormalization concepts", *Polym.,* vol. 48, no. 16, pp. 4883-4899, 2007. [http://dx.doi.org/10.1016/j.polymer.2007.05.080]

[107] F. Schmid, "Self-consistent-field theories for complex fluids", *J. Phys. Condens. Matter,* vol. 10, no. 37, pp. 8105-8138, 1998. [http://dx.doi.org/10.1088/0953-8984/10/37/002]

[108] S.F. Edwards, "The statistical mechanics of polymers with excluded volume", *Proc. Phys. Soc.,* vol. 85, no. 4, p. 613, 1965. [http://dx.doi.org/10.1088/0370-1328/85/4/301]

[109] G.S. Manning, "Counterion condensation theory constructed from different models", *Physica A,* vol. 231, no. 1–3, pp. 236-253, 1996. [http://dx.doi.org/10.1016/0378-4371(95)00452-1]

CHAPTER 15

Interrelation Between Polysaccharides and Different Surfactant Types

M. C. Ortiz-Tafoya and **A. Tecante***

Departamento de Alimentos y Biotecnología, Facultad de Química, Universidad Nacional Autónoma de México, Cd. Universitaria, CdMx, 04510, México.

Abstract: Interactions between polysaccharides and surfactants in aqueous solutions are the topic of this chapter. These components are often found in foods as well as in other systems like cosmetics, paints, detergents, textiles and many more. Therefore, investigation of the interactions between polysaccharides and surfactants has a great practical and fundamental interest. For the sake of clarity, we have divided the discussion into three sections: polysaccharides-nonionic surfactants; polysaccharides-cationic surfactants and polysaccharides-anionic surfactants. The behavior of polysaccharide-surfactant systems depends on many factors, which mainly include: 1) physical properties of both components, like molecular weight, degree of branching, number, and type of charge groups, backbone rigidity, 2) polysaccharide concentrations, 3) surfactant type; ionic or nonionic, polar head, chain length, and concentration. Environmental conditions like pH, ionic strength and presence of salt can partially screen electrostatic interactions, and there are also other factors needed to be considered when working with this type of mixtures. All these factors make it difficult to predict the behavior of polysaccharide-surfactant mixtures. However, in almost all cases, addition of surfactant modifies the behavior of the polysaccharide, regardless of polysaccharide and surfactant, but the modification is particular to the chemical nature of the polysaccharide and surfactant.

Keywords: Alginate, Biopolymers, Carboxymethylcellulose, Carrageenan, Chitosan, Electrostatic interactions, Emulsion, Hydrophobic interactions, Micelles, Polyelectrolyte, Polysaccharide, Surface tension, Surfactant.

INTRODUCTION

Foods are complex systems formed by a great number of components, including but not restricted to, sugars, proteins, polysaccharides, lipids, salts, and surfactants.

* **Corresponding author Alberto Tecante:** Departamento de Alimentos y Biotecnología, Facultad de Química, Universidad Nacional Autónoma de México, Cd. Universitaria, CdMx, 04510, México; Tel/Fax: +52-55-5622-5307; E-mail: tecante@unam.mx

Martin Masuelli & Denis Renard (Eds.)

These components may interact with each other in different ways depending on process conditions such as temperature and pH which make it difficult to control the shelf life and characteristics of the product such as texture and stability.

Surfactant is an abbreviation of surface active agent. A surfactant is a chemical compound with a tendency to adsorb at surfaces and interfaces. An interface is a boundary between two immiscible phases, and the term surface indicates that one of the phases is a gas, usually air [1]. The use of surface-active molecules in prepared foods is quite common. Low molecular weight surfactants, *e.g.* mono and diglycerides, phospholipids, are frequently used in combination with macromolecules like proteins and polysaccharides. In general, polysaccharides do not exhibit surface activity, but proteins can adsorb at interfaces and therefore it is important to know the interfacial properties and the dynamic behavior of all these molecules not only to understand the properties and behavior of complex systems but to develop products for specific purposes [2, 3].

Polysaccharides and surfactants are added to prepared foods basically to improve the stability of a food system. Polysaccharides are usually used to manage the rheological properties, while surfactants are used to handle the interfacial properties. In the case of polysaccharides, the main purpose is to get the desired viscoelastic properties. The interaction between polysaccharides and surfactants is studied in a model system (*i.e.* in aqueous solution), as foods are too complex to conduct fundamental studies. Therefore, most of our knowledge comes from a "simple" system. On the other hand, when polysaccharides and surfactants are used in combination, undesirable effects may occur, and the stability of the system can be affected. For example, incompatibility and formation of polysaccharide-surfactant complexes may eventually lead to phase separation. The interactions between polysaccharides and surfactants depend strongly on the properties of the former, on the type of the latter, on their respective concentrations and proportions in the mixture and on the environmental conditions of the aqueous medium, *e.g.* pH, ionic strength and temperature [4].

Surfactants can self-associate into micelles or micelle-like structures, and these aggregates can, in turn, interact with polymers present in the bulk phase. The concentration of surfactant when the interaction occurs is called the critical aggregation concentration (CAC). This concentration is lower than the critical micelle concentration (CMC) which is eventually reached by increasing the concentration of the surface-active molecule at a constant temperature. Polymers and surfactants can interact in different ways. Electrostatic, hydrophobic and dipolar interactions can be present depending on the nature of the polymer and the surfactant. Neutral polysaccharides can interact with anionic and cationic surfactants. However, the ways these interactions occur are different for each type of surfactant. Charged polymers, *i.e.* polyelectrolytes, interact with oppositely charged surfactants by electrostatic attraction and when the surfactant and the polymer bear the same charge, interaction is still possible if the macromolecule is sufficiently hydrophobic. The behavior of surfactants is also different because anionic surfactants show stronger interaction than cationic ones of comparable chain length [5].

Interactions between surfactants and macromolecules are intensely investigated both for food and non-food applications. Most of the previous studies have been focused on protein and surfactant interaction while the interaction between polysaccharide and surfactant not currently approved for food use, has deserved less attention. Understanding the interactions between polysaccharides and surfactants can be crucial to obtain a given target product. In this chapter, the discussion is focused exclusively on the interaction between polysaccharides with nonionic, cationic and anionic surfactants regardless of the specific application, *i.e.* food or non-food. The chapter is expected to provide insight on the advancement and challenges of this interesting and complex subject.

POLYSACCHARIDES AND NONIONIC SURFACTANTS

Chitosan is one of the most studied polysaccharides and its interaction with different surfactants has been the object of numerous investigations. Mixtures of chitosan with nonionic surfactants like sorbitan monolaurate (Span 20), sorbitan monooleate (Span 80) and sorbitan trioleate (Span 85), have numerous applications in the food and cosmetic industries due to their ability to stabilize emulsions [6]. Although the interaction of chitosan with nonionic surfactants is weak, it is possible to prepare the emulsion, solution or cream with mixtures of various compositions of chitosan and sorbitan ester surfactants. However, some differences are observed depending on the chitosan-surfactant concentration and chain length of the surfactant. For example, the size of surfactant droplets dispersed in the continuous phase of chitosan is essentially the same (*ca.* 2.0 μm) in dilute (0.02 mol/dm^3) polysaccharide-surfactant solution. A ten-fold increase in the concentration of both components leads to higher droplet sizes, but the effect depends on the hydrophobicity of the surfactant; it is more dramatic for Span 80 in which aggregation of surfactant molecules leads to droplets approximately ten times larger than in dilute solutions [7]. On the other hand, confocal microscopy shows chitosan-surfactant aggregates formed by a chitosan-rich shell and a chitosan-poor core with highly ordered micrometric supramolecular structures resulting from high surfactant concentrations. These results indicate the affinity of chitosan, arising from its amphiphilic character to surfactants. This characteristic makes this polysaccharide capable of interacting at the interface of various colloidal systems like micelles, vesicles or emulsion droplets [6].

Viscosity measurement has been used to characterize the interaction between chitosan and octaethyleneglycol mono *n*-dodecyl ether ($C_{12}E_8$) in dilute and semidilute solutions [8]. The interaction, though weak, between chitosan and $C_{12}E_8$, has a strong effect on the viscosity of admixture solutions. Different effects were observed. At surfactant concentrations essentially below or near the CMC, the viscosity is constant regardless of aging the solutions for up to nine days. However, even if viscosity is independent of surface concentration, it decreases with aging time as normally observed in individual chitosan solutions. Above the CMC, a sudden drop in viscosity occurs, but this drop depends on the aging time of the solutions [8]. Therefore, there is a time-dependent effect apart from the time dependency of individual chitosan solutions. The rate of viscosity decreases with time for a given constant chitosan concentration depending on surfactant concentration. For dilute solutions, *i.e.* below CMC, the rate is independent of surfactant concentration. For semi-diluted solutions, a steep increase of the rate with concentration is observed and for concentrated solutions, the rate remains again constant or slightly decreases depending on the aging time of the solutions, *e.g.* up to 16 days. This behavior has been explained by changes in chitosan conformation induced by the presence of the surfactant. When surfactant concentration reaches a certain critical value, C_1, micelles are formed, and association with chitosan chains takes place. This cooperative process leads to conformational changes in the chitosan molecules that induce polysaccharide adsorption at the interface to minimize the electrostatic interactions between chitosan chains. Association continues as long as the surfactant concentration is increased and reaches a given value, C_2, for which every chitosan molecule is bound to one micelle. If the surfactant concentration is further increased, the rate of viscosity increase becomes essentially constant until a third critical surfactant concentration, C_3, is reached. This condition points the existence of interactions only between micelles. At this concentration, chitosan may be adsorbed onto more than one micelle. Under this condition, the system is composed of spherical micelles enclosed by chitosan segments and linked by chitosan coils that coexist with free chitosan chains and free surfactant micelles. A further association is impeded by steric repulsion within the chitosan layer surrounding the micelle surface [8]. The occurrence of time-dependent effects is explained by the presence of very slow conformational changes.

Surface tension studies of individual chitosan and its mixture with a surface-active triblock copolymer of polyethylene oxide-polypropylene oxide-polyethylene oxide (PEO-PPO-PEO) traded under the commercial name Lutrol® F127 and known generically as poloxamer 407, in water and in acetate buffer solutions have demonstrated that pH, ionic strength, chitosan concentration, surfactant concentration, and temperature, play different and significant roles in self-aggregation and adsorption at the air-solution interface [9]. Chitosan dilute solutions (< 0.001%) in water do not show surface activity at 25 (pH 5.06) and 34 °C (pH 5.21). However, pH (and also ionic strength) has an important role in this behavior as it has an influence on the hydrophilic character of the polysaccharide. At pH 5.0, the amino groups of chitosan molecules are protonated, and there are not large hydrophobic groups capable of adsorbing at the interface. Under this condition, the polysaccharide does not exhibit surface activity. As pH increases (≈ 6.0) amino groups are less protonated, and the hydrophilic character of chitosan becomes stronger. The polysaccharide can self-aggregate, and its hydrophilic segments adsorb at the interface. Under this condition, chitosan exhibits slight, but noticeable, surface activity. On the other side, the surface tension of individual surfactant solutions and chitosan-surfactant mixtures decreases linearly with the natural logarithm of surfactant concentration regardless of pH and temperature as shown schematically in Fig. (**1**). Three zones and two break points have been observed in the surface tension isotherm of chitosan-F127 solutions, which means different adsorption behavior at the interface. It is suggested that before the occurrence of the first break point, *i.e.* zones I and II, surfactant molecules can be adsorbed at the interface as 1) individual chains without aggregation, 2) dimers or larger aggregates and 3) chains with a modified configuration. The first break provides the critical aggregation concentration (CAC). The second break point, *i.e.* zones II and III, is attributed to the formation of micelles and therefore is the critical micelle concentration (CMC). Chitosan-F127 solutions in water behave in a different way than solutions in acetate buffer. In the case of water solutions, it has been speculated that before the first brake, complex formation occurs between chitosan and F127. It is shown in Fig. (**1**) that surfactant solution has a steeper slope and higher surface tension than the surfactant-free solution. Therefore, such complexes are less surface active and desorb from the interface. Beyond the CMC, all systems converge to the same surface tension. On the other side, solutions in acetate buffer show steeper slopes, but surface tensions for a given surfactant concentration are lower than chitosan-free solutions. This has been attributed to coadsorption of both components at the interface. The observed effects are the result of multiple interaction processes in which hydrophobic and electrostatic interactions are involved depending on pH, temperature, and ionic strength. As a consequence, the behavior of these systems is complex [9].

The effect of the chemical nature of the head and tail groups of nonionic surfactants on the properties of ethyl cellulose (EC) oleo gels has been studied using differential scanning calorimetry (DSC) and rheometry techniques [10]. The surfactants included sorbitan monostearate (SMS), sorbitan monooleate (SMO), glycerol monostearate (GMS) and glycerol monooleate (GMO). DSC results show a shift to lower sol-gel and gel-sol transition temperatures; the reduction is higher for glycerol-based surfactants, GMS and GMO, than for sorbitan-based surfactants, SMS, and SMO. This behavior is attributed to the plasticizing nature of the head group of glycerol. The addition of surfactants to EC oleo gels results in an increase in gel strength. This behavior can be due to surfactant-EC interactions, and the effect is higher for GMS, probably due to crystallization of the GMS molecules. As a conclusion, the surfactant head group plays an important role in the association between the polysaccharide and the different surfactants [10].

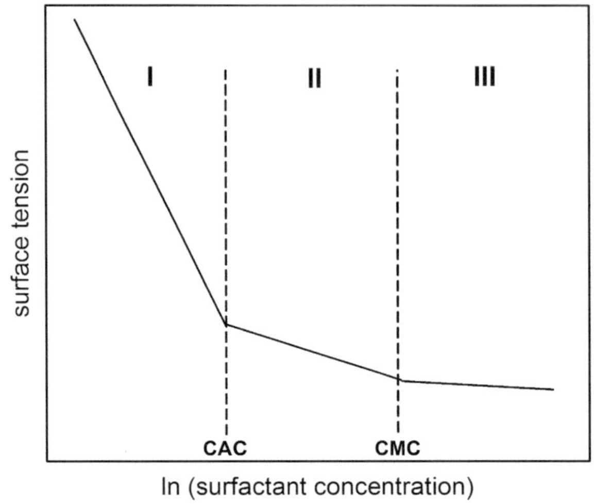

Fig. (1). Surface tension isotherm showing three zones and two break points; CAC and CMC (highly schematic).

Different effects are observed when various polysaccharides and surfactants are mixed. For example, the presence of Triton X-100 can modify the adsorption behavior of carboxymethylcellulose on the solid surface of alumina particles [10]. The most striking effect is that the amount of polysaccharide adsorbed on the particles surface increases when the surfactant is added, but the thickness of the adsorbed layer become smaller compared with that for carboxymethylcellulose-sodium dodecyl sulfate (SDS), an anionic surfactant, mixtures. This has been attributed to differences in polysaccharide conformation, with anionic Triton X-100 carboxymethylcellulose adopting a closely packed conformation [11]. The mechanical strength, rigidity (G'), gelling and melting temperatures of agar gels and viscosity of agar sols decrease when the nonionic surfactants Triton X-100, Brij 35 (polyoxyethylene 23-lauryl ether), polyethylene 10 lauryl ether ($C_{12}E_{10}$), SC893 and Crexlox 3485 are added to the polysaccharide in concentrations near or above the critical micelle concentration (CMC) [12]. The reason for this behavior is still unclear for sols and gels. However, it is believed that in sols interaction between surfactant and water is stronger than that with the polysaccharide and in gels surfactant micelles might interfere with network formation [12]. Addition of TR 8 (polyoxyethylene 7 isododecyl ether), TR 12 (polyoxyethylene 11 isododecyl ether), Brij(R) 56 (polyoxyethylene 10 cetyl ether) and Brij(R) 58 (polyoxyethylene 20 cetyl ether) to alginate, guar and sodium carboxymethyl guar aqueous solutions results in different effects on the steady shear and linear viscoelastic properties of the mixtures depending on the surfactant type and concentration [13]. In general, marked shear-thinning and strong viscoelasticity is observed and only in the case of guar, phase separation attributed to a strong association between the polysaccharide and the surfactant, *via* hydrogen bonding, is observed for high (3%) surfactant concentrations. These results will be valuable in textile printing applications [13]. The release profiles of the hydrophilic and hydrophobic drugs theophylline and tolbutamide, respectively, from agarose gels containing the surfactants Pluronic® F-68, a polyoxyethylene-polyoxypropylene block copolymer, and Tween 80, polyoxyethylene (20) sorbitan monooleate, are remarkably affected by the solubility of drugs and the microstructure of the lyophilized hydrogels. The water uptake of the dehydrated hydrogels increases due to the diminution of the surface tension of the water used to rehydrate the gels. Surfactant micelles, which actuate as a structure support of the agarose network, reduce the hysteresis normally observed after rehydrating and increase the swelling of the dehydrated material. The pore architecture and the polysaccharide-surfactant interactions play an important role in the controlled release of those drugs [14]. The inclusion of Triton X-100 in κ-carrageenan solutions increases the adsorption of the polysaccharide on the surface of alumina particles showing a cooperative behavior. The interaction between carrageenan and alumina particles is electrostatic, while hydrogen bonding is responsible for complexation with Triton X-100. Adsorption of carrageenan on alumina particles is exacerbated when the cationic surfactant hexadecyltrimethylammonium bromide and its equimolar mixtures with Triton X-100 are used [15].

POLYSACCHARIDES AND CATIONIC SURFACTANTS

Carrageenans are present in many food and non-food systems and in most of them oppositely-charged surfactants are also present; so, the interaction between both components is expected. Kogej [16] used a potentiometric technique to study the interactions between the sodium forms of kappa, iota and lambda carrageenan and the cationic surfactants N-dodecyl pyridinium chloride (DPC) and N-cetylpyridinium chloride (CPC) in aqueous solution with the addition of sodium chloride. Kappa and iota carrageenan were studied under non-gelling conditions. The variation in polysaccharide charge density, *i.e.* type of carrageenan, and surfactant hydrophobicity, for example, CPC has a shorter hydrocarbon length than DPC, resulted in noticeable differences attributed to distinct ways of interaction. The critical aggregation concentration (CAC) was essentially independent of sodium chloride concentration for kappa and iota carrageenan but increased with salt concentration for lambda carrageenan. This behavior was observed for both surfactants, but the CAC values for DPC were approximately two orders of magnitude higher than for CPC. The results rationalized through a binding model providing different cooperativity parameters, point to the existence of a lamellar-type ordering of surfactant resulting from the association with polysaccharide chains. Therefore, aggregation of surfactants in the presence of carrageenans is different from the usual micelle formation, and the cooperative effect is due to the hydrophobic interaction between bound surfactant ions.

The interactions between dodecyl ammonium chloride (DDACl) and the sodium form of κ, ι and λ-carrageenan have been studied using dynamic light scattering (DLS), zeta potential and rheometry [15]. The addition of the cationic surfactant to carrageenan solutions leads to the formation of different soluble or insoluble complexes. The collapse of carrageenan chains is revealed by a decrease in the size and charge of the insoluble complexes in addition to the decrease in solution viscosity with increasing DDACl. DLS shows that at low surfactant concentration the size of the insoluble complex particles was slightly lower than the smaller sizes observed in the surfactant-free solutions. The particle size decreases when surfactant concentration increases. This is attributed to a reduction in the intrachain repulsion in carrageenans. Carrageenan solutions show shear-thinning behavior. The addition of surfactant to carrageenan solutions changes the chain conformation of carrageenan and when the DDACl concentration increases the flow behavior starts to be nearly Newtonian. κ-carrageenan shows a higher effect than ι and λ-carrageenans, suggesting that the structure properties of the complexes are mainly determined by the type of carrageenan chains. All surfactant-free solutions show viscoelastic behavior. Progressive addition of DDACl increases the viscous contribution to viscoelasticity, so the elastic character of carrageenan solutions is slightly reduced. This behavior is more evident in all κ-carrageenan mixtures compared with ι and λ-carrageenan. All these changes depend on carrageenan type and DDACl concentration [17].

Isothermal titration calorimetry (ITC), zeta-potential measurements, DLS, transmission electron microscopy (TEM), atomic force microscopy (AFM), and surface tension measurement have been used to study the interactions between κ, ι, λ-carrageenan, alginate and ulvan, and the cationic surfactant glycine betaine (GB) amide [16]. The variation of the interaction enthalpy with surfactant concentration for κ-carrageenan, alginate, and ulvan showed three regions associated with different phenomena dependent on polysaccharides and surfactant concentrations. This clearly indicates the existence of different interactions between polysaccharide and surfactant. At low cationic surfactant concentrations, electrostatic attractions exist between the negative polar group of the polysaccharide and the positive polar head of GB, and as a result, the polysaccharides bind to the surfactant and form a complex. At intermediate cationic surfactant concentrations, unbound micelles dissociate, and interactions are created between the polysaccharide and micelles or particles of surfactant. At high cationic surfactant concentrations, additional micelles appear in the solution because the surfactant concentrations in the aqueous phase exceed the CMC. Only in the case of ι and λ-carrageenans, it is possible to distinguish two stages, which show that the nature of the interactions of GB with various types of carrageenans is different. In addition, higher cationic surfactant concentrations are required to reach neutral interactions. This result is may be due to the difference in carrageenan chemical structures, *i.e.*, the number of sulfate groups along the polysaccharide backbone [18].

The interaction of carboxymethylcellulose in water and in aqueous salt solutions with the cationic surfactants, dodecyl trimethyl ammonium chloride (DTAC), dodecyl trimethyl ammonium bromide (DTAB), tetradecyl trimethyl ammonium bromide (TTAB), tetradecyl triphenyl phosphonium bromide (TTPB), hexadecyl trimethyl ammonium bromide (CTAB), tetradecyl piridinium bromide (TPB) and dodecyl trimethyl ammonium iodide (DTAI) has been studied [19]. The goal of this investigation was to determine the effect of the size of the head group and chain length of the surfactant, different halide counterion, and ionic strength of the medium on the binding between the polysaccharide and surfactant. The variation of surface tension with surfactant concentration shows that the CAC of the mixtures decreases when chain length increases from C_{12} to C_{14}, although no perceptible changes are noticed from C_{14} to C_{16} and decreases with the increase in the size of the head group because the surfactant is more hydrophobic and there is a reduction in counterion binding. The counterion binding also influences the CMC, micelle shape, and phase equilibria in the order $Cl^{-1} < Br^{-1} < I^{-1}$. The CAC for iodide is the lowest of all the halides. This can be attributed to the highest binding capacity of this counterion and as a consequence better packing of surfactant molecules. Therefore, the order for CAC is $I^{-1} < Br^{-1} < Cl^{-1}$. Moreover, the viscosity of surfactant solutions is higher than that of water. At a polymer concentration Cp, the system is in the semidilute regime, which means that polymer chains are extended and can overlap. In carboxymethylcellulose-surfactant solutions, the polysaccharide chains collapse because they cannot longer be overlapped, and as a result, the viscosity drops to that of water. The addition of different salts is necessary to separate the hydrophobic from the electrostatic effects. Viscosity decreases slightly until salt concentration reaches 0.1 mmol/dm^3 because the electrostatic repulsions between polysaccharide charges are screened, which lead to a collapse of the chains. In the presence of surfactants, polysaccharide chains can collapse due to the electrostatic interactions with the surfactant. The same behavior is observed when salt is added.

Differences are not observed for the different salts used $BaCl_2$, NaCl, and KCl; hence the colligative effect dominates over the hydration behavior of the different ions. Furthermore, the size of the polymer-surfactant complex is reduced because of the increment in salt concentration. This can be attributed to the screening effect of the salt in the electrostatic repulsion between the head groups of the surfactant in the bound micelles. As a result, the repulsion between micelles held onto the polysaccharide backbone is reduced too [19].

POLYSACCHARIDES AND ANIONIC SURFACTANTS

Chitosan can display strong interactions with sulfated anionic surfactants like sodium dodecyl sulfate (SDS), sodium lauryl ether sulfate (SLES) and sodium taurocholate (ST) among others. Water insoluble and structured complexes are formed with these surfactants over a wide range of concentrations and mixing ratios. Binding of chitosan to SDS is a cooperative process governed by the hydrophobic interactions with the surfactant alkyl chain. This is also true for other sulfated surfactants. However, strong electrostatic interactions are found between the glucosammonium units of chitosan and the sulfated and sulfonated head groups of surfactants. On the other hand, chitosan also exhibits weak interactions with anionic surfactants like fatty acids. As in the case of sulfated surfactants, chitosan-fatty acids mixtures form water insoluble and structured complexes. The degree of ionization of chitosan and fatty acids and their mutual interaction depend on pH. This yields very complex mixtures. Depending on pH, three regions can be identified: 1) at low pH, chitosan is positively charged and soluble while fatty acids are neutral and insoluble, 2) at intermediate pH, chitosan, and fatty acids are partially charged and soluble and 3) at high pH, chitosan is not charged and insoluble and fatty acids are ionized. These different states should be considered when preparing chitosan-fatty acids mixtures since pH is a very important process parameter [6].

The behavior of chitosan-fatty acids mixtures at the air-water interface is determined by hydrophobic interactions, but hydrogen bonds are also present. When short chain carboxylic acids (C_n, $n \leq 5$) are added to chitosan solutions, complex formation does not occur due to the absence of hydrophobic interactions between aliphatic chains. However, for long chain carboxylic acids (C_n, $n \geq 16$) emulsions are formed even if these acids are not protonated. Finally, aggregates are observed when intermediate chain carboxylic acids are present [6].

The stability to flocculation and coalescence of an oil-in-water emulsion containing sodium stearoyl 2-lactylate (SSL) as the surface active agent can be improved by the addition of chitosan. This is attributed to the formation of a thicker interface consisting of chitosan-SSL complexes with better surface active properties and to the enhancement of bulk viscosity induced by the presence of the polysaccharide [20, 21].

The interactions between maltodextrin (MD) (DE = 10) and SDS have been studied at neutral pH by isothermal titration calorimetry (ITC), surface tension, DSC and turbidity techniques [22]. ITC and surface tension measurements were used to determine the CMC of SDS. Each one of these two techniques yielded different results; 3.4 mmol/dm^3 with ITC and 2.2 mmol/dm^3 with tensiometry. The existence of interactions between both components, at low surfactant concentrations is revealed by exothermic enthalpy changes in heat flow in buffer solutions. As surfactant concentration increases, binding of both components decreases, because fewer sites in maltodextrin molecules are available for interacting with SDS. In addition, it seems that one SDS molecule is bound to two MD molecules. Surface tension measurements indicate insignificant binding between MD and SDS when surfactant concentration is around 0.05 mmol/dm^3. Above this concentration, more complexes of MD-SDS are formed when MD concentration increases. The phase diagram of MD-SDS mixtures obtained from tensiometry and ITC data shows three concentration zones; 1) at low SDS concentrations MD does not interact with SDS, 2) at intermediate SDS concentrations formation of MD-SDS complexes occur and SDS molecules in the complex are in equilibrium with free surfactant molecules, 3) at high SDS concentrations SDS saturates maltodextrin molecules, and the excess surfactant exists as monomers in the aqueous phase.

Binding of starch to the anionic surfactants SDS, SSL and glycerol monostearate (GMS) has been studied by X-ray diffraction, DSC and rheometry [23, 24]. Amylose forms strong complexes with SSL or GMS. Formation of these complexes is favored by the increased mobility of amylose molecules resulting from swelling of starch granules. However, surfactants may also form complexes with amylose molecules present in the swollen granules. Amylose-surfactant complexes are less flexible than individual amylose molecules. This can retard the separation of amylose and amylopectin during heating of starch suspensions. Thus, the formation of the amylose-surfactant complex should explain the ability of some surfactants to retard firmness and retrogradation. Amylose and SDS also form complexes. This surfactant can also form complexes with amylopectin, but this union is more restricted. The number of SDS molecules bound per 1000 glucose units is approximately, 35 for amylose on a 0.2% (w/w) amylose solution and only 6 to 7 for amylopectin on a 3 and 5% (w/w) amylopectin solution. In this case, the addition of SDS has the same effect than SSL or GMS, *i.e.* firmness and retrogradation are retarded [25].

On the other side, the viscosity of 2-hydroxy-3-trimethylammonium propyl starch solutions, *i.e.* cationic starch, decreases when SDS is added [26]. This reduction in viscosity is attributed to the increasing repulsion between charged segments in the polyelectrolyte. Beyond a certain SDS concentration, viscosity drops rapidly because the addition of enough surfactant causes the starch chains to collapse. Viscosity falls until gel particles are formed. The length of the hydrocarbon chain of the surfactant has an effect on viscosity behavior. Viscosity decreases when the surfactant chain length increases, but the end group of the surfactant does not have a significant impact.

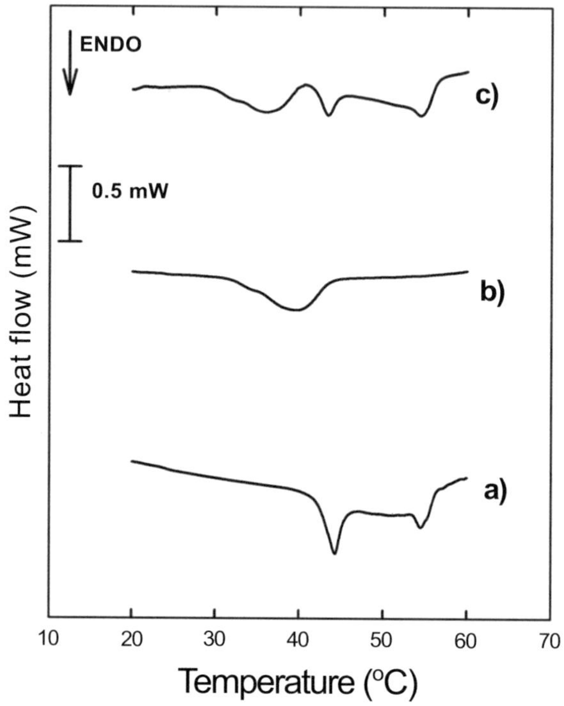

Fig. (2). Heating thermographs of a) 0.5% SSL, b) 0.5% κ-carrageenan in 13 mM KCl and c) 0.5% κ-carrageenan in 13 mM KCl with 0.5% SSL. Cooling thermographs show the same number of peaks at lower temperatures.

The behavior of aqueous semidilute anionic hydroxyethylcellulose (HEC(-)) in the presence of the anionic surfactants SDS or the cationic surfactant CTAB or Gemini(+) was studied by turbidity, small-angle neutron scattering (SANS), rheometry, and DLS measurements [27]. Weak interactions exist between HEC(-) and anionic SDS, while electrostatic attraction between HEC(-) and cationic CTAB or Gemini(+) is present. The key point here is that addition of an oppositely-charged surfactant reinforces the structure of the resulting complexes, flocculation occurs at higher surfactant concentrations and association of the complex leads to the formation of a mixed fragmented network. On the other hand, the addition of a like-charged surfactant promotes the formation of polysaccharide-surfactant micelles.

In our laboratory, we have used DSC and rheometry to study the effect of the nonionic Tween 20 (T20) and the anionic SSL surfactants on the viscoelastic properties and the gel-sol-gel transition of KCl-containing κ-carrageenan gels. The gels were made with 0.5% κ-carrageenan, 1-5 mM T20 or 0.05-0.5% SSL and surfactant concentrations were above the CMC. The storage modulus, G'(ω), of carrageenan-T20 gels is on the average 35% higher than that in surfactant-free gels. On the contrary, G'(ω), of carrageenan-SSL gels is lower by an average of 36% than that in individual carrageenan gels. On the other side, the melting and gelling temperatures of κ-carrageenan-T20 gels are higher with respect to those of surfactant-free gels. As shown in Fig. (**2**) carrageenan-SSL gels show three thermal transitions on melting and gelling; one transition is attributed to κ-carrageenan and the remaining two to the surfactant. The temperatures and enthalpies of melting and gelling of κ-carrageenan gels decrease when the concentration of SSL is increased. The second transition occurs at a constant temperature of about 43 °C on heating and 40 °C on cooling. The third transition decreases further on heating and cooling. The melting and gelling enthalpies in the presence of surfactant show an increase for both transitions. These results suggest that κ-carrageenan gels are strengthened by the presence of the nonionic surfactant possibly by hydrogen bonding with the hydroxyl groups of T20. On the other hand, the negative charges in κ-carrageenan and SSL cause repulsive electrostatic interactions, and this affects the formation of the κ-carrageenan network. As a consequence, the transition temperature and enthalpy decrease upon increasing the SSL concentration.

CONCLUSION

Unlike an individual surfactant, (Fig. (**3**) (top)), in general, the association between polysaccharides and surfactants involves three regions as shown in Fig. (**3**) (bottom). Each region is characterized by the existence of different interactions between both components. In the first region when a small amount of surfactant is added to a polysaccharide solution, surfactant molecules interact with the macromolecule. The interaction depends on the type of surfactant and polysaccharide. Surface tension decreases with the increase in surfactant concentration until a first critical point which is the CAC and corresponds to the onset of micelle formation on the polysaccharide. In the second phase, further addition of surfactant does not change surface tension, because surfactant molecules bind to the polysaccharide. The progressive increment in surfactant concentration saturates the polysaccharide molecules and the second breakpoint appears. The addition of more surfactant after this point causes a decrease in surface tension because the surfactant-polysaccharide complex is desorbed from the surface and surfactant molecules adsorb on it. This point is called the polymer saturation point (PSP) and is close to CMC [28]. Finally, the third region occurs when the surface tension of the polysaccharide-surfactant mixture reaches the surface tension value of the surfactant solution; third critical point. At this point, the surfactant molecules cover the surface, and their concentration is high enough to form free micelles.

On the other hand, the most important interactions between polysaccharides and oppositely-charged surfactants are electrostatic and hydrophobic. Hydrophobic interactions are present when polysaccharides have hydrophobic segments in their structures. If these segments are not present, these kinds of interactions occur due to surfactant-surfactant binding. Hydrophobic interactions are very important in neutral polysaccharides, but hydrogen bonds have an important contribution too.

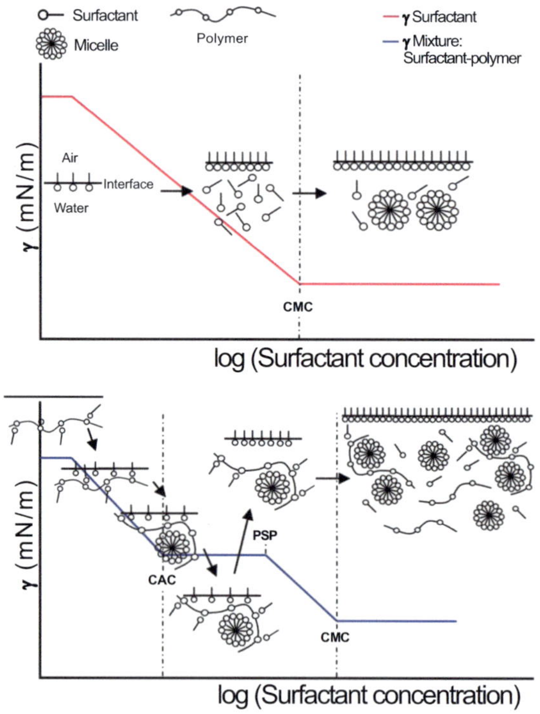

Fig. (3). Simple model of association of surfactant in water (top) and surfactant-polymer in water (bottom).

Important information on polysaccharide-surfactant interactions is currently being provided by a number of techniques, such as tensiometry, rheometry, dynamic light scattering, zeta potential, isothermal titration calorimetry, differential scanning calorimetry, x-ray diffraction, turbidity, small-angle neutron scattering, confocal laser scanning microscopy, electron microscopy and atomic force microscopy. The complexity of the systems justifies the combined use of several of these techniques.

The process conditions of a system containing polysaccharides affect the association of a polysaccharide with another polysaccharide or non-polysaccharide structure-forming molecules, like proteins or surfactants. Physicochemical properties, *e.g.* helix-coil transition temperature, electrical properties, sensitivity to specific monovalent or multivalent ions, are very important and have to be taken into account in order to have a better understanding of their behavior in multicomponent systems [29].

CONFLICT OF INTEREST

The author (editor) declares no conflict of interest, financial or otherwise.

ACKNOWLEDGEMENTS

Author M.C. Ortiz-Tafoya acknowledges CONACyT (Grant 211344) for financial support. Funding from DGAPA-UNAM (Grant IN221915) is also appreciated. Thanks are also due to Mariana Ramírez-Gilly for useful suggestions.

REFERENCES

[1] K. Holmberg, B. Jönsson, B. Kronberng, and B. Lindman, *Surfactants and Polymers in Aqueous Solution.*John Wiley & Sons, Ltd: Surrey, 2002. [http://dx.doi.org/10.1002/0470856424]

[2] C. Carrera-Sánchez, M.R. Rodríguez-Niño, A.L. Caro, and J.M. Rodríguez-Patino, "Biopolymers and emulsifiers at the air–water interface. Implications in food colloid formulations", *J. Food Eng.,* vol. 67, no. 1-2, pp. 225-234, 2005. [http://dx.doi.org/10.1016/j.jfoodeng.2004.05.065]

[3] L.M. Sagis, and E. Scholten, "Complex interfaces in food: Structure and mechanical properties", *Trends Food Sci. Technol.,* vol. 37, no. 1, pp. 59-71, 2014. [http://dx.doi.org/10.1016/j.tifs.2014.02.009]

[4] D. Bais, A. Trevisan, R. Lapasin, P. Partal, and C. Gallegos, "Rheological characterization of polysaccharide-surfactant matrices for cosmetic O/W emulsions", *J. Colloid Interface Sci.,* vol. 290, no. 2, pp. 546-556, 2005. [http://dx.doi.org/10.1016/j.jcis.2005.04.044] [PMID: 15963520]

[5] A. Chatterjee, S.P. Moulik, P.R. Majhi, and S.K. Sanyal, "Studies on surfactant-biopolymer interaction. I. Microcalorimetric investigation on the interaction of cetyltrimethylammonium bromide (CTAB) and sodium dodecylsulfate (SDS) with gelatin (Gn), lysozyme (Lz) and deoxyribonucleic acid (DNA)", *Biophys. Chem.,* vol. 98, no. 3, pp. 313-327, 2002. [http://dx.doi.org/10.1016/S0301-4622(02)00107-2] [PMID: 12128183]

[6] L. Chiappisi, and M. Gradzielski, "Co-assembly in chitosan-surfactant mixtures: thermodynamics, structures, interfacial properties and applications", *Adv. Colloid Interface Sci.,* vol. 220, pp. 92-107, 2015.
[http://dx.doi.org/10.1016/j.cis.2015.03.003] [PMID: 25865361]

[7] J. Grant, J. Cho, and C. Allen, "Self-assembly and physicochemical and rheological properties of a polysaccharide-surfactant system formed from the cationic biopolymer chitosan and nonionic sorbitan esters", *Langmuir,* vol. 22, no. 9, pp. 4327-4335, 2006.
[http://dx.doi.org/10.1021/la060017g] [PMID: 16618183]

[8] D. Jocić, M.R. Juliá, and P. Erra, "The time dependence of chitosan/nonionic surfactant solution viscosity", *Colloid Polym. Sci.,* vol. 274, no. 4, pp. 375-383, 1996.
[http://dx.doi.org/10.1007/BF00654058]

[9] I. Pepić, J. Filipović-Grčić, and I. Jalšenjak, "Interactions in a nonionic surfactant and chitosan mixtures", *Colloids Surf. A Physicochem. Eng. Asp.,* vol. 327, no. 1-3, pp. 95-102, 2008.
[http://dx.doi.org/10.1016/j.colsurfa.2008.06.009]

[10] M. Davidovich-Pinhas, S. Barbut, and A.G. Marangoni, "The role of surfactants on ethylcellulose oleogel structure and mechanical properties", *Carbohydr. Polym.,* vol. 127, pp. 355-362, 2015.
[http://dx.doi.org/10.1016/j.carbpol.2015.03.085] [PMID: 25965494]

[11] E. Grządka, "Influence of surfactants on the structure of the adsorption layer in the system: Carboxymethylcellulose/alumina", *Mater. Chem. Phys.,* vol. 126, no. 3, pp. 488-493, 2011.
[http://dx.doi.org/10.1016/j.matchemphys.2011.01.047]

[12] K. Prasad, A.K. Siddhanta, A.K. Rakshit, A. Bhattacharya, and P.K. Ghosh, "On the properties of agar gel containing ionic and non-ionic surfactants", *Int. J. Biol. Macromol.,* vol. 35, no. 3-4, pp. 135-144, 2005.
[http://dx.doi.org/10.1016/j.ijbiomac.2005.01.004] [PMID: 15811467]

[13] R. Fijan, S. Šostar-Turk, and R. Lapasin, "Rheological study of interactions between non-ionic surfactants and polysaccharide thickeners used in textile printing", *Carbohydr. Polym.,* vol. 68, no. 4, pp. 708-717, 2007.
[http://dx.doi.org/10.1016/j.carbpol.2006.08.006]

[14] T. Marras-Marquez, J. Peña, and M.D. Veiga-Ochoa, "Agarose drug delivery systems upgraded by surfactants inclusion: critical role of the pore architecture", *Carbohydr. Polym.,* vol. 103, pp. 359-368, 2014.
[http://dx.doi.org/10.1016/j.carbpol.2013.12.026] [PMID: 24528741]

[15] E. Grządka, "Interactions between kappa-carrageenan and some surfactants in the bulk solution and at the surface of alumina", *Carbohydr. Polym.,* vol. 123, pp. 1-7, 2015.
[http://dx.doi.org/10.1016/j.carbpol.2015.01.024] [PMID: 25843828]

[16] K. Kogej, "Binding of cationic surfactants by kappa, iota, and lambda carrageenans in aqueous solution in the presence of sodium chloride", *Acta Chim. Slov.,* vol. 55, pp. 985-991, 2008.

[17] M. Vinceković, J. Katona, M. Bujan, and V. Sovilj, "Interactions between dodecylammonium chloride and carrageenans in the semidilute regime", *Colloids Surf. A Physicochem. Eng. Asp.,* vol. 384, no. 1-3, pp. 739-748, 2011.
[http://dx.doi.org/10.1016/j.colsurfa.2011.05.056]

[18] R. Covis, T. Vives, C. Gaillard, M. Benoit, and T. Benvegnu, "Interactions and hybrid complex formation of anionic algal polysaccharides with a cationic glycine betaine-derived surfactant", *Carbohydr. Polym.,* vol. 121, pp. 436-448, 2015.
[http://dx.doi.org/10.1016/j.carbpol.2015.01.001] [PMID: 25659719]

[19] J. Mata, J. Patel, N. Jain, G. Ghosh, and P. Bahadur, "Interaction of cationic surfactants with carboxymethylcellulose in aqueous media", *J. Colloid Interface Sci.,* vol. 297, no. 2, pp. 797-804, 2006.

[http://dx.doi.org/10.1016/j.jcis.2005.11.022] [PMID: 16376917]

[20] K.G. Zinoviadou, E. Scholten, T. Moschakis, and C.G. Biliaderis, "Engineering interfacial properties by anionic surfactant-chitosan complexes to improve stability of oil-in-water emulsions", *Food Funct.*, vol. 3, no. 3, pp. 312-319, 2012.
[http://dx.doi.org/10.1039/c2fo10197a] [PMID: 22298029]

[21] K.G. Zinoviadou, T. Moschakis, V. Kiosseoglou, and C.G. Biliaderis, "Impact of emulsifier-polysaccharide interactions on the stability and rheology of stabilised oil-in-water emulsions", *Procedia Food Sci.,* vol. 1, pp. 57-61, 2011.
[http://dx.doi.org/10.1016/j.profoo.2011.09.010]

[22] A. Wangsakan, P. Chinachoti, and D.J. McClements, "Maltodextrin-anionic surfactant interactions: isothermal titration calorimetry and surface tension study", *J. Agric. Food Chem.*, vol. 49, no. 10, pp. 5039-5045, 2001.
[http://dx.doi.org/10.1021/jf0103471] [PMID: 11600063]

[23] K. Ghiasi, E. Varriano-Marston, and R.C. Hoseney, "Gelatinization of wheat starch. II. Starch-surfactant interaction", *Cereal Chem.*, vol. 59, no. 2, pp. 86-88, 1982.

[24] C.J. Keetels, T. van Vliet, A. Jurgens, and P. Walstra, "Effects of lipid surfactants on the structure and mechanics of concentrated starch gels and starch bread", *J. Cereal Sci.*, vol. 24, no. 1, pp. 33-45, 1996.
[http://dx.doi.org/10.1006/jcrs.1996.0035]

[25] E. Svensson, M. Gudmundsson, and A-C. Eliasson, "Binding of sodium dodecylsulphate to starch polysaccharides quantified by surface tension measurements", *Colloids Surf. B Biointerfaces,* vol. 6, no. 4-5, pp. 227-233, 1996.
[http://dx.doi.org/10.1016/0927-7765(95)01260-5]

[26] J. Merta, and P. Stenius, "Interactions between cationic starch and anionic surfactants 2. Viscosity and aggregate size in dilute solutions", *Colloids Surf. A Physicochem. Eng. Asp.*, vol. 122, no. 1-3, pp. 243-255, 1997.
[http://dx.doi.org/10.1016/S0927-7757(96)03823-X]

[27] N. Beheshti, G.T. Nguyen, A.L. Kjøniksen, K.D. Knudsen, and B. Nyström, "Structure and dynamics of aqueous mixtures of an anionic cellulose derivative and anionic or cationic surfactants", *Colloids Surf. A Physicochem. Eng. Asp.*, vol. 279, no. 1-3, pp. 40-49, 2006.
[http://dx.doi.org/10.1016/j.colsurfa.2005.12.031]

[28] J. Sovilj, J. Milanović, and L. Petrović, "Viscosimetric and tensiometric investigations of interactions between gelatin and surface active molecules of various structures", *Food Hydrocoll.*, vol. 32, no. 1, pp. 20-27, 2013.
[http://dx.doi.org/10.1016/j.foodhyd.2012.12.003]

[29] A. Matalanis, O.G. Jones, and D.J. McClements, "Structured biopolymer-based delivery systems for encapsulation, protection, and release of lipophilic compounds", *Food Hydrocoll.*, vol. 25, no. 8, pp. 1865-1880, 2011.
[http://dx.doi.org/10.1016/j.foodhyd.2011.04.014]

CHAPTER 16

Theoretical Models for Biopolymers

J. Ginés Hernández Cifre and **F. Guillermo Díaz Baños***

Departamento de Química Física, Universidad de Murcia, Campus de Espinardo, 30100, Murcia, Spain

Abstract: Theoretical models play a very important role in experimental practice and are essential for the development of current science and technology. They have been accessible for many years, but their development in recent decades, in parallel with the improvements in computation, has allowed the emergence of an important number of methodologies to complement experimental measurements. In this chapter, we present a brief overview of these methodologies. Among them, bead modeling has put a significant step forward. For rigid molecules, programs like those contained in the HYDRO suite have become very popular tools. For flexible molecules, computer simulation of their behavior in solution, either with a detailed atomic description (molecular dynamics) or based on coarse grain models for longer time scales (Brownian dynamics), is now a usual tool with a variety of accessible programs for use by scientists. Finally, a few examples of recent applications of these methodologies for proteins and polysaccharides are included. With this, we seek to illustrate to non-specialized readers the possibilities available today and, hopefully, to encourage them to find ways to incorporate theoretical models to complement their work.

Keywords: Bead modeling, Brownian dynamics, Flexible models, HYDRO, Molecular dynamics, Monte Carlo, Polysaccharides, Proteins, Rigid models.

INTRODUCTION

There is a very interesting philosophical discussion about modeling [1] and its role in experimental practice and the development of present science and technology. According to Zeidler [2], even when theoretical models can be corroborated empirically, they cannot be treated as representations of real empirical systems. This discussion is not within the scope of this book, but we agree with the statement.

The choice of the right model must follow a rule which has been well described

* **Corresponding author F. Guillermo Díaz Baños:** Departamento de Química Física, Universidad de Murcia, Campus de Espinardo, 30100, Murcia, Spain; Tel: +34 688687394; E-mail: fgb@um.es

Martin Masuelli & Denis Renard (Eds.)

by Carroll [3]: "We must choose the model that is sufficiently accurate for our computational purposes, yet still simple enough that we have some understanding of what the model describes. Otherwise, the model is a black box, and we have no understanding of what it does, perhaps even no idea whether the answers it produces are physically reasonable". For us, theoretical models are those used to represent biomolecules and their processes. They usually have two components: a physical representation of the system and a mathematical description of the internal and external processes that are taking place. Both components can be thought of in very different time and length (size) scales, depending on the natural phenomenon that we are trying to describe.

At present, computing is the basic tool for modeling. Virtually any kind of physical representation can be shown graphically in a computer, and the mathematical description of even the simplest processes of nature in which biopolymers is implied would be nearly impossible without the help of a computer.

Defining What to Model

Biopolymers, like all molecules, are composed of atoms made of electrons, protons and neutrons that follow the rules that quantum chemistry tries to describe. But the characteristics that we are interested in studying are those which come from the fact that these macromolecules contain thousands or even millions of atoms, and, yet their global behavior can be described in terms of a few (from one to hundreds) elements. This declaration immediately has vital consequences in the time scale, because we are not interested in phenomena lasting picoseconds or a few nanoseconds. We want to study things that take from many nanoseconds to microseconds to happen.

Another general statement is about flexibility. We can find biopolymers in the whole range of flexibilities. If a macromolecule needs a large amount of energy to change its conformation, we can predict that the changes in native conditions are going to be very few. Even if changes in small regions take place, for us, its global size and shape are going to be the same all the time (in our time scale). We describe this situation as a rigid molecule. Good examples are globular proteins. On the contrary, if we can easily find the atoms of a molecule in different positions (respect to each other) at the temperatures that we are studying, and we observe that the global shape (and size) is changing, we can say that we are working with a flexible molecule. Examples of this situation are some long linear polysaccharides or denatured proteins. Of course, these are the limiting cases and in nature, we usually find the whole range of possibilities or even mixture of them. That is why, we also define the so-called semi-flexible molecules with different types of flexibility, with worm-like chains (like a long DNA molecule) and those with segmental flexibility (like myosin or many immunoglobulins) being very significant for biopolymers.

Building Models

To build a model, we must define the level of detail in the structure that is going to be modeled. Let us assume that the smallest unit we are going to distinguish is an atom. From this starting point, we can build a model describing the relative positions of all atoms present in a biopolymer. In recent decades, the extraordinary development of X-ray crystallography, structural NMR and cryo-electron microscopy has prompted this possibility for many molecules. Two important limits appear at this point. The first is that not all biopolymers are rigid or suitable for crystallization, and in many cases the positions of part of the structure (or all of it) remain unknown, and the second is that the positions of the atoms are going to change in solution (the usual state of the macromolecule) unless we take the polymer as infinitely rigid. If we handle a model at this level of resolution, we have an atomistic model. Another option is to group (how this is done is another matter) the atoms of a certain region to form a single unit to model the molecule. In most cases this unit is represented by a sphere (or a bead), giving rise to a coarse grain model, which is a very useful option for biopolymers, as we show in this chapter.

Once the structure has been formed (in an atomistic or in a coarse grain model), the size and shape of the model is defined if the molecule is considered rigid. But if any degree of flexibility is included, then connectors that reproduce the dynamic behavior of the molecule must also be included. In this case, the model becomes much more complex and the number of parameters is greatly increased. In addition, a procedure to determine the number of conformations present in the systems in certain conditions must be included.

In many occasions, it is interesting to know not only the characteristics of a molecule in solution at a certain instant, but also the dynamics of a system in which the studied molecule is only a part. When that is the case, computer simulation methodologies of the dynamics of a sample must also be included. In the following pages, we shall try to get an overview of the current situation of all of these aspects.

RIGID MODELS

The size and shape of rigid macromolecules are the only factors which determine their hydrodynamic properties. Scientists have taken advantage of this situation to develop hydrodynamic theories and methods to gain knowledge of the molecular characteristics from experimental measurements. Thus, measurements of overall conformational and hydrodynamic properties in dilute solution (sedimentation and diffusion coefficients, intrinsic viscosity, radius of gyration...) using techniques like analytical ultracentrifugation, viscometry or small angle X-ray scattering (SAXS) have been an important source to obtain the overall size and shape of rigid (and flexible) structures. Classical theoretical approaches were based on elementary "whole-body" models that involve the exact hydrodynamic equations linking hydrodynamic measurements (rotational and translational diffusion, intrinsic viscosity...) with the axial ratios describing the ellipsoidal shape, along with some improvements in solving the "hydration problem" [4]. This simple ellipsoid-based modeling is useful for providing an overall idea of the molecular shape, particularly for those systems where only a limited amount of data is available. The suite of ELLIPS algorithms [5, 6] facilitates the use of this approach.

Most rigid biomolecules cannot be well described by defined geometrical forms like spheres, cylinders or ellipsoids. Illustrative examples are oligosaccharides, like cyclodextrins, and proteins, like hemoglobin, porin, calmodulin or those formed by more than one polypeptide chain. In 1967, Bloomfield *et al.* [7] proposed to model such complex molecules with spherical elements as building blocks, based on an idea first proposed by Kirkwood for flexible entities [8]. This is the bead model concept, according to which the shape of a particle is represented by an array of spherical elements ("beads") from which different properties of the particle can be obtained [9] (see Fig. **1**). Thus, the calculation of hydrodynamic properties of rigid particles can be carried out using methods from fluid mechanics combined with a generalized form of Einstein's theory of Brownian motion. This so-called "hydrodynamic bead modeling" has been shown to be very useful for the peculiar shapes of biomolecules [10, 11]. The ultimate level of detail in a bead model is reached if each atom (usually each extended atom including hydrogen) in the molecule is represented by a bead. Another possibility is to replace each repeating unit (amino acids, nucleotides...) with a bead [12]. A clear advantage of the latter is that the number of elements in the model is much smaller than in bead-per-atom models, and therefore it can be applied to large molecules. Other bead modeling strategies are also possible. Thus, if the rigid macromolecule is regarded as a compact, globular particle with a definite contour, then there is an immediate possibility of building a bead model that consists of filling the volume enclosed by that contour with beads. This is the type of approach implemented in the AtoB procedure [13]. Another efficient strategy is to build a "shell model" in which the molecular surface is modeled by an array of small and tangent beads while ignoring all internal molecular features, since hydrodynamic friction takes place only at the surface [10]. In bead modeling, care must be taken with bead overlap, since it adds problems to the complicated description of hydrodynamic interactions.

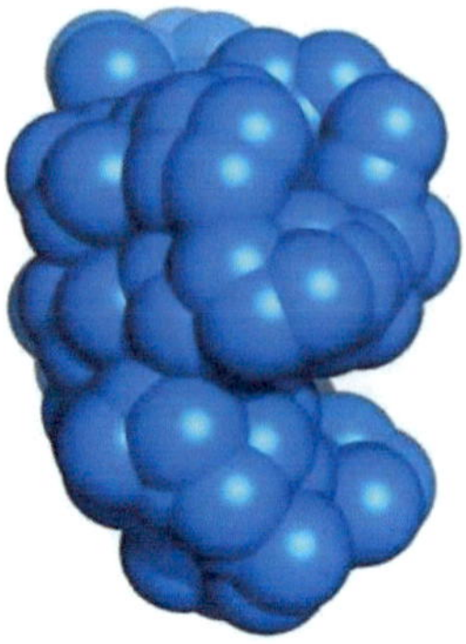

Fig. (1). Bead-per-atom model for the protein lysozyme.

That hydrodynamic bead modeling methodology has been extended and computationally implemented by García de la Torre and co-workers in the public-domain computer program HYDRO [14]. At present, this program has been improved [15] and has evolved to a very complete suite for the study of macromolecules and nanoparticles that is available from our web site [16], including detailed documentation and sample files. Several aspects of the bead modeling strategy are implemented in the HYDRO programs suite. In Table **1**, we show a summary of the components as given by García de la Torre and Harding [17].

In the computer program HYDROSUB [18], the primary model is an array of subunits that have the shape of cylinders (rods or disks) or ellipsoids of revolution. Then, from the solution properties, the dimensions of the equivalent ellipsoids can be determined and the surfaces represented by bead–shell models, which are then assembled *via* a linking hinge region to give a structure (see Fig. **2**). The paradigmatic application of this procedure is the determination of the basic structural features of antibody molecules [19, 20]. This approach has also been successfully used in other applications, like those presented by Zorrilla *et al.* [21], Gapinski *et al.* [22] or Sokolova *et al.* [23]. The HYDROPRO [24] and HYDRONMR [25] programs calculate hydrodynamic properties and NMR relaxation times respectively after building bead models of biomolecules from their atomic coordinates, as taken from PDB (Protein Data Bank) files [26]. HYDROPRO is also able to build residue-level resolution models with one bead per amino acid, thus allowing calculations when atomic resolution is not available or when coarse grain models are preferred [27]. HYDROPRO and HYDRONMR have been used on several occasions to study proteins of different characteristics, as we shall illustrate in the Example 1 section. Another interesting program included in the HYDRO suite is HYDROMIC [28], which predicts solution properties of macromolecular complexes whose volumes have been generated from electron microscopy images. Thus, HYDROMIC reads the density map obtained from cryoEM and builds the hydrodynamic shell model from which the solution properties can be obtained [29, 30]. From other low-resolution techniques, like small-angle neutron scattering (SANS) and X-ray scattering (SAXS), we obtain the angular dependence of the scattering intensity. Analysis programs like DAMMIN [31] and DAMMIF [32] reconstruct a three-dimensional structure which can be employed as input data for HYDROPRO to predict the hydrodynamic properties. Some examples of successful applications have been presented by Scott *et al.* [33], Baldock *et al.* [34] and Patel *et al.* [35].

Table 1. Components of the HYDRO suite

Program	Starting structure/information	Results
HYDRO	User-constructed model Spherical beads: coordinates and radii	1, 2
HYDROPRO	Atomic- or residue-level structure (PDB-formatted list of coordinates)	1, 2
HYDRONMR	Atomic-level structure (PDB-formatted list of coordinates)	3
HYDROMIC	3D density map from electron microscopy Several formats	1, 2
HYDROSUB	User-constructed model Ellipsoids, cylinders: size, position and orientation	1, 2
HYDFIT & HYDROFIT	Experimental data Output from any of the HYDROxxx programs	4
1- Hydrodynamic coefficients (sedimentation, diffusion, rotational times, intrinsic viscosity), radius of gyration, co-volume 2- Angular dependence of scattering intensities and distribution of distances 3- NMR relaxation times 4- Selection of an optimum, best-fitting structure		

Once we have determined, experimentally or from HYDRO calculations, a set of properties like the sedimentation coefficient, intrinsic viscosity, radius of gyration and longest distance, the data can be analyzed by the ancillary programs HYDROFIT [36, 37] and HYDFIT [38] to obtain the structure that best fits these data. As we show in the Example 2 section, we have applied this methodology to study the flexibility of alginate.

In addition to the above mentioned ELLIPS [5] and HYDRO [14] softwares, other computational tools based on the bead modeling concept have appeared over the years, like HI4 [39] and SOMO [40, 41].

An alternative to bead modelling, and equally valid for predicting hydrodynamic properties in the case of rigid structures, consists of representing the macromolecular surface by a set of boundary platelets. The hydrodynamic description is then made using finite elements methods. This strategy is implemented in the program BEST [43, 44], which models the protein surface by an array of triangular platelets.

A very recent comparative study of the methods mentioned above has been presented by Rocco and Byron [45].

Fig. (2). Bead model of a structure with two rigid subunits connected by a flexible chain [42]. Note the shells that represent the ellipsoidal subunits and the string of beads that represent the chain.

FLEXIBLE MODELS

The skeleton of the long and (usually) linear chains that constitutes many biological macromolecules consists of chemical bonds arranged in such a manner that bond lengths and bond angles are nearly constant, but there is an important degree of freedom in the internal rotation around all or some of the bonds in the chain. The result of this conformational variability is that the macromolecule can adopt multiple conformations. Thus, many macromolecules of biological relevance are flexible or semi-flexible entities of non-defined shape. The prediction of the properties of these macromolecules, which are of a statistical nature, requires models that take into account the flexibility and the use of computer simulation techniques.

Two simple mathematical models frequently used to represent the solution conformation of globally linear polymers with uniform flexibility along the chain are the random coil, valid for fully flexible entities, and the wormlike chain, valid for semi-flexible entities. The degree of flexibility of a wormlike chain is tuned by two parameters: the persistence length, L_p, which accounts for the degree of local stiffness, and the contour length, L. Depending on the ratio of those two parameters, the model includes intermediate conformations between the limits of the random coil (large L/L_p value) and the rigid rod (low L/L_p value). The paradigmatic application of the wormlike model is the double-stranded DNA in solution. It is also the proper model to describe the appreciable local stiffness of many polysaccharides, as we will see in the Example 2 section. More details about these and other models for biological macromolecules can be found in textbooks [46]. For macromolecules with more complex and non-uniform flexibility, the development of other theoretical formalisms is necessary. Indeed, concepts like bead models and hydrodynamic interaction (already mentioned in the previous section) were incorporated to understand the solution behavior of flexible polymers [8, 47, 48].

As commented previously, one option to model a molecule is the use of atomic-level models. However, the typical large size of macromolecules and the typically long times of their dynamics usually preclude their use (unless the interest is in small pieces of the chain or short time dynamics). Thus, the prediction of macromolecular properties requires simplified schemes based on coarse grain models [49] whose elements represent large pieces of the polymer chain. Two widely used coarse grain models for flexible polymers are the freely-jointed bead-rod chain and the bead-and-spring chain in which the macromolecule is modeled as an array of spheres, a geometrical form convenient for hydrodynamic calculations, connected by either rigid rods or elastic springs. In both cases, the center of the beads is the place where the frictional drag on the chain is distributed and where the different interactions take place. Physically, the rod in a bead-rod chain corresponds to a Kuhn length (twice the value of the persistence length). Those rods act as a constraint on the system which ensures that adjacent beads in the chain are maintained at a constant separation at all times. However, large flexible polymers are preferably modeled as bead-and-spring chains where, in the classical concept, each spring models a large portion (subchain) of the chain (thus corresponding to several rods of the bead-rod chain, *i.e.* to a large number of Kuhn lengths) [47] (see Fig. **3**). Nowadays, the bead-and-spring chain is also being applied with greater detail to models whose elements represent, for instance, the repeating units (amino acid or nucleotide residues) of biopolymers [50]. Multiple conformations of the chosen model can be generated with the help of some computer simulation technique [51] to build a so-called molecular trajectory. Both conformational properties, like the radius of gyration, and dynamic properties, like the diffusion and sedimentation coefficients, the intrinsic viscosity or the relaxation times, can be obtained from this molecular trajectory.

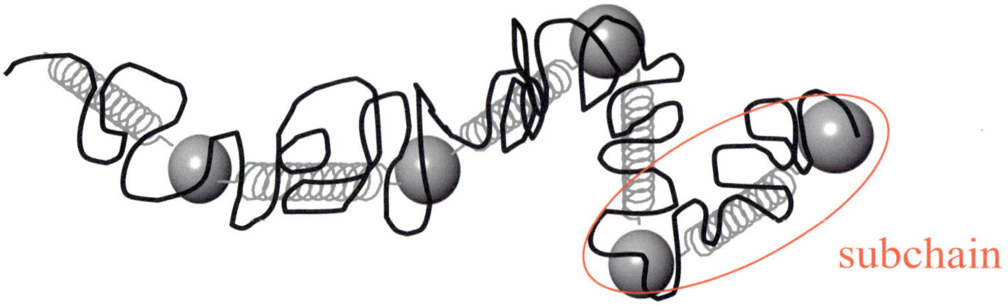

Fig. (3). Bead-and-spring model of a linear polymer chain. The continuous curve is the real polymer that the bead-and-spring chain is meant to represent.

When the computer simulation is restricted to the prediction of conformational, equilibrium properties, as well as some overall hydrodynamic coefficients, Monte Carlo (MC) methods are applicable [51]. Generically speaking, the MC simulation technique aims to generate possible states of a system (in our case, conformations of a flexible macromolecule) in order to obtain observable properties that correspond to averages over all the accessible states. In order to perform MC simulations of general bead-and-spring models, García de la Torre and co-workers have developed the public domain program MONTEHYDRO [52] available from our web site [16], which implements an importance-sampling procedure coupled to the rigid-body hydrodynamics calculations developed for the HYDRO programs suite (see the above section).

When the interest lies in dynamic aspects of flexible macromolecules in solution, such as relaxation processes and non-equilibrium behavior, it is necessary to solve the equation of motion that governs the macromolecular dynamics, *i.e.* Newton's second law. Two typical simulation techniques that carry out this task are molecular dynamics (MD) and Brownian dynamics (BD) [51]. MD explicitly takes into account all the elements constituting the system (in our case the macromolecules and the solvent molecules). It is so widely used that many commercial and public domain MD tools are available to address the different aspects of polymer dynamics [49]. Some popular MD simulation packages which include force fields especially designed for the simulation of proteins, lipids and nucleic acids are AMBER [53, 54], CHARMM [55, 56], which both require a license fee, and GROMACS [57], freely available from the web site [58]. MD is appropriate when carrying out close to atomic-detail simulations aimed at exploring short time and length scales. However, they are not generally adequate for simulating the long time and length scales characteristic of the global dynamics of large molecules. One appropriate alternative is BD, which is a technique to solve the stochastic equation of motion that arises from considering the solvent as a continuum, thus eliminating the solvent degrees of freedom, which allows longer simulation times to be reached. There are fewer publically available computational tools for BD simulations than for MD simulations. Still, several groups have devised BD simulation programs with different scopes or structures [59 - 62]. Some public domain BD simulation packages that handle bead-and-spring models are BD_BOX [61, 63], which efficiently treat both single molecules and multi-component systems, Brownmove [62, 64], and SIMUFLEX [16, 60], developed by García de la Torre and co-workers, which is able to treat a variety of macromolecular topologies and calculate many steady and time-dependent dilute solution properties.

In order to minimize the statistical noise intrinsic to the MC and the BD procedures, many independent molecular trajectories need to be produced and ensemble-averaged properties calculated. Below we provide an overview of the fundamentals of the bead-and-spring model, as well as the MC and BD simulation techniques. The concepts described are general and applied by a number of different MC and BD simulation programs.

Bead-and-spring Model Fundamentals

The bead-and-spring model is a convenient representation of the conformational variability of flexible polymers in solution. Beads are the sites of the model where all the interactions (hydrodynamic, electric, excluded volume…) take place. Springs represent the restoring forces of entropic origin that oppose the chain stretching as well as serving to define the polymer topology (linear, branched, *etc.*). The primary interactions in the model are those between connected (first neighbor) beads. Additional short-term interactions, like bending interactions (between second neighbor beads) and internal-rotation restrictions (between third neighbor beads), are also possible. Finally, there may be long-term interactions (so-called excluded volume interactions) between non-connected beads which are far apart along the chain contour but eventually close in space. Fig. (**4**) is a sketch of a generalized bead-and-spring model. The total potential energy, $V^{(tot)}$, due to the interactions experienced by the N beads of the model, can be decomposed into several contributions:

$$V^{(tot)} = \sum_{conn} V_{ij}^{(conn)} + \sum_{ang} V_{ijk}^{(ang)} + \sum_{tors} V_{ijkl}^{(tors)} + \sum_{EVpairs} V_{ij}^{(EVpairs)} + \sum_{i} V_{i}^{(ext)} \quad \textbf{(1)}$$

The first term of the right hand side in eq. (1) corresponds to the potential of the springs connecting first neighbor beads i and j. The two most employed mathematical expressions for this potential are the Gaussian potential and the

FENE (Finite Extension Non-linear Elongation) potential [65]. The Gaussian potential is that of a Hooke spring with zero equilibrium length ($l_{ij}^{eq} = 0$) and an appropriate value of the spring constant *H*. The general expression for the potential of a Hooke spring reads:

$$V_{ij}^{(conn-Hooke)} = -\frac{1}{2}H\left(l_{ij} - l_{ij}^{eq}\right)^2 \quad \textbf{(2)}$$

where l_{ij} is the instantaneous length between beads *i* and *j*. The FENE potential describes the chain behavior more realistically and reads:

$$V_{ij}^{(conn-FENE)} = -\frac{1}{2}Hl_{\max}^2 \ln\left(1 - \frac{l_{ij}^2}{l_{\max}^2}\right) \quad \textbf{(3)}$$

where the parameter l_{max} is the maximum length attainable by the FENE spring. It is the only choice when the chain is elongated far from equilibrium.

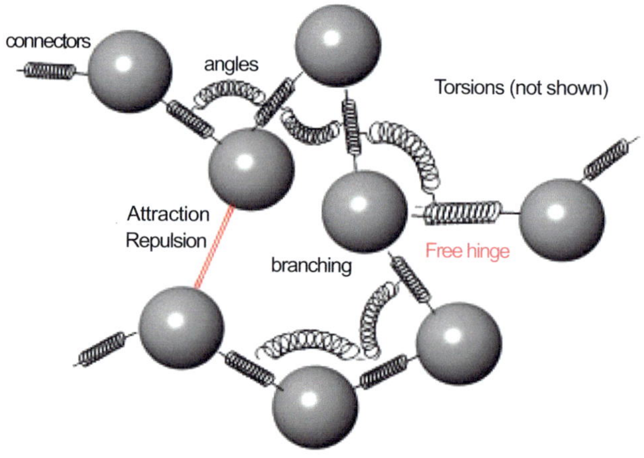

Fig. (4). Pictorial representation of the generalized bead-and-spring model [42].

The second term of the right hand side in eq. (1) corresponds to the potential associated to the instantaneous angle, α_{ijk}, subtended by two successive connector springs joining beads *i* and *j*, and *j* and *k*. A useful potential is the quadratic form:

$$V_{ijk}^{(ang)} = -Q\left(\alpha_{ijk} - \alpha_{eq}\right)^2 \quad \textbf{(4)}$$

where α_{eq} is the equilibrium value of the angle and *Q* is the bending force constant. A discrete version of the wormlike chain can be built as a string of touching beads connected by both hard Hooke springs with non-zero equilibrium length (eq. 2) and angular springs [42]. The degree of flexibility is tuned through the bending constant Q, since it is related to the persistence length [42].

The third term of the right hand side in eq. (1) corresponds to the torsional potentials associated to hindered internal rotations. Expressions for those potentials can be found in the literature [51]. The fourth term takes into account pair-wise excluded volume interactions between beads that do not interact through connector or angular springs. Those interactions prevent bead overlap and represent the solvent quality. One of the most employed potentials is the Lennard-Jones potential:

$$V_{ij}^{(EVpair-LJ)} = 4\varepsilon_{LJ}\left[\left(\frac{\sigma_{LJ}}{r_{ij}}\right)^{12} - \left(\frac{\sigma_{LJ}}{r_{ij}}\right)^{6}\right] \quad \textbf{(5)}$$

where r_{ij} is the instantaneous distance between non-connected beads i and j, and ε_{LJ} and σ_{LJ} are the Lennard-Jones parameters whose values are able to mimic good, theta and bad solvent conditions [66]. We include within this type of potential any other pair-wise interaction that would superimpose over the excluded volume effect. An example is the intra-molecular electrostatic interaction between charged beads, which can be properly described by the screened Coulomb, Debye-Hückel potential [67]. Finally, the fifth term indicates any interaction between a bead i and an external agent (or constraint), like an electric field or a wall. Concerning the size of the beads, this is usually adjusted to reproduce the hydrodynamic aspects of the model correctly [65].

Monte Carlo Fundamentals

In the MC method, equilibrium conformations of the bead-and-spring model are generated using the Boltzmann statistics associated with the total internal energy of the system, $V^{(tot)}$ (eq. (1)). Thus, the probability of a conformation is proportional to the Boltzmann factor, $\exp(-V^{(tot)}/k_BT)$, where k_B is the Boltzmann constant and T the absolute temperature. If the conformations are generated in an absolutely random manner, this factor should be used as a statistical weight in the evaluation of the averages. In more efficient MC algorithms, the conformations are generated so that their statistics already obey the Boltzmann distribution. Thus the averages are merely means of the properties of the individual conformations.

A usual and adequate choice to carry out a MC simulation is the quite simple importance-sampling (or Metropolis) procedure [68]. From a previous conformation of the chain, a new one is generated by making small random displacements of the beads. The potential energy of the new conformation, $V^{(tot)}$, is evaluated and compared to that of the previous conformation, $V_{pre}^{(tot)}$. The new conformation is accepted if $V^{(tot)} < V_{pre}^{(tot)}$. Otherwise a random number u with uniform distribution in (0,1) is generated. Then, if $\exp[(V_{pre}^{(tot)}-V^{(tot)})/k_BT] > u$, the new conformation is accepted and if $\exp[(V_{pre}^{(tot)}-V^{(tot)})/k_BT] < u$, the conformation is rejected and the resulting conformation after the MC step is a copy of the previous one. Other implementations of the MC procedure can be devised for specific cases.

The MC procedure is strictly applicable to equilibrium properties like the radius of gyration and the scattering form factor. For dynamic properties, it is employed in the so-called rigid body approximation [69], in which the generated conformations are treated as if they were instantaneously rigid particles. For each conformation, the hydrodynamic properties are evaluated using the procedures proper to rigid particles and, like for the equilibrium properties, the results are just the sample averages. The rigid-body procedure yields acceptable values for the translational coefficients and the intrinsic viscosity, but it is inappropriate to calculate properties related to the reorientational dynamics of the molecule.

Brownian Dynamics Fundamentals

In the BD method, the effect of the solvent molecules on the chain is taken into account in a mean-field manner by incorporating both a friction force and a stochastic force, in the equation of motion of the bead-and-spring model. Thus, the BD simulation yields the time evolution of a collective of frictional elements (beads) experiencing the deterministic forces associated to the potentials given by eq. (1) along with the random influence of the solvent. Because the friction experienced by the Brownian particles (beads) is usually high compared to their masses, the inertia can be neglected and the Newton's second law for each bead i is written as:

$$\mathbf{F}_i^{(tot)} = \mathbf{F}_i^{(\mathrm{int})} + \mathbf{F}_i^{(ext)} + \mathbf{F}_\mathbf{i}^{(fric)} + \mathbf{F}_\mathbf{i}^{(Brow)} = 0 \qquad \textbf{(6)}$$

The total force vector on a particle, $\mathbf{F}_i^{(tot)}$, is composed of the friction force from the particle moving through the continuous viscous solvent, $\mathbf{F}_i^{(fric)}$, the Brownian force due to the random collisions of the solvent molecules with the particle, $\mathbf{F}_i^{(Brow)}$, and all the deterministic non-hydrodynamic forces which include both the internal forces, $\mathbf{F}_i^{(int)}$ (connector springs, angular springs, torsions and excluded volume) and external forces, $\mathbf{F}_i^{(ext)}$ (*e.g.* an electric field). The generated stochastic differential equation of motion is integrated forward in time to obtain a series of position of the beads along time. In order to carry out the simulation, the procedure is discretized in sufficiently small time steps, Δt, so that the forces do not change appreciably over the step. In this way, from the position vector of a bead i at time t^0, $\mathbf{r}_i^0$, its position vector at time $t^0+\Delta t$, $\mathbf{r}_i$, is evaluated according to the general expression (algorithm details are omitted):

$$\mathbf{r}_i = \mathbf{r}_i^0 + \left[\Delta\mathbf{r}_i\right]^{(forces)} + \left[\Delta\mathbf{r}_i\right]^{(Brow)} + \left(\Delta t\right)\mathbf{v}_i \quad \textbf{(7)}$$

The displacement due to the deterministic forces requires the calculation of a force field and a diffusion tensor of size $3N\times3N$. The Brownian displacement is a Gaussian random number with zero mean and a co-variance that depends on the same diffusion tensor mentioned before [65]. Finally, if there is a flow, there will be term associated to the flow velocity at the point occupied by the bead i, $\mathbf{v}_i$.

The motion of a particle in a solvent exerts a force that perturbs the original velocity field of the solvent. The disturbance velocity changes the viscous drag force exerted on the other particles. This interaction between particles mediated by the solvent is called hydrodynamic interaction (HI) and must be taken into account in order to accurately predict dynamic properties of macromolecules in solution [66, 70]. The HI effect is represented by the so-called HI tensor, which is part of the diffusion tensor [65]. BD algorithms including rigorous, fluctuating HI demand long simulation times which are proportional to N^3. Since the bottleneck of a BD-HI algorithm is the generation of the random displacement (due to the type of manipulations performed on the diffusion tensor), several procedures have been proposed to improve its efficiency [71]. On the other hand, BD simulations without HI sample correctly the equilibrium conformational space. Thus, if the interest is in equilibrium (not dynamic) properties, BD simulations without HI can be perform in which case the CPU time becomes proportional to N.

EXAMPLE 1: PROTEINS

Proteins are macromolecules made of amino acid residues that play key structural, metabolic and regulatory functions in living organisms. Many proteins have a rigid structure in their wild state, as is the case of the globular proteins (lysozyme, bovine serum albumin, *etc.*), whereas others can be quite flexible, as is the case of denatured proteins and intrinsically disordered proteins. Modeling and simulation are powerful tools for understanding their solution behavior and getting insights about their structure and function. These tools are of increasing importance in molecular biology, due to the improvement in modeling methods, the advances in computer technology and the huge amount of experimental data available. For instance, the Protein Data Bank (PDB) [26] collects atomic coordinates of thousands of proteins and protein complexes, obtained by X-ray crystallography. In this section, we give some general ideas about models and computational procedures that are available to elucidate different aspects of the structure and dynamics of proteins, highlighting the relevance of hydrodynamic modeling and dilute solution properties in determining size, shape and structure of proteins (either rigid or flexible).

An important effort in protein dynamics is devoted to understanding the mechanism by which the polypeptide chain folds into its functional, three-dimensional structure from random coil, *i.e.* the so-called protein folding. For this purpose, simplifying assumptions (such as Go models [72]) or coarse grain representations of the protein [73] (such as alpha-carbon only models [74]) are usually employed, or, frequently, a combination of both. Since flexibility is an aspect to consider in protein folding, an appropriate coarse grain model is the bead-and-spring chain. Obviously, appropriate simulation procedures for studying such dynamic phenomenon could be BD and mainly MD [75]. Useful MD packages are those previously mentioned: AMBER [53, 54], CHARMM [55, 56] and GROMACS [57, 58]. On the other hand, an example of suitable BD package is SIMUFLEX [16, 60], which includes several torsion potentials able to represent the protein secondary structure. Besides MD and BD, MC simulations can also be used [76]. For instance, the software package SMMP [77, 78] implements a MC procedure.

Another important area of research in protein dynamics is devoted to protein docking, *i.e.* the study of the formation of protein-protein complexes (or any protein-macromolecule complex). In this case, BD is the appropriate technique to treat the diffusive aspects of the problem efficiently. UHBD [79, 80] is a BD package particularly designed to study protein-protein association. Many of the BD programs conceived for protein docking use rigid-body models because they are designed to study the association of two rigid proteins (*e.g.* globular proteins) undergoing rotational and translational diffusion. Two popular examples of such type of programs are SDA [81, 82] and BrownDye [83, 84]. On the other hand, an example of BD software that handles several types of protein-macromolecule complexes is MacroDox, which is freely available from the web site [85]. Finally, it is worth mentioning that some MC-like approaches can also be used in protein docking studies as it is the case of the software ICM [86, 87].

As already mentioned, conformational and hydrodynamic properties of proteins in dilute solution give key information about their size and shape. Hydrodynamic modeling is, then, a powerful strategy to relate protein structure and hydrodynamic properties. In the case of quasi-rigid, globular proteins, we can use from elementary whole-body models (long rods for fibrous proteins, revolution ellipsoids for globular ones) to very detailed (even atomic) models within the spirit of the bead model concept (protein shape is represented by an assembly of beads) [10], as described in the "Rigid models" section (*e.g.* Fig. **1** illustrates a bead-per-atom model of lysozyme). Clearly, if fine structural details need to be taken into account (as occurs when studying non-globular shapes), the whole-body model is not adequate and the strategy must be the application of the bead model concept with a resolution level (atomic, residue, shell…) appropriate for the case under study [10]. In the case of compact structures, the boundary platelets model [44] is also a good option. Thus, the whole set of computer programs and hydrodynamic approaches shown in the "Rigid models" section are applicable to rigid proteins.

Finally, in the case of (totally or partially) flexible proteins like intrinsically disordered proteins, or multi-subunit structures like antibodies, the most appropriate strategy to predict hydrodynamic properties is the bead modeling. Thus, the flexible part can be well represented by a bead-and-spring model and the rigid part by an assembly of beads (see Fig. **2** as an example of a multi-subunit structure with flexible and rigid domains). An ensemble of conformations can be generated by the MC or BD simulation techniques from which conformational and hydrodynamic properties are computed (for instance, by using the already mentioned rigid body treatment).

Application 1: Solution Properties of Rigid Proteins

Below, we show how the bead modeling strategy is able to adequately predict solution properties of rigid proteins. For this purpose, we use the results of García de la Torre and co-workers, who applied the methodology implemented in the HYDROPRO program to a group of globular proteins [24]. For every protein a primary bead model was built from its PDB atomic coordinates. In that first stage, all of the non-hydrogen atoms are represented by identical beads of radius a. If this radius was taken as the average covalent radius of the atoms, neighbor beads would be nearly tangent (*i.e.* practically free of overlap). However, covalent radii underestimate the effective molecular volume, which is better represented (in a vacuum) by the van der Waals radii. Still, larger values of a are actually used to account for the hydration due to the solvent [24]. At this stage, the model is a compact cluster of overlapping beads (like that shown in Fig. **1**) which would create computational difficulties when calculating hydrodynamic interactions. The solution is to build a shell of tangent beads (*i.e.* a shell model) from that primary model [10] where beads are sufficiently small so as to represent the molecular surface as closely as possible. From that shell model, hydrodynamic properties such as the translational diffusion coefficient, Dt, the rotational diffusion coefficient, Dr, and the intrinsic viscosity, [η], as well as conformational properties, like the radius of gyration, Rg, are calculated. Finally, an extrapolation to zero bead size is carried out in order to get the property value as accurately as possible. Every stage of the model construction as well as the calculations is performed with the HYDROPRO program.

Table 2. % of difference between calculated properties and experimental values.

Protein	Dt	Dr	[η]	Rg
Ribonuclease A	2.8		2.4	2.0
Lysozyme	2.3	8.8	4.1	1.4
Myoglobin	-4.9	0.9	2.2	-4.1
β-Lactoglobulin	0.7	1.5		1.0
Ovalbumin	-4.5		-9.1	
Citrate synthase	-3.0		-10.2	-0.4
GPD	-0.4		0.3	-0.7
Lactate dehydrogenase	-2.9		-0.2	-3.8
Catalase	-0.1		-0.2	0.6

In Table **2** we show the percentage of difference between the calculated and experimental values for a group of proteins. The whole set of tested proteins, the sources of the experimental values as well as details on the calculations and results can be found in ref. [24].

As appreciated, up to four solution properties, Rg, Dt, Dr, [η], of a given globular protein were predicted with errors typically smaller than 5%, a difference within the typical experimental errors. We would mention here that the experimental determination of [η] for globular, almost spherical, proteins is prone to error due to its small value.

Application 2: Solution Properties of Flexible Proteins

Although the rigid-body hydrodynamics could be somehow modified to account for the conformational flexibility of disordered proteins [88], a simple coarse grain model, with non-atomic but sufficient structural detail, along with tools for the simulation of solution dynamics is a valuable strategy for treating long timescale features. One adequate coarse grain model would be the bead-an--spring chain with beads representing amino acid residues, and located at the alpha carbons (Cα). We use the results obtained by García de la Torre and co-workers [89] to illustrate the capability of such a one-bead per amino-acid-residue model [74] to predict solution properties of partially disordered proteins, *i.e.* proteins that contain both flexible and rigid (structured) domains. An example of structure containing two rigid domains linked by a flexible chain is illustrated in Fig. (**2**). The methodology implemented in the SIMUFLEX program [60] was used to carry out BD simulations and calculate proteins solution properties. SIMUFLEX is a suitable software tool because it takes HI properly into account, which is required due to the well known influence of HI on the protein dynamics [70], and because it is able to extract conformational and hydrodynamic properties as well as a number of correlation functions from the BD trajectories. The general physical features of the protein model are (see ref [89]. for further details):

1. **Flexible Chains.** The linear flexible region is modeled with one identical bead for each residue. The equilibrium length between consecutive connected beads is the distance between neighboring Cα atoms. The connector spring potential is like that described by eq. (2). The angle between consecutive connectors has associated a quadratic potential like that defined in eq. (4). No torsional potentials are included in the flexible region. Excluded volume between non-bonded beads is represented by a continuous repulsive potential inspired in the Go model.
2. **Structured Domains.** The Go model is employed [70] to determine and maintain the globular, quasi-rigid domains fixed. The starting point is the atomic structure given in a PDB file and the use of the native contacts concept, which are pairs of residues essential for maintaining the ordered structure. This concept is implemented in the CSU (Contacts of Structural Units) program described in ref [90]. The spring potential between first neighbor beads is the same as for flexible chains. The bending angle potential is also the same but with an angle equilibrium value determined from the atomic (PDB) structure. The Go model includes now a torsional potential in which the equilibrium dihedral angles are those found in the PDB structure. For non-bonded interactions a distinction is made between native and non-native contacts. In the latter case, the same repulsive potential as for flexible chains is used. For native contacts, a modified Lennard-Jones potential is adopted [70].

Below, some results for three partially disordered proteins obtained with the above described methodology are presented. The whole set of results for those and other proteins as well as all the references for the experimental data appear in ref [89], where a more exhaustive comparison between experimental and simulation results including several scattering and relaxation times diagrams can be found.

ZipA is a membrane-anchored protein with 328 amino acids. Its scission produces a soluble protein with 306 amino acids, ZipA(23-328), consisting of a globule and a disordered, flexible domain. The placement and number of residues of the successive domains is 306(N) = 163(L) + 143(G, 1F46) where L and G, indicate the number of residues in the flexible linker and in the rigid globular domain, respectively (with indication of the PDB code for the domain), and N indicates the total number of residues. This notation is used hereafter. The experimental sedimentation coefficient of ZipA(23–328) has been found to be 2.2 S (referred to 20ºC, water). From BD-HI simulations representing the protein dynamics during 15 μs, the translational diffusion coefficient Dt was calculated using the Einstein expression for the square displacements of the center of mass. With that Dt and considering that the protein molecular mass is 36474 Da and its specific volume 0.732 cm^3/g, the predicted sedimentation coefficient was 2.1 S, which is in excellent agreement with the experimental value.

Protein X (PX) is one of the two components of the RNA-polymerase of the Sendai virus. It consists of a globule and a disordered, flexible domain with the following placement and number of residues: 109(N) = 56(L) + 53(G, 1R4G). This protein has been characterized by scattering (SAXS and SANS) and NMR relaxation. Four BD-HI trajectories, each representing 10 μs, were generated. The calculated radius of gyration Rg = 26.7 ± 0.2 Å agrees very well with the results from SAXS (29.7 Å) and SANS (25 ± 1 Å). Elsewhere, experimental NMR relaxation measurements for the residues within the globule give a representative correlation time of 7.3 ns. Information on the internal reorientational dynamics can be extracted from the BD-HI trajectories using time correlation functions. Particularly useful is the second Legendre polynomial ($<P_2(t)>$) of the angle subtended between successive orientations of some characteristic vector of the model (*e.g.* some bond). That correlation function can be fitted to a multi-exponential to obtain a series of relaxation times. The longest one (associated to the slowest relaxation mode of the molecule) is the characteristic correlation time, which was 6.6 ± 1.3 ns, which compares well with the experimental value.

Bruton's tyrosine kinase (BTK) is a protein composed of 659 amino acids, which are arranged in four globular domains and three flexible linkers as follows: 659(N) = 169(G, 1BTK) + 51(L) + 55(G, 1AWX) + 8(L) + 96(G, 2GE9) + 22(L) + 258(G, 1K2P). Experimental results for the radius of gyration (obtained from SAXS) is Rg = 50 Å, and the experimental sedimentation coefficient is 3.3 S (referred to 20ºC, water). BD-HI simulations present a high computational cost owing to the large size of BTK (see the HI issue in the "Flexible models" section). As an alternative, BD-noHI simulations were performed along with the rigid body treatment for the calculation of properties. As mentioned in the "Flexible models" section, BD-noHI works as a MC technique to generate equilibrium conformations which, along with the rigid body treatment, can be used to produce rigorous results for scattering predictions and reasonable results for overall hydrodynamics. Thus, six BD-noHI independent trajectories of 10 μs were conducted. After averaging over those trajectories, we obtain Rg = 49 Å, which practically matches the experimental data. The calculated sedimentation coefficient was 3.9 S which is about 10% above the experimental value (although it should be taken into account that BD-noHI is not the best option to obtain dynamic properties).

EXAMPLE 2: POLYSACCHARIDES

Polysaccharides are formed by long chains of monosaccharides linked by glycosidic bonds. For a long time theoretical models have helped to understand their characteristics through the interpretation of experimental measurements. Thus, the size and shape of polysaccharides in solution can be estimated in a variety of ways with hydrodynamic tools such as sedimentation velocity, sedimentation equilibrium, size exclusion chromatography multi-angle light scattering (SEC-MALS), and viscometry [91]. Persistence length, L_p, (see "Flexible models" section) is a measure of the linear flexibility of the polymer and its estimation can be obtained from various theoretical methods, with the most advanced being some of the mentioned in the previous sections. In this section we illustrate the application of theoretical models to polysaccharides with a study of the flexibility of alginate as a function of the ionic strength [92].

From another perspective, a survey of the literature shows the enormous number of applications of polysaccharides that have been studied. It is also found that theoretical models and simulation techniques are essential for a better knowledge of their molecular characteristics and further improvements in their applications. Although we shall not develop this statement, as an illustration we mention the use of these molecules as drug delivery systems, which has received great attention in recent years. Within this context, molecular dynamics simulation is probably the most popular technique, as can be seen in very recent examples presented by Shan *et al.* [93] or Razmimanesh *et al.* [94].

Flexibility of Alginate as a Function of Ionic Strength

As Schefer *et al.* stated [95], despite the large number of studies in the nucleic acids field, the modulation of molecular conformations as a result of changing ionic environments in other biologically relevant charged macromolecules, such as polysaccharides, has not received much attention.

Alginates are structural linear biopolymers made of guluronate (G) and manuronate (M) which comprise a broad family of polysaccharides. Their natural, rich and renewable sources and non-toxic characteristics, accompanied by their versatility and biocompatibility provide an explanation for the wide interest shown in these molecules. For example, as biomaterials, alginates can easily be formulated into a variety of soft, elastic gels, fibers, foams, nanoparticles, multilayers *etc.* in physiological conditions that ensure the preservation of cell viability and function (see, for example, Andersen *et al.* [96]).

The properties of dilute alginate solutions reflect the flexibility and extension of the chains, which are usually considered to be semi-flexible. Two characteristics of these polysaccharides are essential in explaining their flexibility: the sequential structure of the chain (amount and distribution of G and M monomers) and the polyelectrolyte nature of these molecules. It has been shown that high ionic strength, I, leads to a decrease in the viscosity of the solution compared with salt-free conditions, because when an inert salt is added the screening of the charges that takes place means that the electrostatic interactions decrease and the conformation of the chain becomes more compact [67, 97].

The linear flexibility of polymer chains is represented in terms of the persistence length, L_p, of the equivalent wormlike chains. In practical terms, limits for L_p could range from 1 nm for random coils (*e.g.* pullulan) to 200 nm for an extra-rigid rod (*e.g.* schizophyllan) (see, for example, Tombs and Harding [98]). Data sets of radius of gyration, intrinsic viscosity, sedimentation coefficient and even the translational diffusion coefficient *versus* molecular weight, M, permit simple estimates of the chain conformation type (sphere, rod, coil, *etc.*) from a power law ($P = KM^a$, with P = Rg, [η], s, Dt), *i.e.* from a "Mark–Houwink–Sakurada" (MHS) type of analysis. But a survey of the literature shows that the determination of L_p for polysaccharides in general [99], and for alginates in particular [100], is not trivial because of experimental and theoretical-modeling-data processing difficulties.

García de la Torre and co-workers [92] have used the wormlike model to quantify the flexibility of alginate molecules. Two data treatments (Bohdanecky's equations and the HYDFIT program) give very similar results for L_p. The Multi-HYDFIT program [101] (freely available online [16]) combines all available data sets and minimizes a target (error) function, Δ. Using a combination of the Bohdanecky [102] and Yamakawa and Fujii [103] representations of wormlike coils, this program gives a combined or "global" estimate of the wormlike chain parameters: L_p, M_L (mass per unit length) and d (the chain diameter). The use of HYDFIT is simpler because it is only necessary to introduce a reasonably wide range of d, M_L and L_p. After introducing the same set of input data as used with Bohdanecky's equations we can optimize the target function, Δ, as shown in Fig. (**5**). This analysis program seems to be the best choice at present.

According to our results [92], the MHS power law does not provide a good description of the [η]-M relationship for wide ranges of molecular weights of monodisperse fractions. This could explain some of the differences in the interpretation of the [η]-M data found in the literature. The conclusion is that the wormlike model is not sufficient to describe the [*η*]-M relationship over a wide range of M, an observation that gains importance as I increases. Vold *et al.* [100]

also found that the wormlike chain may not be the most suitable model when the flexibility of alginate molecules increases, as occurred in their work in the absence of excluded volume effects for the most oxidized alginate samples.

The quantitative determination of the parameters for these polyelectrolyte chains using the wormlike model shows two different scenarios. The first appears if we assume that M_L (and hence the contour length and the diameter) do not change on account of the effect of I. In this case the "classical" flexibility mechanism of L_p decreasing with I would be adequate. When I = 0.01, the intrinsic component of L_p would be around 11.3 nm, and the electrostatic one around 6 nm. These values are close to previous results in the literature [100, 104]. In the second scenario, HYDFIT allows us not to impose any restriction on the calculation of M_L or L_p in the wormlike model. Thus a new description of the effect of inorganic ions on alginates (and perhaps other polyelectrolytes) could be offered. In this case, the decrease in [η] with increasing ionic strength would be due to a decrease in contour length and, hence, in M_L. Any changes in the value of L_p which, in our case, is 12 nm, would be slight. Although further investigation is needed, this option offers a new way to understand the complex mechanism of flexibility in polyelectrolytes.

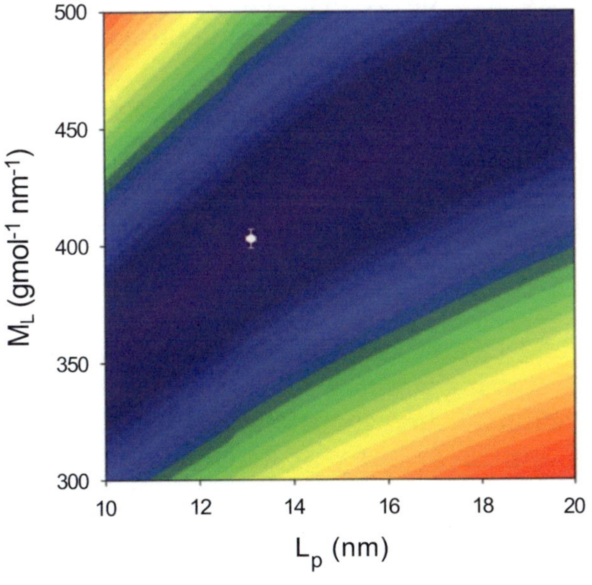

Fig. (5). Contour plot for HYDFIT target function Δ for I = 0.1 with d = 0.69 nm. In this representation the values of Δ are presented by the full color spectrum, from blue (Δ = 0.05) to red (Δ = 25). The minimum (with error bars) is also plotted (white).

A plausible explanation for all these observations could be as follows. When the ionic strength is low, the alginate molecules are well described by the wormlike model, but when concentration of inorganic ions grows, two different effects could appear. On the one hand, short range interactions or screening between COO^- groups of contiguous or very close G and M monomers could shorten the contour length of the molecules as the salt concentration increases, giving rise to shorter (lower $[\eta]$) and more rigid molecules. At the same time, long range interactions or screening between more separated charges in the chain could allow more folded conformations, and, for higher ionic strengths, would also allow aggregation among different molecules. These two effects would therefore be opposite and in competition, and will depend on ionic strength. For shorter molecules, the first factor would be more important while the second would increase strongly with the chain length.

CONCLUDING REMARKS

Theoretical models play a very important role in experimental practice and are essential for the development of present science and technology. We have presented in this chapter a brief overview of methodologies applied to biomolecules which are well developed and easily accessible to the scientific community. With this, we have tried to illustrate to non-specialized readers the current possibilities, and, hopefully, to encourage them to find ways to incorporate the use of theoretical models to complement their work.

CONFLICT OF INTEREST

The author (editor) declares no conflict of interest, financial or otherwise.

ACKNOWLEDGEMENTS

The authors are grateful to Prof. José García de la Torre for enabling them to become part of the fascinating world of science and wish to acknowledge him as their reference in macromolecular modeling and simulation. Financial support was provided by the Spanish Ministry of Economy and Competitiveness under project CTQ2012-33717, including FEDER funds, and by “Fundación Séneca, Agencia de Ciencia y Tecnología de la Región de Murcia” under Projects 19499-PI-14 and 19353/PI/14.

REFERENCES

[1] Edward N Zalta, "The Stanford Encyclopedia of Philosophy", *Available at: http://plato.stanford.edu/.* 2nd June 2015

[2] P Zeidler, "The Epistemological Status of Theoretical Models of Molecular Structure", *HYLE - International Journal for Philosophy of Chemistry,* vol. 6, pp. 17-34, 2000.

[3] F.A Carroll, *Perspectives on Structure and Mechanism in Organic Chemistry.* John Wiley & Sons: Hoboken, NJ, 2010.

[4] S.E Harding, "On the hydrodynamic analysis of macromolecular conformation", *Biophys. Chem,* vol. 55, no. 1-2, pp. 69-93, 1995.
[http://dx.doi.org/10.1016/0301-4622(94)00143-8] [PMID: 17020868]

[5] S.E Harding, J.C Horton, and H Cölfen, "The ELLIPS suite of macromolecular conformation algorithms", *Eur. Biophys. J,* vol. 25, no. 5-6, pp. 347-359, 1997.
[http://dx.doi.org/10.1007/s002490050048] [PMID: 9213555]

[6] "National Centre for Macromolecular Hydrodynamics. University of Nottingham", *Available at: http://www.nottingham.ac.uk/ncmh.* 20th Sept. 2015

[7] V Bloomfield, W.O Dalton, and K.E Van Holde, "Frictional coefficients of multisubunit structures. I. Theory", *Biopolymers,* vol. 5, no. 2, pp. 135-148, 1967.
[http://dx.doi.org/10.1002/bip.1967.360050202] [PMID: 6040712]

[8] J.G Kirkwood, "The general theory of irreversible processes in solutions of macromolecules", *J. Polym. Sci., Polym. Phys. Ed,* vol. 12, pp. 1-14, 1954.

[9] J.G García de la Torre, and V.A Bloomfield, "Hydrodynamic properties of complex, rigid, biological macromolecules: theory and applications", *Q. Rev. Biophys,* vol. 14, no. 1, pp. 81-139, 1981.
[http://dx.doi.org/10.1017/S0033583500002080] [PMID: 7025081]

[10] B Carrasco, and J García de la Torre, "Hydrodynamic properties of rigid particles: comparison of different modeling and computational procedures", *Biophys. J,* vol. 76, no. 6, pp. 3044-3057, 1999.
[http://dx.doi.org/10.1016/S0006-3495(99)77457-6] [PMID: 10354430]

[11] O Byron, "Hydrodynamic modeling: the solution conformation of macromolecules and their complexes", *Methods Cell Biol,* vol. 84, pp. 327-373, 2008.
[http://dx.doi.org/10.1016/S0091-679X(07)84012-X] [PMID: 17964937]

[12] J Antosiewicz, and D Pörschke, "Electrostatics of hemoglobins from measurements of the electric dichroism and computer simulations", *Biophys. J,* vol. 68, no. 2, pp. 655-664, 1995.
[http://dx.doi.org/10.1016/S0006-3495(95)80226-2] [PMID: 7696517]

[13] O Byron, "Construction of hydrodynamic bead models from high-resolution X-ray crystallographic or nuclear magnetic resonance data", *Biophys. J,* vol. 72, no. 1, pp. 408-415, 1997.
[http://dx.doi.org/10.1016/S0006-3495(97)78681-8] [PMID: 8994627]

[14] J García de la Torre, S Navarro, M.C López Martínez, F.G Diaz, and J.J López Cascales, "HYDRO: a computer program for the prediction of hydrodynamic properties of macromolecules", *Biophys. J,* vol. 67, no. 2, pp. 530-531, 1994.
[http://dx.doi.org/10.1016/S0006-3495(94)80512-0] [PMID: 7948671]

[15] J.G de la Torre, Gdel.R Echenique, and A Ortega, "Improved calculation of rotational diffusion and intrinsic viscosity of bead models for macromolecules and nanoparticles", *J. Phys. Chem. B,* vol. 111, no. 5, pp. 955-961, 2007.
[http://dx.doi.org/10.1021/jp0647941] [PMID: 17266248]

[16] "Research Group on Physical Chemistry of Macromolecules", *University of Murcia, Available at: http://leonardo.inf.um/macromol.* 20th Sept. 2015

[17] J García de la Torre, and S.E Harding, "Hydrodynamic modelling of protein conformation in solution: ELLIPS and HYDRO", *Biophys. Rev,* vol. 5, no. 2, pp. 195-206, 2013.
[http://dx.doi.org/10.1007/s12551-013-0102-6] [PMID: 23646070]

[18] J García de la Torre, and B Carrasco, "Hydrodynamic properties of rigid macromolecules composed of ellipsoidal and cylindrical subunits", *Biopolymers,* vol. 63, no. 3, pp. 163-167, 2002.
[http://dx.doi.org/10.1002/bip.10013] [PMID: 11787004]

[19] B Carrasco, J García de la Torre, K.G Davis, S Jones, D Athwal, C Walters, D.R Burton, and S.E Harding, "Crystallohydrodynamics for solving the hydration problem for multi-domain proteins: open physiological conformations for human IgG", *Biophys. Chem,* vol. 93, no. 2-3, pp. 181-196, 2001.

[http://dx.doi.org/10.1016/S0301-4622(01)00220-4] [PMID: 11804725]

[20] Y Lu, S.E Harding, T.E Michaelsen, E Longman, K.G Davis, A Ortega, J.G Grossmann, I Sandlie, and J García de la Torre, "Solution conformation of wild-type and mutant IgG3 and IgG4 immunoglobulins using crystallohydrodynamics: possible implications for complement activation", *Biophys. J,* vol. 93, no. 11, pp. 3733-3744, 2007. [http://dx.doi.org/10.1529/biophysj.107.108993] [PMID: 17704171]

[21] S Zorrilla, T Doan, C Alfonso, E Margeat, A Ortega, G Rivas, S Aymerich, C.A Royer, and N Declerck, "Inducer-modulated cooperative binding of the tetrameric CggR repressor to operator DNA", *Biophys. J,* vol. 92, no. 9, pp. 3215-3227, 2007. [http://dx.doi.org/10.1529/biophysj.106.095109] [PMID: 17293407]

[22] J Gapiński, J Szymański, A Wilk, J Kohlbrecher, A Patkowski, and R Hołyst, "Size and shape of micelles studied by means of SANS, PCS, and FCS", *Langmuir,* vol. 26, no. 12, pp. 9304-9314, 2010. [http://dx.doi.org/10.1021/la100181d] [PMID: 20345164]

[23] A Sokolova, C.S Kealley, T Hanley, A Rekas, and E.P Gilbert, "Small-angle X-ray scattering study of the effect of pH and salts on 11S soy glycinin in the freeze-dried powder and solution states", *J. Agric. Food Chem,* vol. 58, no. 2, pp. 967-974, 2010. [http://dx.doi.org/10.1021/jf902349q] [PMID: 20025226]

[24] J García De La Torre, M.L Huertas, and B Carrasco, "Calculation of hydrodynamic properties of globular proteins from their atomic-level structure", *Biophys. J,* vol. 78, no. 2, pp. 719-730, 2000. [http://dx.doi.org/10.1016/S0006-3495(00)76630-6] [PMID: 10653785]

[25] J García de la Torre, M.L Huertas, and B Carrasco, "HYDRONMR: prediction of NMR relaxation of globular proteins from atomic-level structures and hydrodynamic calculations", *J. Magn. Reson,* vol. 147, no. 1, pp. 138-146, 2000. [http://dx.doi.org/10.1006/jmre.2000.2170] [PMID: 11042057]

[26] "Protein Data Bank", *Available at: http://www.rcsb.org.* 20th Sept. 2015

[27] A Ortega, D Amorós, and J García de la Torre, "Prediction of hydrodynamic and other solution properties of rigid proteins from atomic- and residue-level models", *Biophys. J,* vol. 101, no. 4, pp. 892-898, 2011. [http://dx.doi.org/10.1016/j.bpj.2011.06.046] [PMID: 21843480]

[28] J García de la Torre, O Llorca, J.L Carrascosa, and J.M Valpuesta, "HYDROMIC: prediction of hydrodynamic properties of rigid macromolecular structures obtained from electron microscopy images", *Eur. Biophys. J,* vol. 30, no. 6, pp. 457-462, 2001. [http://dx.doi.org/10.1007/s002490100176] [PMID: 11718300]

[29] N Braun, M Zacharias, J Peschek, A Kastenmüller, J Zou, M Hanzlik, M Haslbeck, J Rappsilber, J Buchner, and S Weinkauf, "Multiple molecular architectures of the eye lens chaperone αB-crystallin elucidated by a triple hybrid approach", *Proc. Natl. Acad. Sci. USA,* vol. 108, no. 51, pp. 20491-20496, 2011. [http://dx.doi.org/10.1073/pnas.1111014108] [PMID: 22143763]

[30] U del Castillo, C Alfonso, S.P Acebrón, A Martos, F Moro, G Rivas, and A Muga, "A quantitative analysis of the effect of nucleotides and the M domain on the association equilibrium of ClpB", *Biochemistry,* vol. 50, no. 12, pp. 1991-2003, 2011. [http://dx.doi.org/10.1021/bi101670s] [PMID: 21309513]

[31] D.I Svergun, "Restoring low resolution structure of biological macromolecules from solution scattering using simulated annealing", *Biophys. J,* vol. 76, no. 6, pp. 2879-2886, 1999. [http://dx.doi.org/10.1016/S0006-3495(99)77443-6] [PMID: 10354416]

[32] D Franke, and D.I Svergun, "DAMMIF, a program for rapid ab-initio shape determination in small-angle scattering", *J. Appl. Cryst,* vol. 42, no. Pt 2, pp. 342-346, 2009. [http://dx.doi.org/10.1107/S0021889809000338] [PMID: 27630371]

[33] D.J Scott, J.G Grossmann, J.R Tame, O Byron, K.S Wilson, and B.R Otto, "Low resolution solution structure of the Apo form of Escherichia coli haemoglobin protease Hbp", *J. Mol. Biol,* vol. 315, no. 5, pp. 1179-1187, 2002.
[http://dx.doi.org/10.1006/jmbi.2001.5306] [PMID: 11827485]

[34] C Baldock, A.F Oberhauser, L Ma, D Lammie, V Siegler, S.M Mithieux, Y Tu, J.Y Chow, F Suleman, M Malfois, S Rogers, L Guo, T.C Irving, T.J Wess, and A.S Weiss, "Shape of tropoelastin, the highly extensible protein that controls human tissue elasticity", *Proc. Natl. Acad. Sci. USA,* vol. 108, no. 11, pp. 4322-4327, 2011.
[http://dx.doi.org/10.1073/pnas.1014280108] [PMID: 21368178]

[35] T.R Patel, R Reuten, S Xion, M Meier, D.J Winzor, M Koch, and J Stetefeld, "Nano-structure of the laminin γ-1 short arm reveals an extended and curved domain", *Matrix Biol,* vol. 31, pp. 135-140, 2012.
[http://dx.doi.org/10.1016/j.matbio.2011.11.004] [PMID: 22210009]

[36] D Amorós, A Ortega, S.E Harding, and J García de la Torre, "Multi-scale calculation and global-fit analysis of hydrodynamic properties of biological macromolecules: determination of the overall conformation of antibody IgG molecules", *Eur. Biophys. J,* vol. 39, no. 3, pp. 361-370, 2010.
[http://dx.doi.org/10.1007/s00249-008-0388-7] [PMID: 19093107]

[37] A Ortega, D Amorós, and J García de la Torre, "Global fit and structure optimization of flexible and rigid macromolecules and nanoparticles from analytical ultracentrifugation and other dilute solution properties", *Methods,* vol. 54, no. 1, pp. 115-123, 2011.
[http://dx.doi.org/10.1016/j.ymeth.2010.12.004] [PMID: 21163355]

[38] D Amorós, A Ortega, and J García de la Torre, "Hydrodynamic properties of wormlike macromolecules: Monte Carlo simulation and global analysis of experimental data", *Macromolecules,* vol. 44, pp. 5788-5897, 2011.
[http://dx.doi.org/10.1021/ma102697q]

[39] R.M Venable, and R.W Pastor, "Frictional models for stochastic simulations of proteins", *Biopolymers,* vol. 27, no. 6, pp. 1001-1014, 1988.
[http://dx.doi.org/10.1002/bip.360270609] [PMID: 3401553]

[40] E Brookes, B Demeler, C Rosano, and M Rocco, "The implementation of SOMO (SOlution MOdeller) in the UltraScan analytical ultracentrifugation data analysis suite: enhanced capabilities allow the reliable hydrodynamic modeling of virtually any kind of biomacromolecule", *Eur. Biophys. J,* vol. 39, no. 3, pp. 423-435, 2010.
[http://dx.doi.org/10.1007/s00249-009-0418-0] [PMID: 19234696]

[41] "Ultrascan SOMO", *Available at: http://www.somo.uthscsa.edu/.* 20th Sept. 2015

[42] J García de la Torre, H.E Pérez Sánchez, A Ortega, J.G Hernández, M.X Fernandes, F.G Díaz, and M.C López Martínez, "Calculation of the solution properties of flexible macromolecules: methods and applications", *Eur. Biophys. J,* vol. 32, no. 5, pp. 477-486, 2003.
[http://dx.doi.org/10.1007/s00249-003-0292-0] [PMID: 12698288]

[43] S Aragon, and D.K Hahn, "Precise boundary element computation of protein transport properties: Diffusion tensors, specific volume, and hydration", *Biophys. J,* vol. 91, no. 5, pp. 1591-1603, 2006.
[http://dx.doi.org/10.1529/biophysj.105.078188] [PMID: 16714342]

[44] S.R Aragon, "Recent advances in macromolecular hydrodynamic modeling", *Methods,* vol. 54, no. 1, pp. 101-114, 2011.
[http://dx.doi.org/10.1016/j.ymeth.2010.10.005] [PMID: 21073955]

[45] M Rocco, and O Byron, "Computing translational diffusion and sedimentation coefficients: an evaluation of experimental data and programs", *Eur. Biophys. J,* vol. 44, no. 6, pp. 417-431, 2015.
[http://dx.doi.org/10.1007/s00249-015-1042-9] [PMID: 26066679]

[46] I.N Serdyuk, N.R Zaccai, and J Zaccai, *Methods in Molecular Biophysics (Structure, Dynamics,*

Function). 1st ed Cambridge University Press: New York, 2007.
[http://dx.doi.org/10.1017/CBO9780511811166]

[47] P.E Rouse, "A theory of the linear viscoelastic properties of dilute solutions of coiling polymers", *J. Chem. Phys,* vol. 21, pp. 1272-1280, 1953.
[http://dx.doi.org/10.1063/1.1699180]

[48] B.H Zimm, "Dynamics of polymer molecules in dilute solution: viscoelasticity, flow birefringence and dielectric loss", *J. Chem. Phys,* vol. 24, pp. 269-277, 1956.
[http://dx.doi.org/10.1063/1.1742462]

[49] S Takada, "Coarse-grained molecular simulations of large biomolecules", *Curr. Opin. Struct. Biol,* vol. 22, no. 2, pp. 130-137, 2012.
[http://dx.doi.org/10.1016/j.sbi.2012.01.010] [PMID: 22365574]

[50] G.S Ayton, W.G Noid, and G.A Voth, "Multiscale modeling of biomolecular systems: in serial and in parallel", *Curr. Opin. Struct. Biol,* vol. 17, no. 2, pp. 192-198, 2007.
[http://dx.doi.org/10.1016/j.sbi.2007.03.004] [PMID: 17383173]

[51] M.P Allen, and D.J Tildsley, *Computer Simulation of Liquids.* 1st ed Clarendon: Oxford, 1987.

[52] J García de la Torre, A Ortega, H.E Perez Sanchez, and J.G Hernández Cifre, "MULTIHYDRO and MONTEHYDRO: conformational search and Monte Carlo calculation of solution properties of rigid or flexible bead models", *Biophys. Chem,* vol. 116, no. 2, pp. 121-128, 2005.
[http://dx.doi.org/10.1016/j.bpc.2005.03.005] [PMID: 15950824]

[53] D.A Pearlman, D.A Case, J.W Caldwell, W.S Ross, T.E Cheatham III, S DeBolt, D Fergurson, G Seible, and P Kollman, "AMBER, a package of computer programs for applying molecular mechanics, normal mode analysis, molecular dynamics and free energy calculations to simulate the structural and energetic properties of molecules", *Comput. Phys. Commun,* vol. 91, pp. 1-41, 1995.
[http://dx.doi.org/10.1016/0010-4655(95)00041-D]

[54] "AMBER", *Available at: http://ambermd.org/.* 20th Sept. 2015

[55] B.R Brooks, R.E Bruccolieri, B.D Olafson, D.J States, S Swaminathan, and M.J Karplus, "CHARMM: A program for macromolecular energy, minimization, and dynamics calculations", *Comput. Chem,* vol. 4, pp. 187-217, 1983.
[http://dx.doi.org/10.1002/jcc.540040211]

[56] "CHARMM (Chemistry at HARvard Macromolecular Mechanics)", *Available at: http://www.charmm.org/.* 20th Sept. 2015

[57] D Van Der Spoel, E Lindahl, B Hess, G Groenhof, A.E Mark, and H.J Berendsen, "GROMACS: fast, flexible, and free", *J. Comput. Chem,* vol. 26, no. 16, pp. 1701-1718, 2005.
[http://dx.doi.org/10.1002/jcc.20291] [PMID: 16211538]

[58] "GROMACS", *Available at: http://www.gromacs.org/.* 20th Sept. 2015

[59] K Klenin, H Merlitz, and J Langowski, "A Brownian dynamics program for the simulation of linear and circular DNA and other wormlike chain polyelectrolytes", *Biophys. J,* vol. 74, no. 2 Pt 1, pp. 780-788, 1998.
[http://dx.doi.org/10.1016/S0006-3495(98)74003-2] [PMID: 9533691]

[60] J.G de la Torre, J.G Hernández Cifre, Á Ortega, R.R Schmidt, M.X Fernandes, H.E Pérez Sánchez, and R Pamies, "SIMUFLEX: Algorithms and tools for simulation of the conformation and dynamics of flexible molecules and nanoparticles in solution", *J. Chem. Theory Comput,* vol. 5, no. 10, pp. 2606-2618, 2009.
[http://dx.doi.org/10.1021/ct900269n] [PMID: 26631776]

[61] M Długosz, P Zieliński, and J Trylska, "Brownian dynamics simulations on CPU and GPU with BD_BOX", *J. Comput. Chem,* vol. 32, no. 12, pp. 2734-2744, 2011.
[http://dx.doi.org/10.1002/jcc.21847] [PMID: 21638295]

[62] U Winter, and T Geyer, "Coarse grained simulations of a small peptide: Effects of finite damping and hydrodynamic interactions", *J. Chem. Phys,* vol. 131, no. 104102, pp. 1-7, 2009.

[63] "BD_BOX", *Available at: http://browniandynamics.org/.* 20th Sept. 2015

[64] "Center for Bioinformatics. Saarland University", *Available at: http://gepard. bioinformatik.uni-saarland.de/services/brownmove.* 21st Sept. 2015

[65] H.C Öttinger, *Stochastic Processes in Polymer Fluids.* 1st ed Springer: Berlin, 1996. [http://dx.doi.org/10.1007/978-3-642-58290-5]

[66] J.G Hernández Cifre, and J García de la Torre, "Steady-state behavior of dilute polymers in elongational flow. Dependence of the critical elongational rate on chain length, hydrodynamic interaction and excluded volume", *J. Rheol. (N.Y.N.Y.),* vol. 43, pp. 339-358, 1999. [http://dx.doi.org/10.1122/1.550987]

[67] J.G Hernández Cifre, and J García de la Torre, "Ionic strength effect in polyelectrolyte dilute solutions within the Debye-Hückel approximation: Monte Carlo and Brownian dynamics simulations", *Polym. Bull,* vol. 71, pp. 2269-2285, 2014. [http://dx.doi.org/10.1007/s00289-014-1186-2]

[68] N Metropolis, A.W Metropolis, M.N Rosenbluth, A.H Teller, and E Teller, "Equation of state calculation by fast computing machines", *J. Chem. Phys,* vol. 21, pp. 1087-1093, 1953. [http://dx.doi.org/10.1063/1.1699114]

[69] B.H Zimm, "Chain molecule hydrodynamics by the Monte-Carlo method and the validity of the Kirkwood-Riseman approximation", *Macromolecules,* vol. 13, pp. 592-602, 1980. [http://dx.doi.org/10.1021/ma60075a022]

[70] T Frembgen-Kesner, and A.H Elcock, "Striking Effects of Hydrodynamic Interactions on the Simulated Diffusion and Folding of Proteins", *J. Chem. Theory Comput,* vol. 5, no. 2, pp. 242-256, 2009. [http://dx.doi.org/10.1021/ct800499p] [PMID: 26610102]

[71] R Rodríguez Schmidt, J.G Hernández Cifre, and J García de la Torre, "J. "Comparison of Brownian dynamics algorithms with hydrodynamic interaction", *J. Chem. Phys,* vol. 135, no. 084116, pp. 1-10, 2011.

[72] H Taketomi, Y Ueda, and N Gō, "Studies on protein folding, unfolding and fluctuations by computer simulation. I. The effect of specific amino acid sequence represented by specific inter-unit interactions", *Int. J. Pept. Protein Res,* vol. 7, no. 6, pp. 445-459, 1975. [http://dx.doi.org/10.1111/j.1399-3011.1975.tb02465.x] [PMID: 1201909]

[73] V Tozzini, "Coarse-grained models for proteins", *Curr. Opin. Struct. Biol,* vol. 15, no. 2, pp. 144-150, 2005. [http://dx.doi.org/10.1016/j.sbi.2005.02.005] [PMID: 15837171]

[74] V Tozzini, W Rocchia, and J.A McCammon, "Mapping all-atom models onto one-bead Coarse Grained Models: general properties and applications to a minimal polypeptide model", *J. Chem. Theory Comput,* vol. 2, no. 3, pp. 667-673, 2006. [http://dx.doi.org/10.1021/ct050294k] [PMID: 19461947]

[75] H.A Scheraga, M Khalili, and A Liwo, "Protein-folding dynamics: overview of molecular simulation techniques", *Annu. Rev. Phys. Chem,* vol. 58, pp. 57-83, 2007. [http://dx.doi.org/10.1146/annurev.physchem.58.032806.104614] [PMID: 17034338]

[76] N Rathore, and J.J de Pablo, "Monte Carlo simulation of proteins through a random walk in energy space", *J. Chem. Phys,* vol. 116, pp. 7225-7230, 2002. [http://dx.doi.org/10.1063/1.1463059]

[77] F Eisenmenger, U.H Hansmann, S Hayryan, and C-K Hu, "[SMMP] A modern package for simulation of proteins", *Comput. Phys. Commun,* vol. 138, pp. 192-212, 2001.

[http://dx.doi.org/10.1016/S0010-4655(01)00197-7]

[78] "SMMP. Computational Biophysical Chemistry. Dept. of Chemistry and Biochemistry University of Oklahoma", *Available at: http://www.hansmann-lab.com/cbpc/smmp/smmp.php*. 24th Sept. 2015

[79] J.D Madura, J.M Briggs, R.C Wade, M.E Davis, B.A Luty, A Ilin, J Antosiewicz, M.K Gilson, B Bagheri, L.R Scott, and J.A McCammon, "Electrostatics and diffusion of molecules in solution: Simulations with the University of Houston Brownian dynamics program", *Comput. Phys. Commun,* vol. 91, pp. 57-95, 1995. [http://dx.doi.org/10.1016/0010-4655(95)00043-F]

[80] T Shen, and C.F Wong, "Brownian dynamics simulation of peptides with the University of Houston Brownian Dynamics (UHBD) program", *Methods Mol. Biol,* vol. 1268, pp. 75-87, 2015. [http://dx.doi.org/10.1007/978-1-4939-2285-7_5] [PMID: 25555722]

[81] M Martinez, N.J Bruce, J Romanowska, D.B Kokh, M Ozboyaci, X Yu, M.A Öztürk, S Richter, and R.C Wade, "SDA 7: A modular and parallel implementation of the simulation of diffusional association software", *J. Comput. Chem,* vol. 36, no. 21, pp. 1631-1645, 2015. [http://dx.doi.org/10.1002/jcc.23971] [PMID: 26123630]

[82] "Heidelberg Institute for Theoretical Studies", *Available at: http://mcm.h-its.org/ sda7/doc/doc_sda7/index.html*. 24th Sept. 2015

[83] G.A Huber, and J.A McCammon, "Browndye: A software package for Brownian dynamics", *Comput. Phys. Commun,* vol. 181, no. 11, pp. 1896-1905, 2010. [http://dx.doi.org/10.1016/j.cpc.2010.07.022] [PMID: 21132109]

[84] "BrownDye", *Available at: http://browndye.ucsd.edu/*. 20th Sept. 2015

[85] "MacroDox Brownian Dynamics Software", *Available at: https://iweb.tntech.edu/ macrodox/macrodox.html*. 20th Sept. 2015

[86] J Fernández-Recio, M Totrov, and R Abagyan, "Soft protein-protein docking in internal coordinates", *Protein Sci,* vol. 11, no. 2, pp. 280-291, 2002. [http://dx.doi.org/10.1110/ps.19202] [PMID: 11790838]

[87] "ICM", *Available at: https://www.molsoft.com/gui/index.html*. 20th Sept. 2015

[88] S-H Bae, H.J Dyson, and P.E Wright, "Prediction of the rotational tumbling time for proteins with disordered segments", *J. Am. Chem. Soc,* vol. 131, no. 19, pp. 6814-6821, 2009. [http://dx.doi.org/10.1021/ja809687r] [PMID: 19391622]

[89] D Amorós, A Ortega, and J García de la Torre, "Prediction of hydrodynamic and other solution properties of partially disordered proteins with a simple, coarse-grained model", *J. Chem. Theory Comput,* vol. 9, no. 3, pp. 1678-1685, 2013. [http://dx.doi.org/10.1021/ct300948u] [PMID: 26587628]

[90] V Sobolev, A Sorokine, J Prilusky, E.E Abola, and M Edelman, "Automated analysis of interatomic contacts in proteins", *Bioinformatics,* vol. 15, no. 4, pp. 327-332, 1999. [http://dx.doi.org/10.1093/bioinformatics/15.4.327] [PMID: 10320401]

[91] G.A Morris, G.G Adams, and S.E Harding, "On hydrodynamic methods for the analysis of the sizes and shapes of polysaccharides in dilute solution: A short review", *Food Hydrocoll,* vol. 42, pp. 318-334, 2014. [http://dx.doi.org/10.1016/j.foodhyd.2014.04.014]

[92] F.G Baños, A.I Díez Peña, J.G Hernánez Cifre, M.C López Martínez, A Ortega, and J García de la Torre, "Influence of ionic strength on the flexibility of alginate studied by size exclusion chromatography", *Carbohydr. Polym,* vol. 102, pp. 223-230, 2014. [http://dx.doi.org/10.1016/j.carbpol.2013.11.023] [PMID: 24507276]

[93] P Shan, J-W Shen, D-H Xu, L-Y Shi, J Gao, Y-W Lan, Q Wang, and X-H Wei, "Molecular dynamics study on the interaction between doxorubicin and hydrophobically modified chitosan oligosaccharide",

RSC Advances, vol. 4, pp. 23730-23739, 2014.
[http://dx.doi.org/10.1039/c4ra01199f]

[94] F Razmimanesh, S Amjad-Iranagh, and H Modarress, "Molecular dynamics simulation study of chitosan and gemcitabine as a drug delivery system", *J. Mol. Model,* vol. 21, no. 7, p. 165, 2015. [http://dx.doi.org/10.1007/s00894-015-2705-2] [PMID: 26044358]

[95] L Schefer, I Usov, and R Mezzenga, "Anomalous stiffening and ion-induced coil-helix transition of carrageenans under monovalent salt conditions", *Biomacromolecules,* vol. 16, no. 3, pp. 985-991, 2015. [http://dx.doi.org/10.1021/bm501874k] [PMID: 25686463]

[96] T Andersen, B.L Strand, K Formo, E Alsberg, and B.E Christensen, "Alginates as biomaterials in tissue engineering", *Carbohydr. Chem,* vol. 37, pp. 227-258, 2012.

[97] R Pamies, R Rodríguez Schmidt, M.C López Martínez, and J García de la Torre, "The influence of mono and divalent cations on dilute and non-dilute aqueous solutions of sodium alginates", *Carbohydr. Polym,* vol. 80, pp. 248-253, 2010. [http://dx.doi.org/10.1016/j.carbpol.2009.11.020]

[98] M.P Tombs, and S.E Harding, *An introduction to polysaccharide biotechnology.* Taylor and Francis: London, 1998, pp. 14-20.

[99] G.A Morris, T.R Patel, D.R Picout, S.B Ross-Murphy, A Ortega, J García de la Torre, and S.E Harding, "Global hydrodynamic analysis of the molecular flexibility of galactomannans", *Carbohydr. Polym,* vol. 72, pp. 356-360, 2008. [http://dx.doi.org/10.1016/j.carbpol.2007.08.017]

[100] I.M Vold, K.A Kristiansen, and B.E Christensen, "A study of the chain stiffness and extension of alginates, in vitro epimerized alginates, and periodate-oxidized alginates using size-exclusion chromatography combined with light scattering and viscosity detectors", *Biomacromolecules,* vol. 7, no. 7, pp. 2136-2146, 2006. [http://dx.doi.org/10.1021/bm060099n] [PMID: 16827580]

[101] A Ortega, and J García de la Torre, "HYDFIT and Related Packages for Linear Molecules", In: *Encyclopedia of Biophysics,* G Roberts, Ed., vol. 2. Springer-Verlag: Berlin, Heidelberg, 2013, pp. 998-1002. [http://dx.doi.org/10.1007/978-3-642-16712-6_296]

[102] M Bohdanecky, "New method for estimating the parameters of the wormlike chain model from the intrinsic viscosity of stiff-chain polymers", *Macromolecules,* vol. 16, pp. 1483-1492, 1983. [http://dx.doi.org/10.1021/ma00243a014]

[103] H Yamakawa, and M Fujii, "Translational friction coefficient of wormlike chains", *Macromolecules,* vol. 6, pp. 407-415, 1973. [http://dx.doi.org/10.1021/ma60033a018]

[104] H Zhang, H Wang, J Wang, R Guo, and Q Zhang, "The effect of ionic strength on the viscosity of sodium alginate solution", *Polym. Adv. Technol,* vol. 12, pp. 740-745, 2001. [http://dx.doi.org/10.1002/pat.97]

CONCLUSIONS

At the end of this ebook, let us recall the tremendous efforts researchers have put on the development of techniques and methodologies for the molecular characterizations and the determination of the physicochemical properties of biopolymers from different sources in order to diversify as such as possible their end-uses for industrial applications. For that, researchers and engineers have adopted a strategy based on the understanding of the relationship between their structure and their macroscopic properties. Because biopolymers are able to generate macromolecular assemblies such as three-dimensional networks, two-dimensional films or one-dimensional monomolecular layers, in hydrated or solid forms, the structure of biopolymers alone or involved in such macromolecular assemblies have to be characterized at different scales, from the nanoscale to the macroscale including mesoscale, in order to deeply screen the structure-function-property relationships. We hope this first objective is reached in the present ebook.

Authors in this ebook have shown, through the diversity of the countries they belong, that biopolymers, from animal, plant, microbial and seaweed sources, can be judiciously extracted and purified according to as much as possible eco-friendly processes, in order to get reliable macroscopic end-use properties. Polysaccharides from plants, with cellulose being the most abundant on Earth, are the most used biopolymers for applications as their useful properties can be largely exploited for end-use applications. Natural gums from plants such as mesquite, tara and arabic gums are also very good candidates to promote economical development in countries where they originate from as they often are good stabilizers and adhesives for food and non-food products. Plant proteins are shown in this ebook as very promising alternatives to petroleum carbon sources for edible coatings and plastic foams.

The ebook displays also a thoroughly state-of-art in the physicochemical methods of characterization of biopolymers from the molecular level, through the exploration of size, polydispersity, chemical nature of repeating units, type of branching, using separation methods coupled to optical detectors but also mass spectrometry and NMR spectroscopy, to the macroscopic scale, through the exploration of the mechanical properties. A special attention is also given to the methods at the mesoscopic scale with the use of spectroscopic and microscopic methods in order to probe the structure covering the nanometric and micrometric scale. Small angle scattering and positron annihilation lifetime spectroscopy are methods highlighted in the ebook to probe nanostructures of single or complex biopolymeric materials.

The continuing development of new and existing biopolymers for end-uses applications are highly documented and we hope readers will find new opportunities for their teaching and research-development activities with this exhaustive list of biopolymers and related applications presented in the ebook. For instance, the design and engineering of innovative bionanocomposites and bioplastics deserve a special attention because it forms a fascinating interdisciplinary area that brings together biochemistry, materials science and nanotechnology. The understanding of the chemistry and structure at the nanometric scale imply an important knowledge for future developments of novel materials based on biopolymers and nanostructured materials. The extraordinary versatility of these new materials comes from the large choice of biopolymers available. These new materials have been elaborated thanks to the developement of new powerful techniques described in the ebook such as electrospinning.

Revolution in biopolymers applications comes from the use of proteins and polysaccharides in nanostructured food and non-food materials with the elaboration of biocompatible and degradable micro and nanoparticles (with a solid or a liquid core) for active delivery applications and bioplastics for packaging applications. Therefore, the toolbox that micro- and nanotechnologies offer provides new opportunities for product and process innovations in the food and non-food industries. The control of the process and functionality at the nanoscale will lead to more sustainable food and non-food production. This approach will allow in food industry for instance the development of nutrient delivery systems with healthy and/or less caloric value nutrients, sensors, and diagnostic devices that can monitor and ensure the safety of food products throughout the food chain. At least, various enhanced packaging concepts will extend the shelf life of fresh products or indicate quality deterioration of the packaged product. Nevertheless, it is imperative to develop a good communication of the applications of biopolymers in the "nano world" that allows the consumers to make an informed decision whether or not they would like to have the benefits of certain applications of biopolymers in nanotechnologies, or whether they do not accept certain risks.

All these driving forces act as stimuli to develop new materials based on biopolymers, and there are many opportunity areas such as industrial, medical, food, consumer products and pharmaceutical applications for which biopolymers act as stabilizers, thickeners, binders, cross-linkers, dispersants, lubricants, adhesives, drug-delivery agents. We sincerely hope the readers will find stimuli-responsive answers to the huge diversity of bioplymers and their related applications in this ebook.

SUBJECT INDEX

Martin Masuelli & Denis Renard (Eds.)

C

N

O

P

Made in the USA
Las Vegas, NV
03 November 2026

57425089R00321